Series Contents

T0258784

CHILDBIRTH

Changing Ideas and Practices in Britain a[nd]
America 1600 to the Prese[nt]

Series Editor

PHILIP K. WILSON

Truman State University

Assistant Editors

ANN DALLY

Wellcome Institute for the Histo[ry of]
Medicine (London)

CHARLES R. KING

Medical College of Ohio

A GARLAND SERIES

VOLUME

4

REPRODUCTIVE SCIENCE, GENETICS, AND BIRTH CONTROL

Edited with introductions by
PHILIP K. WILSON
Truman State University

GARLAND PUBLISHING, Inc.
New York & London
1996

Library of Congress Cataloging-in-Publication Data

Childbirth : changing ideas and practices in Britain and America
  1600 to the present / edited with introductions by Philip K.
  Wilson.
      p.   cm.
  Includes bibliographical references.
  Contents: v. 1. Midwifery theory and practice — v. 2. The
  medicalization of obstetrics: personnel, practice, and instruments — v. 3. Methods and folklore — v. 4. Reproductive science, genetics, and birth control — v. 5. Diseases of pregnancy
  and childbirth.
      ISBN 0-8153-2230-5 (v. 1 : alk. paper). — ISBN 0–8153–
  2231–3 (v. 2 : alk. paper). — ISBN 0–8153–2232–1 (v. 3 : alk.
  paper). — ISBN 0–8153–2233–X (v. 4. : alk. paper). — ISBN 0–
  8153-2234–8 (v. 5 : alk. paper)
      1. Childbirth—United States—History.   2. Childbirth—Great
  Britain—History.   3. Obstetrics—United States—History.   4. Obstetrics—Great Britain—History.   I. Wilson, Philip K., 1961–.
      [DNLM: 1. Obstetrics—trends.   2. Midwifery—trends.
  3. Pregnancy Complications.   4. Reproduction Techniques.
  5. Genetic Counseling.   6. Contraception.   WQ 100C5356   1996]

  RG518.U5C47   1995
  618.4'0973—dc20
  DNLM/DLC
  for Library of Congress   96–794
                                                               CIP

CONTENTS

BIRTH CONTROL

LIST OF ILLUSTRATIONS

SERIES INTRODUCTION

Since "most women are interested in the process of giving birth and all men have been born," it would appear, claimed Johns Hopkins University obstetrician Alan Guttmacher, that the topic of childbirth would above all other topics have "universal appeal." Birth is also one of the most individual moments in each of our lives, but although we all share the experience of being delivered, the processes of delivery have been diverse. The social gathering around the childbed common in earlier times has, for many, been replaced by a more isolated hospital bed. Maternal fears of the pain and peril of procreation have, or so prevalent historiography would have us believe, intensified with the intervention of male midwives and obstetricians bringing along new "tools" of the trade. Markedly divergent beliefs about assisting in labor have created polarized factions of attendants. Some have followed wisdom similar to what Britain's Percivall Willughby first espoused in 1640:

> Let midwives observe the ways and proceedings of nature for the production of her fruit on trees, or the ripening of walnuts or almonds, from their first knotting to the opening of the husks and falling of the nut . . . . These signatures may teach midwives patience, and persuade them to let nature alone perform her work.

Opposing factions adhered to claims similar to that of the early nineteenth-century Philadelphia midwifery professor, Thomas Denman, that belief in:

> labour, being a natural act, . . . not requiring the interference of art for either its promotion or its accomplishment . . . has, from its influence, retarded, more perhaps than any other circumstance, the progress of improvement in this most important branch of medical science.

Other comparisons among midwifery writings suggest that although expectant women may no longer avoid the same "longings and cravings" of pregnancy as did their eighteenth-century forebears, contemporary concern about exposing pregnant women and their fetuses to nicotine, alcohol, known teratogenic agents, and unwarranted stress evokes similar warnings. Indeed, as the works included in this collection illustrate, many similar concerns have been shared by expectant mothers and their labor attendants for centuries.

Although there is a substantial literature on childbirth, it typically lacks the full medical, historical, and social contextualization that these volumes provide to readers. This series attempts to fill the gap in many institutions' libraries by bringing together key articles illuminating a number of issues from different perspectives that have long concerned the expectant mother and the attendants of her delivery regarding the health of the newborn infant. Primary and secondary sources have been culled from British and American publications that focus on childbirth practices over the past three hundred years. Some represent "classic" works within the medical literature that have contributed towards a more complete understanding of pertinent topics. The series draws from historical, sociological, anthropological, and feminist literature in an attempt to present a wider range of scholarly perspectives on various issues surrounding childbirth.

Childbirth: Changing Ideas and Practices is intended to provide readers with key primary sources and exemplary historiographical approaches through which they can more fully appreciate a variety of themes in British and American childbirth, midwifery, and obstetrics. For example, general historical texts commonly claim that childbed (puerperal) fever, a disease that has claimed hundreds of thousands of maternal lives, provoked much fear throughout most of British and American history. In addition to supplying readers with historians' interpretations, *Childbirth: Changing Ideas and Practices* also provides discussions of the causes and consequences of particular cases of childbed fever taken directly from the medical literature of the nineteenth and twentieth centuries, thereby enabling a better understanding of how problematic this disorder initially was to several key individuals who, after first increasing its incidence, ultimately devised specific methods of its prevention.

The articles in this series are designed to serve as a resource for students and teachers in fields including history, women's studies, human biology, sociology, and anthropology. They will also meet the socio-historical educational needs of pre-medical and nursing students and aid pre-professional, allied health, and midwifery instructors in their lesson preparations. Beyond the content of many collections on the history of childbirth, readers frequently need access to the primary sources in order to develop their own interpretive accounts. This five-volume series expands the readily accessible knowledge base as it represents both actual experiences and socio-historical interpretations on select developments within the history of British and American childbirth, midwifery, and obstetrics.

Given the vast and expanding literature on childbirth, it is virtually impossible anymore for any single source to provide a complete coverage of such a broad topic. Selecting precisely what articles to include has been, at times, a painstaking process. We have purposefully excluded works on abortion as many of these articles have recently been reprinted elsewhere. Additionally, we have only touched upon midwifery/obstetrical education, the legal issues surrounding childbirth, marriage, sex, and the family, and genetic engineering since numerous contemporary works in print thoroughly discuss these themes. Seminal articles that are currently available in other edited collections as well as general review articles were, with a few exceptions, not considered for reprinting in this series. There are several areas, including eclampsia, the development and role of the placenta, pregnancy tests throughout history, and Native American childbirth practices, for which suitable articles are wanting. Related topics such as gynecology and gynecological diseases, pediatrics, neonatology, postnatal care and teratology, though of considerable concern to many pregnant women and health care providers, appear beyond the scope of our focus and the interest of our generalist readers. Space did not allow for me to cover childbirth from the viewpoint of what have historically been considered alternative or complementary healing professions such as herbalism, homeopathy, or osteopathy, even though thousands of healthy children have been delivered by practitioners in these professions. The exorbitant permission charges that some journals charge for reprinting their articles has prohibited us from including many important articles. Finally, we have opted not to reprint biographical articles as the typical lengthy accounts of individuals would have precluded addressing more general relevant issues.

Series Acknowledgments

I am grateful to the many individuals who offered their assistance, suggestions, and support throughout the gestation of this project. Foremost, I wish to thank my co-editors, Dr. Charles R. King and Dr. Ann Dally, both highly valued "team players" in what truly became an international collaborative creation. Their medical expertise and historiographical suggestions strengthened the content of this series. Laura Runge, my undergraduate research student and Ronald E. McNair Post-Baccalaureate Achievement Program Scholar, provided exemplary editorial assistance throughout the growth of this project. In addition, she introduced Melissa Blagg-Holcomb to our team, a truly exceptional undergraduate scholar, without whom this project would not have been completed in such a timely manner. Melissa's professional interest in nurse midwifery expanded the scope of the literature we reviewed. Our research would have been impossible without the assistance of many librarians, archivists, and other members of the research staff. In particular, I wish to thank Lyndsay Lardner (The Wellcome Institute, London), Susan Case (Clendening Medical History Library, Kansas City), and Janice Wilson (Hawaii Medical Library, Honolulu; Sterling Medical Library, New Haven, and Pickler Memorial Library, Kirksville) for their exemplary library assistance. The unfailing efforts of Sheila Swafford, in Pickler Library's Reference Department, to secure necessary material are deeply appreciated. The editors also wish to thank Jane Carver, Prof. Mark Davis, Prof. Robbie Davis-Floyd, Nancy Dellapenna, Clare Dunne, Prof. Paul Finkelman, Andy Foley, Dr. Denis Gibbs, Ferenc Gyorgyey, Gwendolyn Habel, Jack Holcomb, Charlene Jagger, Maggie Jones, Carol Lockhart, Barb Magers, Andrew Melvyn, Jean Sidwell, Prof. John Harley Warner, Prof. Dorothy C. Wertz and the staffs of the Library of the Royal Society of Medicine (London), the National Library of Medicine (Bethesda), Pickler Memorial Library (Kirksville) and Rider Drug and Camera (Kirksville) for their assistance in preparing certain parts of this series. Leo Balk of Garland Publishing, Inc., proved to be a stable sounding board during the conception stage of *Childbirth*, a role that Carole Puccino has deftly carried on throughout the later progressions of this work. I also wish to thank my colleagues at the University of Hawaii-Manoa, Yale University, and Northeast Missouri State University (soon to be Truman State University) for their support and critical commentary on this project. Northeast Missouri State University provided a Summer Faculty Research Grant which allowed for the timely completion of this project. Finally, I remain indebted to my wife, Janice, for providing astute critique, able reference library assistance, and continual support and encouragement.

Philip K. Wilson

INTRODUCTION

Childbirth is undoubtedly a natural process. Yet over time, as man
has gained more and more control over the human reproductive
system, the natural process of conceiving, gestating, and bearing a
child has become increasingly structured and scientific. Sex is no
longer necessary for conception. An embryo can be flushed from
one womb and deposited in another. The natural process is manipulated to the point that it is almost *un*natural. Women should
not be slaves to their reproductive systems, but neither should
women's reproductive systems be slaves to technology. At what
point is scientific intervention too great? What dangers lie in
altering the process of nature?

Man has long sought to control fertility through various means—
be it with herbs or magical potions or coitus interruptus—with
varying degrees of success. However, neither government nor
society has fully approved of contraception despite its widespread
use. Birth control information was suppressed for many years. Not
until the 1920s and 1930s did an open movement for an effective
and reliable means of contraception begin.[1] Birth control advocates in Britain and America had various motives for participating
in the crusade for contraception, but one great underlying desire
lay at the heart of the struggle: women wanted control of their
bodies. "Contraception promised the final elimination of women's
only significant biological disadvantage," argued Linda Gordon in
Woman's Body, Woman's Right. "The capacity to reproduce is not a
disadvantage, but lack of control over it is."[2] The ability to bear
children is a very powerful thing, but not when a woman has no
choice in the matter.

Access to safe, effective birth control afforded women more
freedom and autonomy. To a great extent, they had control over
their fertility; they could now choose whether or not, and when, to
have a child. By the late 1950s, oral contraceptive pills were
developed and tested.[3] Today, there are a wide range of contraceptives available, including products such as the Norplant implant
and the Depo-Provera injection, which provide long-term contraception. There is even research on the development of a male pill.[4]

Abortion is an important topic in reproduction and the struggle for control of the woman's body. Due to the proliferation of information and discussion on the subject available elsewhere, we have not included it in this volume.[5]

The eugenics movement, which was founded in London on the ideas of Francis Galton and gained momentum in America around 1900, was intertwined with the movement for birth control. Eugenicists aimed to "improve the genetic condition of the human race"[6] by preventing those individuals deemed "socially inadequate" or physically "defective" from reproducing. Those labeled "socially inadequate" included paupers, criminals, the "feeble-minded," and the blind, among others.[7] Eugenicists supported birth control in order to keep more "defectives" from being born. Margaret Sanger, perhaps America's leading advocate for birth control (she coined the term), even associated herself to some degree with the eugenics movement.

Philip R. Reilly's article in the following pages discusses the extreme attempt to control reproduction through sterilization. Harry H. Laughlin, a leader in the eugenics movement in the early 1900s, was a great proponent of sterilization. Laughlin firmly believed that involuntary sterilization of all "defective" individuals was necessary for the good of society and the human race. He campaigned heartily for sterilization laws and by the end of the 1920s, twenty-four states had enacted them. By 1926, 6,244 individuals had been forcibly sterilized and thousands more would suffer the same fate in the next several years as a result of those laws.[8]

Not surprisingly, the excesses of the early eugenics advocates still arouse public concern over the future of reproductive technology. Medical scientists have simultaneously accumulated tremendous knowledge about reproduction and genetics while generating enormous social controversy. As reproductive science developed, doctors and scientists fought for even more scientific and social control over the reproductive process. For every assumed aspect of the natural process, they sought a detour and often found one. Neither sexual intercourse nor a womb is now necessary for conception. Making babies has, for some, evolved into a complex scientific procedure.

With the discovery of DNA in 1953, the mystery of our genetic make-up began to unravel. Not only did this newly uncoded molecule give scientists insight into the process of heredity, but within a score of years, they were recombining DNA molecules into what were essentially new life forms. In the 1970s, the popular press used Frankenstein-like imagery to provoke the public into urging British and American governments to curtail, regulate, or limit the applications of reproductive science. However, new procedures continued to develop, and now it is even possible to alter the genetic make-up of the unborn. Parents choosing their baby's sex may be but the first step towards creating "designer babies." In 1989, the Human Genome Project was undertaken with the goal to "map and sequence not only the genes but also the non-coding regions of all the DNA contained in the 24 human chromosomes."[9] The project has numerous implications. While it may unlock the secrets of incurable genetic afflictions, it is, for many, frightening to think what could be done with so much knowledge.[10]

What are the implications of this so-called scientific progress? How is the woman's role as childbearer changed? Eventually, women may not even be necessary in the birth process. Babies could be created without the womb. According to scientific reports around the world, it is theoretically possible for a man to have a baby. He could gestate the baby in his abdomen, as many women have done successfully, and have a cesarean birth.[11] A recent movie, *Junior*, features a pregnant Arnold Schwarzenegger as the first man to give birth. How far distant is this vision? The ability to give birth is one form of power and importance that women have exclusively from men. What if this powerful gift is taken away? What will happen to the woman's place in society?

Thousands of years ago, the Goddess was worshipped as the fertile mother of all things. Men revered and even envied women for their mystical ability to produce a child. Then a turning point was reached. Men realized that they, too, played a part in creation, and women lost their ethereal importance. The Goddess was "overthrown" by the God.[12] Now, as science is in the process of achieving complete control of the woman's reproductive system, the Goddess slips away still further. What will happen when all the magic and human warmth dissolves from pregnancy? When cold, hard science replaces the womb, the Goddess could die out altogether and take the woman with her.

Gena Corea claimed in *Mother Machine* that reproductive technologies "are transforming the experience of motherhood and placing it under the control of men."[13] Women gained some control with the advent of effective contraception, but this triumph is now overshadowed by the new technologies that are largely in the hands of male doctors and scientists. Women's bodies are being manipulated in the name of scientific progress, and to what end? Will the future imagined by Andrea Dworkin materialize in the form of "breeding brothels," where women's very wombs are prostituted?[14] Will Aldous Huxley's *Brave New World* vision in which humans are created in high-tech hatcheries finally come to fruition?

To some extent, it is necessary for individuals to have control over their reproductive systems. Overpopulation and other social problems make the option of birth control increasingly important. However, a point could be reached where humans have too much control over nature. Myriad unanswerable ethical questions arise in the application of new reproductive technologies. Certainly, they must be addressed. We have seen frightening visions of the reproductive future in numerous movies and books, such as Huxley's, and many shudder to think this future might not be too far away. Some argue that it is already here.

It is hoped that the articles in this volume incite readers to delve further into the scientific, social, and ethical issues of human genetics and reproduction. The first section of this volume examines both medical and social views of issues including prenatal diagnosis and "test tube" baby technology together with the growth of fields such as genetic counseling, fetal surgery, and surrogate motherhood. The second section is devoted to birth control in Britain and America over the past two hundred years. The practical methods of birth spacing, condoms, sterilization, and male contraceptives are discussed as are the social pressures and propaganda for and against birth control.

Philip K. Wilson

NOTES

1. James W. Reed, *From Private Vice to Public Virtue: The Birth Control Movement in American Society Since 1830* (New York: Basic Books, Inc., 1978).

2. Linda Gordon, *Woman's Body, Woman's Right: A Social History of Birth Control in America* (New York: Penguin Books, 1990), xvii.

3. Gregory Pincus. "Fertility Control with Oral Medication," *American Journal of Obstetrics and Gynecology* 75 (1958): 1333–46.

4. W.J. Bremner and D.M. de Kretser, "Contraceptives for Males," *Signs* 1, no. 2 (1975): 387–96.

5. Maureen Muldoon, *The Abortion Debate in the United States and Canada* (New York and London: Garland Publishing, Inc., 1991).

6. Kenneth M. Ludmerer, *Genetics and American Society: A Historical Appraisal* (Baltimore: The Johns Hopkins University Press, 1972), 2.

7. Philip R. Reilly, *The Surgical Solution: A History of Involuntary Sterilization in the United States* (Baltimore: The Johns Hopkins

University Press, 1991), 59.

8. Reilly, *The Surgical Solution,* 67.

9. Dorothy Nelkin and M. Susan Lindee, *The DNA Mystique: The Gene as a Cultural Icon* (New York: Freeman and Company, 1995), 5. Although each human sexual reproductive cell is composed of 23 chromosomes, there are 22 autosomes plus the x and y chromosomes under investigation.

10. Richard C. Lewontin addressed this concern in *Biology as Ideology: The Doctrine of DNA.* (New York: HarperCollins Publishers, Inc., 1993), 59–83.

11. Andrew Veitch, "How Men Might Be Able to Give Birth," *The Guardian* (May 24, 1983).

12. Gena Corea, *The Mother Machine: Reproductive Technologies from Artificial Insemination to Artificial Wombs* (New York: Harper and Row, 1985), 294.

13. Corea, *Mother Machine,* 289.

14. Dworkin's conceptualized vision in *Right-Wing Women* (New York: Perigee Books, 1983) is expanded in *Mother Machine,* 272–81.

FURTHER READING

Alpern, Kenneth D. The *Ethics of Reproductive Technology.* New York: Oxford University Press, 1992.

Arditti, Rita, et al. *Test-Tube Women: What Future for Motherhood?* Boston: Pandora Press, 1984.

Borell, Merriley. "Biologists and the Promotion of Birth Control Research, 1918–1938." *Journal of the History of Biology* 20, no. 1 (Spring 1987): 51–87.

Braude, Peter R. and Martin H. Johnson. "The Embryo in Contemporary Medical Science." *The Human Embryo: Aristotle and the Arabic and European Traditions,* edited by G.R. Dunstan, 208–21, Exeter: University of Exeter Press, 1990.

Brodie, Janet Farrell. *Contraception and Abortion in Nineteenth-Century America.* Ithaca and London: Cornell University Press, 1994.

Bullough, Vern L. and Bonnie Bullough. "A Brief History of Population Control and Contraception." *Free Inquiry* 14 (Spring 1994): 16–22.

Bullough, Vern L. and Bonnie Bullough. *Contraception: A Guide to Birth Control Methods.* Buffalo, N.Y.: Prometheus Books, 1990.

Chelsea, Ellen. *Woman of Valor: Margaret Sanger and the Birth*

Control Movement in America. New York: Simon and Schuster, 1982.

Corea, Gena, et al. *Man-Made Women: How New Reproductive Technologies Affect Women*. Bloomington: Indiana University Press, 1987.

Corea, Gena. *The Mother Machine: Reproductive Technologies from Artificial Insemination to Artificial Wombs*. New York: Perennial Library, 1985.

Duden, Barbara. *Disembodying Women: Perspectives on Pregnancy and the Unborn*. Cambridge, Mass. and London: Harvard University Press, 1993.

Gordon, Linda. *Woman's Body, Woman's Right: A Social History of Birth Control in America*. New York: Grossman Publishers, 1976. Revised edition, New York: Penguin Books, 1990.

Grant, Nicole J. *The Selling of Contraception: The Dalkon Shield Case, Sexuality, and Women's Autonomy*. Columbus: Ohio State University Press, 1992.

Haines, D.M. "History of Placental Studies." *Obstetrics and Gynecology Annals* 12 (1983): 1–14.

Harris, John. *Wonderwoman and Superman: The Ethics of Human Biotechnology*. Oxford and New York: Oxford University Press, 1992.

Harrison, Michael R.; Mitchell S. Golbus, and Roy A. Filly. *The Unborn Patient: Prenatal Diagnosis and Treatment*. Orlando, Florida: Grune and Stratton, 1984.

Holmes, Helen Bequaert, ed. *Issues in Reproductive Technology: An Anthology*. New York and London: Garland Publishing, Inc., 1992.

Hull, Richard T. *Ethical Issues in the New Reproductive Technologies*. Belmont, Calif.: Wadsworth Publishing Company, 1990.

Johnson, R. Christian. "Feminism, Philanthropy and Science in the Development of the Oral Contraceptive Pill." *Pharmacy in History* 19, no. 2 (1977): 63–78.

Jordanova, L.J. "Gender, Generation, and Science: William Hunter's Obstetrical Atlas." *William Hunter and the Eighteenth-Century Medical World*, edited by W.F. Bynum and Roy Porter, 385–412. Cambridge: Cambridge University Press, 1985.

Katz, Esther. "The History of Birth Control in the United States." *Trends in History* 4, no. 2–3 (1988): 81–101.

Kevles, Daniel J. *In the Name of Eugenics: Genetics and the Uses of Human Heredity*. Berkeley: University of California Press, 1985.

Langer, William. "The Origin of the Birth Control Movement in England in the Early Nineteenth Century." *Journal of Interdisciplinary History* 5 (1975): 669–86.

Laqueur, Thomas. "Orgasm, Generation, and the Politics of Reproductive Biology." *The Making of the Modern Body: Sexuality and*

Society in the Nineteenth Century, edited by Catherine Gallagher and Thomas Laqueur, 1–41, Berkeley: University of California Press, 1987.

Lauritzen, Paul. *Pursuing Parenthood: Ethical Issues in Assisted Reproduction.* Bloomington: Indiana University Press, 1993.

Lillie, F.R. "The History of the Fertilization Problem." *Science* 43 (1916), 39–53.

Martin, Emily. "The Egg and the Sperm: How Science has Constructed a Romance Based on Stereotypical Male-Female Roles." *Signs* 16, no. 3 (1991): 485–501.

Maynard-Moody, Steven. *The Dilemma of the Fetus: Fetal Research, Medical Progress, and Moral Politics.* New York: St. Martin's Press, 1995.

McCann, Carole R. *Birth Control Politics in the United States 1916– 1945.* Ithaca and London: Cornell University Press, 1994.

McLaren, Angus. *Birth Control in Nineteenth Century England.* New York: Holmes and Meier Publishers, Inc., 1978.

McLaren, Angus. *Reproductive Rituals: The Perception of Fertility in England from the Sixteenth Century to the Nineteenth Century.* New York and London: Methuen and Co., 1984.

Reed, James. *From Private Vice to Public Virtue: The Birth Control Movement and American Society Since 1830.* New York: Basic Books, Inc., 1978.

Robertson, John A. *Children of Choice: Freedom and the New Reproductive Technologies.* Princeton: Princeton University Press, 1994.

Rothenberg, Karen H. and Elizabeth J. Thomson, eds. *Women and Prenatal Testing: Facing the Challenges of Genetic Technology.* Columbus: Ohio State University Press, 1995.

Rothman, Barbara Katz. *The Tentative Pregnancy: Prenatal Diagnosis and the Future of Motherhood.* New York: Viking, 1986.

Rowland, Robyn. *Living Laboratories: Women and Reproductive Technologies.* Bloomington: Indiana University Press, 1992.

Shanley, Mary Lyndon. "Surrogate Mothering and Women's Freedom: A Critique of Contracts for Human Reproduction." *Signs* 18, no.3 (1993): 618–39.

Sherman, J. K. "Synopsis of the Art of Semen Banking." *Fertility and Sterility* 24 (1973): 397–412.

Singer, Peter. *Making Babies: The New Science and Ethics of Conception.* New York: Charles Scribner's Sons, 1985.

Spallone, Patricia and Deborah Lynn Steinberg, eds. *Made to Order: The Myth of Reproductive and Genetic Progress.* Oxford: Pergamon Press, 1987.

Spallone, Patricia. *Beyond Conception: The New Politics of Reproduction.* Granby, Mass.: Bergin and Garvey, 1989.

Stanworth, Michelle, ed. *Reproductive Technologies: Gender, Motherhood, and Medicine*. Cambridge: Polity Press, 1987.

Steinbock, Bonnie. *Life Before Birth: The Moral and Legal Status of Embryos and Fetuses*. New York: Oxford University Press, 1992.

Todd, Alexandra Dundas. *Intimate Adversaries: Cultural Conflict between Doctors and Women Patients*. Philadelphia: University of Pennsylvania Press, 1989.

Wertz, Dorothy C. "Ethical and Legal Implications of the New Genetics: Issues for Discussion." *Social Science and Medicine* 35, no. 4 (1992): 495–505.

Reproductive Science
and Genetics

KARYOTYPE

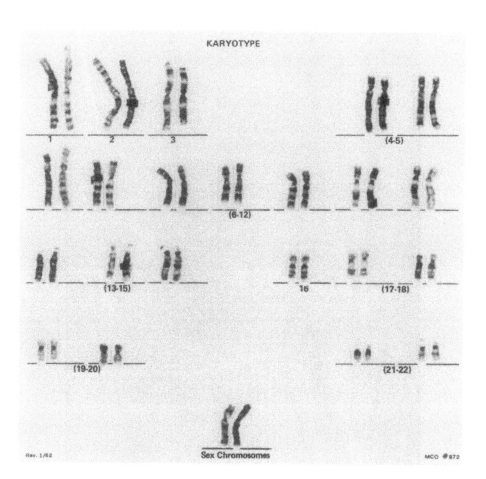

1	2	3		(4-5)

(6-12)

(13-15)   16   (17-18)

(19-20)   (21-22)

Sex Chromosomes

Rev. 1/82

MCO #872

FRONTISPIECE.

O fairest of creation, last and best,
Of all God's works, creature in whom excell'd
Whatever can to sight or thought be form'd
Holy, divine, good, amiable, or sweet!

MILTON.

THE
WORKS
OF
ARISTOTLE.

IN FOUR PARTS.

CONTAINING,

I. His COMPLETE MASTER-PIECE ; displaying the Secrets of Nature in the Generation of Man. To which is added, The FAMILY PHYSICIAN, being approved Remedies for the several Distempers incident to the Human Body.

II. His EXPERIENCED MIDWIFE ; absolutely necessary for Surgeons, Midwives, Nurses, and Child-bearing Women.

III. His BOOK of PROBLEMS ; containing various Questions and Answers relative to the State of Man's Body.

IV. His LAST LEGACY ; unfolding the Secrets of Nature respecting the Generation of Man.

FOURTH EDITION,

EMBELLISHED WITH SEVERAL FINE ENGRAVINGS.

LONDON:

PRINTED FOR THE BOOKSELLERS.

1822.

5

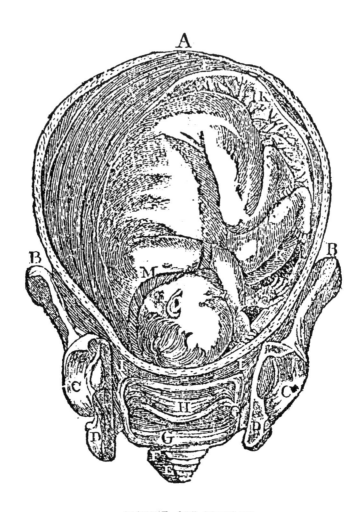

EXPLANATION.

A. The *uterus*, as stretched to near its full extent, containing the *fœtus* entangled in the *funis*.—B B. the superior part of the *ossa ilium*.—C C. The *acetabula*.—D D. The remaining posterior parts of the *ossa ischium*.—E. The *coccyx*.—F. the inferior part of the *rectum*.——G G. the *vagina* stretched on each side.—H. the *os uteri*, stretched to its full extent.—I I. part of the *versica urinaria*.—K K. the *placenta* at the superior and posterior parts of the *uterus*.—L. the *Membranes*.—M. the *funis umbilicalis*.

THE

EXPERIENCED MIDWIFE.

PART I.

A GUIDE TO CHILD-BEARING WOMEN.

INTRODUCTION.

I HAVE given this book the title of The Complete and Experienced MIDWIFE; both because it is chiefly designed for those that profess midwifery, and contains whatever is necessary for them to know in the practice thereof, and also because it is the result of many years experience, and that in the most difficult cases, and is therefore the more to be depended upon. A midwife is the most necessary and honourable office, being indeed a helper of nature; which therefore makes it necesary for her to be well acquainted with all the operations of nature in the work of generation, and instruments with which she works; for she that knows not the operations of nature, nor with what tools she works, must needs be at a loss how to assist therein. And seeing the instruments of operation both in men and women are those things by which mankind is produced, it is very necessary that all midwives should be well acquainted with them, that they may the better understand their business, and assist nature as there shall be occasion. The first thing then necessary, as introductory to this treatise, is an ANATOMICAL DESCRIPTION of the several parts of generation both of men and women: and having designed throughout to comprehend much in a little room. I shall avoid all unnecessary and impertinent matters with which books of this nature are for the most part too much clogged; and which are more than needful. And though I shall be necessiated to speak plainly, that so I may be understood, yet I shall do it with that modesty, that none shall have need to blush, unless it be from something in themselves, rather than from what they shall find here; having the motto of the royal garter for my defence, which is, " *Honi soit qui mal y pense.*" Evil be to him that evil thinks.

7

CHAP. I.

An Anatomical Description of the Instruments of Generation both in Men and Women.

SECT. I. *Of the parts of Generation in Man.*

AS the generation of mankind is produced by the coition of both sexes, it necessarily follows, that the instruments of generation are of two sorts, male and female : the operations of which are by action and passion ; and therein the agent is the seed, and the patient blood ; whence we may easily collect, that the body of man being generated by action and passion, he must needs be subject thereunto during his life. Now, since the instruments of generation are male and female, it will be necessary to treat of them both distinctly, that the honest and discreet midwife may be well acquainted with their several parts, and their various operations, as they contribute to the work of generation. And, in doing this, I shall give the honour of precedence to my own sex, and speak first of the parts of generation in man, which will be comprehended under six particulars, viz. The preparing vessels, the corpus varicolum, the testicles, or stones, the vasa deferentia, the seminal vessels, and the yard. Of each of which in their order.

1. The first are the vasa preparentia, or preparing vessels, which are in number four, two veins, and as many arteries, and they are called preparing vessels from their office, which is to prepare that matter or substance which the stones turn into seed, to fit it for the work. Whence you may note, that the liver is the original of blood, and distributes it through the body, by the veins and not by the heart, as some have taught. As to the original of these veins, the right vain proceedeth from the vena cava, or great vein, which receives the blood from the liver, and distributes it by its branches to all the body ; the left is fro m the emulgent vein, which is one of the two main branches of the hollow veins passing to the reins. As to the arteries, they both rise from the great artery, which the Greeks call that which is indeed the great trunk, and original of all the arteries. But I will not trouble you with Greek derivations of words, affecting more to teach you the knowledge of things and words.

8

2. The next thing to be spoken of is the corpus varicorsum, and this is an interweaving of veins and arteries which carry the vital and natural blood to the stones to make seed of. These, though at their first declension they keep at a small distance the one from the other, yet before they enter the stones they make an admirable intermixture of twisting the one from the other, so that sometimes the veins go into the arteries, and sometimes the arteries into the veins ; the substance of which is very hard and long, not much unlike a pyramid in form, without any sensible hollowness ; the use is to make one body of the blood and vital spirit, which they both mix and change the colour of from red to white, and also the stones may both have a fit matter to work upon, and do their work more easily ; for which reason, the interweaving reacheth down to the stones, and pierceth in their substance.

3. The stones are the third thing to be spoken of, called also testicles ; in Latin, Teste : that is a witness, because they witness one to be a man. As to these I need not tell you their number nor where nature has placed them; for that is obvious to the eye. Their substance is soft, white, spongy, full of small veins and arteries, which is the reason they swell to such a bigness upon the flowing down of the humour in them. Their form is oval ; but most authors are of opinion that their bigness is not equal, but that the right is the biggest, the hottest and breeds the best and strongest seed. Each of these stones, hath muscles called cremaster ; which signifies to hold up, because they pull up the stones in the act of coition, that so the vessels being slackened may the better void the seed.— These muscles are weakened both by age and sickness ; and then the stones hang down lower than in youth and health. These stones are of great use, for they convert the blood and vital spirits into seed for the procreation of man ; but this must not be understood as if they converted all the blood that comes into them into seed for they keep some for their own nourishment. But besides this they add strength and courage to the body ; which is evident from this, that eunuchs are neither so hot, strong, nor valiant, as other men, nor is an ox so hot or valiant as a bull.

4. The next in order are the vasa deferentia, which are the vessels that carry the seed from the stones to the seminal vessels, which is kept there till its expulsion. These are in number two, in colour white, and in substance nervous or sinewy ;

K

and from certain hollowness which they have in them are also spermatic pores. They rise not far from the preparing vessels; and when they come into the cavity of the belly, they turn back again, and pass into the back side of the bladder, between it and the right gut ; and when they come near the neck of the bladder, they are joined to the seminal galls, which somewhat resemble the cells of a honey comb ; which cells contain an oily substance, for they draw the fatty substance from the seed, which they empty into the urinal passage, which is done for the most part in the act of copulation, that so the thin internal skin of the yard, suffers not through the acrimony or sharpness of the seed. And when the vasa deferentia has passed, as before narrated, they fall into the glandula prostrata, which are the vessels by nature ordained to keep the seed, and which are next to be spoken of.

5. The seminal vessels, called glandulum seminale, are certain kennels placed between the neck of the bladder and the right gut, composing about the vassa deferentia, the urethra or common passage for seed and urine, passing through the midst of it, and may properly enough be called the conduit of the yard. At the mouth of the urethra, where it meets with the vasa deferentia, there is a thick skin, whose office is to hinder the seminal vessels, which are of a spongy nature, from shedding their seed against their will : this skin is very full of pores, and through the heat of the act of copulation the pores open, and so give passage to the seed, which being of a very subtle spirit, and especially being moved, will pass through this caruncle or skin as quicksilver through leather ; and yet the pores of this skin are not discernable unless in the anatomy of a man who had a violent running in the reins when he died, and then they are conspicuous, those vessels being the proper seat of that disease.

6. The last of the parts of generation in man to be spoken of is the yard, which has a principal share in the work of generation, and is called penis, from its hanging without the belly ; and it consists of skin, tendons, veins, arteries, sinews, and great ligaments. and is long and round, being ordained by nature both for the passage of the urine, and for the conveying of seed into the matrix. It hath some parts common with it to the rest of the body, as the skin, or the membrana carnosa ; and some parts it has peculiar to itself, as the two nervous bodies, the septum, the urethra, the glands, the four muscles, and the vessels. The skin, which the Latins call cutis, is full

of pores, through which the sweet and fulliginous or sooty black vapours of the third concoction (which concocts the blood into flesh) pass out: these pores are very many and thick, but hardly visible to the eye : and when the yard stands not, it is flaggy ; but when it stands it is stiff : the skin is very sensible, because the nerves concur to make up its being ; for the brain gives sense to the body by the nerves.   As to the cernus membrana, or fleshy skin it is so called, not because its body is fleshy, but because it lies between the flesh, and passeth in other parts of the body, underneath the fat, and sticks close to the muscles ; but in the yard there is no fat at all, only a few superficial veins and arteries pass between the former skin and this, which when the yard stands are visible to the eye : these are the parts common both to the yard and the rest of the body.   I will now speak of those parts of the yard which are peculiar to itself, and to no other parts of the body : and those are likewise six, as has been already said, of which it will also be necessary to speak particularly.   And

1. Of the nervous bodies, there are two, though joined together, and are hard, long and sinewy, they are spongy within, and full of black blood ; the spongy substance of the inward part of it seems to be woven together like a net, consisting of innumerable twigs or veins and arteries.   The black blood contained therein is very full of spirits, and the delights or desire of Venus adds heat to these, which causeth the yard to stand ; and that is the reason that both venereal sights and tales will do it.   Nor needs it be strange to any, that Venus, being a plant cold and moist, would add heat to those parts, since by night, as the Psalmist testifies, Psal. cxxi. 6.   Now this hollow spongy intermixture or weaving was so ordered by nature, on purpose to contain the spirit of venereal heat, that the yard may not fall before it has done its work.   These two side ligaments of the yard, where they are thick and round, arise from the lower part of the share-bone, and at the beginning are separated the one from the other, resembling a pair of horns, or the letter Y, where the urethra or common passage of urine and seed passeth between them.

2. These nervous bodies of which I have spoken, so soon as they come to the joining of the share bone, are joined by the septum lucidum, which is the second internal part to be described, which in substance is white and nervous or sinewy, and its use is to uphold the two side ligaments and the urethra.

K 2

11

2. The third thing in the internal parts of the yard is the urethra, which is the passage or channel by which both the seed and urine is conveyed out through the yard. The substance of it is sinewy, thick, soft and loose, as the side ligaments are ; it begins at the neck of the bladder, and being joined to it, passeth to the glands. It has at the beginning of it three holes, of which the largest of them is in the midst, which receives the urine into it ; the other two are smaller by which it receives the seed from each seminal vessel.

4. The yard has four muscles, on each side two: these muscles are instruments of voluntary motion, without which no part of the body can move itself. It consists of fibrous flesh to make its body, of nerves for its sense, of veins for its nourishment, of arteries for vital heat, of a membrane or skin to knit it together, and to distinguish one muscle from another, and all of them from the flesh ; of these muscles as I said before, the yard has two on each side, and the use of them is to erect the yard and make it stand, and therefore they are also called erectors. But here you must note, that of the two on each side, the one is shorter and thicker than the other ; and these are they that do erect the yard, and so are called erectors ; but the two others being longer and smaller, their office is to dilate the lower part of the urethra, both for making water and emitting seed ; upon which account they are called accelerators.

5. That which is called the glands, is the extreme part of the yard, which is very soft and of a most exquiste feeling, by reason of the thinness of the skin wherewith it is covered ; this is covered with the præputium, or fore-skin, which in some men covers the top of the yard quite-close, but in others it doth not ; which skin moving up and down in the act of copulation, brings pleasure both to men and women. This outer skin is that which the jews were commanded to cut off on the eighth day. This præputium, or fore-skin, is tied to the glands by a ligament or bridle, which is called frænum.

6. The last internal parts of the yard are the vessels thereof, veins, nerves, and arteries. Of these some pass by the skin, and are visible to the eye when the yard stands : others pass by the inward parts of the yard : the arteries are wonderfully dispersed through the body of the yard, much exceeding the dispersion of the veins ; for the right artery is dispersed to the left side, and the left to the right side. It hath two nerves, the lesser whereof is bestowed upon the skin, the greater upon

the muscles and body of the yard. But this much shall suffice to be said in describing the parts of generation in men ; and shall therefore in the next place, proceed to describe those of women, that so the honest and industrious midwife may know how to help them in their extremities.

SECT. II. *Describing the Parts of Generation in Women.*

WHATEVER ignorant persons may imagine, or some good women think, they are unwilling those privy parts which nature has given them should be exposed, yet it is in this case absolutely necessary ; for I do positively affirm, that it is impossible truly to apprehend what a midwife ought to do, if these parts are not perfectly understood by them ; nor do I know any reason they have to be ashamed to see or hear a particular description of what God and nature has given them, since it is not the having these part, but the unlawful use of them, that causeth shame.

To proceed then in this description more regularly, I shall speak in order of these following principle parts ; first, of the privy passage ; secondly, of the womb ; thirdly, of the testicles or stones ; fourthly, of the spermatic vessels.

First, of the privy passage. Under this head I shall consider the six following parts :

1. The lips which are visible to the eve, and are designed by nature as a cover to the fissura magna, or great orifice : these are framed of the body, and have a pretty store of spongy fat : and their use is to keep the internal parts from cold and dust. These are the only things that are obvious to the sight, the rest are concealed, and cannot be seen, unless these two lips are stretched asunder, and the entry of the privities opened.

2. When the lips are severed, the next that appears is the nymphæ, or wings ; they are formed of soft and spongy flesh, and are in form and colour like the comb of a cock.

3. In the uppermost part, just above the urinary passage, may be observed the clitoris, which is a sinewy and hard body full of spongy and black matter within, like the side ligament of the yard, representing in form the yard of a man, and suffers erection and falling as that doth ; and it grows hard, and becomes erected as a man's yard, in proportion to the desire a woman hath in copulation ; for without this a woman hath neither a desire to copulation, nor delight in it, nor can con-

K 3

13

ceive by it. And I have heard that some women have had their clitoris so long that they have abused other women therewith; nay, some have gone so far as to say that those persons that have been reported to be hermaphrodites, as having the genitals both of men and women, are only such women to whom the clitoris hangs out externally, resembling the form of a yard. But though I will not be positive in this, yet it is certain, that the larger the clitoris is in any woman, the more lustful she is.

4. Under the clitoris, and above the neck, appears the orifice, or urinary passage, which is much larger in women than in men, and causes the water to come from them in a greater stream. On both sides the urinary passage may be seen two small membranous appendices, a little broader above than below, issuing forth of the inward parts of the great lips, immediately under the clitoris ; the use whereof is to cover the orifice of the urine, and defend the bladder from the cold air : so that when a woman makes water, she contracts herself so, that she conducts out the urine without suffering it to spread along the privities, and often without so much as wetting the lips : and therefore these small membranous wings are called the nymphæ, because they govern womens' water. Some women have them so great and long that they have been necessiated to cut off so much as has exceeded and grew without the lips.

5. Near this are four caruncles, or fleshy knobs, commonly called caruncles myrtifermes ; these are placed on each side two, and a small one above, just under the urinary passage, and in virgins are reddish, plump, and round, but hang flagging when virginity is lost. In virgins they are joined together by a thin and sinewy skin or membrane, which is called the hymen, and keeps them in subjection, and makes them resemble a kind of rose bud half blown. This disposition of the caruncles is the only certain sign of virginity, it being in vain to search for it elsewhere, or, hope to be informed of it any other way ; and it is from the passing and bruising those caruncles, and forcing and breaking the little membranes, (which is done by the yard in the first act of copulation) that there happens an effusion of blood ; after which they remain separated, and never recover their first figure, but become more and more flat as the acts of copulation are increased ; and in those that have children they are almost totally defeated, by reason of the great distention these parts suffer in the time of their labour. Their use is to straiten the neck of the womb,

to hinder the cold air from incommoding it, and likewise to increase mutual pleasure in the act of coition ; for the caruncles being then extremely swelled, and filled with blood and spirits, they close with more pleasure upon the yard of the man, whereby the woman is much more delighted. What I have said of the effusion of blood which happens in the first act of copulation, though when it happens it is an undoubted sign of virginity, showing the caruncles myrtifermes have never been pressed till then, yet when there happens no blood, it is not always a sign that virginity is lost before ; for the hymen may be broken without copulation by the defluxion of sharp humours, which sometimes happens to young virgins, because in them it is thinner : It is also done by the unskilful applying of pessaries to provoke the terms, &c. But these things happen so rarely, that those virgins to whom it so happens do thereby bring themselves under a just suspicion.

6. The next to be spoken of is the neck of the womb, which is nothing else but the distance between the privy passage and the mouth of the womb, into which the man's yard enters in the act of copulation ; and in women of reasonable stature is about eight inches in length. 'Tis of a membranous substance, fleshy without, skinny and very much wrinkled within, that it both may retain the seed cast into it in the act of copulation, and also that it may dilate and extend itself to give sufficient passage to the infant at its birth. It is composed of two membranes, the innermost of them being white, nervous and circularly wrinkled, much like the palate of an ox, that so it might either contract or dilate itself according to the bigness or length of the man's yard ; and to the end that by the colation, or squeezing, or pressing made by the yard in copulation, the pleasure may be naturally augmented. The external or outward membrane is red and fleshy, like the muscle of the fundament, surrounding the first, to the end the yard may be the better closed within it : and it is by means of this membrane that the neck adheres the stronger both to the bladder and the right gut. The internal membrane in young girls is very soft and delicate, but in women much addicted to copulation it grows harder : and in those that are grown aged, if they have been given much to venery, is almost become grisly.

Secondly, having spoken of the privy passage, I come now to speak of the womb, which the Latins call matrix, yet the only English word is the womb. Its parts are two ; the mouth of the womb and the bottom of it. The mouth is an orifice

at the entrance into it, which may be dilated and shut together like a purse; for although in the act of copulation it be
big enough to receive the glands of the yard, yet after conception it is so close shut, that it will not admit the point of a bod-
kin to enter; and yet again at the time of the woman's delivery it is opened so extraordinary, that the infant passeth
through it into the world; at which time this orifice wholly
disappears, and the womb seems to have but one great cavity,
from its bottom to the very entrance of the neck.   When a
woman is not with child it is a little oblong, and of substance
very thick and close; but when she is with child it is shorten-
ed, and its thickness diminisheth proportionably to its distention: and therefore it is a mistake of some anatomists to affirm, that its substance waxeth thicker a little before a woman's labour; for any one's reason will inform them, that the
more distended it is, the thinner it must be, and the nearer a
woman is to the time of her delivery the greater her womb
must be extended.   As to the action by which this inward orifice of the womb is opened and shut, it is purely nature; for
were it otherwise, there would not be so many bastards begot-
ten as there are; nor would many married women have so
many children were it at their own choice, for they would hinder conception, though they would be willing enough to use copulation; for nature has attended that action with something
so pleasing and delightful, that they are willing to indulge
themselves in the use thereof, notwithstanding the pains they
afterwards endure, and the hazard of their lives which often
follow it; and this comes to pass not so much from any inordinate lust in women, as for that the Great Director of nature, for the increase and multiplication of mankind, and even
all other species in the elementary world, hath placed such a
magnetic virtue in the womb, that it draws the seed to it as
a loadstone draws iron.

The Author of nature has placed the womb in the belly,
that the heat may always be maintained by the warmth of
the part surrounding it; it is therefore seated in the middle
of the hypogastrium, (or lower part of the belly) between the
bladder and the rectum, (or right gut) by which also it is defended from any hurt through the hardness of the bones;
and is placed in the lower part of the belly for the conveni-
ence of copulation, and of a births being thrust out at the full
time.

It is of a figure almost round, inclining somewhat to an oblong, in part resembling a pear, for from being broad at the bottom, it gradually terminates in the point of the orifice which is narrow.

The length, breadth and thickness of the womb differs according to the age and disposition of the body : for in virgins not ripe it is very small in all its dimensions, but in women whose terms flow in great quantities, and such as frequently use copulation, it is much larger ; and if they have had children, it is larger in them than such as have none ; but in women of good stature, and well shaped, it is (as I have said before) from the entry of the privy parts to the bottom of the womb usually about eight inches, but the length of the body of the womb alone does not exceed three inches, the breadth thereof is near about the same, and of the thickness of the little-finger, when the woman is not pregnant ; but when the woman is with child it becomes of a prodigious greatness, and the nearer she is to her delivery, the more is the womb extended.

It is not without reason then that nature (or the God of nature rather) has made the womb of a membranous substance ; for thereby it does the easier open to conceive, and is gradually dilated from the growth of the fœtus, or young one, and is afterwards contracted and closed again, to thrust forth both it and the after-burden, and then to retire to its primitive seat.   Hence also it is enabled to expel any noxious humours which sometimes happen to be contained within it.

Before I have done with the womb, which is the field of generation, and ought therefore to be the more particularly taken care of, (for as the seeds of plants can produce no fruits, nor spring, unless sown in ground proper to waxen and excite their vegetative virtue, so likewise the seed of a man, though potentially containing all the parts of a child, would never produce so admirable an effect if it were not cast into the fruitful field of nature, the womb) I shall proceed to a more particular description of the parts thereof, and the use to which nature has designed them.

The womb then is composed of various similar parts, that is, of membranes, veins, arteries, and nerves. Its membranes are two, and they compose the principal parts of its body ; the outmost of which ariseth from the peritoneum, or cawl, and is very thin, without smooth, but within equal, that it may the better cleave to the womb, as it were fleshy and thicker

17

than any else we met with in the body when a woman is not pregnant, and it is interwoven with all sorts of fibres or small strings, that may the better suffer the extention of the child, and the waters caused during the pregnancy, and also that it may the easier close again after delivery.

The veins and arteries proceed both from the hypogastrics and the spermatic vessels, of which I shall speak by and by ; all these are inserted and terminated in the proper membrane of the womb. The arteries supply it with blood for its nourishment, which being brought thither in too great a quantity, sweats through the substance of it, and distils, as if it were a dew, into the bottom of its cavity ; from whence do proceed both the terms in ripe virgins, and the blood which nourisheth the embryo in breeding women. The branches which issue from the spermatic vessels are inserted in each side of the bottom of the womb, and are much less than those which proceed from the hypogastrics, those being greater and bedewing the whole substance of it. There are yet some other small vessels, which arising the one from the other, and conducted to the internal orifice, and by these, those that are pregnant do purge away the superfluity of their terms when they happen to have more than is used in the nourishment of the infant, by which means nature hath taken such care of the womb, that during its pregnancy, it shall not be obliged to open itself for the passing away of those excrementious humours, which, should it be forced to do, might often endanger abortion.

As touching the nerves they proceed from the brain, which furnishes all the inner parts of the lower belly with them, which is the true reason it hath so great a sympathy with the stomach, which is likewise very considerably furnished from the same part ; so that the womb cannot be affected with any pain but the stomach is immediately sensible thereof, which is the cause of those loathings or frequent vomitings that happen to it.

But beside all these parts which compose the womb, it hath yet four ligaments, whose office is to keep it firm in its place, and prevent its constant agitation by the continual motion of the intestines which surround it, two of which are above and two below : those above are called the broad ligaments, because of their broad and membranous figure, and are nothing else but the production of the peritoneum, which growing out of the side of the loins towards the reins, come to be inserted

In the sides of the bottom of the womb, to hinder the body from bearing too much on the neck, and so from suffering a precipitation, as will sometimes happen when the ligaments are too much relaxed; and do also contain the testicles, and as well safely conduct the different vessels as the ejaculatories to the womb. The lowermost are called round ligaments, taking their original from the side of the womb near the horn, from whence they pass the groin, together with the productions of the peritoneum, which accompanies them through the rings and holes of the oblique and transverse muscles of the belly, which divide themselves into many little branches, resembling the foot of a goose, of which some are inserted into the os pubis, and the rest are lost and confounded with the membranes that cover the upper and interior parts of the thigh; and it is that which causeth the numbness which women with child feel in their thighs. These two ligaments are long, round, and nervous, and pretty big in their beginning near the matrix, hollow in their rise, and all along to the os pubis, where they are a little smaller, and become flat, the better to be inserted in its manner aforesaid: it is by their means the womb is hindered from rising too high. Now, although the womb is held in its natural situation by means of these four ligaments, yet it has liberty enough to extend itself when pregnant, because they are very loose, and so easily yield to its distention. But besides these ligaments, which keep the womb as it were in a poise, yet it is fastened, for greater security, by its neck, both to the bladder and rectum, between which it is situated. Whence it comes to pass, that if at any time the womb be inflamed it communicates the inflamation to the neighbouring parts.

Its use or proper action in the work of generation is to receive and retain the seed, and to reduce it from power to action by its heat, for the generation of the infant, and is therefore absolutely necessary for the conversion of the species. It also seems by accident to receive and expel the impurities of the whole body, as when women have abundance of whites, and to purge away from time to time the superfluity of the blood, as it doth every month by the evacuation of the blood, as when a woman is not with child. And thus much shall suffice for the description of the womb, on which I have been the larger, because, as I have said before, it is the field of generation,

3dly. The next thing to be described in the genitals of women is the testicles or stones, for such women have as well as
men, but are not for the same use, and indeed are different to
those of men in several particulars : as, 1st, in place, being
within the belly ; whereas in men they are without. 2dly. in
figure, being uneven in women, but smooth in men. 3dly, in
magnitude, being lesser in women than in men. 4thly, they
are not fixed in women by muscles, but by ligatures. 5thly,
they have no prostrates or kernel, as men have. 6thly, they
differ in form being depressed or flattish in women, but oval
in men. 7thly, they have but one skin, where men have
four ; for the stones of men being more exposed, nature has
provided for them accordingly. 8thly, their substance is more
soft than in men. And 9thly, their temperature is colder than
men. And as they differ in all these respects, so do they also
in their use, for they perform not the same actions as men, as
I shall show presently. As for their seat, it is in the hollowness of the abdomen, and therefore not extremely pendulous,
but rest upon the ova or egg. 'Tis true Galen and Hippocrates did erroneously imagine, that the stones in women did
both contain and elaborate the seed, as those do in men, but
it is a great mistake ; for the testicles of a woman are as it
were no more than two cluster of eggs, which lie there to be
impregnated by the most spirituous particles, or animating
effluviums conveyed out of the womb through two tubes, or
different vessels ; but however, the stones in women are very
useful ; for where they are defective, generation work is at an
end. For though those little bladders, which are on their su-
perfices contain nothing of seed, yet they contain several eggs,
(commonly to the number of twenty in each testicle) one
of which being impregnated in the act of coition, by the most
spirituous part of the seed of the man, descends through the
oviducts in the womb, and there in process of time becomes
a living child.

4thly, I am now to speak of spermatic vessels in women,
which are two, and are fastened in their whole extent by a
membranous appendix to the broad ligaments of the womb ;
these do not proceed from testicles as in men, but are distant
from them a finger's breadth at least ; and being disposed
after the manner of the miseriac veins, are tainted along this
membranous distance between the different vessels and the
testicles. Their substance is, as it were, nervous and moderately hard ; they are round, hollow, big, and broad enough at

the end, joining to the horn of the womb. Some authors affirm, that by these, women discharge their seed into the bottom of the womb ; but the whole current of our modern authors run quite another way, and are positive that there is no seed at all in their vessels ; but that after the egg or eggs, in the ovaria or testicles, are impregnated by the seed of the man, they descend through these two vessels into the womb, where being placed, the embryo is nourished. These vessels are shorter in women than they are in men, for the stones of a woman lying within the belly their passage must needs be shorter : but their various wreathings and windings in and out make amends for the shortness of their passage. These vessels are not united before they come to the stones, but divide themselves into two branches, whereof the biggest end passes through the testicles, the lesser to the womb, both for the nourishment of itself and the infant in it. I will only observe further that these spermatic veins receive the arteries as they pass by the womb, and so there is a mixture between vital and natural blood, that so the work might be the better wrought ; and that it is so, appears by this, that if you blow up the spermatic vein, you may perceive the right and left vessel of the womb blow up ; from whence also the communion of all the vessels of the womb may be easily perceived.

The deferentia, or carrying vessels, spring from the lower part of the testicles, and are in colour white, and in substance sinewy, and pass not to the womb straight but wreathed with several turnings and windings, as was said of the spermatic vessels, that so the shortness of the way may be likewise recompensed by their winding meanders ; yet near the womb they become broad again. They proceed in two parts from the womb, which resemble horns, and are therefore called the horns of the womb. And this is all that is needful to be known or treated of, concerning the parts of generation both in men and women.

Only since our modern anatomists and physicians are of different sentiments from the ancients, touching the woman's contributing of seed for the formation of the child as well as the man ; the ancients strongly affirming it, but our modern authors being generally of another judgment ; I will here declare the several reasons for this difference and so pass on.

L

SECT. III. *Of the Difference between the ancient and modern Physicians, touching the Woman's contra-united Seed to the formation of the Child.*

I WILL not make myself a party into this controversy, but set down impartially and yet briefly, the arguments on each side, and leave the judicious reader to judge for himself.

Though it is apparent, say the ancients, that the seed of man is the principal efficient, and beginning of action, motion and generation, yet that the woman affords seed, and contributes to the procreation of the child, is evident from hence, that the woman has seminal excrescence; but since nature forms nothing in vain, it must be granted they were made for the use of seed and procreation, and fixed in their proper places to operate, and contribute virtue and efficacy to the seed; and this, say they, is further proved from hence, that if women at the age of maturity use not copulation to eject their seed, they often fall into strange diseases, as appears by young women and virgins: and also it is apparent, that women are never better pleased than when they are often satisfied this way, which argues the pleasure and delight then take herein; which pleasure and delight say they, is double in women to what it is in men; for, as the delight in men in copulation consists chiefly in emission of their seed, so women are delighted both in the emission of their own, and the reception of the man's.

But against this all our modern authors affirm, that the ancients were very erroneous; forasmuch as the testicles in women do not afford seed, but are two eggs, like those of fowls and other creatures, neither have they any such office as men, but indeed are an ovarium, or receptacle for eggs, wherein these eggs are nourished by the sanguinary vessels dispersed through them; and from thence one or more, as they are fœcundated by the man's seed, are conveyed into the womb by the oviducts. And the truth of this, say they, is so plain that if you boil them, their liquor will have the same taste, colour, and consistency, with the taste of bird's eggs. And if it be objected that they have no shells, the answer is easy; for the eggs of fowls, while they are in the ovary, nay after they are fallen into the uterus, have no shell: and though they have one when they are lain, yet it is no more than a fence which nature has provided for them against outward injuries, they being hatched without the body; but those of women being hatched within the body, hath no need of any fence than the womb to secure them.

They also further say, there are in the generation of the fœtus, or young one, two principles, active and passive: the active is the man's seed, elaborated in the testicles, out of the arterial blood and animal spirits; the passive principle is the ovum, or egg, impregnated by the man's seed; for to say that women have true seed (they say) is erroneous. But the manner of conception is this; the most spirituous part of the man's seed in the act of copulation, reaching up to the ovarium, or testicles of the woman,) which contains divers eggs, sometimes more, sometimes fewer) impregnates one of them, which, being covered by the oviducts at the bottom of the womb, presently begins to swell bigger and bigger, and drinks in the moisture that is plentifully sent thither, after the same manner that the seed in the ground sucks the fertile moisture thereof to make them sprout.

But notwithstanding what is here urged by our modern anatomists, there are some late authors of the opinion of the ancients, viz. that women hath both, and emit seed in the act of copulation; and the good women themselves take it ill to be thought merely passive in those wars wherein they make such vigorous encounters, and positively affirm they are sensible of the emission of their seed in those engagements, and that in it a great part of the delight which they take in that act consists: I will not therefore go about to take any of their happiness away from them, but leave them in the possession of their imagined felicity.

Having thus laid the foundation of this work, in the description that I have given of the parts dedicated to the work of generation both in man and woman, I will now proceed to speak of conception, and of those things that are necessary to be observed by women from the time of their conception to the time of their delivery.

CHAP. III.

Of Conception, what it is, the Signs thereof, whether conceived of a Male or Female; how Women are to order themselves after Conception.

SECT. 1. *What Conception is, and the requisites thereto.*

CONCEPTION is nothing else but an action of the womb, by which the prolific seed is received and retained, that

L 2

an infant may be engendered and formed out of it. There are two sorts of conception; the one according to nature, which is followed by the generation of the infant in the womb; the other is false and wholly against nature, in which the seed changes into water, and produces only false conception, moles, or other strange matter. Now here are three things principally necessary in order to a true conception, so that generation may follow; to wit, diversity of sex, congression and emission of seed. Without diversity of sexes there can be no conception: for though some will have a woman to be an animal that can engender of herself, it is a great mistake; there can be no conception without a man to discharge his seed into her womb. What they allege of pullets laying eggs without a cock's treading them is nothing to the purpose; for those eggs, should they be set under a hen, will never become chickens, because they never received any prolific virtue from the male; which is absolutely necessary to this purpose, and is sufficient to convince us, that diversity of sex is necessary even to those animals as well as to the generation of man. But diversity of sex though it be necessary to conception, yet it wont do alone: there must also be a congression of those different sexes; for diversity of the sex would profit little if copulation did not follow. I confess I have heard of some subtle women, who to cover their sin and shame, have endeavoured to persuade some peasant that they were never touched by man to get them with child; and that one in particular pretended to conceive by going into a bath where a man had washed himself a little before, and spent his seed in it, which was drawn and sucked into her womb, as she pretended; but such stories as those are only fit to amuse them that know no better.—Now that these different sexes should be obliged to come to the touch, which we call copulation or coition, besides the natural desire of begetting their like, which stirs up men and women to it, the parts appointed for generation are endowed by nature with a delightful and mutual itch, which begets in them desire to the action; without which, it would not be very easy for a man, born for the contemplation of divine mysteries, to join himself by the way of coition, to a woman, in regard of the uncleanliness of the part, and of the action; and on the other side, if women did but think of those pains and inconveniences to which they are subject by their great bellies, and those hazards even of life itself, it is resonable to suppose they would be affrighted from it. But neither sex make these reflections till after the action is over, considering nothing beforehand but the pleasure of enjoyment. So that it is from this voluptuous itch that nature obligeth both sexes to this congression. Upon which the third thing followeth of course, to wit the emission of seed into the womb in the act of copulation. For the woman having received this prolific seed into her womb, and retained it there, the womb thereupon becomes compressed, and embraces the seed so closely, that being closed, the point of a needle, as saith Hippocrates, cannot enter it without violence ; and now the woman may be said to have conceived ; being reduced by its heat from power into action, the several faculties which are in the seed it contains making use of the spirits with which the seed abounds, and which are the instruments by which it begins to trace out the first lineaments of all the parts ; to which afterwards, making use of the menstruous blood flowing to it, it gives in time growth and final perfection. And thus much shall suffice to shew what conception is.

SECT. II. *The Signs of Conception.*

THERE are many prognostics or signs of conception : I will name some of the chief, which are the most certain, and let alone the rest.

1. If a woman has been more than ordinarily desirous of copulation, and has taken more pleasure than usual therein (which upon recollection she may easily know) it is a sign of conception.

2. If she retain the seed in her womb after copulation ; which she may know, if she perceives it not to flow down from the womb as it was used to do before ; for that is a sure sign the womb has received it into the inward orifice, and there retains it.

3. If she finds a coldness and chillness after copulation, it shows the heat is retired to make conception.

4th. If, after this, she begins to have loathings to those things which she loved before, and this attended with a loss of appetite, and a desire for meats to which she was not affected before, and has often nauseating and vomitings, with sour belchings, and exceeding weakness of stomach.

6. After conception the belly waxeth very flat, because the

L 3

womb closeth itself together, to nourish and cherish the seed contracting itself so as to leave no empty space.

6. If the veins of the breast are more clearly seen than they were wont to be, it is a sign of conception.

7. So it is if the tops of the nipples look redder than formerly, and the breasts begin to swell and grow harder than usual, especially if this be attended with pain and soreness.

8. If a woman has twisting and griping pains, much like those of the cramp in the belly, and about the navel, it is a sign she has conceived.

9. If under the lower eye-lid the veins be swelled, and appear clearly, and the eye be something discoloured, it is a certain sign she is with child, unless she have her menses at the same time upon her, or that she hath sat up the night before. This sign has never failed.

10. Some also make this trial of conception; they stop the woman's urine in a glass or phial for three days, and then strain it through a fine linen cloth, and they will find small living creatures in it, they conclude, that the woman has certainly conceived.

11. There is also another easy trial; let the woman that supposes she has conceived, take a green nettle and put it in her urine, cover it close, and let it remain therein a whole night : if the woman be with child it will be full of red spots on the morrow, but if she be not with child it will be blackish.

12. The last sign I shall mention, is that which is most obvious to every woman, which is the suppression of the terms : for after conception, nature makes use of that blood for the nourishment of the embryo, which before was cast out by nature, because it was too great in quantity. For it is an error, to think that the menstrual blood, simply in itself considered, is bad ; because if a woman's body be in good temper, the blood must needs be good ; and that it is voided monthly is, because it offends in quantity, but not in quality. But though the suppression of the terms is generally a sure sign of conception to such persons as had them orderly before, yet the having them is not always a sign there is no conception ; for as much as many that have been with child have had their terms, and some even till the fifth or sixth month, which happens according to the woman's being more or less sanguine : for if a woman has more blood than will suffice for the nourishment of the embryo, nature continues to void it in the usual way. Whence the experienced midwife may learn there

are few general rules which do not sometimes admit of an exception. But this shall suffice to be spoken of the signs and prognostics of conception.

SECT. III. *Whether Conception be a Male or Female.*

AUTHORS give us several prognostics of this ; though they are not at all to be trusted, yet there is some truth among them : the signs of a male child conceived are,

1. When a woman at her rising up is more apt to stay herself upon her right hand than her left.

2. Her belly lies rounder and higher than when she has conceived of a female.

3. She first feels the child to beat on her right side.

4. She carries the burden more light, and with less pain, than when it is a female.

5. Her right nipple is redder than the left, and her right breast and more plump.

6. Her colour is more clear, nor is she so swarthy as when she has conceived a female.

7. Observe the circle under her eye, which is a pale and bluish colour ; and if that under the right eye be most apparent, and most discoloured, she has conceived a son.

8. If she would know she hath conceived of a son or a daughter, let her milk a drop of her milk into a bason of fair water ; if it spreads and swims at top, it certainly is a boy : but if it sinks to the bottom as it drops in round in a drop, it is a girl. The last is an infallible rule. And in all it is to be noted, that what is a sign of a male conception, the contrary holds good of a female.

SECT. IV. *How a woman ought to order herself after Conception.*

MY design in this treatise being brevity, I shall pretermit all that others say of the causes of twins, and whether there be any such thing as superfœtations, or a second conception in a woman, which is yet common enough, when I come to show you how a midwife ought to proceed in the delivery of those women that are pregnant with them. But having already spoken of conception, I think it necessary to show how such as have conceived ought to order themselves during their pregnancy, they that may avoid those inconveniences which often endanger the life of the child, and many times their own.

A woman after her conception, during the time of her being with child ought to be looked on as indisposed or sick, though in good health ; for the child-bearing is a kind of nine month's sickness, being all that time in expectation of many inconveniences, which such a condition usually causes to those that are not well governed during that time : and therefore ought to resemble a good pilot, who, when sailing in a rough sea, and full of rocks, avoids and shuns the danger if he steers with prudence ; but if not, it is a thousand to one but he suffers shipwreck.   In like manner, a woman with child is often in danger of miscarrying and losing her life, if she is not very careful to prevent those accidents to which she is subject all the time of her pregnancy ; all which time her care must be double, first of herself, and secondly of the child she goes with, for otherwise a single error may produce a double mischief ; for if she receives any prejudice, her child also suffers with her.

Let a woman therefore after conception observe a good diet suitable to her temperament, custom, condition, and quality ; and if she can, let the air where she orderly dwells be clear and well tempered, free from extremes either of heat or cold ; for being too hot, it dissipateth the spirits too much, and causeth many weaknesses : and by being too cold and foggy, it may bring down rheums and distillations on the lungs, and so causeth her to cough, which by its impetuous motions forcing downwards, may make her miscarry ; she ought also to avoid all nauseous ill smells, for some times the stink of a candle not well put out may cause her to come before her time : and I have know the smell of charcoal to have the same effect.   Let her also avoid smelling of rue, mint, penny-royal, castor, brimstone, &c.

But with respect to their diet, women with child have generally so great loathings and so many different longings that it is very difficult to prescribe an exact diet for them.   Only this I think advisable, that they may use those meats and drinks which are to them most desirable, though perhaps not in themselves so wholesome as some others, and may not be so pleasant, but this liberty must be made use of with this caution, that what she so desires be not in itself absolutely unwholesome ; and also that in every thing they take care of excess.   But if a child-bearing woman finds herself not troubled

with such longings as we have spoken of, and in such quantities as may be sufficient for herself and child, which her appetite may in a great measure regulate; for it is alike hurtful
for her to fast too long as to eat too much, and therefore, rather let her eat a little and often, especially let her avoid eating too much at night; because the stomach being too much
filled, compresseth the diaphragms, and thereby causeth difficulty of breathing. Let her meat be easy of digestion, such
as the tenderest parts of beef, mutton, veal, fowls, pullets,
capons, pigeons, and partridges, either boiled or roasted, as
she likes best; new laid eggs are also very good for her; and
let her put into her broth those herbs that purify it as sorrel,
lettuce, succory, and burrage; for they will purge and purify
the blood; let her avoid whatsoever is hot seasoned, especially
pies and baked meats, which being of hot digestion overcharge
the stomach. If she desires fish, let it be fresh, and such as is
taken out of rivers and running streams. Let her eat quinces,
or marmalade, to strengthen the child; for which purpose
sweet almonds, honey, sweet apples, and full ripe grapes are
also good, Let her abstain from all sharp, sour, bitter, and
salt things, and all things that tend to provoke the terms:
such as garlic, onions, olives, mustard, fennel with pepper,
and all spices, except cinnamon, which in the three last
months is good for her. If at first her diet be sparing, as she
increases in bigness let her diet be increased; for she ought to
consider she has a child as well as herself to nourish. Let her
be moderate in her drinking, and if she drinks wine let it be
rather claret than white, which will breed good blood, help
the digestion, and comfort the stomach, (which is always but
weakly during her pregnancy) but white wine, diuretic, or
that which provokes urine, ought to be avoided. Let her have
a care of too much exercise; and let her avoid dancing, riding
in a coach, or whatever puts the body into violent motion,
especially in her first month. But to be more particular, I
shall here set down rules proper for every month for the childbearing women to order herself, from the time she has first
conceived to the time of her delivery.

Rules for the first two Months.

As soon as a woman knows (or has reason to believe) she
hath conceived, she ought to abstain from all motions and exercise, whether to walk on foot or ride on horseback, or in a
coach, it ought to be very gently. Let her also abstain from

29

venery, (to which after conception, she has usually no great inclination lest there be a mole or superfœtation, which is adding of one embryo to another. Let her beware she lift not her arms too high, nor carry too great burdens, nor repose herself on hard and uneasy seats. Let her use moderately, meat of good juice and easy digestion, and let her wine be neither too strong nor too sharp, but a little mingled with water; or if she be very abstemious, she may use water wherein cinnamon is boiled. Let her avoid fastings, thirst, watching, mourning, sadness, anger, and all other perturbations of the mind. Let none present any strange or unwholesome things to her, nor so much as name it, lest she should desire it, and not be able to get it, and so either cause her to miscarry, or the child have some deformity on that account, Let her belly be kept loose with prunes, rasins, or manna in her broth : and let her use the following electuary, to strengthen the womb and child.

"Take conserve of burrage, bugloss, and red roses, each two ounces, of balm an ounce, citron, peel, and shells, mirobolans candied, each an ounce ; extract of wood aloes a scruple ; pearl prepared half a dram ; red coral, ivory, each a dram ; precious stones a scruple ; candied nutmeg two drams ; and with syrup of apples and quinces make an electuary."

Let her observe the following Rules.

" TAKE pearls prepared a dram : and coral prepared and ivory, each half a dram ; precious stones a scruple ; yellow citron, peel, mace, cinnamon, cloves, each half a dram : saffron a scruple, wood aloes half a scruple, ambergrease six drams, and with six ounces of sugar dissolved in rose water, make rolls." Let her also apply strengtheners to the navel, of nutmeg, mace, mastich, made up in bags, or a toast dipped in malmsey, sprinkled with powder of mint, if she happens to desire clay, chalk, or coals, as many women with child do, give her beans boiled with sugar, and if she happens to long for any thing which she cannot obtain, let her presently drink a large draught of pure water.

Rules for the Third Month,

IN this month and the next be sure to keep from bleeding ; for though it may be safe at other times, it will not be so to the end of the fourth month ; and yet if too much blood abound, or some incident disease happen, which requires evacuation, you may use a cuppping-glass with scarification, and a little blood may be drawn from the shoulders and arms, especially if she has been accustomed to bleed. Let her also take care of lacing herself too straitly, but give herself more liberty than she used to do; for by inclosing her belly in too straight a mould, she hinders the infant from taking its free growth, and often makes it come before its time.

Rules for the Fourth Month.

IN this month you ought also to keep the child-bearing woman from bleeding, unless in extraordinary cases : but when this month is past, blood letting and physic may be permitted, if it be gentle and mild ; and perhaps may be necessary to prevent abortion.   In this month she may purge in an acute disease ; but purging may be only used from the beginning of this month to the end of the sixth ; but let her take care that in purging she use not vehement medicine, nor very bitter, as aloes, which is an enemy to the child, and opens the mouth of the vessels ; neither let her use coloquintida, scammony, nor turpith : she may use cassia, manna, rhubarb, agaric, and senna; but dyacidonium purgans is best, with a little of the electuary of the juice of roses.

Rules for the Fifth, Sixth, and Seventh Month.

IN these months child-bearing women are often troubled with coughs, heart-beating, fainting, watching, pains in the loins and hips, and bleeding.   The cough is from a sharp vapour, that comes to the jaws and rough artery from the terms, or from the thin part of that blood gotten into the veins of the breast, or falling from the head to the breast ; this endangers abortion, and strength fails from watching ; therefore purge the humours that fall from the breast with rhubarb and agaric, and strengthen the head as in a catarrh, and give sweet lenitives as in a cough.   Palpitation and fainting arises from vapours that go to it by the arteries, or from the blood that aboundeth, or cannot get out at the womb, but ascends, and oppresseth the heart ; and in that case cordials should be used both inwardly and outwardly.   Watching is from sharp dry vapours that trouble the animal spirits ; and in this case use frictions, and let the woman wash her feet at bed time, and let her take syrup of poppies, dried roses, emulsions of

31

sweet almonds with white poppy seeds. If she be troubled with pains in her loins and hips, as in these months she is subject to be from the weight of her child, who is now grown big and heavy, and so stretcheth the ligaments of the womb, and parts adjacent; let her hold it up with swarthing bands about her neck. About this time also the woman often happens to have a flux of blood, either at the nose, womb, or hemorrhoids from plenty of blood, or from the weakness of the child that takes it not in, or else from evil humours in the blood that stirs up nature to send it forth. And sometimes it happens that the vessels of the womb may be broken either by some violent motion, fall, cough, or trouble of mind; (for any of these will work that effect) and this is so dangerous, that in such a case the child cannot be well; but if it be from blood only, the danger is less, provided it flows by the neck of the womb, for then it prevents pleithory, and takes not away the nourishment of the child; but if it proceeds from the weakness of the child that draws it not, abortion, often follows, or hard travail, or else she goes beyond her time. But if it flows by the inward veins of the womb, there is more danger by the openness of the womb if it comes from evil blood, the danger is alike from cacochimy which is like to fall upon both. If it ariseth from pleithory, open a vein, but with very great caution use astringents, of which this following will do well; "Take pearls prepared, a scruple, red coral two scruples, mace, nutmegs, each a dram; cinnamon, half a dram; make a powder, or with sugar rolls." Or give this powder in broth: " Take red coral half a dram, precious stones half a scruple, red saunders half a dram, bole a dram, scaled earth, tormentil roots, each two scruples with sugar of roses and manus chrissio, with pearl, five drams, make a powder." You must also strengthen the child at the navel; and if there be a cacochimy, after the humours: and if you do it safely, evacuate; you may likewise use amulets in her hand and about her neck. In a flux of hemorrhoids wear off the pain; and let her drink wine with a toasted nutmeg. In these months the belly is also subject to be bound; but if it be without any apparent disease, the broth of a chicken, or of veal sodden with oil, or with the decoction of mallows, or marshmallows, mercury, and linseed, put up in a clyster, will not be amiss, but in less quantity than is given in other cases; to wit of the decoction five ounces, of common oil three ounces, of sugar two ounces, of cassia fistula one ounce,

32

But if she will not take a clyster, one or two yolks of new-laid eggs, or a few pease pottage warm with a little salt and sugar supped up a little before meat, will be very convenient; but if her body should be distended, and stretched out with wind, a little fennel seed and anniseed reduced into powder, and mingled with honey, and sugar, made after the manner of an electuary, will do very well. Also if the thighs and feet swell let them be anointed with exphrodium (which is a liquid medicine made with vinegar and rose water) mingled with a little salt.

Rules for the Eighth Month.

The eighth is commonly the most dangerous, and therefore the greatest care and caution ought to be used, and her diet ought to be better in quality, but not more, nor indeed so much in quantity as before, but as she must abate her diet so she must increase her exercise; and because then women with child, by reason the sharp humours alter the belly, are accustomed to weaken their spirits and strength, they may as well take before meat an electuary of diarrhodon or aromaticum rositum, or diamargarton; and sometimes they may lick a little honey, as they will loathe and nauseate their meat, may take green ginger condited with sugar, or the rinds of citron and oranges condited; and let her often use honey for the strengthening of the infant. When she is not far from her labour, let her eat every day seven roasted figs before meat, and sometimes let her lick a little honey; and let her beware of salt and powdered meat, for it is neither good for her nor the child.

Rules for the Ninth Month.

IN the ninth month let her have a care of lifting any great weights; but let her move a little more to dilate the parts and stir up natural heat. Let her take heed of stooping, and neither sit too much nor lie on her sides; neither ought she to bend herself much, lest the child be unfolded in the umbilical ligament, by which means it often perisheth. Let her walk and stir often, and let her exercise be rather to go upwards than downwards. Let her diet now especially be light and easy of digestion; as damask prunes with sugar, or figs and raisins, before meat, as also the yolks of eggs, flesh and broth

M

of chickens, birds, partridges and pheasants; astringent and roasted meats, with rice, hard eggs, millet and such like things are proper, baths of sweet water, with emolient herbs, ought to be used by her this month with some intermission. And after the bath let her belly be anointed with oil of roses and violets; but for her privy parts, it is better to anoint them with the fat of hen's grease, or ducks, or with the oil of lilies, and the decoction of linseed and fenugreek, boiled with oil of linseed, and marshmallows, or with the following liniment:

"Take of mallows and marshmallows, cut and shred, of each an ounce; of linseed one ounce; let them be boiled from twenty ounces of water to ten; then let her take three ounces of the boiled broth; of oil of almonds, and oil of flower-de-luce, of each one ounce, of deer's suet three ounces; let her bathe with this, and anoint herself with it warm."

If for fourteen days before the birth she do every morning and evening bathe and moisten her belly with muscadine and lavender water, the child will be much strengthened thereby. And if every day she eat toasted bread it will hinder any thing from growing to the child. Her privy parts may be also gently stroked down with this fomentation:

"Take three ounces of linseed; of mallows and marsh-mallows sliced, of each one handful; let them be put into a bag, and boiled immediately;" and let the woman with child every morning and evening take the vapour of this decoction in a hollow stool, taking great heed that no wind or air come to her in any part, and then let her wipe the part so anointed with a linen cloth, that she may anoint the belly and groins as at first. When she is come so near her time as to be within ten or fourteen days thereof, if she begins to feel any more than ordinary pain, let her use every day the following;

"Take mallows and marshmallows, of each one handful, camomile, hard mercury, maiden hair, of each a handful, of linseed, four ounces; let them be boiled in such a sufficient quantity of water as may make a broth therewith." But let her not sit too hot upon her seat, nor higher than a little above her navel; nor let her sit on it longer than about half an hour, lest her strength languish and decay; for it is better to use it often than stay too long in it. And thus have I shown how a child-bearing woman ought to govern herself to each month during her pregnancy: how she must order herself at her delivery shall be shown in another chapter, after I have first shown the industrious midwife how the child is formed in the womb, and the manner of its decumbiture there.

CHAP. III.

Of the parts proper to a Child in the Womb ; how it is formed there, and the manner of its situation therein.

IN the last chapter I treated of conception, showed what it was, how accomplished, it signs, and how she who had conceived ought to order herself during the time of her pregnancy. Now before I come to speak of her delivery, it is necessary that the midwife be first acquainted with the parts proper to a child in the womb, and also that she be shown how it is formed, and the manner of its situation and incumbiture there ; which are so necessary to her, that without the knowledge thereof, no one can tell how to deliver a woman as she ought. This therefore shall be the work of this chapter. I shall begin with the first of these,

SECT. I. *Of the Parts proper to the Child in the Womb.*

IN this section I must first tell you what I mean by the parts proper to a child in the womb ; and they are only those that either help or nourish it, whilst it is lodged in that dark repository of nature, and that help to clothe and defend it there, and are cast away as of no more use, after it is born ; and these are two, to wit the umbilicurs, or navel vessels, and the secundinum : by the first it is nourished, and by the second clothed and defended from wrong. Of each of these I shall speak distinctly ; and, first,

Of the Umbilicurs, or Navel Vessels.

These are four in number, viz. one vein, two arteries, and the vessels which is called urachos : 1. The vein is that by which the infant is nourished, from the times of its conception to the time of its delivery ; till being brought into the light of this world, it has the same way of concocting its food that we have. This vein ariseth from the liver of the child, and is divided into two parts when it hath passed the navel ; and these two are again divided and subdivided, the branches being upheld by the skin called chorion, (of which I shall speak by and by) and are joined to the veins of the mother's womb,

M 2

35

from whence they have their blood for the nourishment of the child. 2. The arteries are two on each side, which proceed from the back branches of the great artery of the mother, and the vital blood is carried by these to the child, being ready concocted by the mother. 3: A nervous or sinewy production is led from the bottom of the bladder of the infant to the navel, and this is called urachos; and its use is to convey the urine of the infant from the bladder to the alantois. Anatomists do very much vary in their opinions concerning this, some denying any such thing to be in the delivery of women, and others on the contrary affirming it; but experiencece has testified there is such a thing. For Bartholomew Carbrolius, the ordinary doctor of anatomies to the college of physicians at Montpelier in France, records the history of a maid, whose water being a long time stopped, at last issued out through her navel. And Johannes Fernelius speaks of the same thing that happened to a man of 30 years of age, who having a stoppage in the neck of the bladder, his urine issued out of the navel many months together, and that without any prejudice at all to his health which he ascribes to the ill lying of his navel, whereby the urachos was not well dried. And Vulchior Coitus quotes such another instance in a maid of 34 years of age at Nuremburg in Germany. These instances, though they happen but seldom, are very sufficient to prove that there is such a thing as an urachos in men. These four vessels before mentioned, viz. one vein, two arteries, and the urachos, do join near to the navel, and are united by a skin which they have from the chorion, and so become like a gut or rope, and are altogether void of sense; and this is that which the good women call the navel-string. The vessels are thus joined together, that so they might neither be broken, severed, nor entangled; and when the infant is born are of no use, save only to make up the ligament which stops the hole of the navel, and some other physical use, &c.

Of the Secundine, or After-birth.

Setting aside the name given to this by the Greeks and Latins, it is called in English by the name of secundine, after-birth, and after-burden, which are held to be four in number.

1. The first is called placentia, because it resembles the form of a cake, and is knit both to the navel and chorion, and makes up the greatest part of the secundine or after-birth.

The flesh of it is like that of the melt, or spleen, soft, red, and tending something to blackness, and has many small veins and arteries in it ; and certainly the chief use of it is for containing the child in the womb.

2. The second is the chorion. The skin, and that called the amnios, involve the child round, both above and underneath, and on both sides which the alantois doth not : this skin is that which is most commonly called the secundine, as it is thick and white, garnished with many small veins and arteries, ending in the placentia, before named, being very light and slippery. Its use is not only to cover the child round about, but also to receive and safely bind up the roots, and the veins and arteries, or navel vessels before described.

3. The third thing which makes up the secundine is the alantois, of which there is a great dispute among anatomists, some say there is such a thing, and others that there is not, those that will have it to be a membrane, say it is white, soft, and exceeding thin, and just under the placentia, where it is knit to the urachos, from whence it receives the urine : and its office is to keep it separate from the sweat, that the saltness of it may not offend the tender skin of the child.

4. The fourth and last covering of the child is called amnios and it is white, soft, and transparent, being nourished by some very small veins and arteries. Its use is not only to enwrap the child round, but also to retain the sweat of the child.

Having thus described the parts proper to the child in the womb, I will now proceed to speak of the formation of the child therein, as soon as I have explained the hard terms in this section, that those for whose help this is designed may understand what they read. There is none, sure can be so ignorant as not to know that a vein is that which receives blood from the liver, and distributes it in several branches to all the parts of the body. Arteries proceed from the heart, are in a continual motion, and by their continual motion quicken the body. Never is the same with sinew, and is that by which the brain and sense and motion to the body. Placentia properly signifies a sugar-cake ; but in this section it is used to signify a spongy piece of flesh, resembling a cake full of veins and arteries, and is made fo receive the mother's blood appointed for the infant's nourishment in the womb. Chorion is the outward skin which compasseth the child in the womb. The amnios is the inner skin which compasseth

M 3

37

the child in the womb. The alantois is the skin that holds
the urine of the child during the time that it abides in the
womb. The urachos is the vessel that conveys the urine from
the child in the womb to the alantois.—I now proceed to

Sect. II. Of the Formation of the Child in the Womb.

TO speak of the formation of the child in the womb we
must begin where nature begins : and that is, at the act of
coition in which the womb having received the generative
seed, without which there can be no conception, the womb
immediately shuts up itself so close that not the point of a
needle can enter the inward orifice ; and this it does partly to
hinder the issuing out of the seed again and partly to cherish
it by an in-bred heat, the better to provoke it to action ; which
is one reason why women's bellies are so lank at their first
conception. The woman having thus conceived, the first thing
which is operative in the conception is the spirit, whereof the
seed is full, which nature quickening by the heat of the
womb, stirs it up to action. The seed consists of very different parts, of which some are more and some are less pure.
The internal spirits thereof separateth those parts that are less
pure which are thick, cold, and clammy, from them that are
more pure and noble. The less pure are cast to the outside,
and with them the seed is circled round, and with them the
membranes are made, in which that seed which is the most
pure is wrapped round and kept close together, that it may
be defended from cold and other accidents, and operate the
better.

The first thing that is formed is the amnois, the next the
chorion ; and they enwrap the seed round as it were in a cur-
tain. Soon after this, (for the seed thus shut up in a woman
lies not idle) the navel vein is bred, which pierceth these skins
being yet very tender, and carries a drop of blood from the
veins of the mother's womb to the seed : from which drop is
formed the liver, and from which liver there is quickly bred
the vena cava, or chief vein, from which all the rest of the
veins that nourish the body spring ; and now the seed hath
something to nourish it, whilst it performs the rest of nature's
work, and also blood administered to every part of it to form
flesh.

The vein being formed, the navel arteries are soon after
formed, then the great artery, of which all others are but

branches and then the heart : for the liver nourisheth the arteries with blood to form the heart, the arteries being made of seed, but the heart and flesh of blood. After this the brain is formed and then the nerves, to give sense and motion to the infant. Afterwards the bones and flesh are formed, and of the bones, first the vertebræ or chine bone, and then the skull, &c.

As to the time in which this curious part of nature's workmanship is formed, physicians assign four different seasons wherein this microcosm is formed, and its formation perfected in the womb: the first is immediately after coition ; the second time of forming, say they, is when the womb by the force of its own innate power and virtue makes a manifest mutation to coagulation in the seed, so that all the substance thereof seems congulated flesh and blood, which happens about the twelfth or fourteenth day after copulation ; and though this concretion of fleshy mass abound with spirits, yet it remains undistinguishable without any form, and may be called a rough draught of the fœtus or embryo. The third time in which this fabric is come to some further maturity, is when the principal parts may be in some measure distinguished, and one may discern the liver, umbilical veins, arteries, nerves, brain, and heart: and this is about eighteen days after conception. The fourth and last time assigned by physicians for the formation of the child, is about the thirtieth day after conception for a male, but for a female, they tell us forty-two or forty-five days are required, though for what reason I know not, nor does it appear by the birth ; for if the male receives its formation fifteen days sooner than the female, why should it not be born so much sooner too ? but, as to that, every day's experienced shows us to the contrary ; for women go to the full time of nine months both with male and female. But as to this time of thirty days (or some will have it 45) the outward parts may be also seen exquisitely elaborate, and distinguished by joints ; and from this time the child begins to be animated, though as yet there is no sensible motion, and has all the parts of the body, though small and very tender, yet entirely formed and figured, although not longer in the whole than one's middle finger ; and from thenceforward, the blood flowing every day more and more to the womb, not by intervals like their courses, but it continually grows bigger and stronger to the end of nine months, being the full time of a woman's ordinary labour.

Very great have been the disputes among both philosophers and physicians about the nourishment of the child in the womb, both as to what it is, and which way it receives it. Almæon was of opinion, that the infant drew in its nourishment by its whole body, because it is rare and spongy, as a sponge sucks in water on every side ; and so he thought the infant sucked blood, not only from its mother's veins, but also from the womb. Democritus held, that the child sucked in the nourishment at its mouth. Hippocrates affirms, that the child sucks in both nourishment and breath by its mouth from the mother, for which he gives two reasons : 1. that it will suck as soon as it is born, and therefore must have learned to suck before. 2. Because there are excrements found in the guts as soon as it is born. But nothing of these reasons are sufficient to prove his assertions : for as to the first, " That the child will suck as soon as it is born," it is from a natural instinct : for take a young cat that never saw her dam catch a mouse, and she will catch mice herself as soon as she is able. And as to his second reason, it is a sufficient answer to say, that the excrements found in the guts of an infant new born are not excrements of the first concoction, which is evident because they do not stink, but are the thickest part of the blood, which is conveyed from the vessels of the spleen to the guts. Having therefore said enough to confute the opinion of the child's receiving its nourishment by the mouth, I do affirm that the child receives its nourishment in the womb by its navel ; and that it shall be so, is more consonant to truth and reason ; which being granted, it will easily follow, that the nourishment the child receives is by the pure blood conveyed into the liver by the navel vein, which is a branch of the vena porta, or great vein, and passeth the small veins of the liver. Here this blood is made more pure, and the thicker and rawer part of it is conveyed to the spleen and kidney, and the thick excrement of the guts, which is that excrement found there so soon as they are born. The pure part is conveyed to the vena cava, and by it distributed throughout the body by the small veins, which, like so many small rivulets, pass to every part of it. This blood is accompanied (as all blood is) with a certain watery substance, the better to convey it through the passage it is to run in, which, as in men, is breathed out by sweating, and contained in the amnios, as I have already said.

40

SECT. III. *Of the manner of the Child's lying in the Womb.*

I COME now to show after what manner the child lies in the womb ; a thing so essential for a midwife to know, that she can be no midwife who is ignorant of it, and yet even about this authors extremely differ : for there is not two in ten that agree, what is the form that the child lies in the womb or in what manner it lies there : and yet this may arise in a great measure from the different figures that the child is found in, according to the different times of the woman's pregnancy ; for near the time of its deliverance out of these winding chambers of nature, it oftentimes changes the form in which it lay in before for another. Hippocrates affirms that the child is so placed in the womb as to have its hands, its knees, and its head bent down towards it feet, so that it lies round together, its hands upon both its knees, and its face between them ; so that each eye toucheth each thumb, and nose betwixt its knees ; and Bartholinus was also of the same opinion. Columbus describes the posture of the child thus : " The right arm bowed, the fingers whereof under the ear and above the neck ; the head bowed down so that the chin toucheth the breast, the left arm bowed above both breast and face, and the left arm is propped up by the bending of the right elbow ; the legs are lift upwards, the right of which is so lifted up, that the thigh toucheth the belly, the knees, the navel, the heel, the left buttock, and the foot is turned back and covereth the secrets, the left thigh toucheth the belly, and the leg is lifted to the breast, the back lying outward." And this much shall suffice touching the opinion of authors.

I will now show the several situations of the child in the mother's womb, according to the different times of pregnancy, by which those that are contrary to nature, and are the chief cause of all ill labours, will be the more easily conceived by the understanding midwife : it ought therefore, in the first place, to be observed, that the infant, as well male as female, is generally situated in the midst of the womb ; for though sometimes to appearance a woman's belly seemeth bigger on one side than the other, yet it is so with respect to her belly only, and not of her womb in the midst of which it is always placed.

But in the second place, a woman's great belly makes different figures, according to the different times of pregnancy; for when she is young with child the embryo is always found of a round figure, a little oblong, having the spine moderately turned inwards, the thighs folded, and a little raised, to which the legs are so joined that the heels touch the buttock, the arms are bending, and the hands placed upon the knees, towards which the head is inclining forwards, so that the chin toucheth the breast: in which posture it resembles one sitting to ease nature, and stooping down with the head to see what comes from him. The spine of its back is at that time placed towards the mother's, the head uppermost, the face forwards, and the feet downwards; and proportionably to its growth, it extends its members by little and little, which were exactly folded in the first month. In this posture it usually keeps, till the seventh or eighth month, and then by a natural propensity and disposition of the other parts of the body, the head is turned down towards the inward orifice of the womb, tumbling as it were over its head, so that when the feet are uppermost, and the face towards the mother's great gut, and this turning of the infant in this manner, with his head downwards, toward the latter end of a woman's reckoning, is so ordered by nature that it may thereby be the better disposed for its passage into the world at the time of its mother's labour, which is not then far off; (and indeed several children turn not at all until the very time of their birth) for in this posture all its joints are most easily extended in coming forth; for by this means the arms and legs cannot hinder the birth, because they cannot be bended against the inward orifice of the womb; and the rest of the body being very supple, passeth without any difficulty after the head, which is hard and big, being past the birth. 'Tis true, there are divers children that lie in the womb in another posture, and come to the birth with their feet downwards, especially if they be twins; for then by their different motions they so disturb one another, that they seldom come both in the same posture at the time of labour, but one come with the head, and another with the feet, or perhaps lie across; and sometimes neither of them will come right. But however the child may be situated in the womb or in whatever posture it presents itself at the time of birth, if it be not with its head forwards, as I have before described, it is always against nature; and the delivery will occasion the mother more pain and danger, and requires greater care and skill from the midwife than when the labour is more natural; of which the following scheme will give a great demonstration, which is

The FORM of a FEMALE CHILD in the WOMB.

Ready for the Birth, naked and disrobed of its Tunacles, proper and common.

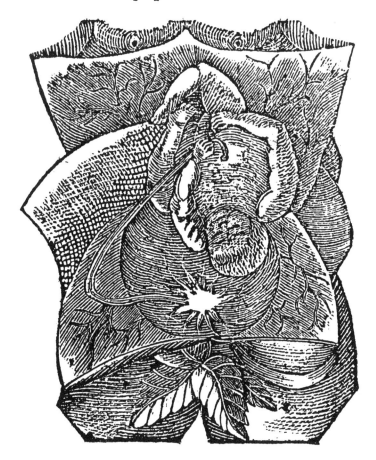

lady in question, was about 32 years of age, and at the period of her death had been married only four months, and, therefore, it is to be presumed that the fœtus found in the uterus after death was a fœtus between the third and fourth month ; the size of the fœtus is itself a satisfactory evidence in support of this fact. The dissection of the uterus was made with great care ; it was perfectly healthy, and evidently performing all its functions in the most satisfactory manner ; the parts were preserved for the museum with the fœtus enclosed within the amnios, otherwise we should have here added its measurement, weight, &c.

CONTRIBUTIONS

TO THE

HISTORY OF THE CORPUS LUTEUM,

HUMAN AND COMPARATIVE,

BY

ROBERT KNOX, M.D.,

Corresponding Member of the French Academy of Medicine, Feb. 1840.

PART I.

THE following brief history of facts and opinions respecting the *corpus luteum*, and the circumstances attending its origin, development or growth and decay, was commenced some years ago on the occasion of my receiving from an esteemed friend the human uterus and its appendages, which seemed to me to throw considerable light on the obscure history of this interesting body. Shortly after I had examined the uterus in question a very valuable work touching on this matter appeared from the pen of Dr. Montgomery, and this was soon afterwards followed by a " Memoir" on the structure of the corpus luteum, to which is attached the name of one of the most accurate and talented observers of the present day.* As I found it impossible to arrange the facts I had observed, so as to suit the theoretical views supported by these gentlemen, I have thought it might be agreeable to the readers of THE LANCET, and to physiologists generally, to publish in its columns the observations themselves, nearly as they stand in my note-book.

I much fear that all the labour desirable may not have been bestowed on the following memoir, and that contradictory opinions are supported, which, however, on being pointed out, I shall be the first to acknowledge; but I foresee that unless availing myself of the few days of leisure afforded by the Christmas holidays, the publication of these notes may be put off indefinitely.

In the month of March, 1835, the gravid uterus of a most respectable woman, who died suddenly of perforation of the stomach, was put into my hands. Mrs. B.——, the

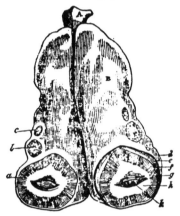

Description of Fig. 1.

Right ovarium section along the free margin of the organ, dividing nearly to its base or fixed margin.

A *proper* ligament of the ovary.

B cut surface of the ovary.

a section of a *corpus luteum* at between the third and fourth month of pregnancy.

b section of another body resembling *a*, in miniature.

c section of a third body still smaller than but resembling the two former.

d external cut margin of the ovarium.

e cut surface of the stroma of the ovarium.

f cavity out of which the corpus luteum has been partially dissected.

g plicated yellow body.

h white albuminous-looking substance in the interior of the yellow body.

k fold traversing the centre of the corpus luteum.

On the surface of the right ovarium*

---

* R. Lee, M.D., F.R.S., " Med. Chirurg. Trans.," vol. xxii., p. 329.

---

* This ovarium weighed 128 grs.; greatest length, 2 inches 9 lines ; greatest depth over the corpus luteum, 10 lines, and its thickness here equalled 8 lines. The ovarium of a woman, aged 30, who had had a family, and died in full health, weighs 65 grs., being little more than one-half that of the ovarium

was observed several short fissures or grooves, and towards its distal part a general fulness, immediately over which there existed a well-marked fissure. These facts led to the belief that it contained a corpus luteum, and, accordingly, on dividing the ovarium along its free margin, in the axis of its greatest length, but avoiding the fissure above alluded to, the appearances as seen in the engraving, figure 1, presented themselves. *a*, marks the large corpus luteum which the knife had bisected longitudinally, and above this corpus luteum two other bodies, marked *b c*, precisely resembling the first in miniature, presenting really no other discernible differences but that of size. On the exterior of this ovarium, as we have already remarked, there are several fissures, and it is just possible that there may exist other small bodies besides those marked *b c*, but it was deemed unadvisable further to bisect the preparation. No Graafian vesicles were observable, either on the external or cut surfaces of the ovary.

In the left ovarium* a similar section in the long axis of the organ displayed a single corpus luteum, about one-third the size of that marked *a*, fig. 1 ; and in this ovarium (the left) may be readily observed numerous vesicles (Graafian ?).

The dissection of these ovaria was then unavoidably delayed, and the preparations were deposited in alcohol, in which they remained for a considerable time, before another opportunity occurred of returning to their examination. The careful and deliberate scientific examination was resumed, partly with the view of verifying the above brief description, but chiefly to examine with more care into the intimate structure of the largest of these bodies (marked *a*, fig. 1). Upon the right segment the corpus luteum was dissected, and turned partially out of its position in connection with the substance of the ovarium ; a loose cellular texture, with numerous vessels, was found to be the only connecting medium between the *plicated yellow body* and the stroma of the ovary, and this delicate cellular texture was the only structure investing the exterior of the yellow body in the shape of a *capsule*. This investing cellular texture sent numerous processes into the yellow body like partitions, and evidently giving rise to that plicated appearance which the section presents. Moreover, the substance of the yel-

low body admitted of being separated with great ease from its investing cellular tissue, and when so divested presenting the strongest possible resemblance to a gland. We have already mentioned that the ovaria had both been divided in the long axis, the section having apparently intersected the entire structure of the large *corpus luteum* (marked *a*, fig. 1). Of the characteristic plicated yellow part, marked *g*, we need say nothing further at present ; it encloses on each section a smooth, dull white, membranous-looking structure, marked *h* in the figure, which seems to send processes outwards into the plicated yellow body, and to the inner surface of which it everywhere adheres intimately ; lastly, quite in the centre of this there runs a fold marked *k*, elevated from the surface, and presenting the appearance of lateral branches ; the centre and the fold itself being again, so far as we are able to observe, strictly analogous to, if not positively a portion of, the yellow plicated body. To this fold Dr. Montgomery and others (*if we understand their descriptions*) have given the name of *cicatrix*, but to which it does not bear the slightest resemblance, and in no shape can be called a *cicatrix*. The fold occupies the centre of a structure, which seems to be, and as we are indeed quite satisfied it is, nothing but the plicated yellow body itself, as if the whole albuminous-looking membrane or substance (*h*), which the knife had evidently bisected, either did not occupy the very centre of the plicated yellow body, or was so thin as to have exposed, by the section, a portion of the yellow plicated body itself. The object of this first part of our contributions not being in any shape critical in respect to the observations of others, I shall here avoid all remarks of that nature, but shall only observe that they seem altogether at variance with what has been said about "*fibrinous deposits, cavities filled with blood*, and *cicatrices* as indicating true corpora lutea from false," &c. Before quitting this ovarium I shall now describe the other appearances which the section presents, and which appearances have been usually described under the name of *false* corpora lutea. *b*, fig. 1, presents all the appearances of the large corpus luteum, marked *a*, but in miniature ; the absence of the fold seems attributable to the comparative smallness of the body bisected ; *c* resembles, upon one segment, *a* in miniature, but, upon the opposite section, the plicated yellow structure scarcely surrounds the whole of the white albuminous centre, in the midst of which there is a very obvious cavity, containing what appears to be a small quantity of grumous blood. How far these appearances, as we have described them, are reconcilable with the opinions and theories of late writers, and which of these are *true* corpora lutea, and which *false*, or, to use a more precise language, *from which*

delineated in our plate. It may be observed that the ovarium weighing 65 grs. presents two very distinct Graafian vesicles, but no remains even of corpora lutea. The preparation is preserved in our private museum.

* This ovarium weighed 112 grs., or 16 grs. less than the right ; its greatest length is 1 inch 9 lines ; depth near distal end, 9 lines ; breadth near distal end, 6 lines.

Q 2

of these bodies sprung the fruitful ovum? are questions which we shall endeavour to discuss on their simple merits, without mixing up either our own theories or those of others, in the third part of these contributions; they may, indeed, be considered as purely theoretical questions, there being no inquiries before the public adequate to their solution.

The left ovarium, which had also been bisected in its long axis, and along its free margin, presents a well-marked plicated yellow body, with its albuminous centre; the absence of the central fold I am inclined to attribute to the comparative smallness of this body, although, in all other respects, there is really no difference in its structure from the large one marked *a* in fig. 1.

The second case to which I shall allude is that of a lady, who died two days after delivery at the full term. In one ovarium was found a *corpus luteum* rather more than one-half the size of the largest described in fig. 1; the remains of two miniature *corpora lutea* were also distinctly observable, and one pretty large cyst, with several smaller cysts, but these latter I considered pathological. Previous to the last child the lady had been delivered of several others at the full period.

I examined, with Dr. William Campbell, Lecturer on Midwifery in Edinburgh, the uterus (which had been several years in alcohol) of a person who had died in the fourth month of her pregnancy; the right ovarium* upon being mesially bisected in its long axis, presented a *corpus luteum* about 9 lines long, and 7 lines broad; there was also a central cavity about 3 lines long, and of the same breadth, but instead of occupying precisely the centre it was within less than a line of the surface of the ovary, with a corresponding decrease in the thickness of the plicated yellow texture around it.

The left ovary, when cut into, showed first a cavity similar in position and size to that just described in the right ovary; a blueish-white substance surrounded it, but no yellow plicated texture. In another part of the same organ is a cavity, close to the outer tunic, which, with the membranes enclosing it, resembles a Graafian vesicle, beneath which, and in immediate contact with it, is a sac much resembling the central cavity of a Graafian vesicle, surrounded for about two-thirds of its circumference by a structure, proving the whole, in our opinion, to be the remains of a fruitful Graafian vesicle, and the remains of a corpus luteum.

I may now briefly mention, that by the kindness of my medical friends in town I have had every opportunity of examining the

* This ovarium measured 1 inch 8 lines in length, and 9 lines in breadth; its weight was not ascertained.

corpora lutea in six cases of persons dead of acute diseases, and thus I have been able to observe the condition of the recent ovaria at almost every stage of pregnancy, from a period so early that an ovum could not be detected in the uterus, although the decidua was present, up to the full term, and likewise after delivery, from a period of two days to about six months.

Now, without troubling the reader with the details of these cases, I may here remark that the impression left on my mind was, that, co-existing with conception in the human female, there would be found a *corpus luteum* of considerable magnitude, but whether this corpus luteum was the *effect* or *cause* of conception, had not been determined. That the size of the body generally supposed to be the one from which the fœtus in the womb has proceeded, had not been determined, as to its different periods, with accuracy, but that upon the whole it seemed to have a rapid growth; that the periods of decay of this body, and the changes it undergoes, had not been described in such a way as to entitle the observations to be held as facts either by the physiologist or medical jurist; and that, lastly, the whole question of those miniature corpora lutea found in the ovaries of women, under a variety of circumstances, was one accompanied with the greatest difficulties, in solving which neither physiological nor obstetrical works, hitherto published, afford adequate data. Indeed, it seemed to me that the only practical conclusion which could be drawn from a careful examination of the whole facts and opinions to which I had access, was, that co-existing with a fœtus in the uterus, there would always be found a body of some magnitude, and of the distinctive characters of a corpus luteum; but that this conclusion would scarcely prove of any utility to the physiologist or medical jurist, in consequence of the numerous difficulties which beset the whole subject of its formation, development, and decay; and if this be true in respect to the single corpus luteum, which, from its size and regular character, the anatomist would be disposed to call the *true* one, that is, *the one from which the fœtus in the uterus at the time had come*, every practical anatomist knows that the difficulties are immeasurably greater when a number of such bodies are found in the ovaria of the same person scarcely differing from the so-named *true* one, in any circumstance, but that of size, and sometimes (as we shall afterwards show) not even in that.

In concluding our Part I., it may be useful to draw the attention of the reader to the difficult questions which arise out of a single case. These difficulties have not been observed merely of late years; they were clearly apprehended by Santorini, Morgagni, Hunter, and others; Haller also had given them much of his attention. A detailed history of the views of these great men would be useful to the physiologist; but I shall here allude only to the opinions of Hunter. And as I propose discussing these views at greater length, I shall here advert to them very briefly. Sir E. Home states in the preface to his work on "Comparative Anatomy," that this work, though published under his name, may be considered merely as a descriptive catalogue of Hunter's "Museum." I presume, therefore, that it contains the most of Hunter's opinions; but if any doubt had remained on this point, such doubt was entirely removed by Mr. Clift's evidence before the Committee of the House of Commons, who added, that besides being a descriptive catalogue of Mr. Hunter's "Museum," the greater part, or nearly the whole of the text of Sir E. Home's "Lectures on Comparative Anatomy," were merely extracts from the MSS. of Mr. Hunter, which MSS. Sir E. Home afterwards burned; and that these extracts, or lectures, or by whatever other name they may be called, did most unquestionably contain by far the greater part of Mr. Hunter's theoretical opinions on most physiological subjects.

Since these facts are so, I hope I may venture to draw the following conclusion, viz., that the lecture on Generation, contained in Sir E. Home's work on "Comparative Anatomy," may be viewed as a memoir compounded of copious extracts from Mr. Hunter's MSS., and of a few hasty observations by Sir E. Home himself, interpolated with but little art amongst Mr. Hunter's observations and theories.

Now, judging of Case I., detailed in the commencement of these Contributions, and applying his theory to it, conclusions directly opposed to the ones generally received would be arrived at: for, *First*, The largest corpus luteum found in the right ovarium, would have been viewed by Mr. Hunter, *not* as the one from which had come the fruitful ovum found in the uterus after the death of the patient, but rather as the one from which was to come the next foetus, had the lady lived, and become again pregnant. *Secondly*, The smaller corpora lutea (marked *b c* in the same drawing) would have been viewed by Mr. Hunter as appertaining, the one to the foetus then existing in the uterus, and the other as one from which no fruitful ovum had ever come. The left ovarium also contained a well-marked and distinct corpus luteum. *Thirdly*, We are warranted, I think, in inferring, from the same work by Sir E., that Mr. Hunter viewed the corpus luteum as a body whose production and growth were altogether *independent* of any connection with the male,—that its function is to form or secrete the ovarian ovules, and that these are formed by, and will be found in virgin ovaria of all mammals.* But, from the imperfect notices of Sir E. Home, I cannot be quite certain whether Mr. Hunter thought that corpora lutea formed in females that had not only no connection with, but that had never even seen the male. These theoretical views, for they are strictly so, it is my intention to examine in Part III. of these Contributions.

---

* We cannot perceive that Mr. Hunter includes the human species. We are of opinion that great and important differences exist between the human structure and that of the lower animals, and which we shall endeavour to show in Part II. of these Contributions.

ON THE CONSTRUCTION

OF THE

HUMAN PLACENTA.

AN HISTORICAL SKETCH.

BY FRANCIS ADAMS, LL.D., M.D.

PUBLISHED BY
A. BROWN AND COMPANY,
BOOKSELLERS, ABERDEEN.

48

PREFACE.

I AM induced to re-publish the following Papers, which appeared originally about ten years ago in the LONDON MEDICAL GAZETTE, in consequence of my attention having been of late forcibly directed anew to the subject on which they treat, by discussions that have recently taken place in the Medico-Chirurgical Society of Edinburgh anent the important physiological questions,—Whether the mucous membrane of the Uterus be cast off at the moment of conception, to form an envelope for the ovum and a bond of connection between the embryo and its parent: and, whether it be fairly torn off during the process by which the full-grown fœtus is ejected from the seat of its incubation ? or, Whether this membrane remains uninjured and unconnected with the fœtal appendages during the whole process of Utero-gestation ?

It cannot but appear singular that any doubt should exist on these simple questions, in an age so remarkable for scientific progress, and in one of the most learned cities on the face of the earth. Yet, unfortunately, it is so ; for, if the subject was previously dark and perplexed, it must be admitted that it has not been sufficiently illumined by the new light which has been recently cast upon it. Indeed, one is almost tempted to apply to this disquisition the words of an ancient Father in the Church with regard to a puzzling dogma in Theology, that " the more he read on it the less he understood it, and, the more he wrote on it, the greater difficulty did he find in expressing his meaning."

I cannot but think that one cause of the unsatisfactory results of all such inquiries has been the fragmentary manner in which they have been taken up and conducted. For, no sooner do we plunge into any one of these discussions than we find facts stated and principles assumed which we demur in recognising as settled ; and, consequently, when we hesitate in admitting the premises, we are equally unprepared to grant the conclusions. The object of the present Historical Sketch is to do away, at any rate, with this objection, by presenting to the reader a comprehensive and continous exposition of the whole subject, from the earliest dawn of Physiological Science up to the present date. Some good, I flatter myself, will necessarily result from this method of investigation.

Another and a still greater cause of the slowness with which this, and indeed all new truths in science gain ground in this country, is the pertinacity with which the people of Great Britain, and more especially of Scotland, cling to every opinion which has obtained the sanction of what are now looked up to and worshipped as "Great Names." This is so much the case, that I am not afraid to maintain, that scarcely did our forefathers, three hundred years ago, submit their judgment more slavishly to the dogmas of Anselm and Dun Scotus, and to those of Aristotle and Galen, than their descendants of the present day submit theirs to Knox and Chal-

49

mers in Theology,—and to Newton and the Hunters in Science. To give one example :—perhaps the most brilliant discovery in Natural Philosophy of which the present century can boast is Dr. Young's Undulation Theory of Light. Now, when this new opinion was first propounded by its most ingenious author about forty years ago, it was immediately pounced upon, and, I may say, trampled in the dust by the sages of the Edinburgh Review ; and all this for no other apparent reason but because, forsooth, the new doctrine ran counter to a dogma of the "Immortal Newton." At all events, so completely was the heresy suppressed for a time, that it was never more heard of on this side the Channel until it was re-cognised as a great truth by a high continental authority, Arago ; when, as a matter of course, it rose into honour. May I be permitted then, reve-rently and respectfully, to address my learned friends in Edinburgh and say—" Ye men of Modern Athens, I perceive that in many things ye are too superstitious " ! that is to say, you are too prone to Hero-worship, and to refuse an impartial hearing to any one who professes strange doctrines not recommended by what you now regard as respectable authority.

For this mental condition the only true remedy, as far as I know, is the diligent study of the History of Science ; for when the student comes to learn from examples, that those whom he has idolized as the " Demigods of Fame" squabble among themselves, like the " ancient Divinities of Olym-pus," and manifest not a few of the weaknesses of ordinary mortals, he learns to moderate his devotion to any one of them, and to cultivate the habit of relying more upon his own judgment, and less upon theirs.

By such as are disposed to take up the investigation of our present subject in the spirit now indicated, I confidently flatter myself, that this little work, albeit it does not carry with it the *prestige* of a " Great Name," and has not emanated from any of the celebrated seats of Learn-ing but from the Alpine region of the Far North,

aut laudatus erit aut excusatus.

F. A.

BANCHORY, March 29, 1858.

ON THE

CONSTRUCTION OF THE PLACENTA,

AND THE MODE OF COMMUNICATION BETWEEN THE MOTHER AND THE

FŒTUS IN UTERO.

IT cannot but appear remarkable that, considering the industry and success with which Fœtal Anatomy and Physiology have been cultivated, both in ancient and modern times, there should still be the greatest discrepancy of opinions among anatomists regarding the construction of the human Placenta, and that physiologists should be equally at variance as to the mode by which the fœtus draws nourishment from its parent. The Hunters, indeed, were long supposed to have finally settled both these questions; but of late years it has been admitted by many original inquirers who have re-examined the subject, that a considerable portion of the Hunterian hypothesis is based on erroneous principles,—so that it must either be rejected altogether, or be subjected to important modifications. Still, however, no well-defined theory has been established in the place of it, and descriptions have been recently given by eminent authorities on anatomy, containing the most contradictory statements as to the facts of the case ; so that altogether the subject of the Placenta may be justly pronounced to be the great *opprobrium medicorum*. What makes this state of matters the more to be lamented is, that indisputably the subject is one of the utmost importance, as affecting the practical views of the physician and surgeon in urgent cases of almost daily occurrence: for example, the rules for the management of uterine hæmorrhage as laid down by our highest authorities in midwifery during the last sixty or seventy years, are all founded upon the Hunterian hypothesis; and if it be now admitted to be untenable, surely it becomes the duty of every obstetrical practitioner to reconsider his principles treatment, and abandon such as are based on a false doctrine. I think myself called upon, therefore, to state explicitly the grounds upon which my own opinions are founded, and shall embrace the present opportunity of reviewing the literature of the whole subject, from the earliest times down to the present period. This seems to me to be the only legitimate way of getting the question set at rest ; for, when all the views which have been ever entertained on it are fairly before us, there are certainly some admitted tests of truth which will enable us to decide what opinions are true and what are based on error. I shall endeavour to execute my undertaking.

51

within as narrow bounds as possible, by confining my attention to such doctrines as possess importance from having been extensively received, —dealing only with the larger facts of the case, and avoiding all prolixity of detail, or the use of terms not properly defined. I purpose, then, to divide the literature of the subject into five periods, giving—1st, The opinions which prevailed anciently, and down to the discovery of the circulation by Harvey ; 2d, The opinions held by Harvey and his followers ; 3d, The opinions held by the physiologists of the eighteenth century, and more especially by Haller and the Hunters ; 4th, The opinions held by Dutrochet, Velpeau, and others ; 5th, The opinions held by the advocates of the Cell theory at the present time. I shall then draw my conclusions from the whole.

Period I.—*On the opinions which prevailed anciently and down to the discovery of the circulation by Harvey.*

The opinions advanced in several of the Hippocratic treatises, as *De Natura Hominis, de Natura Pueri, de Septimestri, de Octomestri,* &c. &c., I shall not dwell upon at any length, since it is now generally acknowledged that these works are not the genuine productions of the great Hippocrates, and the physiological doctrines contained in them are crude and not very well defined. I may just mention that the original of the embryo is there assumed to be the male semen,[*] but that the earliest appearance of the embryo in the uterus is pretty accurately described, as being enclosed in its chorion, which is said to be a membrane like that which encloses the egg within the shell, and through which membrane blood is absorbed for the nourishment of the fœtus (p. 386, 387, ed. Kühn). The fœtus is further said to breathe by means of its umbilical cord (p. 388). The opinions advocated by Aristotle on this subject are of a general nature, and the text in the passage containing the fullest exposition of them appears to me to be corrupt (*de generat. animal.* ii. 7.) He would seem to have derived his information on this subject mostly from observations made, either by himself or his predecessor Democritus, on sheep and cows, as is evident from his description of the cotyledons, a term which, like many others, he misapplies to the human placenta, and represents them as being the instrument by which the fœtus derives a sanguineous pabulum from the mother.[†] The appearance of the chorion he describes with considerable accuracy (*l. c.*) We shall pass on, however, to Galen, who is the great ancient authority on physiology, the principles of which he may be said to have fixed during fourteen succeeding centuries, that is to say, during the remainder of the period of

* I may just mention that this system of embryology, although now entirely exploded, found an able advocate towards the end of the 17th century in the celebrated Anthony Leeuwenhoek, who strenuously attacked the doctrine of Harvey and De Graaf, and thought he had proved, from microscopical observation, that the male semen, and not the female ovum, is the original of the embryo in all animals. His treatise on the subject, entitled " Anatomia et Contemplationes," is undoubtedly the best defence of the Hippocratic hypothesis which has appeared in modern times.
† The term *cotyledon* occurs in the Hippocratic collection. It is often misunderstood and misapplied both in ancient and modern works on Fœtal Anatomy. It cannot, strictly speaking, be properly applied to any other animal except the ewe. The term is derived from κοτύλη, which was originally applied to the acetabulum, or cavity in which the head of the femur is rotated, See Homer, *Iliad.* v. 305.

which we are now treating. He devotes an entire treatise to the investigation " Of the Formation of the Fœtus in Utero," (vol. iv. p. 652), and also treats of it at considerable length in his great work " On the Uses of the Parts," and in other of his works. Now it is beyond all doubt that he holds there is a direct communication between the mother and the fœtus, by means of an artery and vein running from the uterus into the placenta. The rationale of his doctrine on this point will be readily apprehended by any one who rightly understands his general theory regarding the arteries and veins, which was briefly this,—that every part of the animal frame is supplied with spiritual or äerated blood by means of arteries, and with alimentary blood by veins—the one to maintain its innate heat, and the other to provide its pabulum. Galen, then, adopting the Hippocratic dogma, that the male semen is the original of the fœtus, naturally enough supposed that it must derive both its spiritual and alimentary blood from its mother through the usual channels by which the parts of the adult body are supplied with them. He states distinctly that the chorion is not connected with the uterus at any other point, except where the blood-vessels of the mother enter. He further compares the vessels which unite to form the umbilical cord, to the roots of a tree; and those which proceed to the liver of the fœtus, to its branches.

The treatises referred to above contain also many curious observations, correct descriptions of parts, and ingenious speculations in philosophy; but all these we must pass over as not bearing directly on the point which we have more particularly under consideration. From what we have stated, then, it will be clearly seen that Galen decidedly maintained that there is a vascular connection between the mother and fœtus in utero. Galen being, as it were, the autocrat on all professional subjects, during many succeeding ages, it would be vain to look elsewhere for anything original, either in observation or speculation, during the whole of this period. We may just mention that the same descriptions and the same hypotheses, with little or no variation, are given by Aëtius, one of his more immediate successors, who appears to hold very decidedly that the vessels of the cotyledous are formed from the prolongations of the uterine artery and vein into the semen, and that from the re-union of them the umbilical cord is constituted.* This hypothesis was held as late as the beginning of the 17th century by Fabricius of Aquapendente, the celebrated master of the still more celebrated Harvey, who keenly defends it against the strictures of Arantius, the only physiologist, as far as I am aware, during the whole of the period we are treating of, that had ventured to call in question the dogma of Galen. Fabricius thinks he silences Arantius most effectually by a mode of argumentation, which, at all times carries great weight with ordinary minds, who find it extremely convenient to escape from the labour of reflection and observation by taking shelter under the authority of great names. "Am I," he says, "to embrace this opinion along with a single individual in preference to so many learned men who have maintained the contrary?" It is but justice, however, to the memory of Fabricius, to state that even

* Opera xiii, 2, 3.—The original of this part of the works of Aetius has never been published, but through the kindness of a learned friend in Oxford, I obtained a copy of the two chapters referred to above, from the MS. belonging to the Bodleian Library.—*Cod. MS. Bodl. Canon,* Gr. 109, f. 330.

at the present day there is scarcely a work more replete with original matter on the subject now under consideration than his treatise *de formato fœtu.* In it he has described and delineated most faithfully the appearances of the fœtal and maternal apparatus connected with gestation in a great many animals; so that, if he missed the truth himself, he deserves credit for having forwarded the progress of others in the search of it; and, as by his descriptions and demonstrations of the valves of the veins he brought his pupil to the very verge of his discovery of the circulation, so by his labours on fœtal anatomy and physiology he had the merit of leading Harvey and his followers to the adoption of those very sound views which, as we shall presently see, they entertained respecting the connection between the mother and the fœtus in utero. And at the same time we may draw a useful lesson from his mistakes; as, for example, the following is a memorable instance of the influence which preconceived opinions have in distorting the mental vision, and making a man of even a cultivated mind believe that he sees things exactly as he fancies he should see them. Fabricius affirms, as if from personal observation, that in the bitch, the ewe, and the cow, the vascular connection between the uterus and fœtus can be readily recognised, and he infers that this connection must exist in all other animals of the same class, although it cannot be so easily demonstrated in some of them.* How egregiously Fabricius deceived himself on this point we shall have an opportunity of proving demonstrably in the next section.

PERIOD II.—*Opinions held by Harvey and his followers.*

Harvey having disturbed the established opinions on so important a subject as the functional office of the arteries and veins, it was naturally to be expected that many other received doctrines in physiology which hinged on the old theory would be destined to undergo a corresponding modification. This was accordingly the case with regard to the subject now under investigation, which, moreover, was one to which Harvey had devoted most particular attention, as being connected with his investigations "On Generation." This work accordingly abounds with most interesting observations on the connection between the mother and fœtus, and, also, evinces a very respectable acquaintance with polite literature and the higher philosophy of the ancients; so that, even at the present day, it may be read with much interest and advantage. The only thing which prevented Harvey from taking a correct view of this subject was his scepticism with regard to the discovery of the *lacteals,* which had been demonstrated by Aselli a few years previous to the announcement of his Theory on the circulation.† His views, then, are in some respects not so complete and accurate as those of his immediate successors who admitted this discovery—I mean De Graaf,‡ Ruysch,§ Needham,‖ Swammerdam,¶ Drelincourt,** Bartholinus,†† Steno,‡‡ Hoboken,§§ and Malpighi,‖‖—illustrious names!—whose works every person who would wish to understand perfectly the physiology of

---

* Op. cit. ii. 3.
† See the English edition lately printed by the Sydenham Society, pp. 604-5.
‡ De Mulier. organis, 15. § Observatiuncula de Ove, &c. ‖ De formato fœtu.
¶ Miraculum Naturæ. ** De conceptione adversaria. †† De Ovariis Mulier.
‡‡ Acta Medica Haffniensia, vol. i. p. 210. §§ De secund. vitulin. ‖‖ De Utero, &c.

generation will find it to his delight and improvement—

" Nocturna versare manu, versare diurna "

These distinguished inquirers take the only method which ever can lead to a correct and satisfactory view of the structure and functional office of the human placenta—namely, by studying its analogues through the whole class of Mammals, from its lowest and simplest form up to Man. No one who is at all familiar with their works can fail to recognise the advantage of thus taking a general view of the whole subject, and tracing a certain type of structure through all the orders of the class. My necessary limits on the present occasion preclude me from doing anything like justice to the enlightened views contained in their works, and I must content myself with giving such an outline of their opinions as, I trust, will serve to make them intelligible to the reader. I shall, at least, without further preamble, now attempt an abridged exposition of what they have written regarding the connection between the mother and fœtus, in the orders *Ruminantia, Rodentia, and Bisulca*, taking under each head the guidance of the author whom I look upon as being most successful in his method of illustrating it. On the ruminants, I shall principally follow Hoboken, whose treatise " de Anatomia Secundinæ Vitulinæ" I regard as being the most complete monograph on a single subject in physiology with which I am acquainted. According to him, in the earlier months of gestation the embryo is entirely nourished by means of a gelatinous juice, which is secreted by the womb, and imbibed by the pores of the chorion. At some period of the process, not stated by Hoboken, and never exactly determined by myself, the parts of the vaccine secundines which are the analogues of the human placenta, are formed in the following manner :—The umbilical vessels immediately after issuing from the navel of the fœtus, divide into from sixty to eighty ramifications, which radiate to the circumference of the ovum, and then attach themselves to as many points on the chorion, where they gradually are formed into those protuberances called the *fœtal cotyledons* by the ancient physiologists, and *carunculæ* or *placentulæ* by Hoboken, Needham, and other physiologists of that age. Corresponding to, and in intimate connection with these placentulæ, there are an equal number of cellular protuberances on the inner surface of the uterus, called *glandulæ* and *uterine cotyledons* by the physiologists, these being all formed by the increased developement of the uterine structure, in consequence, no doubt, of the stimulus of impregnation imparted by the placentulæ. It is of the utmost importance, then, to remark that Hoboken has clearly described and delineated the chorion as surrounding the whole of the secundines, including these *placentulæ*, which last are therefore completely separated from the maternal parts in contact with them by a fold of the chorion ; so that neither do the umbilical vessels of which they are principally composed protrude through this membranous envelope, nor do any of the uterine vessels enter it from without : in a word, Hoboken, De Graaf, Malpighi, and all the others referred to above, are agreed in holding that there is no vascular connexion between the cow and her calf, and that the latter is nourished not by the blood of its mother, but by an alimentary juice secreted from it. The process by which this nutritious fluid is separated from the maternal blood, and is transmitted into the fœtal apparatus of vessels composing the placentulæ, is compared by Hoboken to filtration and transcolation, although at the same time, he does not fail to intimate that these terms are used in a figurative sense. In this account of the cotyledons, fœtal and maternal, and their functional offices, Hoboken is supported by all his contemporaries, and most especially by Malpighi, who gives a very accurate description of these parts in his work " De Utero et Viviparorum Ovis." The *glandulæ* or *maternal cotyledons*, he says, are liberally supplied with arteries and veins which secrete " a copious juice like to ptisan." He also describes most graphically the *placentulæ*, and the appearance they put on if slowly separated in water from the *glandulæ* to which they are connected, when the radicles of the placentulæ may be seen divided into numerous capillaries, which present the beautiful appearance of a forest. These radicles or tufts, he distinctly says, are every one of them supplied with branches of blood-vessels from the fœtus. Hoboken, in like manner, states, that these vessels of the placental tufts consist of an artery and a vein. Entertaining the views which we have described of the complete separation of the placentulæ of the fœtus from the adjoining parts of the mother, he approves highly of Aristotle's comparison of the former, to the roots of a plant, which imbibe nourishment to the trunk through its pores (spongioles) without the aid of any vessels from without. With regard to the Ruminants, then, these physiologists regard it as being proved by actual observation, that there is no vascular connection between the mother and the fœtus, and that the latter is nourished by means of a nutritious liquor secreted from the blood of the mother. And here I may be permitted to observe that, having had ample opportunities of verifying these statements, I am quite satisfied of their correctness, and, in particular, I can have no doubt, from my own personal observation, that the *placentulæ* are inclosed in a fold of the chorion, and the *glandulæ* covered by the lining membrane of the uterus ; that, consequently, there is a most complete separation of the maternal and fœtal parts, and therefore that there is no rupture of vessels at the separation of the secundines of the ruminants.

We shall next briefly examine the case of the *rodents*, and in the present instance, shall take for our guide, De Graaf, who has given the fullest and most accurate description I have ever seen of the process of fœtification in the rabbit. For the first seven or eight days after impregnation, the ovum draws all its nourishment from the fluid secreted by the uterus, and yet, as De Graaf remarks, it is wonderful to see how full it is of liquor on the seventh day. This juice he shows to be albumen, from its coagulating by heat. During the early stage, then, it is obvious that the ovum or rudimentary fœtus draws all its nourishment through the pores of the chorion. The first traces of a placenta are visible on the ninth day ; when it is more fully developed, it is found to consist of two portions, a white and a red, both of which, as De Graaf remarks, are clearly formed within the chorion, and come away along with it. It has evidently no vascular connection with the uterus ; this I can affirm from personal observations to be clearly shown towards the end of gestation, although it is not so evident towards the middle of the term, when the placenta is so closely agglutinated to the uterus as might lead a careless inspector to believe that there is a vascular connection. The uterus, too, I may further mention, from my own observation, although it has no prominence on its inner surface corresponding exactly to the *glandulæ* of the ruminants, is much thickened where the placenta is attached, and its blood-vessels there are enlarged.   De Graaf, and all the other physiologists of that age, arrive at the conclusion that, in the case of the rabbit, the placenta is altogether a portion of the fœtal apparatus, and that at birth there is no rupture of any vascular connection between the mother and fœtus.   Needham adds, that, with some slight differences, the same type of structure prevails in the hare, the shrew, the Indian sow, the mouse, the mole, and the hedge-hog.

In the ewe the construction of the different parts is very similar to that of the cow, as already described ; that is to say, the fœtal placentulæ are nearly one hundred in number, and are inserted into as many protuberances on the surface of the uterus, having the appearance of the vinegar-cruets of the ancients, and hence they were called *cotyledones*, or *acetabula;* for it was from observations upon sheep that the term *cotyledon* in fœtal anatomy took its origin.   Now, it is distinctly affirmed, in particular by Steno, but is also assumed and acknowledged as an indisputable fact by the others, that between the fœtal and maternal parts there is no vascular connection, and that the nourishment of the 'fœtus is derived from a milky liquor secreted in the cotyledons of the mother.   Here, again, I beg leave to add my own testimony in favour of the statements just quoted from Steno and his contemporaries ; for, from actual inspection, I can positively affirm that the parts of the fœtal lamb, are not connected to the mother by any vessels, and the fœtus is evidently supported, not by blood but by a thick mucus derived from the cotyledons of the mother.

As stated by Needham, the same mode of construction, with scarcely any perceptible difference, exists in the goat.

In the sow, as Needham and several of the others have correctly remarked, there is no appearance of any distinct placenta or placentulæ from beginning to end of gestation, there being merely a general thickening of the chorion all around, and the connection between the mother and fœtus particularly loose.   In the mare, too, the fœtus is supported for the first half of the period, solely by an alimentary liquor imbibed by the pores of the chorion ; and, even in the latter months, the small tubercles which form on it cannot properly be compared to the placentæ or placentulæ of other animals.   The physiologists of the 17th century then came to the conclusion, that in all the inferior orders of Mammals there is no vascular connection between the mother and fœtus ; that the latter is nourished by an alimentary juice which percolates through the pores of the lining membrane of the uterus, and is imbibed through those of the investing membrane of the fœtal secundines. These enlightened inquirers do not fail, likewise, to point out the analogy which here prevails between the oviparous and viviparous animals ; but their observations on this head we have neither time nor space to give, and shall proceed to expound their views with regard to the *human placenta.*

I may mention, then, in the first place, that Harvey does not hesitate to aver, that he decidedly agrees with Arantius in denying that there is an inosculation between the vessels of the uterus and fœtus, and does not scruple to declare that it was either envy towards Arantius or undue veneration for the ancients, that made Fabricius controvert this doctrine.   He calls the placenta *hepar uterinum* and *mamma uterina,* that is to say, he held it to be analogous to these parts in the adult.   It is to be borne in mind that, as stated above, Harvey did not believe in Aselli's discovery of the lacteals, and accordingly he most probably agreed with the ancients in holding the liver to be the great organ of sanguification. One can readily see, then, what led him to compare the placenta to the liver. His comparison of it to the mammæ, of course implies that he held it to be an organ for preparing an alimentary fluid for the fœtus. Caspar Bartholinus, and Wharton, held very decidedly and distinctly opinions in the main coinciding with those of Harvey. Wharton in particular expresses himself with great precision on this subject, contending that the vessels of the fœtus terminate in the placenta, and those of the uterus in that portion of it which is in contact with the placenta; that there is no inosculation of vessels between them, nor any rupture of the uterine vessels at the separation of the placenta. He further holds that it is a nutritious fluid like the albumen of an egg, and not blood, which is attracted from the uterus by the placenta (Adenographia, Sec. 35.)* Reyner de Graaf, with great ability and force of argument, contends in like manner that there is no vascular connection between the uterus and the placenta, and that the latter organ is an appendage of the chorion. Among other facts which he adduces in proof of this position, he calls attention to the circumstance that in extra-uterine pregnancy a placenta is not wanting, as it certainly would be if it were not a portion of the fœtal apparatus. He holds that the vessels of the placenta suck in a nutritious fluid resembling milk from the mother, in the same manner as the meseraic veins absorb chyle from the intestine in the adult. He further compares the extremities of the umbilical vessels to the roots of trees, and decidedly maintains that at the expulsion of the secundines there is no rupture of any vascular connection (De Mulier. Org. 15). The same theory is defended and espoused by Drelincourt, Swammerdam, Needham, and the other physiologists of the 17th century, without one dissentient voice. as far as I am aware, and including in their number the great father of microscopical anatomy, MALPIGHI. I could have wished to give Malpighi's description of the placenta entire in this place, but it is too long for my limits; and, standing by itself, it would not be readily understood by those who are unacquainted with the Latinity of that age. I must content myself, therefore, with giving a short abstract of his views.

He sets out with admitting the difficulty which he found in describing the uterine parts connected with gestation, but says he will give the results of his own observations *after repeated dissections of women who had died immediately after delivery, or about the seventh month of pregnancy.* After giving a general description of the enlarged structure of the uterus, and more especially of its sinuses, he states, that, on the separation of the chorion and placenta there are discovered " certain pellicles which, during gestation, adhered to the inner surface of the uterus, but several of which are attached to the chorion and placenta:" he adds, " they are soft, mucous, and easily torn." (These pellicles, by the way, are evidently the *decidua* of the Hunters, to which so much importance has been attached

* Ruysch expresses himself in language particularly strong and precise to the same effect, namely, that no blood-vessels pass from the uterus to the placenta, but merely a juice percolated through the glandular body of the uterus, and that the fœtus forms its own blood, from which the placenta, in a great measure at least, derives its growth. Clerici et Manget, Bibl. Anat.) i. 551.

by them and their followers.)* The placenta itself he describes as consisting of a contexture of umbilical vessels, supported by a substance peculiar to itself. He says, that, as the cotyledons (of the ruminants) are possessed of a peculiar structure, which gives support to their vessels and serves the purpose of a sieve or strainer to them, so in like manner the placenta, which is, as it were, an aggregate of cotyledons, namely, of parts which enter the vaginulæ of the uterus, is composed of a congeries of the umbilical vessels. He says that the surface of the placenta by which it is attached to the uterus is unequal, consisting of appendices, which enter the sinuses and cavities of the uterus, like cotyledons. He concludes his treatise with stating that the placenta absorbs a juice secreted by the uterine vessels.† It thus appears that the first person who described the placenta with the aid of the microscope found it to consist entirely of ramifications, of the fœtal vessels along with a peculiar structure for giving them support. Malpighi, therefore, concurs with his contemporaries in holding the placenta to be a portion of the fœtal appendages. I cannot leave this part of my subject without giving the opinion of that ingenious physiologist, Mayow, as delivered by him, in his celebrated treatise, " De Respiratione Fœtus." He says : " As the lacteal vessels deriving their origin from the membranes of the intestines receive a nutritious juice, as it were, by a process of straining through their membranes, and convey it to the mass of the blood, so also in the egg, and in other conceptions, it is to be supposed that a nutritious juice, properly concocted, reaches the mouths of the umbilical vessels only by percolation through these membranes."

And here I would request the reader to remark the evidence which these extracts cursorily given furnish that the physiologists of the 17th century had very correct ideas regarding secretion and absorption through membranes—that is to say, of the processes now denominated *endosmose* and *exosmose*; that they did not hold that the lymphatics are the only absorbents in the animal frame ; and further, that they were aware that these absorbent vessels, whether sanguineous or lymphatic, do not terminate by patulous orifices on the free surface of membranes. Setting out, then, from these accurate premises, it is not so much to be wondered at that these eminent men arrived at correct conclusions on the question now on hand, more especially considering that they took an enlarged view of the whole subject, beginning with the more simple modes of structure, and ascending to the more complex. What these conclusions were, I shall briefly recapitulate before concluding the present section.

They held, then, that, in all the orders of mammals, the fœtus is nourished by an alimentary juice secreted from the blood of its mother, and not by blood itself ;—that in the earlier stage of gestation this liquor is absorbed through the simple pores of the chorion, or external envelope of the ovum ;—that in the lower orders of the class this mode of conveying the nourishment to the fœtus continues to the end of the process,—as, for example, in the sow ; while in others, as the mare, it prevails during the first five or six months ;—that all the higher orders have placentæ, either

* See Appendix B.
† De Utero et Viviparorum ovis Dissertatio.

solitary, or divided into a greater or smaller number of separate masses ;—that these placentæ and placentulæ are altogether portions of the fœtal apparatus, and are mere appendages to the chorion, *within a fold of which they are enclosed* ;—that the maternal structures subservient to the nourishment of the fœtus are entirely separate from the placenta and the rest of the fœtal secundines, and consequently that the fœtal and maternal parts are not united together by any vascular connection.

Such were the tenets of our forefathers in the profession 200 years ago ; and, most assuredly, I think, all must admit that if these, their opinions, are not true, they wear, at least, the semblance of truth, and are in conformity with what is acknowledged as such in all the departments of natural science. Here we seem to see developed a rule of which every department of the microcosm and the macrocosm furnishes parallel examples, namely, simplicity in the general plan and diversity in the particular application of it. Take, as an example of the former, the construction of the respiratory organ in the whole class of mammals, and the planetary motions throughout our system of the other, and the same diversified uniformity will be discovered in both. In short, we no where find in nature anything anomalous ; there is no object throughout all its works which stands by itself, is regulated by peculiar laws, or is constructed upon a particular model.

It thus appears that the opinions held by the illustrious physiologists of the 17th century seem to possess the genuine stamp of truth whether tried by the direct evidence of the facts upon which they are founded, or by the general tests of what is acknowledged to be truth in other sciences. But perhaps it will be objected that, if these their opinions had been really true and well-founded, they ought to have stood the test of time, and not have yielded, as we shall presently see they did, to the prevalence of a different hypothesis. But, to this objection I would reply, that, although I in so far agree with the Roman philosopher, that, "opinionum commenta delet dies, Naturæ judicia confirmat," I must protest that it is only the long, the very long lapse of time, that can be reasonably recognised as a test of what is true and what is false in scientific hypothesis. Thus, for example, the true system of the universe, as now universally acknowledged, was taught by Philolaus more than 300 years before the Christian era ; was admitted by Aristippus, and probably by Plato, his contemporaries ; but immediately afterwards was coldly received by Aristotle, and after the lapse of some 400 years, was rejected by Ptolemy, who propounded a false hypothesis, which superseded the true system for more than twelve centuries. And when the true system was again espoused by Copernicus, in the beginning of the 16th century, did it meet with a general and ready reception? Far the contrary ; most of the lights of the learned world, during nearly 100 years, refused to adopt it. Bacon rejected it, George Buchanan thought he had refuted it, (V. de Sphærâ) Milton hesitated in deciding whether it be the earth or the sun that moves ; Tycho Brahe propounded a new Hypothesis, and Galileo, after demonstrating the truth of the Copernican system, recanted, and professed himself to have been in error. *

* The following references relative to the ancient authorities who advocated the true system of the universe may be interesting to the reader. Diogenes Laertius, *in Vita Philolai* ; Cicero, *Acad. Quæst. IV.* ; Plutarchus, *De Facie in Orbe Lunæ* ; Archimedes, *Arenarius.* I need not give references to the modern authorities—See Whewel's Hist. of the Inductive Sciences.

So difficult is the discovery of truth, and so slow its progress in the world, obstructed as it is by the ignorance, the contending interests, and the prejudices of mankind !   I have indulged in these general reflections in order to prevent the reader from too hastily drawing the conclusion that the opinions which we have just described cannot have been well founded, otherwise. as we shall presently see, they would not have been lost sight of for a time, and been superseded by a different hypothesis.   As it was, however, from the time of Harvey to the end of the 17th century,—that is to say, during more than 50 years, they constituted the acknowledged creed of the profession on this subject.*   In the succeeding century, however, as will presently appear, they did not long remain unchallenged.

PERIOD III.—*Opinions held by the Physiologists of the 18th century, and more especially by Haller and the Hunters.*

One of the first steps towards the overthrow of the theory, which we have now been expounding, was taken by the anatomist Cowper, when he announced that the lacteals all terminate in the free surface of the intestinal canal by patulous mouths (osculis hiantibus.)†   Following up these new but mistaken views which he had adopted, he straightway pretended to show, by mercurial injections, that the umbilical vessels in the placentula of a cow could be injected from the uterine artery.‡   Both these doctrines are advocated by Manget in his "Theatrum Anatomicum," published at Geneva, A. D. 1717, (vol i. p. 306. v. ii. p. 139.)   He does not hesitate to declare it as his opinion that there is a transfusion of blood from the vessels of the mother into those of the foetus, and a mutual circulation of the same between them.   Noortwyk,§ Roederer,‖ Gibson of Leith,¶ and others, concurred in advocating these views.   There were not wanting eminent men, however, who still stood up for the theory of Harvey, and denied the force of the objections recently stated to it by the authorities just mentioned.   Among these the most celebrated was Dr. Alexander Munro *primus*, the father of academical medicine in Scotland, who, in his most interesting treatise "On the Nutrition of the Foetus," decidedly contends that there is no vascular connection between the uterus and placenta, and shows most satisfactorily, how Noortwyk had fallen into the mistake of supposing that such a connection does exist.   He states that he had examined the bodies of five women who had died with child, and that the following were the results : "In all of them I found a thick, fungous, succulent, cellular substance between the muscular part of the womb

* I may mention in proof of this statement, that, in the great anatomical work of Clerc and Manget, published at Geneva in 1699, there is not a single author who disputes the truth of the theory of De Graaf and Malpighi.

† I am not sure, however, that Cowper was the first to make this announcement—indeed, I suspect W. Cockburn must have preceded him.   Cockburn at all events, distinctly describes the lacteals as arising by open mouths in the intestinal canal ; he says, "Lacteorum vero *orificia in intestina hiare*, licet non sint satis spectabilia docet contenti succi color Albicans, &c."—*Œconomia Animalis.*

‡ See Drape's Anthrop.

§ Uteri Anatom. § 6, No.

‖ Icones Uteri humani, &c. 1759.

¶ Edinburgh Medical Essays, vol. i.

and its villous coat, through which numerous. thin-coated vessels passed, and in this cellular substance the sinuses were. Excepting its sinuses it resembled the internal coat of the intestines.* I was ignorant of this structure when I began the dissection of the first woman, and, therefore, when I cut through the firm muscular part of the womb, and saw this fungous substance, I imagined it to be the placenta. I was surprised to find the cohesion of this supposed placenta to the womb so firm; but persisted to separate the muscular part of the womb from it, till having torn a little of the fungous substance, *I observed the smooth, tense chorion, from which the fungous substance separated most easily, as it did likewise from the placenta,* by only gently pressing the ovum with one hand, and raising the womb with the other, without the assistance of any other instrument. I avoided this mistake in dissecting the other four impregnated uteri, which I had occasion to examine afterwards, and then had the villous coat entire, *and the smooth chorion spread over all the secundines.*" He goes on to show, from Dr. Noortwyk's own account of his dissections, that " the doctor must have persisted in the error which I committed in dissecting the first impregnated uterus, which I had occasion to examine, and brought off the internal cellular substance and sinuses of the womb with the ovum, in which case all the appearances would be exactly as he has described them." He distinctly describes the placenta as *" being covered on the side next to the womb, with a membranous continuation of the chorion."* To Slade, who had pretended to show the communication between the vessels of the *glandulæ* and *placentulæ* in cows, he answers "that having tried the experiment variously, he had come to the very contrary conclusion, and that he could not be more certain of any thing than that there is no anastomosis or continuity of these vessels in cows."

The following passage will explain his physiological views: " Were I allowed to illustrate the communication between a mother and her child in the womb, by a gross comparison, I would say that the uterine sinuses are to a fœtus what the intestines are to an adult; the uterine blood poured into the sinuses being analogous to the recent ingesta of food and drink. The liquors sent from the umbilical arteries to be mixed with the uterine blood, resemble the bile, pancreatic juice, and other liquors separated from the mass of blood. The umbilical veins, and those on the surface of the chorion, take up the finer part of this compound mass, as the lacteal and meseraic veins do from the contents of the guts: but the grosser parts of the blood in the sinuses are carried back by the veins of the womb, as the excrements of the guts are discharged at the anus."

Whether or not any attempt was made to answer the arguments contained in Monro's essay, or how they could have been answered, I am alike ignorant, but certain it is that the opinions of Noortwyk soon afterwards found an advocate of such celebrity as raised them to a degree of importance which they would not, otherwise have attained.† I allude to

---

* I beg the reader to remark how analagous these appearances are to what is seen in the glandulæ of the cow.

† The Medical Essays and Observations published by a society in Edinburgh, in and about the year 1737, contain certain papers which show, in a very interesting manner, the unsettled state of professional opinion in Scotland, regarding the placenta, about that period. See in particular " Observations on the Placenta," by Dr. Thomas Simpson of St. Andrews.

Haller, who about this period was rising to be one of the most eminent authorities in physiology and medicine. Haller, it is true, admits, in a letter to his pupil, Dr. Donald Munro, son of the aforesaid Dr. Alexander Munro, that he himself had never made any original observations on the placenta; but, notwithstanding this admission, there can be no doubt that the authority of his name would go far to sanction any opinions which he advocated. Haller, then, decidedly held that all the absorbent vessels terminate on free surfaces by *patulous mouths* (osculis patentibus), and that there is a mutual connection of vessels between the mother and fœtus in utero. (See *Primæ Lineæ*, § 769 and 891). I have seen it stated, indeed, in certain recent publications, that Haller did not believe in an anastomosis between the vessels of the mother and the fœtus in the placenta*; but I cannot conceive any language more express than that which he uses in support of this doctrine. He says—"Respondentibus ex utero in placentæ venas arteriis exhalantibus, tum placentæ arteriis ad uteri magnas venas hiantibus."—*l. c.* He does not hesitate to declare it as his opinion, that in the manner now described, by means of the placenta, a humour of a milky nature is at first transmitted to the fœtus, *and ultimately blood.* In proof of this doctrine that the fœtus is supplied with blood from its mother, he appeals to the well-known phenomena of menstruation and uterine hæmorrhage,—to injections " lately performed by distinguished men," whereby mercury and wax had been thrown from the uterine vessels into those of the placenta,—and to the fact that a fœtus had been known to be formed without a heart, in which case, as he argues, the whole of the blood must have been furnished by the mother.† Here, then, we find described a most extraordinary mode of communication between the vessels of the mother and the fœtus, the veins of the fœtus sucking in blood from the arteries of the mother, and the veins of the latter sucking it back from the arteries of the fœtus! Thus were the arteries and veins supposed to be locked together in each other's embraces! No wonder that the author calls this mode of communication by the name of " commercium." How the arteries and veins, deriving their origin from such distant points as the hearts of the mother and fœtus, contrive to find out one another's mouths " in the dark chamber," he does not attempt to explain. The illustrious Boerhaave, in his work "Œconomia Animalis," gives a sort of equivocal assent to this doctrine, although, at the same time, it is apparent that he did not feel at all satisfied with it, for he admits that appearances

Some of his statements and opinions are very inaccurate, and in that respect form a strong contrast to those contained in the writings of the physiologists of the 17th century. For example, he holds that in the Ruminants the *placentula* are produced on the chorion by contact with the *glandulæ*, and denies that in extra-uterine conceptions any placenta can exist, vol. iv. p. 93. I need scarcely remark that the existence of a placenta in extra-uterine pregnancies is a fact which cannot now be controverted. See Turnbull's case, and many others, published of late years.

* Edin. Med. and Sur. Jour., No. 118,

† The case of the child without heart and lungs to which Haller here alludes, is no doubt the one related by Mery, *Mem. de l' Acad. des Sciences*, 1720. It is noticed by Munro in his essay referred to above. The case, however, is not at all in point; for Mery himself states that this monster was twin to a perfect child, whose *funis umbilicalis* sent off the small navel string of the monster, which was thus supplied with blood from its perfect brother. The injections " lately performed by distinguished men," to which he alludes, were probably those of Cowper and Drake, which Munro shows to be utterly inconclusive.

are certainly in so far against it, since no mouths of vessels can be detected on the lining membrane of the uterus, *nor on the uterine surface of the placenta, which he describes as being smooth, and covered by the chorion* (§ 678, 679). The theory of Haller is the one which is adopted by Fleming in his *Introduction to Physiology* —a work which it is well-known was long used as a text-book on the subject it treats of by the students of medicine in this country. He says —" I stick not to believe that red blood as such is brought from the mother to the fœtus, and transmitted from the fœtus to the mother, as it were, in a circle."

Such, then, was the state of established opinions about the middle of the 18th century; some with Haller, contending for a singular sort of vascular connection between the mother and fœtus; and others, with Munro, still adhering to the theory of Harvey, and denying all such connection. It was not long after this time that the HUNTERS rose to so great celebrity that their opinions became, as it were, laws, which no one was at liberty to dispute. Now it is to be borne in mind that the Hunters acquired a mighty reputation by finally establishing (as was supposed) what had been often maintained and denied before—that the lymphatic vessels perform exclusively the function of absorption in the organism and that these all terminate *by open mouths* on the free surfaces of membranes. This doctrine regarding the absorbents it was long considered their greatest glory to have established beyond all possibility of controversy, by injected preparations contained in their museums. In particular (as has been stated by Professor Goodsir), Dr. W. Hunter, and his colleague Mr. Cruickshank, had in their collection two preparations which were considered to exhibit most distinctly the openings of the lacteals in the intestinal canal. But, as the tale is highly interesting and instructive, I must give it in Mr. Goodsir's own words :—" Mr. Cruickshank, in treating of the orifices of the lacteals and lymphatics, states that he and Dr. Wm. Hunter observed the openings by which the lacteals communicated with the cavity of the gut in portions of the intestine of a woman who died after eating a hearty supper. The two preparations of the intestine on which these anatomists made their observations came into the possession of the College of Surgeons in Edinburgh, as part of the Collection of the late Sir Charles Bell. * * Repeated examinations satisfied me that Dr. W. Hunter and Mr. Cruickshank were quite correct in describing and figuring radiating lacteals within the villi, but that they were led into error in describing those vessels as opening on the free surface of the gut, partly by imperfect instruments and methods of observations,—partly by general prejudice of the period in favour of absorbent orifices."—*Anatomical and Pathological Observations*, No. 2). It thus appears that Hunter's preparations, which were supposed to prove the truth of his doctrines respecting the absorbents, as far at least as regards their terminations by *open mouths*, prove nothing of the kind, and that the Hunters were led into the mistake of *fancying they saw what it is clear they did not actually see*,—"partly by imperfect instruments and methods of observation, partly by the general prejudices of the period in favour of absorbent orifices" —but also, I would add, in a very considerable measure by their inordinate ambition to raise themselves a name above all other names in the profession, and to construct an original system, both of physiology and medicine,

—quod nec Jovis ira neque ignis,
Nec poterit ferrum nec edax abolere vetustas.

In such a frame of mind, and with these prejudices, the two brothers, in 1780, took up the subject of the placenta; and a most unfortunate undertaking it must have proved to both, as regards their happiness, seeing the rivalry engendered by the simultaneous publication of their supposed discovery created a breach in that brotherly affection which had so long subsisted between them.

In the first place I would call attention to the remark, that the doctrines of the Hunters, regarding the absorbents, and Harvey's Theory of the Placenta, are evidently incompatible with one another, since the latter exhibits an instance of absorption through membrane and performed without the aid of lymphatics. Now I do not mean to insinuate that this consideration led the Hunters to palm deliberately on the scientific world a false statement of facts, to conceal the weak points of their own hypothesis, but I do not hesitate to say that I believe prejudices in favour of their own doctrines on absorption made them see the phenomena connected with the placenta through a distorting medium, and in such a light as suited their peculiar views and prepossessions.*

I find it difficult to convey to any reader who is not familiar with the subject, a distinct view of the description of the placenta given by the Hunters; more especially restricted, as I necessarily am, within very narrow limits. It must be apparent to every one who is at all conversant with the literature of the fœtal structures, that the Hunters mystified it in a remarkable degree by combining hypothesis with their descriptions of those "soft and gelatinous pellicles" formerly described by Malpighi, to which W. Hunter gave the very incongruous name of "*deciduæ*," from a mistaken idea that they consist of the inner membrane of the uterus, which he fancied to be cast off like "the slough" of a serpent. I need not say how much the perplexity has been increased by the preposterous ingenuity of their followers, so that it has now become a perfect puzzle to comprehend the *decidua vera, decidua reflexa, decidua parietina, decidua serotina, &c.* To me it appears that the importance of the *deciduæ*, in connection with the literature of embryology, has been absurdly exaggerated, and that they are merely flocculent films or pellicles, formed by the concretion of a gelatinous exudation from the vessels of the uterus; for they are all evidently devoid of structure and vascularity, and have none of that firmness by which the true membranes of the organ are characterised.† And now, with regard to J. Hunter's celebrated description of the dissection of an injected uterus, contained in his "Animal Economy,"

---

* The classical scholar will here recollect the old adage, "credunt quia credere cordi est, " or, as it is otherwise expressed, "quod volunt, id credunt homines."

† It will be understood, I trust, that I apply this character merely to the films formed by exudations from the mother. 1 of course except what has latterly been called the *decidua vera*, by which is meant the lining membrane of the uterus, thickened and expanded like all the other structures of the organ, by the stimulus of conception. 1 need scarcely remark what a gross misnomer *decidua vera* is when thus applied. *Decidua falsa* would be a more appropriate term! It would be most desirabl ethat microscopical physiologists should pay some attention to logic, as much, at least, as to teach them the importance of using properly defined terms. See Appendix B.

suffice it to say, that he describes arteries "*about the size of a crow-quill*," passing from the surface of the uterus into the placenta, and terminating there "in a very fine spongy substance; and that the veins originating from this same spongy substance pass obliquely through the decidua, and communicate with the proper veins of the uterus." The following is his description of the appearances remarked by him on recently expelled placentæ :—"Soon after this time, Dr. Hunter and I procured several placentæ, to discover if, after delivery, the termination of the veins and the curling arteries could be observed; *they were discernible almost in every one :* and by pushing a pipe into the placenta, we could fill, not only its whole substance, but also the vessels on that surface which were attached to the uterus, with injection." The views of W. Hunter are quite similar, and from these two originals, all the descriptions of the placenta which occur in our treatises of anatomy, and works on midwifery, during the next fifty years, are, with very slight modifications, entirely borrowed. In order to place the subject in a more distinct point of view, I shall here quote the description of the placenta and the fœtal circulation, given in a Treatise on Anatomy, which was a common text book in the schools of medicine at the time when I was engaged in my professional studies, and from which I formed my first conceptions of the structure and functional office of the organ :—"In the placenta are to be observed, on the side next the child, vessels, forming the principal part of its substance; on the side next the mother, the ramifications of the umbilical *branches of the uterine arteries, almost of the size of crow-quills*, passing in a convoluted manner between the uterus and placenta, and terminating in the latter. Veins corresponding with these arteries, but flat, and of a good size, running obliquely from the placenta to the uterus ; and, in the substance of the placenta, an appearance, which has been supposed by many authors to be the common cellular tissue, and easily ruptured by injection, but which is considered by late authors *as a regular spongy substance similar to that in the body of the penis*." "The placenta receives blood from the uterus, and, according to the opinion of modern anatomists, *purifies the blood, as the lungs do in the adult*, for the nourishment of the fœtus."* The blood is sent by the

---

* That the placenta is a respiratory organ, was no new doctrine ; indeed, as hinted above, it is as old as Hippocrates, but still, in my opinion, it is quite untenable. Needham properly remarks that it is only by a play upon the term that the functions of the placenta can be assimulated to respiration, since it evidently is so situated as to want the requisites of a respiratory organ. That it is, to a certain extent, a depuratory organ, may be readily admitted, but so, in fact, are the vessels on the inner surface of the intestinal tube, and even the spongioles of trees, for both these excrete recrementitious matters from the organisms to which they also absorb alimentary matters. But this is evidently a very different process from respiration, the primary end of which would appear to be the preservation of the innate heat of animals. We may venture then, I think, to state it as an universal law, that no animal in the state of embryo stands in need of respiration, seeing that in viviparous animals the fœtus derives heat from the vitals of its mother, and in the oviparous it is supplied with the same from without., It has, therefore, always appeared to me an extremely improbable hypothesis, that in incubation the allantoid is for the purpose of respiration ; indeed, the fact that urinary deposits are often found in the allantoid, towards the end of incubation, seems to me to prove the contrary, most decidedly, and that in oviparous animals, as well as in the ruminantia and bisulca, it is rather an appendage of the urinary organs. See Wagner's Physiology, English Ed., p. 196 ; and Mr. Town's experiments, which are very similar to those reported to have been performed by the father of Dr. A. Munro *primus*, and appear to me to have never been fairly met. I observed, indeed, a very interesting paper on the other side of the question, in the Microscopical Journal, for

arteries of the uterus to the substance of the placenta, from which, according to the opinion of many of the ancient anatomists, it passes to the umbilical vein by a direct communication of branches,—or, according to that of the greater part of modern authors, by absorption. From the iliac arteries it is conveyed by the umbilical branches to the substance of the placenta, where one portion of it returns by corresponding veins to the fœtus, *the rest going to the uterus in the manner it was discharged from the uterine arteries to the branches of the umbilical veins.*"—*Fyfe's Compendium of Anatomy*, vol. ii. pp. 295, 296, 307, 308; Edin. 1812.

From the above descriptions it would appear that, according to the Hunters and their followers, the placenta is constructed in the following way :—The great mass of it next to the fœtus is formed from the umbilical vessels, and the more external layer, next its uterine surface, from the curling arteries of the size of crow-quills and corresponding veins of great size ; while its uterine and fœtal portions are connected together by means of a spongy substance, namely the placental cells. Where the cells are situated I am not aware that any one has accurately defined, but I think it probable from the descriptions both of J. Hunter and his copyist Fyfe, as given above, that they were supposed to be much nearer the uterine than the fœtal surface of the organ.*

From what we have stated, it will be remarked that the views of Harvey, Haller, and Hunter, differ in the following respects. Harvey and his followers held that the placenta consists entirely of ramifications of umbilical vessels, with a peculiar substance to support them, enclosed in a fold of the chorion, so that the placenta is completely separated from the uterus, and all connexion between the maternal and fœtal parts is through their investing membranes. Haller held the same views with Harvey as to the vessels which compose the placenta, but he supposed the arteries and veins of the fœtus to terminate on the uterine surface of the placenta by blunt extremities, and there to inosculate with the veins and arteries of the mother. The Hunters and their followers held that the placenta consists mainly of fœtal vessels, but partly also of maternal, united together by means of certain cavities called placental cells, through which an interchange of blood takes place between the mother and fœtus. It is to be further remarked, that the peculiar views of the Hunters rest entirely, 1st, *upon a single dissection of an injected uterus* made by J. Hunter ; 2d, upon certain anatomical preparations preserved in their museums ; and 3dly, upon the appearances remarked on the uterine surface of fresh placentæ. It is not my intention at present to enter upon the consideration of the merits of the case thus made out by the Hunters in favour of this hypothesis, further than to remark, that I believe it will turn out, as we shall see by and by, that injections of so tender an organ as the placenta are extremely fallacious;

Feb. 1846, by Mr. J. Dalrymple, but I cannot admit that his injections, however carefully devised and executed, fairly counterbalance Mr. Town's experiments, and the fact I have alluded to of urinary deposits being found in the allentoid.

* See further Burns' *Principles of Midwifery*, 1832, p. 200. This author certainly appears to understand the construction of the placenta to be as I have represented above. He distinctly describes the placenta as consisting of two portions, which can be separated from one another during the first three months, and which can at all times be injected, separately,—the one from the fœtal and the other from the uterine side, the space called the placental cells being left between them.

and that the appearances described by the Hunters as being seen on the uterine surface of the placenta, are very different from those which Malpighi, Munro, and Boerhave had described, and *from those which I myself have observed.*

The doctrine of the Hunters, however, under the authority of their great name, soon superseded all others, and for more than half a century it may be pronounced to have been the established creed of the profession. Indeed, before Dr. R. Lee, who published his objections to the Hunterian hypothesis in the *Philosophical Transactions* for the year 1832, I am not aware that a single individual had ventured to call it in question. Of his remarkable paper "On the Construction of the Placenta," I have now to give a very brief abstract. He states that his observations are the result "of the examination of *six gravid uteri and many placentæ,* expelled in natural labour, which seem to demonstrate that a cellular structure does *not* exist in the placenta, and that *there is no connection between this organ and the uterus by great arteries and veins.*" He declares that, on detaching the placenta carefully from the uterus, "there is no vestige of the passage of any great blood-vessel, either artery or vein, through the intervening decidua from the uterus to the placenta, *nor has the appearance of the orifice of a vessel been discovered even with the help of a magnifier, on the uterine surface of the placenta";* and further, " *that no cells are discernible in its structure by the minutest examination.*" He argues against a vascular connection between the uterus and placenta, from the surface of the latter appearing "uniformly smooth, and covered with the deciduous membrane, which could not be the case did any large vessels connect it with the uterus; and from the circumstance that in the majority of cases it is separated with the least possible force, and without hæmorrhage." He further gives an analysis of all the preparations in the Hunterian Museum at Glasgow, which were supposed to demonstrate a connection between the uterus and placenta, and shows on the testimony of two intelligent friends, who had examined them for this purpose at his request, that none of them warrant the inference which the Hunters and others had drawn from them. He therefore holds that "the facts stated warrant the conclusion that the human placenta does not consist of two parts, maternal and fœtal: that no cells exist in its substance; and that there is no communication between the uterus and placenta by large arteries and veins."

Such were the conclusions respecting the structure of the placenta which Dr. Lee had arrived at in 1832. But in 1842, he declares that "the discovery of the circulation of the maternal blood in the placenta, made by the Hunters, which throws so much light upon the whole economy of the fœtus, especially the processes of respiration and nutrition, will be regarded in all future ages as one of the greatest that has ever been made in human anatomy, and as second only to the discovery of the circulation of the blood by Harvey" (MED. GAZ. vol. xxxi.) From the following extract it will be apparent that he could then, that is to say, in 1842, see, as he thought, the extremities of ruptured blood-vessels on the uterine surface of a fresh placenta, and could satisfy himself of the existence of the placental cells. "If you keep this (uterine) surface of the placenta convex, you can see numerous small tortuous arteries in the decidua filled with maternal blood.

Their open mouths are visible at the surface of the membrane, and they soon disappear, after making, as John Hunter describes, ' a twist or spiral turn upon themselves.' These decidual arteries soon terminate in the cavernous structure of the placenta, &c." (Ibid.)

I cannot help thinking that when Dr. Lee published this recantation, it was incumbent upon him to explain how, in his six dissections of the bodies of pregnant women performed ten years before, he had missed seeing "the curling arteries, of the size of crow-quills," passing between the uterus and placenta,—how he then failed to detect the ruptured extremities of the same on the uterine surface of fresh placentæ, and also the placental cells; and by what process he arrived at the discovery of all these parts at the latter period. To me it would have been peculiarly gratifying and instructive if he had done so, for I must say that in 1847 I can perceive no openings of blood-vessels on the uterine surface of the placenta, nor can I discover those cavities which have been called placental cells.

I cannot leave this part of my subject without mentioning that Dr. Lee's paper quoted above contains a letter to him from Mr. Owen, giving a description of dissections made by him confirmatory of the views then held by Dr. Lee. He, too, has since explained all this away, and professed himself satisfied that Mr. Hunter's general views are correct in the main. Contrary even to what, as we shall presently see, is the established opinion of the microscopists, he seems to admit the existence of placental cells.

> " Can such things be,
> And overcome us like a summer cloud
> Without our special wonder ! "

Period IV.—*Opinions advanced by Dutrochet, Velpeau, and others.*

We ought to entertain a stronger feeling of gratitude than I suspect we generally do towards our Gallic neighbours for *compelling* us to see and acknowledge the errors into which the Hunters had led us respecting the absorbents, and for explaining to us the nature of imbibition and transudation through membranes, at a time when we had fairly lost all knowledge of the true nature of these processes. I agree, indeed, with Liebig (*Animal Chemistry*, 3d edit. p. 165,) that *endosmose* and *exosmose* are little else than different names for *filtration*; but undoubtedly Dutrochet has great merit for having recalled attention to these phenomena in the animal frame, and for having investigated their laws at a time when complete forgetfulness of them had led physiologists into errors of the most serious description. Light having been thus generally diffused over physiological subjects, Dutrochet, Breschet, and Velpeau naturally thought of reconsidering the prevalent doctrines regarding the construction of the placenta, and the mode of communication between the mother and fœtus. Velpeau's " Embryologie " is one of the most important works ever published on this subject, and therefore I have to regret that I cannot give a proper exposition of his peculiar views without the plates, which constitute its greatest value. I am confident that no one can rise from an examination of these plates without coming to the conclusion that if they are carefully and faithfully executed, as there is every reason to suppose they are, it is impossible to resist the conclusion that the placenta is formed by the extension of the villi of the chorion, and consequently that this organ must belong exclusively to the fœtal apparatus. I would refer the reader particularly to plate vii. fig. 1 ; plate ix. fig. 3 ; plate xi. fig. 2; and plate xii. fig., I.

The principal arguments by which Velpeau combats the doctrines of the Hunters, that the placenta is supplied with blood-vessels from the mother for the purpose of conveying blood to the fœtus, are the following :— 1st. In extra-uterine pregnancies such an arrangement is impossible. 2d. The placenta at first does not exist, and even until the third month it consists of agglomerated filaments only, and consequently no sinuses can exist between its lobules. 3d. A regular-formed placenta has been found in connexion with a fibrous polypus and hardened portion of the womb. 4th, Velpeau has seen the uterine surface of a recently delivered female, hard, leathery, and without orifices. (See further, Edinburgh Med. and Surg. Journ. No. 118, p 174.) Velpeau gives two figures from the great work of W. Hunter, in illustration of the peculiar views of the latter, with some interesting observations of his own. (See Embryol. plate ix. fig. 5, 6.) Of these the one, as he says, "is formed entirely from the imagination ;" and in the other, what Hunter gives as the natural orifices of vessels on the surface of the decidua lining the placenta, he holds to be mere lacerations, —"sont de simples lacerations, au lieu de constituer des orifices naturels comme le croit l'auteur."

To this head may be referred the description of the placenta given by Mr. John Dalrymple, of London, which is highly deserving of notice, as being perhaps the most lucid, precise, and accurate description of the organ, as usually presented to us at the full period of gestation, which is to be found in the whole compass of medical literature. The following extracts will enable the reader to comprehend his general views :—"The umbilical arteries, after dividing and passing on in a convoluted and serpentine form over the fœtal surface of the placenta, dip at various intervals into its substance, there dividing and subdividing infinitely. The trunks are covered on the surface of the organ by the fœtal membranes, and each branch, as it dips into the thickness of the tissue, *carries before it a fold of the chorion.*" " The whole mass of the placenta is made up of the innumerable ramifications of the arteries terminating in beautiful coiled and convoluted capillaries, which form tufts or bouquets at various intervals ; these finally become continuous with the minute origins of the umbilical vein, which returns to the fœtus in the same direction that the arteries left, viz., coiled and twisted in the umbilical cord." " The chorion constitutes by division into processes true villi, and each villus contains a tortuous capillary, which, entering from the arterial side, leaves by the venous." "The uterine surface of the placenta is covered by decidua." " *There are no distinct or defined cells constituting a maternal portion of the placenta.* * * The interstices between the villi have been usually but improperly called the cells of the placenta." " In the placenta must go on a double interchange of fluids, for the blood returned to this organ by the arteries is unfitted for a second circulation through the embryo;—at least, this is true in part. if not entirely. Hence, when the blood, or nutrient material of the blood, brought by the uterine arteries, and previously aërated by the mother, enters by *endosmose* the absorbent capillaries of the fœtal villi, that portion of the fœtal blood which requires the action of oxygen escapes by *exosmose,* and returns by the uterine sinuses and veins to the maternal part."*

This description is so remarkably clear, so devoid of mystification, and, I may be allowed to add, is so much in accordance with my own observations made with the microscope, as to satisfy me that the placenta consists entirely of a congeries of the umbilical vessels, strengthened by some fibrous matter, and enveloped in a fold of the chorion, and having nothing maternal in its structure further than a thin pellicle called the decidua, formed no doubt from a gelatinous exudation poured out by the vessels of the uterus. It agrees so well, moreover, with the above-mentioned descriptions given by Malpighi and Munro, the one two hundred, and the other at least one hundred years ago, that no one can avoid the conclusion that this coincidence cannot be otherwise accounted for but upon the supposition that each of these eminent anatomists described appearances as he had remarked them, and, unlike too many others who have handled the same subject, did not allow his mental vision to be perverted by the mists of prejudice and hypothesis.

And now we find that our investigations have brought us back again at the end of this, the fourth stage of our progress, to exactly the same conclusions as those we had arrived at, at the end of the second stage, namely, that the placenta is altogether a fœtal organ, and that there is no vascular connexion between it and the uterus.

PERIOD V.—*Opinions held by the advocates of the cell-theory at the present time.*

There never occurred, I am inclined to think, within so brief a period, such a revolution in any physical science as has been produced in physiology lately by the celebrated hypothesis of Schwann and Sleiden regarding the functional office of the cell in the formation of all organic substances. A scholar who is conversant with the old atomic theory of Democritus and Epicurus,† who taught that all things are originally formed of atoms, might fancy he saw it revived when he finds the microscope actually shows that all organic substances are composed of molecules surrounded by thin films, namely, *nucleated cells.* Few persons who have any pretensions to an acquaintance with natural science can require to be told that all the structures of the animal and vegetable world are now held to be originally formed from these cells.

---

* I would also beg to call attention to the following observations contained in the same paper :—" It has been observed by some anatomists that the uterine veins are filled by injections thrown in by the umbilical arteries. The explanation of this phenomenon is sufficiently easy ; the tufted villi are very delicate, and it not unfrequently happens that the injection bursts the covering of the chorion, and so escapes into the interstices between the villi, which have been usually but most improperly called the cells of the placenta. If the injection so escapes it will easily find its way, after distending the spongy mass, into the uterine sinuses, and thus fill the uterine veins. On the other hand, coloured fluid thrown into either of the uterine arteries or veins will distend the placenta or spongy interspaces, and if the fœtal tufts be lacerated by the distension or force of the manipulation, some of it will enter the broken extremities of the fœtal vessels," &c. There is an interesting paper in No. 86 of the Edinburgh Medical and Surgical Journal, on the Maternal Fœtal Circulation, by Dr. Williams, from which it appears how very fallacious all attempts to ascertain the minute structure of the placenta by means of injections have generally proved. Dr Munro's Essay, referred to above, likewise contains many statements and remarks all leading to the same conclusion.

† See in particular Diogenes Laertius, *in Vita Democriti ;* and Lucretius, *De Rerum Natura.*

It is also well known with what enthusiasm this doctrine was received, and with what eagerness all subjects connected with animal and vegetable physiology have been re-explored, with the hopes of deriving additional illustration to them from the lights generally diffused by this brilliant discovery.* It was not to be supposed that the placenta would be overlooked in the general survey, and accordingly it will be found that the Teutonic microscopists were not slow in announcing to the scientfic world the new discoveries which they had made in this interesting field of investigation. The foremost to distinguish themselves in this way were Weber, Wagner. and Baer, whose descriptions of the placenta certainly form a most extraordinary contrast to those which we have just been considering. I must now endeavour to convey to the reader, with as much brevity as I can, a distinct notion of the views lately promulgated by Weber with regard to the construction of the human placenta. (See Wagner's Physiology, by Dr. R. Willis, Note, p. 200—206.

He illustrates his idea of the placenta by the following comparison of it to a sponge: "The fibrous tissue of the sponge corresponds with and represents the branching subdivisions of the chorion, and their uniting medium derived from the decidua; the cavities and interspaces of the sponge, however, represent the passages in which the blood of the mother flows. * * The arteries and veins of the uterus open at once into the spongy substance of the placenta." He says elsewhere (p. 201), "that the arteries and veins of the uterus, the channels of the mother's blood, *penetrate in great numbers into the placenta*, and are distributed through its substance in such wise, that every one of its minutest lobules has a canal carrying the blood of the mother, and so comes in contact with the vessels in which the blood of the embryo is flowing. The umbilical vessels of the embryo divide in the manner of a tree, into very numerous and minute branches, which finally turn round, forming loops and anastomoses, and again collect into larger and fewer branches, which at length unite into a single trunk, and form the umbilical vein. *The whole placenta, and therefore every individual lobule, consists of two distinct parts, the one a continuation of the chorion and vessels of the embryo, and the other a continuation of the membrana decidua, and vessels of the uterus.*"† Baer's description is to the same effect: he says "by the growth of the vessels of the uterus into the *decidua serotina* this is transformed into the placenta. *That vessels pass from the walls of the uterus into the placenta has been long known and admitted:* but in regard to the form and mode of this passage or transference, opinions still vary. *It was long believed with Hunter that they passed into cavities.* In more recent times there appeared a growing disposition *to regard these spaces as enlarged veins with extremely thin walls, a structure which is assigned*

---

* After an acquaintance with the cell Theory of Schwan and Sleiden, now extending to upwards of twelve years, I will venture to say of it that it is *un grand peut-etre*,—a splendid speculation—but that it has no pretensions to the rank of a Scientific Theory. If science be correctly defined by Aristotle to be " an immutable opinion," (Topic. vi. 8), surely an hypothesis which has been and is constantly undergoing changes does not deserve to be regarded in this light.—A.D. 1858.

† Wagner's own account of the origin of the placenta is much the same. He clearly derives the origin of it from the *decidua*, that is to say from the uterus (p. 199.)

*to them by many others, and particularly by E. Weber.** He professes himself inclined to agree with Weber, but owns that since he had become acquainted with Dr. R. Lee's views "he had no opportunity of appealing to nature for a solution of the question." Weber, in like manner, although he delivers his extraordinary account of the placenta in a strain of the greatest self-confidence, artlessly lets out that, " *Seiler believed himself authorised to conclude that no vessels from the mother penetrate the placenta but that the maternal vessels only come into contact with the surface of the placenta where it is bounded by the uterus.*" Since, then, it would appear that one portion of the German physiologists hold that the placenta is mainly formed from the vessels of the mother, while others maintain that the uterine vessels do not enter into the structure of it at all, it must surely be admitted that professional opinions in Germany, on this important subject are altogether speculative.! We shall turn our attention, then, to the examination of what has been doing in this department, of late years, by the microscopists of our own country.

Dr. J. Reid has the merit of being one of the first in this country to describe the placenta with the aid of the microscope, after the use of this instrument in physiological investigations had been revived. His views amount to this,—that the placenta consists of a congeries of umbilical vessels, *which terminate in blunt extremities on its surface,* and that it is divided on its uterine surface into a multitude of "tufts," which enter into the sinuses of the mother, these tufts being covered externally by a thin membrane derived from the mother, "consisting of a reflection of the inner coat of the venous system of the mother, or, *at least, of a membrane continuous with it.*" He seems to hold that, at delivery, these tufts are often broken off, and portions of them left behind upon the surface of the womb. Some parts of his description I am not sure that I fully understand, and it will now be generally admitted that he was mistaken in supposing that the umbilical arteries and veins terminate by "blunt extremities." Mr. Dalrymple and Professor Goodsir point out the mistake into which he has fallen in this matter. He further speaks of having satisfied himself, "but not without considerable difficulty, of the existence of the utero-placental vessels described by the Hunters." Now I am at a loss to comprehend how so excellent an anatomist as Dr. Reid could have experienced any difficulty in satisfying himself whether or not "vessels of the size of a crow-quill" pass from the uterus into the placenta, as represented by the Hunters. From some parts of his description one would be led to suppose that he considers the placenta to be formed altogether from foetal vessels, but in other parts he seems to suppose the existence of maternal vessels in it. The functional office and structure of the placenta he clearly holds to

‡ The following experiment is the only positive proof stated by Weber in support of his hypothesis : " If the uterine surface of a very fresh placenta, that has not been put into water, be moistened with a strong solution of corrosive sublimate, in alcohol, in order to coagulate and prevent the escape of the blood still contained in it, and the whole placenta be then soaked in a weaker solution of the same kind, the whole of the maternal blood that remained in the spaces between the divisions of the chorion will be found coagulated ; even in the larger lacerated veins which have just passed from the uterus into the placenta, coagulated blood will be found ; and the manner in which these veins open into the inter-spaces mentioned, will be seen, and the course of the maternal blood during life be found indicated." I have repeated this experiment with the utmost possible care, but have failed to detect the appearances described by Weber.

be analogous to that of the branchia or gills of certain aquatic animals. He says "the placenta is therefore not analogous in its structure to the lungs, but to the branchial apparatus of certain aquatic animals."* According to this hypothesis, then, the fœtus in utero is in a state analogous to the tadpole or young frog. It must be obvious therefore, that, if the structure and functional office of the placenta be as Dr. Reid represents, it is not formed in the human subject upon the type of structure which prevails generally throughout the class of mammals. No person, for example, who examines the fœtal apparatus in the cow, the ewe, or the sow, would say that in them it bears the least analogy to branchia or gills. And besides it is clearly an organ of nutrition, and not of respiration which the fœtus stands in need of.

The description of the placenta given by Dr. Knox† is held to be confirmatory of Weber's views, but I must say that to me it is by no means so clear as that given by his original. On one important point, however, he is decided—namely, that there are no cells or cavities in the placenta, and consequently he rejects *in toto*, the Hunterian theory, which is utterly untenable, provided it be shown that the supposed cells have no existence. He speaks of the decidua being interposed between the placenta and uterus, but admits "that it is obscure in its real nature." He further speaks of the placental vessels penetrating through this decidua until they reached the surface of the uterus, where they floated in one of the venous sinuses of the uterus. This would appear to me to be much at variance with Weber's views as stated above. Altogether, however, I must use the liberty of saying that I desiderate clearness of ideas in Dr. K's exposition of his views.

The elaborate description of the placenta lately given by Professor Goodsir, of Edinburgh, must be regarded as the one which at present gives the tone to professional opinion in this country, and therefore it now demands our serious attention. I must express my regret, however, at the outset, that my prescribed limits preclude me from giving so full an exposition of it as would be necessary to render the views therein advanced intelligible to any one who is wholly unacquainted with them. But as the work is now widely circulated, it is to be presumed that few readers of this communication can be entire strangers to it. I must here, then, refer the reader to Figg. 19 and 20 in Professor Goodsir's Essays, as I find difficulty in getting them correctly copied. I shall only remark beforehand that these figures may be conceived as representing "a tuft" or a single point of the placenta where it comes into immediate contact with the uterus. *Ex uno disce omnes.*

"Fig. 19.—The extremity of a placental villus. *a* The external membrane of the villus, the lining membrane of the vascular system of the mother. *b* The external cells of the villus, cells of the central portion of the placental decidua. *c c* Germinal centres of the external cells. *d* The space between the maternal and fœtal portions of the villus. *e* The internal membrane of the villus, the external membrane of the chorion. *f* The internal cells of the villus, the cells of the chorion. *g* The loop of umbilical vessels."

* Edinburgh Med. and Surg. Journal, 1841, p. 1—12.
† Medical Gazette, Oct. 30, 1840, p. 209,

" FIG. 20.—This drawing illustrates the same structures as the last, and has been introduced to show the large space which occasionally intervenes between the internal membrane and the external cells. It would appear that into this space the matter separated from the maternal blood by the external cells of the villus is cast before being absorbed through the internal membrane by the internal cells. This space, therefore, is the cavity of a secreting follicle, the external cells being the secreting epithelia, and the maternal blood-vessel system the capillaries of supply. This maternal portion of the villus and its cavity correspond to the glandular cotyledons of the ruminants, and the matter thrown into the cavity, to the milky secretion of these organs."

It will here be perceived at a glance that the fœtal parts are represented as being enclosed within the internal membrane of the villus, *and that this membrane is held to be the external membrane of the chorion.* On this most important point, then, in the anatomy of the placenta, Mr. Goodsir, is completely in accordance with De Graaf, Malpighi, Munro, Boerhave, Velpeau, and Dalrymple. It will also be seen that the fœtal and maternal parts are represented as perfectly separate, and involved each in its own proper membrane, as further held by the physiologists of the 17th century; that, as also maintained by them, the fœtus is held to be nourished by means of an alimentary fluid secreted by the maternal portion, and absorbed by the fœtal; and, as likewise held by them, that the fœtal parts correspond to the *placentulæ,* and the maternal to the *glandulæ* of the ruminants. And moreover it will be seen that, like these physiologists, Professor Goodsir compares the function of the placental villi to that of the villi of the intestines, and to the spongioles of plants. In so far, then, one might, at first sight, be disposed to think that Mr. Goodsir's object was to revive the theory of Harvey and De Graaf, and to show it to be in unison with his own favourite cell theory. But on looking more narrowly into his " Essay on the Construction of the Placenta," one meets with opinions which it is difficult to reconcile with what I have stated to be the general bearing of it. It is impossible not to recognise in it a marked disposition to keep on terms with the followers of the Hunters. Accordingly he frequently speaks of the " maternal portion of the placenta,', although it must be obvious that if, as he himself holds, this maternal portion be analogous to the *glandulæ* of the ruminants, it is no portion of the placenta at all. Another of his conclusions wherein the same *animus* is manifest, is this: "The placenta, therefore, not only performs, *as has been always admitted,* the function of a lung, but also the function of an intestinal tube." Now Mr. Goodsir, upon reflection, must be aware that it has by no means been always admitted that the placenta performs the function of a lung—nay, more, there is nothing in his own paper to show that there is the least analogy between a placenta and a lung, for he himself makes out an exclusive case in favour of its analogy to an intestine. His drift however, would appear to have been, to reconcile in so far the conflicting theories of Harvey and Hunter. To prejudices in favour of the Hunterian doctrines may be also ascribed his comparison of " the uterine cotyledons of the ruminant and other mammalia to a permanent decidua vera, and the milky juice interposed between them and the fœtal cotyledons to a decidua reflexa in its primitive and simplest form." Now, the former of these comparisons is, no doubt, in so far appropriate ; only I cannot see how the uterine cotyledons of the ruminants can be called *permanent*, since, as any person may see upon examining the carcase of a cow two or three months after calving, the *glandulæ* or *uterine cotyledons* disappear in the same manner that the decidua vera, or internal membrane of the uterus abnormally developed, does in women. And the milky juice of the ruminants is evidently analogous to the alimentary juice or liquor imbibed by the placenta, and not to the decidua reflexa, to which it can bear no analogy. I also find it a defect in Mr. Goodsir's description of the placenta, that he does not point out the place where, according to his views, the natural separation takes place between the maternal and fœtal parts. It might be supposed that he represents it to be at the union of what he calls the internal and external membranes, since these two membranes are represented as being separated from one another by the alimentary fluid (more especially in fig. 20); and there is no vascular connection between them. The latter, at all events, is merely what is called in the nomenclature of the microscopists a basement membrane, that is to say, " a pellicle of such extreme delicacy that its thickness scarcely admits of being measured," (*Carpenter's Physiology*, 206.) Whether this very fine film is to be usually found on the uterine surface of a newly-expelled placenta, although I have made attempts to satisfy myself as to the fact with the aid of a powerful micros-cope, I am neither competent to affirm nor deny, and, as Mr. Goodsir has not spoken out decidedly, I am at a loss to say what his views expressly are on this point. At all events, it is evidently an insignificant affair, being, as he himself represents it, wholly devoid of all structure. That this is the outer membrane which envelopes the maternal parts must be conceded, I suppose, to so great an authority, although I must confess, it appears to me hard to believe how such an immense sinus comes to be formed external to the lining membrane of the uterus. In the ruminants, with the fœtal anatomy, of which I am most familiar, the latter membrane—that is, the lining mem-brane of the uterus—is the envelope of the maternal parts or *glandulæ*, and all the sinuses are within it. One cannot help thinking that this would have been the most natural construction in the case of the human subject, but I readily admit that I am not competent to dispute this point with so eminent an authority, in microscopical anatomy, as Professor Goodsir, in whose opinions on all these matters I would place the most unbounded reliance, were it not that he evidently betrays throughout the whole of this paper a strong disposition to find all the phenomena connected with the placenta in accordance with the principles of the cell-theory. This bias has led him to adopt certain opinions regarding the placenta which I venture to predict he will find it necessary to modify at no very distant day. For example he holds that " the nucleated cells " are the sole instru-ments of absorption, both in the intestinal villi, and in those of the placenta ; thus refusing to admit to simple imbibition, or *endosmose* and *exosmose*, any operation at all in this case. I perceive that Professor Matteucci has already pointed out this part of Mr. Goodsir's theory as being overstrained. (See *British and Foreign Medical Review*, No. 46, p. 378.)

Dr. Carpenter, in his " Manual of Physiology," professes to borrow a considerable part of his description of the placenta from Mr. Goodsir, but, upon the whole, he would seem to incline rather to the theory of Weber than of Harvey. Thus, treating of the nature of the fœtal and maternal portions of the placenta, he says, that in the human subject "the two elements are mingled together through its whole substance:" and he further speaks of the blood being "conveyed into the *cavity of the placenta* by the curling arteries," a description which I am not sure that I understand aright; indeed, I must freely confess that in the course of my numerous dissections of the placenta, I have never been able to detect either "the cells" of Hunter, or "the cavity" of Carpenter.

I am aware that I may subject myself to the charge of presumption for thus venturing to criticise the opinions of two contemporaries so distinguished for their contributions to physiology as Professor Goodsir and Dr. Carpenter. But whoever undertakes the investigation of a subject like our present one, will find that at every stage of his progress he will have to encounter such discordant views and conflicting statements, advanced under the sanction of GREAT NAMES, as must soon compel him either to abandon the pursuit altogether in despair, or exercise his own judgment manfully in discriminating between truth and error. With much respect, therefore, for the two individuals I have just named, and fully sensible how much physiology owes to the one as an original inquirer, and to the other as a diligent expounder of the discoveries of others, I must still be permitted to state, that I too feel conscious of having cultivated this department of physical science with so much diligence, that I am not afraid to claim for myself the privilege of exercising an independent judgment in every case; I must venture to say with Correggio, when rousing himself to contend with the great Masters in his art, "*ed io anche sono pittore.*"

RETROSPECT.—Having thus finished my historical sketch of all the opinions which have been entertained on the subject in question, I will now briefly recapitulate the results, and state the conclusions which I hold may be legitimately drawn from the same.

I. It would appear that all the opinions which have ever been held regarding the construction and functional office of the placenta, may be referred to the following heads:—1st. Physiologists in ancient times, and down to the days of Harvey, holding that the male semen is the original of the embryo, believed that the uterine vessels penetrate into it, and thus directly furnish blood to the fœtus. 2nd. Harvey, and all the physologists of the 17th century after him, held that there is no vascular connection between the mother and fœtus, and that in the human subject, as in all the other orders of mammals, the fœtus and its appendages are inclosed in a proper membrane of their own, through which they imbibe an alimentary juice, which constitutes the pabulum out of which the blood of the fœtus is formed. This theory likewise accords in the main with the views entertained by Monro *primus*, in the middle of the 18th century, by Dutrochet and Velpeau, by Seiler, Dalrymple, and other authorities of the present day. 3rd. Certain physiologists in the 18th century, including the illustrious Haller, held that a mutual inosculation of the foetal and maternal vessels takes place upon the uterine surface of the placenta, and that in this way the uterine arteries supply pure blood to the placenta, while the uterine veins remove impure blood from it. 4th. The Hunters and their followers held that the placenta is formed partly from the ramifications of the uterine vessels, and partly from those of the umbilical, and that these two distinct portions are united together near its uterine side, and interchange blood by means of certain cavities named the placental cells. 5th. Weber, Baer, Wagner, and other advocates of the cell-theory in Germany, have lately propounded the doctrine that the whole mass of the placenta is composed of a double set of vessels, the maternal and umbilical ; that these run along, side by side, through the whole structure of the placenta, the latter corresponding to the fibrous tissue of a sponge, and the former to the cavities or interstices of the same ; that the maternal vessels are far larger than the foetal, the latter being mere capillaries, whereas the former are represented to be " far too large to be spoken of as capillaries." 6th. We have shown above that all the views lately propounded by British microscopists, are modifications of the theories of Harvey, Hunter, and Weber.

II. It will be readily admitted that the first of these theories, being based on an erroneous assumption as to the original of the embryo, and also being formed in ignorance of the true functional office of the blood-vessels, must be entirely rejected ; and that the third is, if possible, still less deserving of any serious consideration, being founded on an erroneous hypothesis as to the termination of the absorbent vessels, and in ignorance of the difference which it is now well known there is between the globules of the maternal and fœtal blood.

III. With regard to the second, I hold that the following conclusions, may be legitimately drawn from the facts and arguments stated in the preceding sketch. 1st. It is a *priori* highly probable as assumed by the physiologists of the 17th century, that the human placenta should be formed upon the same type of structure that prevails through the other orders of the class *Mammalia;* and these physiologists have shown, most satisfactorily that in all the inferior orders, the maternal and fœtal parts connected with gestation are entirely distinct, and that the fœtus is nourished by means of an alimentary liquor, secreted from the maternal blood and imbibed by the fœtal secundines through their outer membrane, the chorion, which envelopes the whole secundines, including the placenta when this organ exists. 2d. The analogy between the appa·ratus for the support of the fœtus in the ruminants and the human subject is so striking as to have been pointed out by physiologists in all ages. Now it is clearly shown by the physiologists of the 17th century—and as to the fact I cannot entertain the slightest doubt—that in the cow the maternal and fœtal cotyledons otherwise called the *glandulæ* and *placentulæ*, are quite distinct, and consequently there can be no vascular connection between the mother and the fœtus. The analogy of the ruminants, therefore, leads strongly to the probable conclusion that in the human subject the maternal and fœtal parts must be entirely separate. 3. The analogy of oviparous animals in like manner leads to the conclusion that it is an universal law in nature that animals in the embryonic state have no vascular connection with their parent. 4. The analogy of the placental tufts to the intestinal villi and the spongioles of trees is so striking as to have been particularly adverted to by physiologists in all ages, and it also leads to the same inference. 5. The appearances presented when separating the placenta from the womb after death, as given by Malpighi, Munro, Dr. Lee in 1832, and others, are so striking, as to preclude the supposition of there being any vascular connection between the uterus and placenta,

(See Appendix B.)   John Hunter's assertion that he detected " arteries of the size of crow-quills, and veins of a large size," passing between the mother and the placenta, has not been confirmed by the observations of any trustworthy inquirer since his time.   6. As is remarked by De Graaf, Dr. Lee, and others, if the placenta were connected to the uterus by blood-vessels of considerable size, it is impossible that the separation of the placenta could ever take place without hæmorrhage.   7. The last remark is still more striking in the case of inversion of the womb.   8. The appearances on the uterine surface of the placenta as described by Malpighi, Munro, Boerhaave, Dr. Lee in 1832, J. Dalrymple, and others, have been confirmed by repeated observations made by myself, and, as far as I can see, they entirely preclude the supposition of any vessels from the mother entering the placenta :—*The uterine surface is covered with a fold of the chorion, and no vestige of any blood-vessel can be detected on it even with the aid of a microscope.*   9. As remarked by De Graaf, Velpeau, and others, if the placenta did not belong exclusively to the fœtal apparatus, this organ could not possibly be formed in extra-uterine pregnancies. 10. The figures of the ovum given by Velpeau in his *Embryologie*, if correct, put it beyond a doubt that the placenta is an appendage of the chorion, and that no maternal vessels enter into its structure.

IV. With regard to the Hunterian hypothesis, the following are the results of our preceding investigation, and the conclusions which I think must necessarily follow from them :—1. That the Hunters took up the consideration of this subject with very erroneous impressions in regard to the absorbent system, and looked upon the phenomena connected with the placenta under a strong bias in favour of their preconceived views. 2. That they allowed themselves to be imposed upon by fallacious appearances in certain injected preparations contained in their museums, as Professor Goodsir has shown that in like manner they were deceived by similar preparations illustrative of the mode by which the absorbents terminate in the intestinal canal.   3. That certain of the figures in W. Hunter's work on the Gravid Uterus are drawn from the imagination, and exhibit appearances purely ideal.   4. That when J. Hunter described " vessels of the size of crow-quills " running between the uterus and the placenta, he most probably committed the mistake which Monro shows Noortwyk to have fallen into—namely, that of confounding the portion of the maternal apparatus which is analogous to the *glandulæ* of the ruminants with the placenta itself; the fact of the matter being, that this fungous substance in the human subject is as entirely separate from the placenta as it is in the ruminants.   5. That, as we have shown to have often happened to other inquirers of equal eminence, the Hunters most probably allowed themselves to be imposed upon by preconceived notions when they described the openings of arterial vessels said to be seen by them on the uterine surface of the placenta, no such openings being actually visible.   6. That it seems to be now pretty generally conceded that there is no such structure of the placenta as the cavities described by the Hunters under the name of placental cells.   7. That, according to the Hunterian hypothesis the formation of the placenta in extra-uterine conceptions is utterly inconceivable. 8. That the human placenta, if constructed on the plan represented by the

Hunters, would be a perfect anomaly in nature, as it cannot be shown that the same organ is formed upon the same type in any other animal whatever. 9. That, in particular, the human placenta, as described by the Hunters, is analogous to the lungs in the adult, whereas in all other animals its corresponding part is analogous to the intestinal tube : in other words, the placenta is evidently an organ of nutrition, and not of respiration as held by the Hunters.

V. Respecting Weber's hypothesis, I hold,—1st, That most of the objections stated above to the Hunterian theory of the placenta apply also to that of Weber, as far, at least, as concerns a vascular connexion between the uterus and placenta. 2. That no competent proof is offered of the placenta being formed on the ideal plan he describes. 3. That the German physiologists are utterly at variance as to the facts upon which this hypothesis is founded ; thus, for example, Escricht, of Copenhagen, agrees with Weber on many points, while on others he proclaims his dissent, whereas Seiler entirely denies all vascular connexion between the placenta and the uterus,—all this implying that the hypothesis has been hastily concocted from assumed facts of a very questionable stamp, and not from original observation. 4. That this hypothesis is irreconcileable with a prominent fact than which there is no fact connected with the subject established upon a greater concurrence of high authority, namely, that the uterine surface of the placenta is covered by a smooth and firm reflection of the chorion. 5. That if tried by the acknowledged tests of evidence, the conflicting statements of alleged facts in support of the hypotheses of Hunter and of Weber nullify one another. According to the one the utero-placental vessels are few in number, and of considerable size,—according to the other, they are numerous, and of small size ; according to the one these vessels immediately after entering the placenta are lost in the cavities called the placental cells,—according to the other the placental cells have no existence, and these vessels retain their vascular form throughout the whole structure of the placenta.

Finally.—Seeing, then, that there are so many formidable objections to the other hypotheses, it seems impossible not to recognise the second as being the only true theory of the structure and functional office of the placenta, inasmuch as we have seen that observation, analogy, and reasoning from the undoubted facts of the case, all lead to the same conclusion,—namely, that the human placenta is formed upon the same type as its analogues in all the genera and orders of the class of mammals ;—that it is a portion of the fœtal appendages, having no connection with the maternal parts but by imbibition through its investing membrane ; and that its functional office is analogous to absorption by the intestinal tube and bears no analogy whatever to the process of respiration in adult animals.

APPENDIX A.

EXAMINATION OF THE PREPARATIONS OF THE GRAVID UTERUS IN THE HUNTERIAN MUSEUM AT GLASGOW

I FIND myself called upon to resume this subject, in consequence of having been obliged to leave one portion of my argument incomplete at the time I composed my papers on the Construction of the Placenta, which appeared in the MEDICAL GAZETTE during the months of July, August, and September, 1847. At that time, having never had an opportunity of inspecting the anatomical preparations in the Hunterian Museum at Glasgow, I was under the necessity of passing judgment upon those which relate to the construction of the placenta entirely at second hand, and principally upon the report of the authorities quoted by Dr. Lee, in his paper contained in the Philosophical Transactions for the year 1832. I have since had it in my power to inspect them carefully for myself, and consequently it becomes my duty to state publicly the results of my examination, as far as they bear upon the opinions advanced by me in my former communications. I think the present, moreover, a most fitting occasion to re-direct the attention of the profession to this question, as I find that a zealous advocate of the Hunterian hypothesis, in a late number of the MEDICAL GAZETTE, rests his defence of it on the appearances presented by these preparations : his words are—" Many years ago the Hunters demonstrated that vessels passed (pass ?) from the uterus into the placenta, and the beautiful injections left behind them still remain to certify to this fact. Since then several attempts, have been made to repeat these injections, *but without success*, and thus incontrovertible evidence seemed to be afforded in favour of the opinion that the placenta was (is ?) entirely fœtal. The injections, and the doctrines founded upon them, were considered to be equally fallacious, &c." It thus appears that the evidence in support of the Hunterian hypothesis is now made to rest entirely upon the anatomical preparations contained in the Hunterian Museum. Whether or not the gentleman whose words I have just quoted has ever actually inspected these preparations, I have no means of knowing; but if he has, I must say that either he or I have been looking at the same objects through a coloured medium, and have drawn very different conclusions from the same data. Indeed, I may mention, that when I entered the Hunterian Museum, for the purpose I have stated,—although I must admit that I did so under the impression that the Hunterian hypothesis is at variance with a great law of the animal economy,—I did expect to meet with appearances by which I should be staggered, and fancied to myself that my mind for a time would be in such a state of suspense, as the Roman poet professes to have been, when called upon to pronounce judgment on the justice of a cause, which had the Gods on the one side, and Cato on the other !

" Victrix causa Diis placuit sed victa Catoni."

I was not a little surprised then,—I may almost say disappointed,—to find that, notwithstanding the imposing titles which certain of the preparations on the gravid uterus bear in the catalogue, there is not a single one of them which, when impartially examined, would warrant the inference drawn from them by the Hunters—namely, that arteries " of the size of crow-quills, and veins of a considerable size," pass between the uterus and the placenta.

I now proceed to give literally the remarks which I wrote on these preparations at the time I made my examination of them. It would serve no good purpose, however, to detail my observations on all the preparations of the gravid uterus which I examined, as a very large proportion of them do not at all bear upon the question at issue ; and therefore I shall be content with selecting what I consider to be a sufficent number for forming a general judgment on the whole :—

No. 31, s.—A portion of the uterus at the place where the placenta adhered : the orifices of the torn veins full of plugs of coagulated blood : very remarkable. [*Very unsatisfactory : no certainty that what are here represented as vessels arc vessels. The substance on the inner surface of the uterus evidently a portion of the maternal cotyledon much torn.*]

33, s.—A portion of the uterus in which the arteries had been injected red, the veins yellow : shows inside surface and the torn orifices of the veins filled with the yellow injection. [*Pieces of red wax, certainly having some resemblance to vessels, are to be seen, but they prove nothing as to the construction of the placenta.*]

34.—Ditto, shows ditto. [*Nothing certain can be made of this preparation ; very unsatisfactory.*]

96, s.—A portion of the placenta and its membranes : on the surface which adhered to the uterus may be seen some very small curling arteries injected red, and veins injected black, which are going to the cells of the placenta. [*I cannot make anything of this preparation. Substance of the placenta a mass of red wax.*]

100, s.—A small section of placenta with part of the membranes : the cells of the placenta have been filled from the veins of the uterus, and vice versa ; the cells are not very bare ; on the side which adhered to the uterus the veins may be seen very distinctly. [*Difficult to say what is meant here by the cells of the placenta. Quite an indistinct preparation.*]

106.—A section of uterus with membranes turned partly down, and showing a double layer of decidua. [*A beautiful preparation. What is called decidua, a mere film, seemingly devoid of regular structure.*]

118, s.—A section of uterus with placenta partly adhering and partly detached : showing in the angle the mode of adhesion. [*Very interesting, but no appearance of vessels at the angle. Quite at variance with the Hunterian hypothesis, as not exhibiting the utero-placental vessels.*]

124, s.—A small portion of the placenta and uterus, where the cells of the placenta have been iejected from the veins of the uterus ; the veins are seen very large, entering into the substance of the uterus : injection green. [*The green pieces of wax here taken for veins passing between the placenta and uterus, are as large as the femoral vein of an adult. Quite out of the question that this can be a correct preparation : evidently the result of laceration.*]

145, s.—A portion of the uterus with placenta adhering injected red : the cells of the placenta injected from the uterus. [*Difficult to make out what it meant by the cells : altogether the placenta is a confused mass.*]

147, s.—A portion of placenta with the cells apparently filled with fine injection of a red colour ; less distinct than when coarse injection is employed; the vessels of the navel-string are quite empty, although the vessels of the cells had been very minute, proving no communication. [*The entire mass of the placenta is here seen injected, except the cord ; consequently the injection must have burst the vessels, even according to the Hunterian hypothesis.*]

149, t.—A portion of uterus and placenta ; the arteries injected of a dark colour, and veins green: both vessels are seen entering into the substance of the placenta. [*Pieces of wax to be seen on the uterine surface of the placenta, but no reason to suppose that they are vessels.*]

158, t.—A portion of uterus and placenta; the placenta being partly detached, shewing veins injected green from the uterus, going into the posterior surface of placenta ; the placenta itself injected with a different injection. [*Certainly no inference as to the construction of the placenta can be drawn from this preparation. The green substances are taken for vessels, but in all probability they are lacerations ; the wax has burst the vessels.*]

160, s.—A placenta injected from the navel-string red; to great minuteness, most entirely unravelled, showing a most beautiful shag of vessels : it has been hardened by spirits of wine probably, and put into oil of turpentine. [*A curious preparation, but shows nothing in regard to the construction of placenta. Indeed, it seems at variance with the Hunterian hypothesis, for the whole mass of the placenta is injected from the umbilical cord.*]

(?) s. t.—A portion of uterus with placenta adhering ; the vessels of the uterus injected red and black : the cells of the placenta are filled with a different injection, and therefore not from the vessels of the uterus, but must have been previously filled from the spongy surface of the placenta itself. [*What is here said about the cells is quite imaginary: here the mass of the placenta would seem to be injected from the uterine vessels. Preparation quite unsatisfactory.*]

176, s.—Section of uterus with placenta adhering : the cells of the placenta are injected from the vessels of the uterus. [*The centre is filled with a red injection from the uterus, but no appearance of vessels passing between the uterus and placenta.*]

178, s.—A small section of the uterus, with the veins injected green, and broken off where they are entering the placenta. [*Green pieces of wax are to be seen on the surface of the uterus, but no reason to suppose them truncated vessels.*]

From what I have now stated, it will be readily understood that, in my opinion, the preparations in the Hunterian Museum at Glasgow do not at all warrant the inference that there is any connection by arteries and veins between the uterus and placenta, and that the appearances of connection which they exhibit may all be reasonably supposed to be the result of laceration. At all events, as the collection exhibits the most contradictory appearances, it is indisputable that one is not warranted in founding any theory upon them. For example, No. 147 exhibits a placenta wholly injected from the uterus, while No. 160 is a placenta entirely injected from the umbilical vessels. Now most assuredly it will be admitted that one or other of these preparations must be incorrect, seeing they lead to incompatible and contradictory inferences. Then, again, who for a moment can believe that vessels of the size of a femoral vein pass between the uterus and placenta, as exhibited in No. 124.? And, to give another example, when 'masses held to be vessels are exhibited in No. '124, how does it happen that the said vessels do not appear in No. 118, which exhibits an uterus with a piece of placenta partially detached from it ?

I repeat, then, after a careful, and I conscientiously believe, an impartial inspection of the preparations of the gravid uterus contained in the Hunterian Museum, I do not fear to declare it as my decided opinion, that they do not at all warrant the inferences which the Hunters drew from them of a vascular connection between the uterus and placenta.

How the Hunters came to entertain these erroneous notions regarding the placenta I have partly explained in a former part of my communication ; namely, that it was owing to their minds having been occupied by strong prepossessions in favour of the termination of the absorbent vessels in patulous mouths, and their prejudices against the doctrine of imbibition through membranes. It is a melancholy instance how a superior mind may be blinded by prejudices, that Dr. W. Hunter professed to have actually seen distinctly the terminations of the lacteals in an intestinal villus, and that the Museum of the Hunters contained preparations which were held to show decidedly the patulous orifices of these vessels.* With such unfounded prejudices and mistaken views, it was morally impossible that the Hunters could have solved the problem as to the mode of communication between the mother and fœtus in utero. How the opinions of the Hunters on this subject should still command authority in this country, can also admit of a ready explanation, when we advert to the extraordinary veneration in which their names have been held for the last sixty or seventy years ; this is so much the case, that Mr. Samuel Lane, in his excellent paper on the Lymphatic and Lacteal System, in the Cyclopædia of Anatomy, complains that he found the minds of professional men had not yet freed themselves from the influence of the Hunterian views with respect to the parts performed by the lymphatic vessels, and that we are still allowing ourselves to be misled by these impressions. It is now at least thirty years since our Gallic brethren overturned the doctrines of the Hunters regarding the Lymphatics, and yet we stuck to them down almost to the present date. We were long behind our neighbours, also, in admitting the possibility of absorption by veins and through membranes; but now, all must allow that on these points the Hunters were greatly in error. To allow, then, that they had also deceived us on the subject of the placenta appears altogether monstrous in the eyes of these ardent worshippers who are not yet prepared to cast off the Hunters as their professional Indigetes. What adds much to the tenacity with which their hypothesis on the placenta is still defended, is the circumstance that it is intimately connected with the art of midwifery, and that many of our standard authorities in this line, are already strongly committed on this subject, and naturally feel reluctant to believe and to confess that they have long been propagating erroneous doctrines on points of the most vital importance, as regards the lives of their fellow-creatures.

Vel quia nil rectum, nisi quid placuit sibi, ducunt,
Vel quia turpe putant parere minoribus, et quæ
Imberbes didicere, senes perdenda fateri.

* See Mr. Lane's paper on the Lymphatic and Lacteal System, in the Cyclopædia of Anatomy, and Goodsir's Anatomical Essays.

This feeling, then, so well expressed by the poet in the verses just quoted, has operated powerfully in all ages, and for the reason which I stated it weighs very strongly with the obstetrical authorities at the present time. Hence some of them obstinately cling to the Hunterian hypothesis, while at the same time they admit facts bearing upon the question, which, to any unprejudiced mind, must appear quite decisive against their own opinions on the question as to the supposed vascular connection between the uterus and placenta. For example one of them lately made the following candid statement of facts: " The uterine surface of the placenta is covered by a delicate membrane, and seems to be so applied to the walls of the uterus as to close the venous openings on its surface without any direct connection with them. *The placenta may be peeled from the uterus more easily than the rind from an orange: no vessels seem to be broken.* The natural inference from these facts would be, that the placenta belongs altogether to the fœtus; that no maternal blood passes into it; and that any interchange between the blood of the child and the mother takes place only at the surface of the uterus, to which the placenta is applied like a cake of unbaked dough."—MED. GAZ:, No. 1094. p. 826. On this remarkable passage I shall only remark, that the two facts here distinctly admitted appear to me quite decisive of the question at issue; for if the placenta can be peeled from the surface of the uterus more easily than the rind from an orange, without any vessels being seen to be broken, and if no vessels can be detected on the membrane which lines the uterine surface of a separated placenta, we may rest assured that the so-called utero-placental vessels are altogether ideal.

In the course of my examination of the preparations in the Hunterian Museum, I was much struck with one of them, and with the title it bears in the catalogue :—" No. 320, s.—A portion of gravid uterus from the cow, showing the oval fungus of the maternal part of the placenta, resembling in its surface pretty much a cauliflower. This and the foregoing preparation show that in many quadrupeds the maternal and fœtal parts of the placenta are quite distinct in structure from each other, and may throw light on the human placenta, where there is a more intimate connection between the fœtal and maternal portions."

Here then it would appear that W. Hunter had before his eyes a specimen of a placenta constructed upon a totally different type from what he conceived the human placenta to be—namely, with a complete separation between the maternal and placental portions, and where of course nutrition must take place by absorption through membrane. Strange ! that it should not have occured to his acute mind, that if absorption through membrane can take place in one of the mammalia, there is every reason from analogy to suppose that the same vital process must operate in its congeners, and more especially in the highest genus of the class; and that if the secundines be entirely separated from the uterus in one of the genera, that there is every reason from analogy to infer the same of the others. But whatever their blind worshippers may say to the contrary, the minds of the Hunters, and especially of John Hunter, were not of a logical cast nor capable of entertaining any very enlarged general views on professional or scientific subjects. Had they been well trained in tracing the structural analogies in the animal kingdom, and in drawing legitimate inferences from them, they could not have failed to arrive at the conclusion, that every other organ in the bodies of man and the ox are constructed on similar types—as, for example, the lungs, the heart, the liver, the kidneys, the bladder, the womb; nay, if even with regard to the contents of the cranium itself, which is as it were " the dome of thought and palace of the soul," every particular part in the brains of the two animals is formed on the same fundamental type;* it is contrary to all analogy to suppose that such crude structures as the secundines should be constructed on entirely different types, and that their functional office should be essentially different

And here I cannot deny myself the pleasure of introducing a quotation in which this train of thought appears to me very conclusive on the argument which I am now enforcing. " In all the principles of his internal structure, in the composition and function of his parts, man is but an animal. The lord of the earth, who contemplates the eternal order of the universe, and aspires to communion with his invisible Maker, is a being composed of the same materials, and framed on the same principles, as the creatures which he has tamed to be the servile instruments of his will, or slays for his daily food. The points of resemblance are innumerable; they extend to the most recondite arrangements of that mechanism which maintains instrumentally the physical life of the body,—which brings forward its early development, and admits, after a given period, its decay,—and by means of which is prepared a succession of similar beings, destined to perpetuate the race."—(Pritchard's Natural History of Man, p. 2.

From what has been here stated, I trust that it will be now generally admitted that I am warranted in drawing the following inferences:—

1. That a careful inspection of the preparations of the gravid uterus, in the Hunterian Museum at Glasgow, gives no support to the hypothesis advanced by the Hunters and their followers regarding the construction of the placenta, and the mode of communication between the mother and fœtus in utero.

2. That these preparations, by appearing to prove too much in regard to a vascular connection between the mother and fœtus, lose all claim to be held as competent evidence on the question at issue: since, for example, some of them exhibit vessels of the size of the femoral vein, passing between the uterus and placenta; and in others the whole substance of the placenta is injected from the uterus. *Qui probat nimium probat nihil.*

3. That considering how close an analogy subsists between the respective organs in the bodies of the ruminants and the human subject, it is highly improbable that their secundines should be composed upon totally different types.

4. That since no one pretends to say that there is an utero-placental circulation in any other animal, it is contrary to all analogy to suppose that such a process takes place in the human subject.

5. That the human placenta, if constructed in the manner represented by the followers of the Hunters, that is to say, if composed partly of fœtal and partly of maternal vessels, all blended together into one compact structure, would be an absolute monstrosity, without a parallel in the whole works of Nature.

* See Teidemann on the Fœtal Brain, *passim.*

6. That the shape of the fœtal globules indicates that they have not been derived direct from the blood of the mother.

7. That it being now universally admitted that the placenta can be peeled from the surface of the womb more easily than the rind from an orange, without any vessels seeming to be broken, and that there is no appearance of vessels on the uterine surface of an expelled placenta, it is impossible any longer to contend that the so called utero-placental vessels have any existence.

Lastly. That in the human subject, as in all other animals, the secundines are altogether a fœtal structure, and that no maternal vessels can possibly be lacerated at the separation of the placenta in natural labour.

APPENDIX B.

ON HUNTER'S PLATES OF THE GRAVID UTERUS.

Before concluding I think myself obliged to notice the attempt lately made to bolster up the Hunterian Hypothesis by an appeal to the elegant Plates of the GRAVID UTERUS, published by Dr. W. Hunter, and of which a reprint was issued a few years ago by the Council of the Sydenham Society.

In the first place let me state decidedly, that I have no great faith in knowledge of anatomy acquired from plates, having seen many proofs of the fallaciousness of such delineations even when executed under the directions of the highest authorities. I request the reader's particular attention to the following confirmation of this statement taken from Mr. Guthrie's admirable work on Hernia. Mr. Guthrie gives four distinct engravings of the inguinal ring by the highest authorities in this line, along with the anatomical descriptions of these parts by Sir Astley Cooper, Cloquet, Blandin, and Velpeau, and then makes the following striking remarks :—

" The reader cannot fail to be surprised at the great difference which exists between these different versions of the same thing, and that a plain matter of fact and not of imagination ; and a student in anatomy and surgery, on trying to reconcile them by an actual examination of the parts will find considerable difficulty in making his dissection correspond with any one of the descriptions which have been quoted. * * No student can look at the four engravings appended to this paper, and believe that they are intended to represent the same parts in the same stage of dissection, without drawing much on his imagination ; yet they are really intended for that purpose. * * I dare hardly venture to give the reason which, in my mind, has led to the great apparent discrepancy of opinion which exists between so many able men on so plain a matter of fact. *It is possible that it may have arisen from the great minuteness with which it has been attempted to describe parts which scarcely deserve it, especially the fascia transversalis.*" p. 10. He adds, " If the student is taught to consider the fascia transversalis as a sheet of condensed cellular membrane divisible in some parts into two layers he will readily understand it." p. 11.

Now, I will here venture to affirm, that if the student of Fœtal Anatomy will compare the Hunterian engravings with one another, and with the preparations in the Hunterian Museum, and further will compare both with the appearances which he will discover on actual dissection, he will find himself extremely puzzled, and come at last to the same conclusion respecting the *deciduæ*, as Mr. Guthrie does regarding the Fascia Transversalis, namely, "that this great discrepancy of opinion may have arisen from the minuteness with which it has been attempted to describe parts that scarcely deserve it":—for that, after all, these *membranæ deciduæ* are neither more nor less than concretions formed from the glutinous cement or mucous tractus by which the *chorion* is glued, as it were, to the inner membrane of the uterus I am fully persuaded. This has been my decided opinion for a good many years, and it is a great satisfaction to find all my opinions amply confirmed by the following descriptions of dissections lately made by Dr. Meigs, of Philadelphia, under so favourable circumstances as seemingly to preclude all suspicion of mistake.

"In a necroscopy made in the presence of Dr. Yardley and Dr. Wallace, I detached the whole of the placenta from the womb, after the careful injection made of the aorta by Dr. Wallace. Neither I, nor those gentlemen, upon the most minute and careful search, aided by good lenses, could verify the existence of even a single vessel passing from the womb to the placenta. We arose from the dissection equally and unanimously convinced that we had not seen a single vessel broken off, or pulled out in the slow, gentle, and most careful divulsion of the two utero-placental surfaces.

"During the epidemic of cholera here, I examined a womb within a very few hours after the death of the woman, in company with the late Dr. J. Hopkinson, then prosector at the University of Pennsylvania. He though an able anatomist, was unable, as I was, to detect anything broken, *save mucous tractus*, though the light and the glasses were good, and the most scrupulous care was used. A similar opportunity was enjoyed a few years since, in the Pennsylvanian Hospital, in a womb gravid with twins. *Here also I detected nothing but mucous tractus.* Another very fine specimen, at the seventh month, was afforded me by Professor Pancoast, at the Jefferson College. In this case many medical students observed the devulsion of the surfaces without detecting any vessels."—Treatise on Obstetrics, p. 178.

He thus states his conclusions:—"As I must confide in my own, rather than in other men's senses, I find it impossible, under my own observations, to adopt the views of the Hunters, and I prefer the opinions of Seiler and of Velpeau."

It thus appears that Historical research and the most recent observation lead to the same conclusion, namely:—

THAT THE HUMAN PLACENTA IS ALTOGETHER A FŒTAL STRUCTURE HAVING

NO VASCULAR CONNECTION WITH THE MOTHER.

KNOWLEDGE OF THE OVULATORY CYCLE AND COITAL FREQUENCY AS FACTORS AFFECTING CONCEPTION AND CONTRACEPTION

ROBERT G. POTTER, JR.[1], PHILIP C. SAGI[1], AND CHARLES F. WESTOFF[2]

INTRODUCTION

ONLY during a relatively short period, approximately in the middle of the menstrual cycle, does a woman have an appreciable chance of conceiving. The average duration of this "fertile period" is not known precisely but almost certainly averages under 48 hours.[3] Since the time period is so brief, coital frequency has a bearing on conception ease. This relevance is the more assured because of evidence, furnished primarily by the studies of MacLeod and Gold,[4] indicating that coital frequency can be increased without jeopardizing virility, except possibly in cases where the increase is to very high levels or the male is oligospermatic. Any relationship of coital frequency to contraceptive effectiveness is more speculative. Strong libido, reflected in high levels of sexual activity, may contribute to less effective contraception by generating more occasions for chance-taking and, in the case of rhythm, by making continence more difficult to maintain during the unsafe period.

Having a less direct but potentially important bearing on conception ease and contraceptive effectiveness is knowledge of the fertile period with reference to its timing within the

[1] Office of Population Research, Princeton University.
[2] New York University.
The writers are indebted to A. J. Coale, F. W. Notestein, and C. Tietze for their helpful comments.

[3] See C. Tietze, Probability of Pregnancy Resulting from a Single Unprotected Coitus, *Fertility and Sterility*, XI (September–October, 1960), 485–488 and R. G., Potter, Length of the Fertile Period, Milbank Memorial Fund *Quarterly*, XXXIX (January, 1961), 132–162.

[4] Of special relevance, among several pertinent articles by these authors, is J. MacLeod and R. Z. Gold, The Male Factor in Fertility and Infertility: Semen Quantity in Relation to Age and Sexual Activity, *Fertility and Sterility*, IV (January–February, 1953), 10–33. For brief comment on this work and the related work of E. J. Farris, see R. G. Potter, Length of the Fertile Period, *op. cit.*, p. 140, fn. 18.

monthly cycle. Efforts to hasten pregnancy are not likely to succeed unless the couple have realistic ideas about when to increase their sexual activity above customary levels. Then, too, beliefs about the timing of the fertile period govern a couple's judgment about when they may omit contraception with greater or lesser risk of pregnancy. The correctness of such beliefs should be especially crucial for rhythm users and for users of chemical or mechanical methods who regularly omit contraception when they feel it safe to do so.

It is the purpose of this paper to test the above hypotheses on the basis of data collected in the Family Growth in Metropolitan America study—a longitudinal survey being conducted at the Office of Population Research,[5] under the sponsorship of the Milbank Memorial Fund with grants from the Population Council.

SAMPLE

In this survey, 1,165 wives were initially interviewed shortly after their second delivery and were reinterviewed approximately three years later. The original panel constituted a probability sample of native-white, once-married couples residing in one of the seven largest Standard Metropolitan Areas and having a second birth in September of 1956.[6] At this time the wives averaged 26 years of age and a little longer than 5 years of marriage.

Compared with the total panel, wives sucessfully reinterviewed three years later tended to be slightly older, married at a later age, more concentrated in the white-collar occupational class, of higher education, and included proportionately fewer Catholics. The importance of these biases is modified by the fact that this paper will be concerned with relationships rather than means or distributions.

[5] The results of the first interviews are reported in C. F. Westoff, R. G. Potter, P. G. Sagi, and E. G. Mishler, FAMILY GROWTH IN METROPOLITAN AMERICA, (Princeton: Princeton University Press, 1961).
[6] For details of the study design, see C. F. Westoff, *et al.*, FAMILY GROWTH IN METROPOLITAN AMERICA, chapter 2.

DATA

In both interviews detailed data were collected on conception delays and contraception. In addition, during the second interview, information was obtained about such topics as:

I. current frequency of marital intercourse.
II. belief about the timing of fertile and infertile periods in the menstrual cycle,
III. practical use made of this belief either to hasten or to avoid pregnancy,
IV. time when belief was acquired, and
V. frequency of skipping contraception.

Data on the last item are available only for the 673 respondents practicing contraception at the time of the second interview, whereas reports on all other items embrace the entire sample.

FREQUENCY OF COITUS

The coital frequencies reported by the respondents are tabulated in Table 1.[7] As in most other studies, these frequencies decline with advancing age. The correlation with wife's age is − .21, to be compared with a corresponding correlation of − .33 obtained by Terman.[8] Within an age control, the present data correspond fairly well to re-

Table 1. Coital frequency at time of second interview.

RATE PER WEEK	FREQUENCY
Less Than Once per Week	48
Once	140
One and a Half Times	142
Twice	254*
Two and a Half Times	81
Three Times	117
Three and a Half Times	60
Four Times	37
Four and a Half or More	26
TOTAL	905

* 31 indeterminate frequencies are assigned to this modal category.

[7] Roughly one-third of the women indicated monthly frequencies and these frequencies were multiplied by 7/23 to convert them to a weekly rate. Twenty-three days was viewed as a typical intermenstrum—i.e., 28 days minus 5 days for flow. No strong bias appears to have resulted from this procedure inasmuch as a nonsignificant difference is found between the distributions of converted frequencies and those not requiring conversion.

[8] L. M. Terman, PSYCHOLOGICAL FACTORS IN MARITAL HAPPINESS, (New York: McGraw-Hill, 1938), p. 271.

sults from the female sample of Kinsey *et al.*[9] Another point of agreement with the latter study is the absence of significant class-religious differences with respect to coital rates, a result also obtained by Terman.[10]

BELIEFS ABOUT THE OVULATORY CYCLE

A series of questions were used to elicit opinions about the timing of the fertile period in the monthly cycle. After preliminary queries about typical flow and cycle length, the respondent was asked to indicate, on a chart representing her typical monthly cycle, the cycle days she believed to be especially fertile. Often more than one fertile period was alleged and each of these periods might include one day or several.

All responses have been classified into three categories: "correct," "incorrect," and "don't know." Replies designating a single fertile period in the middle of the cycle were rated as correct. More specifically, this fertile period must overlap with at least one of the three most fertile days as judged by the Ogino-Knaus and/or Farris rhythm calculations, predicated on the cycle length which the respondent reports as most typical for her.

Altogether, 49 per cent of the replies have been labeled correct, 23 per cent "don't know," and 27 per cent incorrect. Thus about half the sample have a realistic view of the ovulatory cycle; another quarter hold inaccurate opinions; and the remaining quarter do not pretend to know. Class-religious differentials exist (see Table 2) with persons of higher socio-economic status possessing more knowledge on the average and, within class, Catholics having an advantage over non-Catholics. This religious difference is related to the Catholics' greater emphasis upon the rhythm method. As will be shown later, most rhythm users have correct opinions about the timing of the fertile period.

[9] A. C. Kinsey, *et al.*, SEXUAL BEHAVIOR IN THE HUMAN FEMALE. (Philadelphia: W. B. Saunders Company, 1953), p. 394.
[10] *Ibid.*, pp. 355 and 360; Terman, *op. cit.*, p. 275.

CLASS AND RELIGION	NUMBER OF COUPLES	KNOWLEDGE ABOUT THE FERTILE PERIOD			
		Correct	Incorrect	Don't Know	Total
White-collar Catholic	179	60.3	21.2	18.4	99.9
White-collar Protestant	191	53.9	22.5	23.6	100.0
Jewish	110	48.2	20.9	30.9	100.0
Blue-collar Catholic	244	44.7	˙32.0	23.4	100.1
Blue-collar Protestant	181	40.9	32.0	27.1	100.0
TOTAL	905	49.4	26.5	24.1	100.0

Table 2. Information about the positioning of the fertile period, by class and religion.

USE OF INFORMATION ABOUT THE OVULATORY CYCLE

Only 18 per cent of the sample report ever trying to hasten pregnancy by deliberately increasing coital frequency during particular times of the month. In this connection it should be kept in mind that in the absence of contraception or after deliberately interrupting it, a majority of couples conceive within six months and a very large majority within a year. No important class-religious differences are found. On the other hand, as one might expect, couples who report a past problem in becoming pregnant are much more likely also to report that they tried to hurry at least one pregnancy. About half these latter couples have sought to hasten a conception, as compared to 15 per cent among the remainder who remember no conception problems.

Women in the present sample report using their beliefs about the timing of the fertile period more often to avoid pregnancy than to hasten it. All told, 35 per cent of the respondents mention avoiding intercourse at particular times of the month as a precaution against pregnancy. These 35 per cent do not coincide exactly with the 39 per cent who report ever using rhythm. Some of the latter combine rhythm with another method by practicing no method during supposedly safe days and the supplementary method during unsafe days. Class-religious differentials are found with respect to avoiding intercourse at particular times of the month and follow the pattern of relative emphasis upon the rhythm method.

INFLUENCE UPON CONTRACEPTIVE EFFECTIVENESS

To what extent does low coital frequency and correct information about the safe and unsafe periods of the menstrual cycle facilitate effective contraception? Since little or no association exists between coital frequency and information about the ovulatory cycle, the two factors may be considered independently. A crude measure of contraceptive effectiveness will be used, namely the per cent successful in any single pregnancy interval. By 'successful' is meant deliberately interrupting contraception in order to conceive and by 'failure,' experiencing pregnancy either while practicing contraception or omitting it to take a chance. This crude index does not take into account time with contraception preceding success or failure and therefore becomes liable to bias if either coital frequency or knowledge of the ovulatory cycle are appreciably correlated with intended length of pregnancy postponement. However, there is little reason to surmise such links.

Interestingly enough, coital frequency fails to discriminate contraceptive success or failure when all methods of contraception are considered jointly, and even when use of rhythm is taken alone. An important reason for this unexpected result is that coital frequency is only weakly correlated, if at all, with frequency of chance-taking, contrary to our original hypothesis. Of the 673 respondents using contraception at the time of sec-

Table 3. Proportions successful with the rhythm method used alone, by accuracy of information about the fertile period.*

CONTRACEPTIVE EXPERIENCE	KNOWLEDGE OF THE FERTILE PERIOD			
	Correct	Incorrect	Don't Know	Total
Failure	73	25	14	112
Success	81	24	9	114
TOTAL	154	49	23	226

* The associated significance level falls between .3 and .5.

ond interview, the 211 wives who acknowledge skipping contraception "once in a while" or "quite often" do not differ significantly in their mean coital rate from the 462 wives who claim never to omit contraception.

Worth examining next is the relationship between knowledge about the ovulatory cycle and the contraceptive effectiveness of rhythm. Use of rhythm alone (Table 3) is distinguished from use of rhythm in combination with another method (Table 4). To increase sample size the experiences of first, second, and third pregnancies preceding second birth have been pooled. As a result, one couple may contribute experience from as many as three intervals, so that the observations are not strictly independent and the accompanying chi square values are only approximate.

As noted earlier, only a minority of the respondents citing use of rhythm have incorrect or no opinion about the ovulatory cycle. Partly for this reason, the differences obtained in Tables 3 and 4, although in the expected direction, are not statistically significant to a decisive degree. Some of the women who currently possess correct information about the ovulatory cycle acquired it after one or more intervals of rhythm use. Elimination of these intervals does not materially bolster the association in Table 3, but does slightly strengthen the association observed in Table 4.[11] Another reason why the effectiveness of

Table 4. Proportions successful with rhythm used in combination with another method, by accuracy of information about the fertile period.*

CONTRACEPTIVE EXPERIENCE	KNOWLEDGE OF THE FERTILE PERIOD			
	Correct	Incorrect	Don't Know	Total
Failure	26	14	5	45
Success	28	4	2	34
TOTAL	54	18	7	79

* The associated significance level falls between .05 and .10.

[11] In 124 intervals of rhythm use preceded by acquisition of correct information, successful contraception is claimed in 67 instances, unsuccessful in 57. When these

(Continued on page 53)

rhythm is not more strongly conditioned by accuracy of opinion about the ovulatory cycle is that chance-taking, acknowledged by roughly one-third of the current users of rhythm, appears to be equally common in the three groups representing correct, incorrect and no opinion about the unsafe period.

When attention is shifted to the other methods of contraception besides rhythm used alone or in combination with another method, no association whatever is found between beliefs about the timing of the unsafe period and contraceptive effectiveness. Presumably, for the women involved here, a failure to mention rhythm as a supplement to the principal method of contraception means that any omissions of contraception are not usually guided by rhythm considerations; and therefore the accuracy of beliefs about the unsafe period becomes largely irrelevant.

INFLUENCES UPON CONCEPTION DELAY

If the fertile period is less than 48 hours, then doubling coital frequency at mid-cycle, say from 2 to 4 times per week, should reduce mean conception delay by a factor of one-third to one-half, oligospermatic husbands always excepted.[12] In a previous study by Stix,[13] the empirical relationship between coital frequency and conception delay approaches this theoretical standard. A strong association is also found by Gold and MacLeod, although their data is presented in a form which precludes precise comparison with Stix's results.[14] In a third study by Stix and Notestein,[15] the relationship between reported

figures are substituted into the first column of Table 3, the associated chi square is raised from 1.5 to 1.8 but still fails to reach statistical significance. With respect to Table 4, there are 38 intervals in which rhythm used in combination with another method is preceded by acquisition of correct information. Here, successful contraception is claimed 22 times and unsuccessful contraception admitted 16 times. So revised, Table 4 yields a barely significant chi square between the 2 and 5 per cent level.

[12] Potter, The Length of the Fertile Period, *op cit.*, pp. 149–156.

[13] Regine K. Stix, Birth Control in a Mid-Western City, Part I. Milbank Memorial Fund *Quarterly*, XVII (January, 1939), p. 82.

[14] MacLeod and Gold, The Male Factor in Fertility and Infertility, *op cit.*, p. 29.

[15] Regine K. Stix and Frank W. Notestein, CONTROLLED FERTILITY, (Baltimore: The Williams & Wilkins Co., 1940), p. 11.

COITAL FREQUENCY (PER WEEK)	FIRST CONCEPTION		SECOND AND THIRD CONCEPTION	
	No Contraception Used During Interval	Contraception Deliberately Interrupted	No Contraception Used During Interval	Contraception Deliberately Interrupted
Less than Twice	11.0 mos. (158)	7.1 (111)	21.7 (91)	7.0 (185)
Twice	7.1 (144)	4.5 (122)	19.8 (82)	5.2 (197)
Three or More	6.6 (106)	4.4 (85)	12.9 (49)	5.8 (146)
TOTAL	8.5 (408)	5.4 (318)	19.0 (222)	6.0 (528)

Table 5. Mean conception delay as related to coital frequency, by pregnancy interval and contraceptive status.

coital frequency and conception delay proved negligible for reasons that are not entirely clear.

The results of the present study are tabulated in Table 5. Care has been taken to distinguish first conceptions from subsequent ones and, within these two classes, to separate delays following deliberate interruption of contraception from those occurring before first use of contraception. Thus four trials of the relationship are provided. Nine classes of coital frequency are combined into three to give more stability to the mean conception delays. A consistent tendency for conception delay to shorten as coital frequency increases is evident in three of four trials, though the associations are weaker than the one observed in Stix's data.

Doubtless one factor operating to weaken relationships in the present sample is the reporting of coital frequency at a time point several years after the conception delays being correlated with it. Unfortunately there is no ready way to gauge the importance of this factor.[16]

It remains to consider whether endeavors to hasten preg-

[16] The conception delays occurring since second birth are the most concurrent and for that reason might be expected to yield the highest correlation with reported coital frequency. However there are only 46 completed pregnancy intervals since second birth in which no contraception was practiced. In another 146 pregnancy intervals since second birth, contraception was deliberately interrupted but the ensuing conception delays are strongly selected for brevity, only 6 showing durations greater than 6 months.

EFFORTS TO HASTEN PREGNANCY	FIRST CONCEPTION		SECOND AND THIRD CONCEPTION	
	No Contraception Used During Interval	Contraception Deliberately Interrupted	No Contraception Used During Interval	Contraception Deliberately Interrupted
Yes	30.9 (23)	11.3 (46)	34.0 (23)	9.1 (90)
No	7.1 (384)	4.4 (272)	17.3 (199)	5.3 (438)

Table 6. Mean conception delay as related to deliberate efforts to hasten pregnancy, by pregnancy interval and contraceptive status.

nancy are effective and whether this effectiveness is qualified by knowledge of the ovulatory cycle. Contrary to what one might first expect, couples who strive to speed pregnancy take longer on the average to conceive than those who report no special efforts. According to Table 6, this unexpected relation holds in the initial as well as the following two pregnancy intervals (all preceding second birth) irrespective of whether contraception is deliberately interrupted or not used at all. In 3 out of 4 instances, the relationship is significant beyond the .005 level despite heavy losses of information entailed by the choice of significance test.[17] Obviously this outcome is related to the fact, documented earlier, that women who have experienced trouble conceiving are much more likely to take action toward hastening pregnancy. Doubtless many of these couples did not initiate their special efforts to conceive until several months without pregnancy had elapsed.

Given this selective process, it is impossible to measure directly the effectiveness of deliberately increasing sexual activity as a device to hasten pregnancy. However, women reporting such efforts may be classified according to the correctness of their present opinions about the timing of the fertile period and a check made whether accurate information on this score constitutes an advantage. The differences contained in Table 7 do suggest an advantage. Special attention has been paid to a comparison of conception delays between possessors of correct

[17] To cope with the highly skewed distributions of conception delay, an extension of the median test based on chi square was utilized. Cf. S. Siegel, NONPARAMETRIC STATISTICS, (New York: McGraw-Hill, 1956), pp. 179–184.

INFORMATION ABOUT THE FERTILE PERIOD	FIRST CONCEPTION		SECOND AND THIRD CONCEPTION	
	No Contraception Used During Interval	Contraception Deliberately Interrupted	No Contraception Used During Interval	Contraception Deliberately Interrupted
Correct	17.4 (9)	9.4 (32)	29.0 (8)	6.4 (50)
Incorrect	40.9 (9)	21.4 (8)	30.4 (10)	11.1 (24)
Don't Know	39.5 (14)	8.2 (6)	49.2 (5)	14.6 (16)

Table 7. Mean conception delay among couples trying to hasten pregnancy, by pregnancy interval and accuracy of information about the fertile period.

and incorrect information, with respondents claiming no knowledge set aside. Although all differences are in the expected direction, the number of cases involved are so small that only two of the four relationships are significant at beyond the .05 level.[18]

DISCUSSION

Because of its practical importance, the relationship between coital frequency and conception delay deserves more comment. Table 5 leaves no doubt that higher coital rates generate shorter conception delays *on the average.* Nevertheless the linear correlation between conception delay and reported coital frequency is very low. The four correlation coefficients corresponding to the data of Table 5 are all in the region of − .10.

The basis of this low correlation may be conjectured. Here it is convenient to think of each couple as having, in the absence of contraception, a typical monthly chance of conception, or fecundability. This fecundability is responsive to level of sexual activity and may be raised during a single pregnancy interval if, for instance, the couple suddenly embark on special efforts to hasten pregnancy. However the level of a couple's fecundability is also affected, and perhaps primarily determined, by a host of other, mostly physiological factors.[19] For

[18] The Mann-Whitney U-test has been used, *ibid.,* pp. 116–126. A third of the four tested relationships proved statistically significant at between .1 and .2.

[19] Evidence that fecundabilities vary widely among couples is afforded by two

(Continued on page 57)

example, let us imagine that because of reproductive impairments (such as partially occluded tubes, cystic ovaries, etc.) a particular group of couples have only one chance in 20 of conceiving any month, given their customary coital frequencies. By augmenting their sexual activity at mid-month, they manage to double their fecundability from .05 to .10. Yet they still remain at a disadvantage with respect to the majority of couples whose fecundabilities are well above .10. Furthermore, individual conception delays are subject to chance variation, the magnitude of which tends to increase as fecundability decreases.[20] Thus it is plausible to think that coital frequency is only moderately correlated with fecundability which in turn is only moderately, perhaps only weakly, correlated with individual conception delays.

The practical implication for couples wanting a pregnancy and of low fecundability is obvious. By increasing marital intercourse at mid-month, many such couples can materially reduce their expected conception delay but they do not, thereby, assure themselves of prompt conception. They remain subject to a large element of chance and depending on their luck will conceive promptly or only after a frustratingly long wait.

SUMMARY

The data of this analysis have come from the Family Growth in Metropolitan America study, a longitudinal survey in which 1,165 couples were interviewed approximately six months after their second birth and then 905 were reinterviewed three years later. During both interviews, detailed histories were collected about contraception and conception delays. In addition, during the second interview, questions were asked concerning beliefs about the timing of the fertile period in the menstrual cycle

models designed to reproduce a criterion series of conception delays: C. Tietze: Differential Fecundity and Effectiveness of Contraception. *The Eugenics Review*, Vol. L (January, 1959), pp. 231–237; and R. G. Potter, Length of the Observation Period as a Factor Affecting the Contraceptive Failure Rate, Milbank Memorial Fund *Quarterly*, XXXVII (April, 1960), pp. 141–144.

[20] This point is justified mathematically by reference to the geometric distribution. Cf. R. G. Potter, Some Physical Correlates of Fertility in the United States, [Mimeographed Paper # 12] International Population Conference, 1961.

and use made of these beliefs either to hasten or delay pregnancy.

About half the responses about the timing of the fertile period were classified as correct; the remainder were fairly equally divided between classifications of "incorrect" and "don't know." Thirty-five per cent of the respondents report ever avoiding intercourse during particular times of the month as a means of avoiding pregnancy and most of these couples are correctly informed about the timing of the unsafe period. Only 18 per cent report ever trying to hasten pregnancy by deliberately increasing coital frequency.

Coital frequency is not correlated with contraceptive effectiveness, partly for the reason that greater sexual activity does not appear to occasion more frequent chance-taking. Coital frequency is moderately correlated with fecundability (a couple's typical monthly chance of conception in the absence of contraception) but only weakly associated with individual conception delays.

Correctness of information about the ovulatory cycle does not appear to affect contraceptive effectiveness except in the case of the rhythm method. The gains in conception ease derived from deliberately increasing coital frequency are difficult to gauge because there are disproportionately many subfecund couples in this group who make special efforts to hurry conception. However, among couples reporting special efforts to hasten pregnancy, those correctly informed about the fertile period average shorter conception delays than those who are misinformed.

ROLE OF AMNIOCENTESIS IN THE INTRAUTERINE DETECTION OF GENETIC DISORDERS*

Henry L. Nadler, M.D., and Albert B. Gerbie, M.D.

Abstract One hundred and sixty-two transabdominal amniocenteses were performed between the thirteenth and eighteenth weeks of fetal gestation as part of the management of 155 "high-risk" pregnancies. Successful cultivation of amniotic-fluid cells led to the intrauterine detection of Down's syndrome (10 cases), Pompe's disease (one case), lysosomal acid phosphatase deficiency (one case) and metachromatic leukodystrophy (one case). The risk of this procedure is low since neither fetal nor maternal complications were demonstrated in this series of patients. Cultivation of amniotic-fluid cells obtained by transabdominal amniocentesis early in the second trimester of pregnancy provides a method that enables parents at "high risk" for having offspring with certain serious genetic disorders to have children without risk of such a defect.

A MNIOCENTESIS has been used as a diagnostic aid since the early 1930's.[1] Since the demonstration of its value in the management of Rh isoimmunization, the technic of transabdominal amniocentesis has gained widespread acceptance.[2] This procedure has been performed over 10,000 times after the twentieth week of pregnancy and maternal or fetal morbidity or mortality reported in less than 1 per cent of cases.[3-7] In adverse effects that have been reported fetal mortality appears to be greater than maternal, with fetal deaths reported due to abruptio placentae, amnionitis and fetal hemorrhage.[5-7] Puncture of the fetus has been reported.[7-9] The maternal morbidity includes amnionitis, maternal hemorrhage, abdominal pain and peritonitis. More recently transabdominal amniocentesis has been performed early in the second trimester of pregnancy. In most cases, amniocentesis has been performed immediately before pregnancy was interrupted, making it difficult if not impossible to define the risks to either fetus or mother accurately.[10,11] Transvaginal amniocentesis has been shown to carry an appreciable risk of spontaneous abortion when performed early in pregnancy.[12]

During the past few years, sex-chromatin analysis

*From the departments of Pediatrics and Obstetrics and Gynecology, Northwestern University Medical School, and the Genetic Clinic, Children's Memorial Hospital, Chicago (address reprint requests to Dr. Nadler at the Children's Memorial Hospital, 2300 Children's Plaza, Chicago, Ill. 60614).

Supported by grants from the National Institutes of Health (HD-04339), the National Foundation — March of Dimes and the Chicago Community Trust.

and cultivation of amniotic-fluid cells obtained by transabdominal amniocentesis early in pregnancy have resulted in the detection of a number of genetic defects of the fetus, including X-linked recessive disorders,[12] Down's syndrome,[13,14] mucopolysaccharidosis,[13,15] Pompe's disease,[16] X-linked uric aciduria,[17] cystic fibrosis[18] and lysosomal acid phosphatase deficiency.[19]

The purpose of this study was to evaluate the usefulness of transabdominal amniocentesis and cultivation of amniotic-fluid cells in the management of pregnancies in which there is an appreciable risk that the child will be affected with a serious genetic disorder.

METHODS AND MATERIALS

Transabdominal Amniocentesis

One hundred and sixty-two transabdominal amniocenteses were performed between the thirteenth and eighteenth weeks of fetal gestation in 142 patients during 155 pregnancies. Transabdominal amniocentesis was performed by an obstetrician as an outpatient procedure after thorough explanation of the risks and with signed permission of the pregnant woman and her husband. After the patient voids, her skin is cleansed with alcohol and benzalkonium (Zephiran) solution. Strict aseptic conditions are observed throughout the procedure. A local anesthetic, 1 per cent lidocaine (Xylocaine) solution, is injected into the proposed puncture site. A 22-gauge, 5-inch spinal needle with stylet is inserted through the abdominal wall in the midline, directed at a right angle toward the middle of the uterine cavity. After puncture, the stylet is removed, and a sterile plastic syringe is used to withdraw 10 ml of amniotic fluid, after which the needle is swiftly withdrawn.

Cultivation of Amniotic-Fluid Cells

The amniotic fluid is placed in either a sterile siliconized glass or a plastic tube and transported at ambient temperature to the laboratory. The fluid is centrifuged at 100 × g for 12 minutes, the supernatant removed and frozen, and the cell pellet suspended in 0.5 ml of 100 per cent fetal calf serum. The cells are placed in five small Falcon Petri dishes, immobilized under a glass cover slip, 2 to 3 ml of F-10 nutrient medium (BBL) supplemented with 15 per cent of fetal calf serum containing antibiotic-antimycotic mixture is added, and the cultures placed in 5 per cent carbon dioxide at 37°C. The medium is changed every other day until a number of colonies of cells are seen under the cover slips, usually seven to eighteen days. The medium is removed from the dish, and the cover slip turned cell surface up and placed in another Falcon dish. Two milliliters of medium is added to both the original dish and the new dish. The cultures may now be used directly for chromosome analysis or

subcultured and used for chromosome or biochemical analysis.[15,16,19-22]

Chromosome Analysis

Twenty hours after the cover slip is inverted or after subculture onto a cover slip, 0.2 ml of diacetylmethyl colchicine (Colcemid), 0.01 mg per milliliter, is added. Four hours later the medium is removed, the cover slip rinsed gently with hypotonic solution (4 parts distilled water to 1 part medium) and then incubated in this hypotonic solution at 37° for 30 minutes. The cover slip is gently rinsed in a fixative (1 part glacial acetic acid to 3 parts absolute methanol) and placed in fresh fixative for 20 minutes. The cover slip is rinsed in 50 per cent acetic acid and placed cell side up on a slide. A few more drops of 50 per cent acetic acid are added to the cover slip, which is gently passed about 5 inches above a flame, permitting some of the acetic acid to evaporate. The excess acetic acid is drained into a blotter or tissue. The cover slip is placed in methanol for five minutes and stained for 10 minutes with Giemsa reagent.

RESULTS

Transabdominal amniocentesis was successfully performed in 160 of 162 cases. Repeat amniocentesis was required in seven. Amniotic fluid was not obtained in the initial amniocentesis in two cases, and in the remaining five, either contaminated amniotic fluid or inadequate cell growth necessitated a repeat amniocentesis.

Successful cultivation of amniotic-fluid cells was accomplished in 155 of 160 amniocenteses – that is, in all 155 pregnancies. The time between amniocentesis and successful chromosome analysis ranged from three to 28 days, with a mean of 14.2 days, as compared to a range of 15 to 40 days, with a mean of 30.1 days, for biochemical analysis. Successful cultivation of amniotic-fluid cells for more than three subcultures was accomplished in 75 per cent of all cases. In all but one the predicted sex was confirmed after delivery by examination of the external genitalia, and after therapeutic abortion by chromosome analysis of the cultured abortion material. In three cases when hypertonic saline was used for therapeutic abortion, chromosome analysis of the abortus was not possible, but repeat amniotic-fluid analysis demonstrated similar karyotypes. In one case the predicted sex of the baby was not confirmed at delivery. Chromosome analysis showing 46, XX, was completed four days after amniocentesis and confirmed at six days. The patient gave birth to normal male twins. Re-examination of the chromosome analysis of the amniotic-fluid cell culture, which had been frozen, after 45 days of growth, demonstrated a 46 XY karyotype. In this case and in another case prenatal chromosome analysis did not predict the presence of twins.

TABLE 1. *Indications for Amniocentesis in 155 Pregnancies.*

INDICATION	NO. OF PREGNANCIES	OUTCOME OF PREGNANCY
Chromosomal	132	
Translocation carrier	22	7 with Down's syndrome (therapeutic abortion); 15 normal.
Maternal age > 40 yr	82	2 with Down's syndrome (therapeutic abortion); 80 normal.
Previous trisomic Down's syndrome	28	1 with Down's syndrome; 27 normal.
Familial metabolic	23	
Carrier of X-linked recessive disorder	7	2 males (therapeutic abortion); 5 females (normal).
Pompe's disease	8	1 with Pompe's disease (therapeutic abortion); 6 normal. 1 normal (spontaneous abortion)
Lysosomal acid phosphatase deficiency	2	1 affected (therapeutic abortion) 1 normal
Metachromatic leukodystrophy	1	1 affected (therapeutic abortion)
Mucopolysaccharidosis	2	2 normal
Generalized gangliosidosis	1	1 normal
Maple-syrup-urine disease	2	2 normal

Table 1 lists the indication for the amniocentesis and the outcome of the pregnancy. In 10 patients Down's syndrome was detected in utero. Six pregnancies were terminated by hysterotomy, and three by intra-amniotic instillation of hypertonic saline. The diagnosis of Down's syndrome was confirmed in these nine abortions by examination or culture or both. In one case the parents elected to continue the pregnancy, and an infant with Down's syndrome was delivered at term. Two mothers who had previously borne children with X-linked muscular dystrophy were found to be carrying a male fetus. In each the pregnancy was terminated by intra-amniotic saline injection. In previously reported cases Pompe's disease[16] and lysosomal acid phosphatase deficiency[19] was detected in utero. The pregnancies were interrupted by hysterotomy, and the diagnosis confirmed by biochemical examination of the fetus. In one case aryl sulfatase A activity was deficient in cultivated amniotic-fluid cells obtained from a woman who had previously delivered a child with metachromatic leukodystrophy (65 nmoles of product released per hour per milligram of protein as compared to six controls with values of 825 to 1400 nmoles of product released per hour per milligram of protein). The pregnancy was terminated by hysterotomy, and aryl sulfatase A activity in the liver of the fetus was less than 5 per cent of normal. Enzyme activity in cultivated amniotic-fluid cells from the remaining pregnancies at risk for Pompe's disease, lysosomal acid phosphatase deficiency, generalized gangliosidosis and maple-syrup-urine disease was readily detectable, and normal children, presumably two thirds of whom may be carriers, were delivered. Cultivated amniotic-fluid cells obtained in two pregnancies from a patient at risk for Hurler's syndrome had no detectable metachromatic granules, and accumulation of mucopolysaccharide labeled with radioactive sulfate was within normal limits. The children are now six and 18 months of age, with no evidence of Hurler's syndrome either on physical or on laboratory examination.

No evidence of maternal morbidity was encountered. In one case, one month after transabdominal amniocentesis a spontaneous abortion occurred. Examination of the fetus and placenta revealed evidence of recent minimal chorioamnionitis (Dr. Kurt Benirschke). The cause of the abortion was listed as incompetent cervix. Seven children born to mothers past the age of 40 years were noted to have the following malformations: spina bifida, one; horizontal palmar crease, one; cleft soft palate, one; pilonidal dimple, three; congenital heart disease (patent ductus), one; and café-au-lait spots, one. In one mother who had previously given birth to a child with Down's syndrome a child with dislocated hips was delivered. None of these congenital malformations could have been detected with the methods used in this study. Seven infants were delivered between 34 and 37 weeks of pregnancy with birth weights below 4 lb, 10 oz (3 lb, 2 oz, 3 lb, 6 oz, 3 lb, 9 oz, 4 lb, 4 lb, 1 oz, 4 lb, 3 oz, and 4 lb, 6 oz).*

DISCUSSION

Successful cultivation of amniotic-fluid cells for chromosome analysis was achieved in 97 per cent of cases with one amniocentesis — yielding 10 ml of amniotic fluid. Repeat amniocentesis resulted in successful cultivation in the five cases in which inadequate growth or contaminated specimen had not permitted successful chromosome analysis. Chromosome analysis was accurate in all but one case, in which, as well as in one reported by Uhlendorf,[23] the analysis was performed much earlier than usual because of the rapid cell growth. These cells, presumably maternal macrophages, die after approximately a week, and repeat examination of the cultures after two weeks will permit accurate chromosome analysis. In a case reported by Macintyre[23] one of five cultures from one patient had XX cells whereas the remaining cultures were XY. On the basis of these observations it is suggested that chromosome analysis be performed on at least two cultures at different times of cultivation. Another source of error is the inability to detect the presence of twins of similar or possibly even different sex on the basis of chromosome analysis.

Limited information is available concerning the reliability of intrauterine detection of biochemical disorders. In only one inborn error of metabolism,

*The summary of each patient is available upon request from the authors.

Pompe's disease, has a large enough number of pregnancies been studied to evaluate the reliability of this approach.[16] At present the major limiting factor appears to be the difficulty in obtaining an adequate number of cells for biochemical analysis.

The experience in this study suggests that transabdominal amniocentesis early in the second trimester of pregnancy carries minimal risks to mother and fetus. In addition, to the 155 patients in this study, 132 have been monitored in three other centers, without evidence of fetal or maternal complications.[23] Complications will undoubtedly occur but the risk will probably be about the same as that of transabdominal amniocentesis after 20 weeks of pregnancy, – that is, less than 1 per cent.

The ability to detect a genetic defect in the fetus not only gives new precision to genetic counseling but also creates many moral, legal and medical problems. At present, since effective treatment of genetic defects is limited, the question of interruption of pregnancy is raised. Interruption of pregnancy on the basis of fetal abnormality is legal in relatively few states, and in many countries it must be performed before 20 weeks of gestation. One medical problem that must be evaluated is the safety of therapeutic abortion performed in the second trimester of pregnancy. The two procedures most often used for interruption of pregnancy in the second trimester are hysterotomy and intra-amniotic instillation of hypertonic solutions. The maternal morbidity and mortality of intra-amniotic hypertonic saline for therapeutic abortion include infection, intravascular injection of saline resulting in hypernatremia, post-partum hemorrhage, cervical lacerations, uterine rupture, cerebral infarction and sudden death after shocklike symptoms or vascular collapse.[24] However, the risk of this procedure can be kept to a minimum if careful attention is given to technical details. These include knowledge of the potential complications, a uterus of more than 15 weeks' gestational size and instillation of not more than 200 ml of 20 per cent saline.

The selection of patients for whom intrauterine monitoring for detection of a genetic disorder in the fetus is warranted will depend upon accurate assessment of the risk of the procedure and the reliability of diagnosis as compared to the risk of an affected fetus. The results of this study demonstrate the usefulness of intrauterine monitoring of a number of groups of "high-risk" pregnancies.

We are indebted to Sally Lee, Anita Messina, Cathy Ryan, Elvira Shannon and Marilyn Swae for technical assistance, to the many physicians who helped in obtaining this material, and to Drs. A. E. Emery, C. B. Jacobson, J. W. Littlefield, M. N. Macintyre and B. W. Uhlendorf for sharing their experience with us and for suggestions.

REFERENCES

1. Menees TO, Miller JD, Holly LE: Amniography: preliminary report. Amer J Roentgen 24:363-366, 1930
2. Liley AW: The use of amniocentesis and fetal transfusion in erythroblastosis fetalis. Pediatrics 35:836-847, 1965
3. Freda VJ: Recent obstetrical advances in the Rh problem: antepartum management, amniocentesis, and experience with hysterotomy and surgery in utero. Bull NY Acad Med 42:474-503, 1966
4. Queenan JT: Amniocentesis and transamniotic fetal transfusion for Rh disease. Clin Obstet Gynec 9:491-507, 1966
5. Burnett RG, Anderson WR: The hazards of amniocentesis. J Iowa Med Soc 58:130-137, 1968
6. Liley AW: The technique and complications of amniocentesis. New Zealand Med J 59:581-586, 1960
7. Creasman WT, Lawrence RA, Thiede HA: Fetal complications of amniocentesis. JAMA 204:949-952, 1968
8. Berner HW Jr: Amniography, an accurate way to localize the placenta: a comparison with soft-tissue placentography. Obstet Gynec 29:200-206, 1967
9. Wiltchik SG, Schwarz RH, Emich JP Jr: Amniography for placental localization. Obstet Gynec 28:641-645, 1966
10. Jacobson CB, Barter RH: Intrauterine diagnosis and management of genetic defects. Amer J Obstet Gynec 99:796-807, 1967
11. Nadler HL: Medical progress: prenatal detection of genetic defects. J Pediat 72:132-143, 1969
12. Fuchs F: Genetic information from amniotic fluid constituents. Clin Obstet Gynec 9:565-573, 1966
13. Nadler HL: Antenatal detection of hereditary disorders. Pediatrics 42:912-918, 1968
14. Valenti C, Schutta EJ, Kehaty T: Prenatal diagnosis of Down's syndrome. Lancet 2:220, 1968
15. Fratantoni JC, Neufeld EF, Uhlendorf BW, et al: Intrauterine diagnosis of the Hurler and Hunter syndromes. New Eng J Med 280:686-688, 1969
16. Nadler HL, Messina AM: In-utero detection of Type-II glycogenosis (Pompe's disease). Lancet 2:1277-1278, 1969
17. DeMars R, Sarto G, Felix JS, et al: Lesch-Nyhan mutation: prenatal detection with amniotic fluid cells. Science 164:1303-1305, 1969
18. Nadler HL, Swae MA, Wodnicki JM, et al: Cultivated amniotic-fluid cells and fibroblasts derived from families with cystic fibrosis. Lancet 2:84-85, 1969
19. Nadler HL, Egan TJ: Deficiency of lysosomal acid phosphatase: a new familial metabolic disorder. New Eng J Med 282:302-307, 1970
20. Porter MT, Fluharty AL, Kihara H: Metachromatic leukodystrophy: arylsulfatase-A deficiency in skin fibroblast cultures. Proc Nat Acad Sci 62:887-891, 1969
21. Yarborough DJ, Meyer OT, Dannenberg AM Jr, et al: Histochemistry of macrophage hydrolases. 3. Studies on beta-galactosidase, beta-glucuronidase and aminopeptidase with indolyl and naphthyl substrates. J Reticuloendothel Soc 4:390-408, 1967
22. Dancis J, Hutzler J, Cox RP: Enzyme defect in skin fibroblasts in intermittent branched-chained ketonuria and in maple syrup urine disease. Biochem Med 2:407-411, 1969
23. Emery AEH, Jacobson CB, Macintyre MN: Personal communication
24. Schiffer MA: Induction of labor by intra-amniotic instillation of hypertonic solution for therapeutic abortion or intrauterine death. Obstet Gynec 33:729-736, 1969

1. Severe disorder — high genetic risk: autosomal recessive conditions such as mucopolysaccharidoses and lipidoses (for example, Hurler's and Tay-Sach's diseases), cystinosis, Pompe's disease, usual type of maple-syrup-urine disease, and certain other metabolic disorders; familial chromosome translocations; and X-linked conditions such as Lowe's, Lesch–Nyhan and Hunter's syndromes, chronic granulomatous disease, and certain muscular dystrophies.

2. Severe disorder — moderate genetic risk: chromosome disorders in pregnancy of women 40 years of age or older.

3. Severe disorder — low genetic risk: ? chromosome disorders in pregnancy of women 35-39 years of age; ? recurrence of trisomy 21 after one affected child.

4. Treatable disorder — high genetic risk: ? autosomal recessive conditions such as galactosemia and pyridoxine-responsive homocystinuria; and ? X-linked conditions such as hemophilia, nephrogenic diabetes insipidus, and Bruton form of agammaglobulinemia.

THE PREGNANCY AT RISK FOR A GENETIC DISORDER

In this week's issue of the *Journal* Nadler and Gerbie report the first large series of amniocenteses done around the sixteenth week of fetal gestation for prenatal genetic diagnosis — a new technic heralded in these columns a year ago.[1] These authors have performed 162 amniocenteses in 155 pregnancies, and have collected 132 more cases from three other centers. First of all, this combined experience is reassuring because no complications were apparent in any mother or fetus. Although unlikely long-range effects must be excluded, it appears that the risk of amniocentesis at 16 weeks will be very low, as has been true later in pregnancy. Nadler and Gerbie were able to culture cells from all 155 pregnant women, and all their chromosomal or enzymatic diagnoses subsequently proved correct, although they missed twins twice and wisely caution concerning this possibility. The accuracy of sex prediction from amniotic-fluid cells was also confirmed; only two or three cases of probable contamination of the sample with maternal cells have been encountered, and it is likely that this confusion can be avoided in the future. In two cases a male fetus was detected in a mother who had previously borne a child with X-linked muscular dystrophy; both pregnancies were terminated, since there was a 1:2 chance that the fetus was affected.

With these results amniocentesis for prenatal genetic diagnosis emerges as a legitimate procedure, which should be widely available. It becomes important to identify pregnancies appropriate for examination. Of course, it goes without saying that the ultimate decision for this procedure, as well as for any subsequent intervention, will be made by the families concerned, and will vary with their views, composition and aspirations, as well as consideration of local abortion laws and practices. The doctor should tell them that the likelihood of complications from amniocentesis seems very low, but that complications from the current methods for therapeutic abortion in the second trimester are not negligible. He must be able to inform them fully and accurately of the genetic risk that a child will be affected, and of the nature, available treatment and prognosis of the disorder. A tentative classification of pregnancies appropriate for amniocentesis can be formulated with these variables in mind:

In the "Severe disorder — high genetic risk" category it is likely that the appearance of the disorder in a child will call attention to the family. For the autosomal recessive disorders listed, the risk of recurrence is 1:4 for each subsequent pregnancy. Informed of this risk, such families will usually forego further children, but with the use of amniocentesis, such denial is not necessary. There probably would be widespread agreement that amniocentesis is indicated in subsequent pregnancies in these families, and intervention appropriate if the fetus is found to be affected. The risk of recurrence is almost as high in families carrying a chromosomal translocation, who produce 1 per cent of all cases of trisomy 21, for example. Finally, a distinction between affected and unaffected male fetuses can be made for the Lesch–Nyhan and Hunter's syndromes, while in the other severe X-linked disorders all male fetuses must be aborted if an affected child is to be avoided.

For the "Severe disorder — moderate genetic risk" group there seems increasing agreement about the wisdom of offering amniocentesis to all older pregnant women. In Massachusetts, for example, a pregnant woman 40 to 44 years of age has a 1:80 risk of a child with trisomy 21,[2] and other trisomies[3] and certain X-chromosome disorders (XXY and XXX)[4] are more likely as well; the total risk may be about 1:50. This situation is probably due to increased meiotic nondisjunction in the older woman, rather than to retention to term of trisomic fetuses, which are often aborted spontaneously.[4] Over half the amniocenteses performed by Nadler and Gerbie were done on women over 40, with this possibility of chromosome disorders in mind.

From the purely economic point of view, it would actually be sensible for Massachusetts to encourage and pay for amniocenteses in pregnant women 35 years of age and older[5] — a possible indication falling in the "Severe disorder — low genetic risk" category. Such screening would be many times more productive than that for disorders of amino acid metabolism, and of course from the practical point of view, it would be more useful if done prenatally rather than postnatally, as in various current programs. Also in the "Severe disorder — low genetic risk" group are women who have borne one child with trisomy 21, and are quite concerned that they will have another. Eighteen per cent of the patients of Nadler and Gerbie were examined for this reason, as a number of our own have been. But it needs emphasis that empirically the risk of recurrence is only about twice that for other women of similar age,[6] and is therefore quite low if the mother is young. For example, the risk of recurrence is about 1:700 for the group from 20 to 29 years of age, which is equal to the overall risk for all women, and doubtless less than the risk of amniocentesis itself.

Within the "Treatable disorder — high genetic risk" group the success of treatment is variable, and there will be disagreement about the propriety of amniocentesis and subsequent abortion for an affected fetus. The views of individual families will differ, as do those of both professional and nonprofessional people in general. Of course amniocentesis should not be undertaken unless the family is committed to subsequent intervention if appropriate.

It is one thing to say that a technic should be widely available, and another to make it so. Laboratories competent to perform the specialized assays for various rare recessive conditions need to be designated and supported around the country. Probably the most frequent indication for amniocentesis will be concern for a chromosome disorder, mainly trisomy 21. New facilities for chromosome analysis will become necessary, and perhaps can be centralized and automated in part, but first the efficient use of existing laboratories should be encouraged. Regional co-operation in prenatal genetic diagnosis, and then in other present and future genetic counseling activities, would bring many advantages in improved patient care, lower costs, greater quality control, co-ordinated registries and better doctor education. Here is a challenge in public health well worth the attention and support of private and governmental agencies.

JOHN W. LITTLEFIELD, M.D.

REFERENCES

1. Littlefield JW: Prenatal diagnosis and therapeutic abortion. New Eng J Med 280:722-723, 1969
2. Fabia J: Illegitimacy and Down's syndrome. Nature (London) 221:1157-1158, 1969
3. Magenis RE, Hecht F, Milham S Jr: Trisomy 13 (D₁) syndrome: studies on parental age, sex ratio, and survival. J Pediat 73:222-228, 1968
4. Court Brown WM, Law P, Smith PG: Sex chromosome aneuploidy and parental age. Ann Hum Genet 33:1-11, 1969
5. Littlefield JW: Introductory remarks for the session on chromosomal studies, Symposium on Down's Syndrome, New York, November 24-25, 1969. New York, New York Academy of Sciences (in press)
6. Penrose LS, Smith GF: Down's Anomaly. Boston, Little, Brown and Company, 1966, p 176

110

Making babies
—the new biology
and
the "old" morality

LEON R. KASS

Good afternoon ladies and gentlemen. This is your pilot speaking.
We are flying at an altitude of 35,000 feet and a speed of 700 miles
an hour. I have two pieces of news to report, one good and one
bad. The bad news is that we are lost. The good news is that we
are making excellent time.   —*Author unknown*

THOUGHTFUL men have long
known that the campaign for the technological conquest of nature,
conducted under the banner of modern science, would someday
train its guns against the commanding officer, man himself. That
day is fast approaching, if not already here. New biomedical technologies are challenging many of the formulations which have
served since ancient times to define the specifically human—to
demarcate human beings from the beasts on the one hand, and from
the gods on the other. Birth and death, the boundaries of an individual human life, are already subject to considerable manipulation.
The perfection of organ transplantation and especially of mechanical
organs will make possible wholesale reconstructions of the human
body. Genetic engineering, a prospect already visible on the horizon,
holds forth the promise of a refined control over human capacities

and powers. Finally, technologies springing from the neurological and psychological sciences (e.g., electrical and chemical stimulation of the brain) will permit the manipulation and alteration of the higher human functions and activities—thought, speech, memory, choice, feeling, appetite, imagination, love.

The advent of these new powers for human engineering means that some r. en may be destined to play God, to re-create other men in their own image. This Promethean prospect has captured the imagination of scientist and layman alike, and is being hailed in some quarters as 'ie final solution to the miseries of the human condition. But this optimism (not to say *hybris*) has been tempered by the dim but growing recognition that the use of these new powers will raise profound and difficult moral and political questions—and precisely because the objects on which they are to operate are human beings. In this essay, I consider some of these moral and political questions in connection with one group of new technologies: the technologies for making babies.

Why make babies?

Why would anyone want to provide new methods for making babies? A major reason given is that, in many instances, the "old"[1] method is not possible. Despite greatly increased abilities to diagnose and to treat the causes of infertility in recent years, some couples still remain involuntarily childless. Thus, paradoxically, while the need to limit fertility becomes ever more apparent, some scientists and physicians have taken it as their duty to satisfy the natural desire of every couple to have a child, by natural or artificial means.

Some rather large questions arise here. Physicians have a duty to treat infertility by whatever means only if patients have a right to have children by whatever means. But the "right to procreate" is an ambiguous right, and certainly not an unqualified one. Whose right is it, a woman's or a couple's? Is it a right to carry and deliver (i.e., only a woman's right) or is it a right to nurture and rear? Is it a right to have your own biological child? Even if involuntary sterilization imposed by a government would violate such a right, however de-

---

[1] This awkward use of "old" calls attention to the subtle traps laid for us by the abuse of language. In a time and place where novelty and originality are considered cardinal virtues, and when faddishness has replaced tuberculosis as the scourge of the intellectual classes, one should vigorously resist the tendency to make things attractive simply by emphasizing their newness. Is the "old" way of beginning life *merely* old, simply traditional and conventional?

fined, is it "violated" or denied by sterility not imposed from without but due to disease? Is the inability to conceive a disease? Whose disease is it? Can a couple have a disease? Does infertility demand treatment wherever found? In women over seventy? In virgin girls? In men? Can these persons claim either a natural desire or natural right to have a child, which the new technologies might or must provide them? Does infertility demand treatment by any and all available means? By artificial insemination? By *in vitro* fertilization? By extracorporeal gestation? By parthenogenesis? By cloning—i.e., "xeroxing" of existing individuals by asexual reproduction?[2]

Simply posing these questions suggests that both the language of rights and the language of disease could lead to great difficulties in thinking about infertility. Both point to possessions or properties of single individuals, for it is an individual who bears rights and diseases. Yet infertility is a relationship as much as a condition—a relationship between husband and wife, and between generations too. More is involved than the interests of any single individual. Ultimately, to consider infertility (or even procreation) solely from the perspective of individual rights can only undermine—in thought and in practice—the bond between childbearing and the covenant of marriage. And in a technological age, to view infertility as a "disease," one demanding treatment by physicians, automatically fosters the development and encourages the use of all the new technologies mentioned above.

A second reason given for seeking new methods for making babies is that sometimes the old method is thought to be undesirable or inadequate, primarily on eugenic grounds. A diverse—and ultimately incompatible—collection of champions are presently in bed together under this rationale: patient-centered physicians and genetic counsellors seeking to prevent the transmission of inherited diseases to prospective children of carrier parents, species-centered pessimists concerned to combat the alleged deterioration of the human gene pool, and zealous optimists eager to engineer "improvements" in the human species. The new methods called for include the growth of early embryos in the laboratory with selective destruction of those who do not pass genetic muster; directed mating with eugenically selected eggs, sperm, or both; and asexual replication of existing "superior"

---

[2] Those who seek to submerge the distinction between *natural* and *unnatural* means would do well to ponder these questions, and reflect on what they themselves mean when they speak of "a natural desire to have children" or "a [natural human] right to have children." One cannot speak of natural desires or natural human rights or, indeed, about disease, without having some notion of "normal," "natural," and "healthy" for human beings.

individuals. But serious questions can be raised with respect to these ends as well. For example, we may know which diseases we would wish not to have inflicted upon ourselves and our offspring, but are we wise enough to act upon these desires? In view of our ignorance concerning why certain genes survive in our populations, can we be sure that the eradication of genetic disease (or of any single genetic disease) is biologically a sensible goal? Might it not have unanticipated genetic consequences?

The species-centered goals are even more problematic. Do we know what constitutes a deterioration or an improvement in the human gene pool? One might well argue that, at least under present conditions, the crusaders against the deterioration of the species are worried about the wrong genes. After all, how many architects of the Vietnam war have suffered from Mongolism? Who uses up more of our irreplaceable natural resources and who produces more pollution—the inmates of an institution for the retarded or the graduates of Harvard College? It is probably as indisputable as it is ignored that the world suffers more from the morally and spiritually defective than from the genetically defective. Thus, it is sad that our best minds are busy fighting our genetic shortcomings while our more serious vices are allowed to multiply unmolested.

Perhaps this is too harsh a judgment. Certainly, our genetic inheritance is entrusted to us for safekeeping and not for abuse or neglect. Perhaps a case could be made for the desirability and wisdom of certain negative or even positive eugenic goals. Still, as in the treatment of infertility, we shall also have to consider which means, if any, can be justified in the service of *any* reasonable goals.

Thirdly, there are scientific goals which themselves generate new beginnings in life. In other words, there is a limit to what can be learned about the nature and regulation of fertilization, embryonic development, or gene action from lives begun in the old, undisturbed, natural manner. This is no doubt true. But if the goal is scientific knowledge of these processes for its own sake, there is little need to develop new beginnings in *human life*. Embryological experimentation in a wide range of mammals, employing all the new technological possibilities, would yield the basic understanding. There is at present no reason to believe that the fundamental mechanisms of differentiation differ in monkeys and in man. Until extensive animal studies show otherwise, the human experiments can only be given a technological and not a purely scientific justification. (Indeed, it is the philanthropic foundations interested in finding new drugs for abortion or contraception who are supporting much of the work on the laboratory growth of human embryos. For example, the work of R. G. Edwards and his colleagues in Cambridge, England, is supported by the Ford Foundation.) These technological purposes and activities (and others, such as the use of early embryos in culture to test for muta- tion- or cancer-producing chemicals and drugs, or to work out tech- niques for genetic manipulation) may well be desirable, but they need to be so identified and distinguished from the quest for knowl- edge simply. Adequately to assess the desirability of any specific means, and properly to weigh alternative means, requires a clear understanding of which ends are being served.

Finally, new methods for making babies are being sought pre- cisely because they are new and because they can be sought. While not praiseworthy reasons, these certainly are important reasons, and all-too-human ones. Drawn by the promise of fame and glory, driven by the hot breath of competitors, men do what can be done. Bio- medical scientists are no less human than anyone else. Some of them will be unable to resist the lure of immortality promised the father of the first test tube baby. Moreover, regardless of their private motives, they are encouraged to pursue the novel because of the widespread and not unjustified belief that their new findings will probably help to alleviate one form or another of human suffering. They are also encouraged by that curious new breed of techno-theo- logian who, after having pronounced God dead, discloses that God's dying command was that mankind should undertake its limitless, no- holds-barred self-modification, by all feasible means.

So much then for reasons why some have called for and helped to promote new beginnings for human life. But what precisely is new about these new beginnings? Such life will still come from pre- existing life; no new formation from the dust on the ground is being contemplated, nothing as new—or as old—as that first genesis of life from non-living matter is in the immediate future. What is new is nothing more radical than the divorce of the generation of new human life from human sexuality, and ultimately from the confines of the human body, a separation which began with artificial insemina- tion and which will finish with ectogenesis, the full laboratory growth of a baby from sperm to term. What is new is that sexual inter- course will no longer be needed for generating new life. (The new technologies provide the corollary to the pill: babies without sex.) This piece of novelty leads to two others: There is a new co-progen- itor, the embryologist-geneticist-physician, and there is a new home for generation, the laboratory. The mysterious and intimate processes of generation are to be moved from the darkness of the womb to the bright (fluorescent) light of the laboratory, and beyond the shadow of a single doubt.

But this movement from natural darkness to artificial light has the most profound implications. What we are considering, really, are not merely new ways of beginning individual human lives but also— and this is far more important—new ways of life and new ways of viewing life and the nature of man. Man is partly defined by his origins; to be bound up with parents, siblings, ancestors, is part of what we mean by "human." By tampering with and confounding these origins, we are involved in nothing less than creating a new conception of what it means to be human.

Consider the views of life and the world reflected in the following different expressions to describe the process of generating new life. The Hebrews, impressed with the phenomenon of transmission of life from father to son, used a word we translate "begetting" or "siring." The Greeks, impressed with the springing forth of new life in the cyclical processes of generation and decay, called it *genesis*, from a root meaning "to come into being." (It was the Greek translators who gave this name to the first book of the Hebrew Bible.) The pre-modern, Christian, English-speaking world, impressed with the world as given by a Creator, used the term pro-creation. We, impressed with the machine and the gross national product (our own work of creation), employ a metaphor of the factory, re-production. And Aldous Huxley has provided "decantation" for that technology-worshipping Brave New World of tomorrow.

"In vitro" fertilization—state of the art

The first technological development I shall discuss is an accomplishment recently reported from several laboratories around the world, most publicly from that of Dr. R. G. Edwards in Cambridge, England: the fertilization, in the test tube, of human egg by human sperm, and the subsequent laboratory culture of the young embryo. The major technical difficulty overcome was to obtain mature, functional eggs. To surmount this difficulty, Edwards and his obstetrician colleague, Dr. P. C. Steptoe, have devised a surgical method, known as laparoscopy, to obtain matured eggs directly from the ovaries prior to ovulation. From a single woman as many as three or four eggs can be recovered at one operation. Upon addition of sperm, fertilization occurs with a small but significant fraction of these eggs. Kept in culture medium, a majority of the fertilized eggs begin to divide, and a small fraction reach the blastocyst stage

117

(i.e., age of about 7-8 days), the stage at which the early embryo normally implants itself in the wall of the uterus. Successful implantation of laboratory-grown embryos has been reported in rabbits and in mice, but not in humans. The physical transfer of the embryo into the uterine cavity poses no problem, but implantation may be difficult to achieve with regularity, since this process is poorly understood. It *is* known that the uterine lining is receptive to implantation only for a short portion of the menstrual cycle, and that only embryos at a certain stage are capable of implantation. Thus, the timing of transfer is likely to be critical. The results in mice can be considered to be somewhat encouraging. A recent article reports that nearly half of the transferred blastocysts developed into full-term, apparently normal progeny; however, the success rate over all stages was low, with only four per cent of the initial number of eggs giving rise at the end to viable mice. No gross abnormalities have been noted in any of the animals born alive following blastocyst transfer. And although some researchers would prefer to learn more about the control of implantation in animals before proceeding further in the human work, others are inclined to go ahead in humans on a trial and error basis. In fact, Edwards and Steptoe hope to accomplish transfer and subsequent growth into a baby sometime in the next year or two.

This summarizes the current state of the art on new beginnings in *human* life. But there is much work being done with other mammals which will provide knowledge and techniques someday applicable to humans. Some of these developments deserve mention (although a detailed treatment is beyond the scope of this essay). Considerable progress has been made in growing older mammalian embryos in the laboratory. Dr. D.A.T. New and his colleagues in Cambridge, England, have been successful in growing rat and mice embryos bathed in blood serum medium for about one third (the middle third) of the whole gestation period. As the embryo approaches term-size, it can no longer be maintained bathed in media, but requires more efficient circulation and exchange of nutriments, gases, and wastes. Various artificial pump and perfusion techniques, analogous to the artificial kidney machine, are being studied in an effort to design an artificial placenta. Finally, a long-standing early barrier to extensive laboratory culture, located just after the blastocyst stage, has recently fallen. Dr. Yu-Chih Hsu of Johns Hopkins University has reported the successful culture of mouse embryos from the blastocyst stage to a stage having a differentiated and beating heart. Thus, from both ends and from the middle, researchers are gradually closing in on the possibility of complete extra-corporeal gestation. It should be stressed that these techniques are being pursued primarily to make possible a better understanding of the full scope of embryonic development. However, even though no scientists at present appear to be interested in going from fertilization to birth entirely in the laboratory,[3] the technology to do so is gradually being worked out piece by piece.

Techniques to predetermine the sex of unborn children may also be just around the corner. Since the sex is determined solely by the X- or Y-chromosome content of the sperm, techniques for physical separation of X-carrying from Y-carrying sperm would make sex control possible through artificial insemination. Attempts to effect such a separation have all met with failure, but new efforts can be expected, partly because of recently discovered methods for detecting the Y-chromosome in cells, methods which may serve as an assay for successful separation. A second method of sex control, already successfully demonstrated in rabbits, involves the sexing of embryos (prior to implantation) by cell-staining techniques or chromosome analysis. Embryos of the desired sex could then be transferred to the recipient females. Though accurate, this sexing technique is not without its problems, since embryonic tissue needs to be removed for testing. In the rabbit work, only about one in five embryos sexed by these methods developed to full term, and there was at least one monstrous birth. Even if these technical difficulties were ironed out, it is doubtful that many will accept the costs and inconvenience of *in vitro* fertilization or even artificial insemination merely to control the sex of their offspring—unless perhaps they are known carriers of a sex-linked genetic disease such as hemophilia. Less cumbersome methods, e.g., a chemical method that would selectively destroy either the X- or the Y-carrying sperm in the man's body or soon after intercourse, are conceivable but as yet unreported, even in animals.

Finally, the generation of man-animal hybrids or "chimeras" has been predicted by some reputable scientists. These might be produced by the introduction of selected non-human genetic material into the developing human embryos. Fusion of human and non-human cells in tissue culture has already been achieved.

---

[3] Although no scientist *appears* to be, at least one *is* interested, according to a quotation in Albert Rosenfeld's book, *The Second Genesis* (Englewood Cliffs, N.J.: Prentice-Hall, 1969): "'If I can carry a baby all the way through to birth *in vitro*,' says an American scientist who wants his anonymity protected, 'I certainly plan to do it—though obviously, I am not going to succeed on the first attempt, or even the twentieth'" (p. 117).

"In vitro" fertilization—ethical questions

At the end of a recent popular review article in *Scientific Ameri-can* summarizing the work on human embryos, Drs. Edwards and Fowler offer the following conclusion: "We are well aware that this work presents challenges to a number of established social and ethical concepts. In our opinion the emphasis should be on the rewards that the work promises in fundamental knowledge and in medicine." Here, we are told that we *should* "emphasize" promised rewards in knowledge and power at some future time at the expense of established (don't they really mean "establishment" or "conven-tional"?) ideas of right. But this is itself a judgment of value—un-tenable as a *general* proposition for experimentation on human sub-jects—whose soundness in this particular case cannot be determined until the full range of ethical, social, legal, and political implications is carefully studied and understood.

Let us consider first what is probably the least controversial and likely to be the most popular use, at least initially, for *in vitro* fertilization, the provision of their own child to a childless couple, where oviduct disease in the woman obstructs the free passage of egg and sperm, and hence also fertilization.[4]

At first glance, the intramarital use of *artificial fertilization* re-sembles, ethically, *artificial insemination (husband)*. The procedure simply provides for the union of the wife's egg and the husband's sperm, circumventing the pathological obstruction to that union. But there is at least this difference. There is an alternative treatment for infertility due to tubal obstruction, namely surgical reconstruc-tion of the oviduct, which, if successful, permanently removes the cause of infertility (i.e., it treats the underlying disease, not merely the desire to have a child). Moreover, it does so without need for manipulation of embryos. At present, the success rate for oviduct reconstruction is only fair, but with effort and practice this is bound to improve. This therapeutic surgery for the woman is without possi-ble moral objection or adverse social consequences. Therefore, it is to be preferred over artificial fertilization, both in principle (namely,

---

[4] There are many infertile women so affected, crude estimates suggesting as many as one per cent of all women. Approximately 10 per cent of couples are infertile, and in more than half of these cases the cause is in the female. Blocked or abnormal oviducts account for perhaps 20 per cent of the female causes of infertility. However, not all such women—perhaps only a minority—are suitable candidates for the intramarital use of *in vitro* fertilization. Many women who have blocked tubes also have associated disordered ovaries, making it difficult or impossible for a doctor to obtain eggs from them. For this reason, only some of these women are likely to be able to provide their own eggs for fertilization.

one should use the least objectionable means to achieve the same unobjectionable end) and in practice, should both options be feasible and available.

A sufficient reason for this preference is related to the most important intrinsic difference between artificial insemination and artificial fertilization. In the latter, fertilization occurs in the laboratory, as do the earliest stages of embryonic development. Considerably more manipulation is involved, and, unlike the procedures of artificial insemination, there is necessarily deliberate manipulation of the embryo itself. Serious questions can be raised about the effects of the manipulations on the child who is eventually produced by this procedure. These medical questions about safety and "normality" lead to a perplexing moral question: Does the parents' desire for a child entitle them to have it by methods which carry for that child an unknown and untested risk of deformity or malformation?

How unknown are the risks? The leading researchers appear to disagree. Drs. Edwards and Steptoe are reported ready to proceed with implantation if tests on the embryos can rule out the presence of genetic or other defects. Apparently, they are both concerned about the risks and ignorant of their likelihood. In their judgment, "the normality of embryonic development and efficiency of embryo transfer cannot yet be assessed."[5]

In contrast to Edwards and Steptoe, Dr. Landrum Shettles at Columbia University does not talk about the need for further tests, and appears ready to proceed, having thrown the gauntlet and the burden of proof to those who might be more cautious or who might raise other objections. In a recent article he states that "the grossly normal blastocyst" was not transferred to the patient for the single reason that she had recently undergone uterine surgery. He adds: "Otherwise, there was no discernible contraindication for a successful transfer *in vitro* [*sic;* he means *in utero*] and continued development. This is scheduled for patients with ligated or excised fallopian

---

[5] Perhaps it was a concern for possible risks to the offspring that accounts for the unexplained judgment, quite surprising in the light of his later words and deeds, rendered by Edwards in his *Scientific American* article written only five years ago: "If rabbit and pig eggs can be fertilized after maturation in culture, presumably human eggs grown in culture could also be fertilized, *although obviously it would not be permissible to implant them in a human recipient.* We have therefore attempted to fertilize cultured human eggs *in vitro*" (emphasis added). The last sentence indicates that Edwards had no qualms about the "permissibility" of doing the fertilizations themselves. If taken seriously, the entire quotation displays that curious form of technocratic logic in which a course of action is deduced simply from the possibility of action: "Presumably human eggs can also be fertilized, *therefore* we have tried to fertilize them." Is a similar logic now at work in regard to implantation?

tubes who may want a child, with the ova obtained by culdoscopy or laparoscopy."[6]

The truth is that the risks are very much unknown. Although there have been no reports of gross deformities at birth following successful transfer in mice and in rabbits, the number of animals so far produced in this way is much too small to exclude even a moderate risk of such deformities. In none of the research to date has the question of abnormalities been systematically investigated. No attempts have been made to detect defects which might appear at later times or lesser abnormalities apparent even at birth. In species more closely related to humans, e.g., in primates, successful *in vitro* fertilization has yet to be accomplished. The ability regularly to produce normal monkeys by this method would seem to be a minimum prerequisite for using the procedure in humans.[7]

Laboratory testing of the human embryos themselves, prior to transfer, cannot provide enough information about "normality," and might itself do damage. Ordinary microscopic observation of the early embryo can disclose gross abnormalities, but it provides too crude a measure of normality. Most genetic tests cannot be done on a given embryo without damaging it; morever, there are few genetic tests presently available for the doing. Furthermore, damages could be introduced during the transfer procedure, after the last inspection is completed. Conceivably, the manipulations may even make possible the implantation of some abnormal embryos which would have been spontaneously aborted had they been generated under natural conditions.

---

[6] The tone of the recent scientific articles and the reports of this research in the scientific and popular press suggest to the ordinary reader that a race may be on to do the first embryo transfer. If such a race is on, it is likely that the swift will abandon caution to the more sober, and will trust to luck that his victory in the race does not issue in a deformed or retarded child.

[7] This view is shared by some of the more cautious—and hence, less publicly known—researchers, including Dr. Luigi Mastroianni, chairman of obstetrics and gynecology at the University of Pennsylvania: "It is my feeling that we must be very sure we are able to produce normal young by this method in monkeys before we have the temerity to move ahead in the human." Dr. Mastroianni adds: "In our laboratory, our position is, 'Let's explore the thing thoroughly in monkeys and establish the risk.' Then we can describe that risk to a patient and obtain truly informed consent before going ahead. We must be very careful to use patients well and not be presumptuous with human lives. We must not be just biologic technicians." I would dissent from this fine statement only to suggest that both doctor and prospective parents would be "presumptuous with human lives" in proceeding unless there were known to be no risks. Such confidence cannot be provided by the monkey experiments alone. Monkey experiments would neither rule out nor establish the risk of mental retardation for children resulting from the experiments in humans. Unfortunately, as is often the case, only humans can provide the test system for assessing the risks of using the procedure in humans.

In sum, there is at present no way of finding out in advance whether or not the *viable* progeny of the procedures of *in vitro* fertilization, culture, and transfer of human embryos will be deformed, sterile, or retarded. Even if we would wish to practice abortion on all the misbegotten foetuses, we are not and will not be able to identify (by prenatal diagnosis) many if not most of them. Neither can we count on "nature" to spontaneously abort all of them for us.

I have dwelled at length upon the problem of risks and mishaps that accompany the experimental phase of this new technology, first, because it is a problem widely and remarkably ignored, and, second, because it provides a powerful moral objection sufficient to rebut the implantation experiments. Moreover, this moral objection can and should be widely shared; it does not rest upon arguments about the will of God or about natural right. It rests instead upon that minimal principle of medical practice, "Do No Harm." In these prospective experiments upon the unborn, it is not enough not to know of any grave defects; one needs to know with confidence *that there will be no such defects*—or at least no more than there are without the procedure. Professor Paul Ramsey, in his book *Fabricated Man,* puts the matter quite forcefully and, I think, correctly: "The decisive moral verdict must be that we cannot rightfully *get to know* how to do this without conducting unethical experiments upon the unborn who must be the 'mishaps' (the dead and the retarded ones) through whom we learn how."[8]

It may be objected that all new medical technologies are risky, and that the kind of ethical scrupulosity I advocate would put a halt to medical progress. But such an objection ignores an important distinction. It is one thing voluntarily to accept the risk of a dangerous procedure for yourself (or to consent on behalf of your child) *if the purpose is therapeutic.* Some might say this is not only permissible, but obligatory, in line with a duty to preserve one's own health. It is

---

[8] The matter of mishaps is completely ignored by Edwards and Sharpe (the latter, an American professor of law), in their recent article in *Nature* surveying some of the social and ethical issues attending research in human embryology. Feigning neutrality, these authors reveal their prejudices—and their misconception of medical ethics—when they invoke the principle "Do No Harm" to justify, rather than to oppose, the use of *in vitro* techniques for making babies: "The beginning of medical ethics, however, is *primum non nocere;* this permits alleviation of infertility, and has been stretched to cover destruction of foetuses with hereditary defects, but would it permit the more remote techniques like modifying embryos?" "Do No Harm" is a principle which can only justify *omitting* a medical intervention; it does not permit, let alone justify, committing an intervention. Rightly understood, it can enter medico-moral argument only *in opposition to* risk-filled technologies for making babies.

quite a different thing deliberately to submit a child to hazardous procedures which can in no way be considered therapeutic for him, and are "therapeutic" for you only in that they "treat" your desires (albeit unobjectionable ones). This argument against non-therapeutic experimentation on children applies with even greater force against experimentation "on" a hypothetical child (whose conception is as yet only intellectual). One cannot ethically choose for him the unknown hazards he must face and simultaneously choose to give him life in which to face them. This judgment could be set aside only under a strongly pro-natalist ethic—much more pro-natalist that Roman Catholicism ever was—which would hold that parents have either a preeminent duty or a preeminent right to have their own biological child by whatever means.

The question of "informed consent"

While on the subject of the ethics of experimentation, let me add a few comments concerning the adult participants. Most of the scientific reports on human embryo experimentation are strangely silent on the nature of the egg donors and on their understanding of what was to be done with their eggs. This is surprising considering the growing sensitivity of the scientific community to the requirement of informed consent, and especially surprising given the kind of experiments here being performed. Who are these women and how did they come to "volunteer"? In the report describing the first successful fertilization and cleavage of human eggs obtained via laparoscopy, there are only several passing references to "patients," and the one-sentence abstract of the paper only increases our confusion by its use of the word "mother": "Human oocytes have been taken from the mother before ovulation, fertilized *in vitro* and grown *in vitro* to the eight- or sixteen-celled stage in various media." If the women were indeed patients for infertility, then "the mother" is surely the one thing that they are not. In the recent *Scientific American* article by Edwards and Fowler there is this solitary comment: "Our patients were childless couples who hoped our research might enable them to have children." We are not told, and can therefore only guess as to what these women were in fact told. From the report that the women and their husbands had hopes, we can surmise that they considered themselves to be *patients*. But for the present, they are *experimental subjects*. One wonders if they were told this.

Only one of the many scientific articles, that describing the use of the laparoscopic surgery to recover human oocytes, tells us anything more about the persons used as experimental subjects (in this case, for perfecting the laparoscopy technique), and about how they were informed: "The object of the investigations was fully discussed with the patients, including the possible clinical applications to relieve *their* infertility" (emphasis added). Though welcome, this statement leaves many questions unanswered. Were the couples also told that the much more likely possibility was that it would be future infertile women, rather than they themselves, whose infertility might be "relieved"? Were they told about alternative possibilities, such as surgery on the blocked oviduct, or adoption? Since the same article tells us that three out of 46 "infertile" women became pregnant—by the "old," customary method—*during the first month after the laparoscopy,* we can only wonder about the criteria used for subject selection. Were all other possibilities exhausted before bringing these couples into this uncertain program of experimentation? Finally, we are left to wonder how the discussions were conducted, especially in the light of the following quotation attributed to R. G. Edwards: "We tell women with blocked oviducts, 'Your only hope of having a child is to help us. Then maybe we can help you.'" (Remarks made at a scientific meeting in West Berlin, as reported in *Medical World News,* April 4, 1969.)

It is altogether too easy to exploit, even unwittingly, the desires of a childless couple. It would be cruel to generate for them false hopes by inflated publicity of the sort that some of these researchers have promoted. It would be both cruel and unethical falsely to generate hope, for example, by telling women that they themselves, rather than future infertile women, might be helped to have a child, in order to secure their participation in experiments. That this may have already occurred is suggested by the following extract from a news report by Patrick Massey of Reuters:

> Dr. Patrick Steptoe, who heads the team of doctors working on the experiment, disclosed on television that he had extracted an ovum from a 34-year-old housewife and fertilized it with her husband's sperm. The woman, Mrs. Sylvia Allen . . . said *she hoped the fertilized ovum would be implanted in her womb in the next two to six weeks,* meaning that the world's first baby conceived in a test tube could be born by the end of 1970. (Story in *The Washington Post,* March 3, 1970; emphasis added.)

The implantation was never performed.

Let me also raise some questions concerning experimentation done with eggs obtained from women undergoing ovarian surgery for clinical reasons. Do and should these women know what is going to be done with their eggs? To whom belong the rights governing ordinary tissue removed at surgery? Is reproductive tissue a special case? Surely, if the eggs were going to be implanted in another woman, one would think that the donor's permission should be obtained. Then what about their simple fertilization? If the woman from whom the eggs are taken has religious or other objections against *in vitro* fertilization which would lead her to refuse permission if asked, is she wronged by not being asked or informed? No matter how worthy the research and how well-intentioned the investigator, *ethical* experimentation on human subjects and their tissues requires that persons be treated as ends and not merely as means, and hence that their wishes be considered and respected.

The case of the surplus embryos

So far, I have discussed ethical questions surrounding attempts to generate a normal child by transferring a laboratory-grown human embryo into the uterus of an infertile woman. But what about all the embryos that are not so implanted? Dr. Donald Gould, editor of the British journal, *The New Scientist,* has asked: "What happens to the embryos which are discarded at the end of the day—washed down the sink? There would necessarily be many. Would this amount to abortion—or to murder? We have no law to cope with this kind of situation."

I don't wish to mislead anyone. At this state of the art, the largest embryo we are talking about is a blastocyst, barely visible to the naked eye. But the moral question does not turn on visibility, any more than it does in the case of murder committed by a blind man. The embryos are clearly biologically alive, even at the blastocyst stage. At some future date, improved techniques will permit their growth to later stages, someday even viable stages. Before then, however, the question of discarding will have to be faced. Since there is a continuity of development between the early and the later stages, we had better face the question now and draw whatever lines need to be drawn.

When in the course of development does a living human embryo acquire *protectable humanity?* This is a familiar question which I shall not belabor. But the situation here, though similar, is not identical to that of abortion. For one thing, we don't start with a foetus already *in utero,* which one reluctantly destroys, hopefully only for good reasons. Here, nascent lives are being deliberately created despite certain knowledge that many of them will be destroyed or discarded. (Several eggs are taken for fertilization from each woman, the extra ones being available for experimentation). The foetuses killed in abortion are unwanted, usually the result of so-called accidental conception. The embryos discarded here are wanted, at least for a while; they are deliberately created, used for a time, and then deliberately destroyed. Moreover, unlike abortion, the continued life of the laboratory embryo is in conflict with no one; no claim or right of its "mother" can be invoked to justify overriding its claim or right to life. Even if there is no intrinsic wrong done by discarding at the blastocyst stage—and I am undecided on this question—there certainly would be at later stages. (Those who disagree should at least be concerned about the effects on the attitude toward and respect for human life engendered in persons who are engaged in these practices.)

There is a second, related difference between abortion and discarding laboratory-grown embryos. Who decides what are the grounds for discard? What if there is another recipient available who wishes to have the otherwise unwanted embryo? Whose embryos are they? The woman's? The couple's? The geneticist's? The obstetrican's? The Ford Foundation's? If one justifies abortion on a paramount right of the woman to decide about her family size and spacing, or even on that unbelievable ground that a woman has a paramount right to do what she wishes with her body,[9] whose rights are paramount here? Shall we say that discarding laboratory-grown embryos is a matter solely between a doctor and his plumber?

But the unimplanted embryos raise even more profound problems. Bad as it may be to discard them, it would be far worse to perpetuate them in their laboratory existence—especially when the technology arrives that can bring them to "viability" *in vitro*. Will they then still be considered fit material for experimentation? Or are they then to be released from the machinery and admitted into the human fraternity or, at least, into the premature nursery? Who will the children be, and who their parents, and what their social definition? The need for a respectable boundary "defining" protectable human life can not be overstated. The current boundaries, gerrymandered for the

---

[9] Even if such a right exists, it does not govern actions involving the foetus, because the foetus is simply not a mere part of a woman's body. One need only consider whether a woman can ethically take thalidomide while pregnant to see that this is so. It is distressing that seemingly intelligent people would sincerely look upon a foetus whose heart was beating, and which had its own EEG, as indistinguishable from a tumor of the uterus, a wart on the nose, or hamburger in the stomach.

sake of abortion—namely, birth or viability—may now satisfy both Women's Liberation and the United Methodists, and may someday satisfy even a future pope, but they will not survive the coming of more sophisticated technologies for making babies.

One thing leads to another

Having discussed so far only one serious moral objection to *implantation* of embryos and having raised some questions about the *discarding* and *perpetuation* of unimplanted ones (and deliberately neglecting the ethics of *creating* the embryos in the first place), I suspect that I have persuaded no one not originally opposed to, or, at least, doubtful about, *in vitro* fertilization. Furthermore, the first objection may become a vanishing objection. Since the experimentation will in all likelihood proceed, the problem of risks and mishaps may eventually be eliminated as the technique is perfected. The discarding issue will hardly get a fresh hearing in a society which has so recently converted to foeticide.[10] Moreover, apart from these questions I can find no *intrinsic* moral reason to reject the intramarital use of *in vitro* fertilization and implantation—at least no reason that would not also rule against artificial insemination (husband). But the argument does not stop here, for we must consider both the likely other uses and abuses of this procedure, and also the other and more objectionable procedures that this one makes possible.

Some may object to my making an argument based upon likely or possible misuses and abuses. After all, there are few if any powers, technological and non-technological, that cannot be abused. Many of us would prefer arguments from principle concerning intrinsic rightness or wrongness, arguments that abstract from the difficult task of predicting and of weighing consequences, often quite remote and intangible ones. Nevertheless, we can ill afford such intellectual purism, especially for technologies that touch the foundations of man's biological nature. No technology exists autonomously or in isolation; each arises in the context of other technologies and, more importantly, in a complex and heterogeneous world of men whose proclivity for mischief and folly we cannot in good conscience ignore. We would ourselves display such folly were we to

---

[10] Some may object to my use of the term foeticide, but I am opposed to hiding behind euphemisms. If we are going to be brave enough to practice abortion, let us at least not be cowardly in describing it. Even if foeticide were made legal everywhere, and even if it were morally justified, it would still be killing (though not necessarily murder, which is "wrongful killing").

justify the introduction of each new technology simply because *some* good use can be found for it.[11]

Once introduced for the purpose of treating intramarital infertility, *in vitro* fertilization can then be used for any purpose. There is no reason why the embryo need be implanted in the same woman from whom the egg was obtained. An egg taken from one woman (the biological mother) could be donated to another woman (the gestational mother), either before or after fertilization—the former has been termed "artificial inovulation (donor)," the latter, "embryo transfer (donor)" or "prenatal adoption." Since obtaining eggs for donation is more difficult than obtaining sperm, and requires surgery on the donor, this might not seem a likely occurrence except on a small scale. However, procedures for freezing and storing eggs or young embryos will circumvent this program. (In the past six months, a method has been reported for freezing and storing mouse embryos, with little loss of viability on thawing.) Egg and embryo banks will almost certainly be established—as are sperm banks today—partly to avoid having to do repeated operations on the same woman, and also because there is money to be made. Indeed, enterprising financiers and scientists, with the blessing of the new techno-theologians, are perhaps at present organizing under one corporate canopy, The Chaste Rational Bank and Union: The Logical Semenary.

There are enough women whose infertility is due to reasons other than blocked oviducts to make it extremely likely that donation of eggs and embryos will be attempted, i.e., that the technique will not be confined to those intramarital cases in which it was first used. Clinical use of *in vitro* fertilization will probably rely mainly on donor eggs in a manner closely analogous to donor insemination. And why stop at couples? What about single women, widows, or lesbians? If adoption agencies now permit these women to adopt, are they likely to be denied a chance to bear and deliver?

The converse possibility will also follow, namely the use of one woman simply to incubate and deliver another woman's child. If the previous practice might lead to new business ventures advertised under "eggs for sale," this practice might lead to one advertised under "wombs for rent." Women with uterine abnormalities which preclude normal pregnancy may seek surrogate gestational mothers,

---

[11] I doubt if many readers would find acceptable as a reason sufficient to justify the development of chemical and biological weaponry the fact that it may be more humane to use non-lethal gas on an enemy soldier than to kill him, without first asking whether we can foreclose the further consequences to human civilization and the human race of introducing such military technology.

as may women who don't want pregnancy to interfere with the skiing season. There are certainly enough poor women available to form a caste of childbearers, especially for good pay. The public is already being introduced to this prospect, as the following excerpt from a *Washington Post* news story indicates: "A prominent British embryologist, Dr. Jack Cohen, suggested that the process of impregnation, once perfected, might lead to a system of volunteer 'host' mothers, who would bear other people's children for a fee. Cohen suggested that 2,000 pounds ($5,000) a birth might be a reasonable sum, especially for an actress eager for the joys of motherhood without the cares of interrupting her career or distending her figure during pregnancy."

The more sentimentally minded will point out that these twin forms of foster pregnancy can be humanely and respectfully practiced. A woman may wish to donate an egg or an embryo to her sister, or may agree out of generosity to gestate a friend's embryo. It can be argued that no one should stand in the way of such acts of love.[12] But it is simply naive to think the practice would be limited to these more "innocent" cases. Moreover, there are psychological and ethical reasons for thinking that these cases may not be so innocent. What are the psychological consequences for the womb-lending sister (and the others) of giving birth to her nephew? Will she feel like giving him up? Whose child is he? Confusion and conflict would seem to be almost inevitable. If the donor of sperm has no claim over a child born after artificial insemination, why should the donor of ova, especially if there is a later dispute between the two women? These acts of love cannot be kept anonymous; the visible and continued presence of the progeny might chronically stimulate such conflict. Also, might not female relatives be under intolerable pressures to donate eggs or to lend wombs, once the first such acts of "kindness" are well publicized (just as relatives now often feel constrained to serve as transplantation donors of kidneys)? And is a person morally justified in allowing her body to be used as a "hot house," as a human incubator? Indeed, is not a decisive

[12] The current sentimentality which endorses all acts done lovingly because lovingly done leads to some strange judgments. A teacher friend recently asked one student who was having difficulty appreciating the crimes of Oedipus, what she would think if she discovered that her sister was having an affair with their father. The girl replied that, although she was personally disgusted by the prospect for herself, she thought that there was probably nothing wrong with it "provided that they [sister and father] had a good relationship." It is unlikely that an individual or a whole society that is unable to find reasons (other than genetic ones) for rejecting incest will be willing to appreciate any of the questions raised in this paper.

objection to the extramarital use of these techniques that it requires and fosters, both in thought and in deed, the exploitation of women and their bodies? And is not a second decisive objection that it fosters the notion that children are property, and encourages the practice of child buying and selling? (The further question concerning the separation of procreation from sexual love and marriage will be treated later.)

Use of these technologies need not be confined—nor is it likely to be confined—to the scale of individual couples making private decisions, nor to treatment of infertility. Indeed, several proposals for additional uses have already been placed before the public. As suggested in these proposals, and as I noted at the start, these techniques could serve eugenic purposes. Artificial insemination with semen from selected donors, the (positive) eugenic proposal of Herman Muller, could now be supplemented with the selection of ova as well. For many people, these prospects raise the fear of directed breeding programs under the dictates of a totalitarian regime. Such programs need not be coercive, since the desired donors of egg and sperm, as well as the foster mothers of the regime, might be handsomely paid and highly honored. But, perhaps perversely, in a time when suspicion and fear of governmental abuse of power are high and growing, I am not very worried about government-directed breeding by these methods. The eugenic advantages of this method, if there are any, are also available—and more cheaply too—simply by directed sexual intercourse. A regime could more easily compel or induce this less troublesome, more enjoyable practice. While those who hold to the demonic theory of politics may think me naive, I expect that artificial fertilization and embryo culture would add very little to the already large arsenal of those who would practice mischief and evil.

We stand in much greater danger from the well-wishers of mankind, for folly is much harder to detect than wickedness. The most serious danger from the widespread use of these techniques will stem, not from desires to breed a super-race, but rather from the growing campaign to prevent the birth of all defective children, in the name of population control, "quality of life," and the so-called "right of every child to be born with a sound physical and mental constitution, based on a sound genotype." Thus continues the retiring (but not reticent) President of the American Association for the Advancement of Science, geneticist Bentley Glass, in his presidential address (1971): "No parents will in that future time have a right to burden society with a malformed or a mentally incompetent child." These are not the words of a dictator, but of a gentle biologist. Even granting the desirability of his end—optimum children of no burden to society (except from their "perfection")—just consider what it would require in the way of means. This perfect condition is to be accomplished not by infanticide, not just by prenatal diagnosis and abortion of defectives, but by the laboratory growth and implantation of human embryos:

> The way is thus clear to performing what I have called 'prenatal adoption,' for not only might the selected embryos be implanted in the uterus of the woman who supplied the oocytes, but in that of any woman at the appropriate time of her menstrual cycle. Edwards cautiously limits the application of his developing techniques to the provision of a healthy embryo for a woman whose oviducts are blocked and prevent descent of the egg. It should be obvious that the technique can be quickly and widely extended. The embryos produced in the laboratory might come from selected genotypes, both male and female. Preservation of spermatozoa in deep frozen condition could permit a high degree of selectivity among the sperm donors, who so far have been limited to the husbands of the women donors of the oocytes. Sex determination of the embryos is possible before implantation; and embryos with abnormal chromosome constitutions can be discarded. By checking the sperm and egg donors with a battery of biochemical tests, matching of carriers of the same defective gene can be avoided, or the defective embryos can themselves be detected and discarded.

I leave it to the reader to consider the ethical, social, legal, and political implications of Professor Glass's proposal, and to elaborate his own favorite objections. My point here is simply to show that even before Edwards or Shettles opens Pandora's box, there are well-meaning, decent men already at work to find good uses for its contents.

A similar camel's-nose-under-the-tent argument was advanced by opponents of artificial insemination. Ironically, some of the same people who made light of these arguments in defending artificial insemination are now defending the camel's neck while again dismissing the camel's nose argument. It is true that the practice of artificial insemination has thus far been confined to the treatment of infertility, and has not, to my knowledge, been malevolently, despotically, or even frivolously used. But I am no longer talking about the problem of misuse or abuse of a given technique, but rather about the fact that one technical advance makes possible the next, and in more than one respect. The first serves as a precedent for the second, the second for the third—not just technologically, but also in moral arguments. At least one good humanitarian ground can be found to justify each step. For these reasons, we must try to see more than a few feet in front of us before we set forth.

It was this kind of foresight which prompted Professor James D. Watson, co-discoverer of the structure of DNA, to bring before the public his concern over one technological prospect, the cloning of human beings, which Edwards's work makes very much more possible. In his very sober and careful testimony (before the Panel on Science and Technology, Committee on Science and Astronautics, United States House of Representatives), Professor Watson concluded a discussion of the work on *in vitro* fertilization as follows:

> Some very hard decisions may soon be upon us. For it is not obvious that the vague potential of abhorrent misuse should weigh more strongly than the unhappiness which thousands of married couples feel when they are unable to have their own children. Different societies are likely to view the matter differently and it would be surprising if all come to the same conclusion. We must, therefore, assume that techniques for the *in vitro* manipulation of human eggs are likely to be general medical practice, capable of routine performance throughout the world within some ten to twenty years.
>
> The situation would then be ripe for extensive efforts, either legal or illegal, at human cloning. . . .
>
> Moreover, given the widespread development of the safe clinical procedures for handling human eggs, cloning experiments would not be prohibitively expensive. They need not be restricted to the super-powers —medium-sized, if not minor countries, all now possess the resources needed for eventual success. There furthermore need not exist the coercion of a totalitarian state to provide the surrogate mothers. There already are such widespread divergences as to the sacredness of the act of human reproduction that the boring meaninglessness of the lives of many women would be sufficient cause for their willingness to participate in such experimentation, be it legal or illegal. Thus, if the matter proceeds in its current nondirected fashion, a human being—born of clonal reproduction—most likely will appear on the earth within the next twenty to fifty years, and conceivably even sooner, if some nation actively promotes the venture.

I now turn to consider this second new method for making babies, asexual reproduction, or cloning.[13]

Cloning, or asexual reproduction—state of the art

In genetic terms, asexual reproduction is distinguished from sexual reproduction (whether practiced in bed or in the test tube) by the

---

[13] The discussion which follows is adapted from another essay, devoted entirely to cloning, entitled "Freedom, Coercion and Asexual Reproduction," which will appear in *Freedom, Coercion, and the Life Sciences,* edited by Daniel Callahan and Leon R. Kass (to be published by the Harvard University Press in 1972).

following two characteristics: The new individuals are, first, derived from a single parent, and second, are genetically identical to—are identical twins of—that parent. Asexual reproduction occurs widely in nature, and is the normal mode of reproduction of bacteria, many plants, and some lower animals. By means of a technique known as nuclear transplantation (also called nuclear transfer), experimental biologists have artificially achieved the asexual reproduction of organisms which naturally reproduce only sexually (so far, frogs, salamanders, fruit flies). The procedure is conceptually simple. The nucleus of a mature but unfertilized egg is removed (by micro-surgery or by irradiation), and replaced by a nucleus obtained from a specialized somatic cell of an adult organism (e.g., an intestinal cell or a skin cell). Since almost all the hereditary material (DNA) of a cell is contained within its nucleus, the renucleated egg and the individual into which it develops are genetically identical to the organism which was the source of the transferred nucleus. Thus, the origin of the new individual is not the chance union of egg and sperm, with the generation of a new and unique genetic arrangement or genotype, but rather the contrived perpetuation into another generation of an already existing genotype.

An unlimited number of identical individuals, all generated asexually from a single parent—that is, a clone—could be produced by nuclear transplantation. An adult organism comprises many millions of cells, all genetically identical, each a potential source of a nucleus for cloning. In addition, techniques for storage and subsequent laboratory culture of animal tissues permit the preservation and propagation of cells long after the deaths of the bodies from which they were removed. There would thus be the possibility of a virtually unlimited supply of genetically identical nuclei for cloning.

The extension of nuclear transplantation to mammals has not yet been achieved, although several people have been trying for a few years. The difficulties are technical; there is no *theoretical* reason to believe that cional reproduction is not possible in mammals, including man. The technical problems when this work began included the following: (1) obtaining of mature mammalian eggs; (2) removal of the egg nucleus; (3) insertion of the donor nucleus; (4) transfer and implantation of the renucleated egg into the uterus of a female at the right stage in her menstrual cycle. As a result of the work on *in vitro* fertilization, the first and fourth problems have been solved for rabbits and mice, and will be solved for humans when and if Edwards or someone else succeeds. Recently, chemical methods have been perfected to remove the nucleus from mammalian cells in tissue culture, methods which can probably be used to enucleate egg cells. The only serious difficulty which remains is the introduction of the donor nucleus. And this difficulty may also be short-lived, since there are now very simple methods for fusing almost any two cells to produce a single cell containing the combined genetic material of both original cells. Fusion of an enucleated egg cell with a cell containing the donor nucleus might provide the method for getting the nucleus into the egg. In fact, Dr. Christopher Graham at Oxford has already succeeded in fusing mouse eggs with adult mouse cells. The fused egg divides several times but has thus far not gone on to form a blastocyst. Given the rate at which the other technical obstacles have fallen, and given the increasing number of competent people entering the field of experimental embryology, it is reasonable to expect the birth of the first cloned mammal sometime in the next few years. This will almost certainly be followed by a rush to develop cloning for other animals, especially livestock, in order to propagate in perpetuity the champion meat or milk producers.

With the human embryo culture and implantation technologies being perfected in parallel, the step to the first clonal man might require only a few additional years. Within our lifetime, possibly even as early as 1980, it may be technically feasible to clone a human being.

Among sensible men, the ability to clone a man would not be sufficient reason for doing so. Nevertheless, the apologists and the titillators have been at work, and the laundry list of possible applications keeps growing, in anticipation of the perfected technology: (1) Replication of individuals of great genius or great beauty to improve the species or to make life more pleasant. (2) Replication of the healthy to bypass the risk of genetic disease contained in the lottery of sexual recombination. (3) Provision of large sets of genetically identical humans for scientific studies on the relative importance of nature and nurture for various aspects of human performance. (4) Provision of a child to an infertile couple. (5) Provision of a child with a genotype of one's own choosing—of someone famous, of a departed loved one, of one's spouse or oneself. (6) Control of the sex of future children (the sex of a cloned offspring is the same as that of the adult from whom the donor nucleus was taken). (7) Production of sets of identical persons to perform special occupations in peace and war (not excluding espionage). (8) Production of embryonic replicas of each person, to be frozen away until needed as a source of organs for transplantation to their genetically identical twin. (9) To beat the Russians and the Chinese, and thus to prevent a "cloning gap."

Cloning—some ethical questions

Some of the ethical and social questions raised in connection with *in vitro* fertilization apply also to cloning: questions of experimenting upon the unborn, discarding of embryos, problems of misuse and abuse of power, questions concerning the camel and the tent. I will not repeat what has gone before, except to call special attention to the point about the ethics of experimentation. A significant number of grossly abnormal creatures has resulted from the frog experiments, and there is no reason to be more optimistic about the early attempts in humans. If the attempts to clone a man result in the production of a defective "product," who will or should care for it, and what status and rights will it have? If the offspring is subhuman, are we to consider it murder to destroy it? The twin issues of the production and disposition of defectives provide sufficient moral grounds for rebutting any first attempt to clone a man. Again, there is no ethical way for us to get to know whether or not human cloning is feasible.

There are, however, other questions which apply specifically to cloning and not to the techniques discussed earlier. Among the most important are questions concerning identity and individuality. One problem can be illustrated by exploiting the ambiguity of the word "identity": The cloned person may experience serious concerns about his identity (distinctiveness) because his genotype, and hence his appearance, stand in a relationship of identity (sameness) to another human being.

The natural occurrence of identical twins in no way weakens the argument against the artificial production of identical humans; there are many things which occur accidentally that ought not to be done deliberately. In fact, the problem of identity faced by identical twins should instruct us and enable us to recognize how much greater the problem might be for someone who was the "child" (or "father") of his twin. I cannot improve upon Paul Ramsey's reflections on this subject:

> Growing up as a twin is difficult enough anyway; one's struggle for selfhood and identity must be against the very human being for whom no doubt there is also the greatest sympathy. Who then would want to be the son or daughter of his twin? To mix the parental and the twin

136

relation might well be psychologically disastrous for the young. Or to look at it from the point of view of parents, it is an awful enough responsibility to be the parent of a son or daughter as things now are. Our children begin with a unique genetic independence of us, analogous to the personal independence that sooner or later will have to be granted them or wrested from us. For us to choose to replicate ourselves in them, to make ourselves the foreknowers and creators of every one of their genetic predispositions, might well prove to be a psychologically and personally unendurable undertaking. For an elder to teach his "infant copy" is a repellent idea not because of the strangeness of it, but because we are altogether too familiar with the problems this would exponentially make more difficult.

Perhaps this issue can be pressed even farther, beyond such concerns for undesirable psychological consequences. Does it make sense to say that each person has a right not to be deliberately denied a unique genotype? Is one inherently injured by having been made the copy of another human being, independent of *which* human being? We should not be deterred by the strangeness of these questions, a strangeness due largely to the fact that the problem could not have arisen before.

Central to this matter is the idea of the dignity and worth of each human being. The question we must ask is this: Is individual dignity undermined by a lack of genetic distinctiveness? One might argue, on the contrary, that indistinctiveness in appearance and capacity might produce a greater incentive to be distinctive in deed and accomplishment. Certainly the latter are more germane to any measure of individual worth and self-esteem than the former. On the other hand, our personal appearance is, at the very least, symbolic of our individuality. Differences in personal appearance, genetically determined, reinforce (if not make possible) our sense of self, and hence lend support to the feelings of individual worth we seek in ourselves and from others. Some put it more strongly and argue that a man not only *has* a body, but *is* his body. By this argument, a man's distinctive countenance not only makes possible his sense of self, but is in fact at one with that self. Membership in a clone numbering five to ten would no doubt threaten one's sense of self; membership in a clone of two might also. To answer the question posed above: We may *not* be entitled, in principle, to a unique genotype, but we *are* entitled not to have deliberately weakened the necessary supports for a worthy life. Genetic distinctiveness would seem to me to be one such support.

A second and related problem of identity and individuality is this: The cloned individual is not simply denied genetic distinctiveness; he is saddled with a genotype that has already lived. He will not be fully a surprise to the world; people are likely always to compare his performance in life with that of his alter ego. He may also be burdened by knowledge of his precursor's life history. Imagine living with the knowledge that the person from whom you were cloned had subsequently developed schizophrenia or suffered multiple heart attacks before the age of 40. For these reasons, the cloned individual's belief in the openness of his own future may be undermined, and with it, his freedom to be himself. Ignorance of what lies ahead is a source of hope to the miserable, a spur to the talented, a necessary support for a tolerable—let alone worthy—life for all.

But is the cloned individual's future really determinable or determined? After all, only his genotype has been determined; it is true that his environment will exert considerable influence on who and what he becomes. However, isn't it likely that the "parents" will seek to manipulate and control the environment as well, in an attempt to reproduce the person who was copied? For example, if a couple decided to clone a Rubinstein,[14] is there any doubt that early in life young Artur would be deposited at the piano and "encouraged" to play? It would not matter to the "parents" that the environment in which the true Rubinstein blossomed can never be reproduced or even approximated. Nor would it matter that no one knows what is responsible for the development of genius, or even for the appearance of ordinary talents and traits for which other people might have elected to clone. Such ignorance would not deter the "parents." Why else did they clone young Artur in the first place?[15]

Thus, although the cloned individual's future is probably not determinable according to his "parents'" wishes, enough damage is done by leading him to believe otherwise and by their believing otherwise. His own potential will in all likelihood be stunted and his outlook warped as he is forced into a mold he neither fits nor wants. True, some parents are already guilty of the same crime, but many more are

---

[14] Frequently mentioned candidates for cloning are musicians and mathematicians, such as Mozart, Newton, and Einstein, whose genius is presumed to have a large genetic component. But all such suggestions ignore the wishes of these men. I suspect that none of them would consent to having themselves replicated. Indeed, should we not assert as a principle that any so-called great man who *did* consent to be cloned should on that basis be disqualified, as possessing too high an opinion of himself and of his genes? Can we stand an increase in arrogance?
[15] These reasons make it clear that it will probably prove impossible to keep the knowledge of his origin from the child, especially in the early cases when the geneticist's need to claim credit for his success will also be a factor.

restrained by their impotence in determining the raw material. The opportunity to clone would not only remove this restraint, but would openly invite and encourage more outrageous efforts to shape our children after our own desires.

Although these arguments would apply with even greater force to any large scale efforts at human cloning, I find them sufficient to reject even the first attempts at human cloning. It cannot be repeated too often that these are human beings upon whom these eugenic or merely playful visions (shall I say hallucinations?) are to be worked.

Thus far, I have dealt separately with two technological prospects —one now upon us, the other on the horizon and fast approaching— in an effort to reason about and evaluate each piece of technology one at a time. I have been at pains to analyze the morally relevant features of each, in order to show that real and important distinctions can and should be drawn among different technologies, that the practice of one should not *ipso facto* justify the introduction of another. I am far more concerned that this approach be found reasonable and useful than that of any of my specific arguments be found convincing.

Yet despite its practical utility, this piece-by-piece approach has grave deficiencies. It ignores the great wave upon which each of these techniques is but a ripple. All of the new technologies arise from and are part of the great project of modernity, the "conquest of nature for the relief of man's estate." They go beyond many earlier techniques in that they seek to relieve man's estate by directly changing man himself. We must therefore raise some broader questions concerning this project as these questions arise in connection with the technologies discussed above. Here, I am far more concerned that my arguments be found convincing than that they be found useful.

Questions of power

Though philosophically debatable, the Baconian principle "knowledge is power" is certainly correct when applied to that knowledge which has been sought under that principle. The knowledge of how to begin human life in new ways is a human power to do so. But the power rests only metaphorically with humankind; it rests in fact with particular men—geneticists, embryologists, obstetricians. The triumphant proclamation of "man's" growing power over nature obscures the troublesome reality that it is individual men who wield

139

power. What we really mean by "man's power over nature" is a power exercised by some men over other men with the knowledge of nature as their instrument.

While applicable to technology in general, these reflections are especially pertinent to the technologies of human reproduction and genetic manipulation with which men deliberately exercise power over future generations. Ultimately, as C. S. Lewis points out in *The Abolition of Man,*

> If any one age really attains, by eugenics and scientific education, the power to make its descendants what it pleases, all men who live after it are patients of that power. They are weaker, not stronger: For though we may have put wonderful machines in their hands we have preordained how they are to use them. . . . The real picture is that of one dominant age . . . which resists all previous ages most successfully and dominates all subsequent ages most irresistibly, and thus is the real master of the human species. But even within this master generation (itself an infinitesimal minority of the species) the power will be exercised by a minority smaller still. Man's conquest of nature, if the dreams of the scientific planners are realized, means the rule of a few hundreds of men over billions upon billions of men. There neither is nor can there be any simple increase in power on man's side. Each new power won *by* man is a power *over* man as well.

Please observe that I am not yet speaking about the problem of the misuse and abuse of power. My point is rather that the power which grows is willy-nilly the power of only some men, and that the number of these powerful men tends to grow fewer and fewer as the power increases.

I suggest that this is also true with respect to our topic. Recall that there is a new partner in these new procedures for making babies—the scientist-physician. The obstetrician is no longer just the midwife, but also the sower of seed. Even in the treatment of intramarital infertility, the scientist-physician who employs *in vitro* fertilization and laboratory culture of human embryos has acquired far greater power over human life than his colleague who simply repairs the obstructed oviduct. He presides over many creations in many patients. And once he goes beyond the bounds of marriage, he is not simply the Fertilizer General but the Matchmaker as well —as in the practice of artificial insemination (donor), where a small number of physicians have already arranged for the fathering of several hundred thousand children, many of them, nepotistically, by their professional offspring, the medical students. I am not at present questioning this practice; my point is rather to illustrate how the new technologies lead to the concentration of power.

Both a cause and an effect of the growing power of biomedical technologists is the growing complexity of scientific knowledge, and the related fragmentation of disciplines and extreme specialization of their practitioners. Science understands and explains the world in ideas and language which the layman cannot understand. I am not speaking now about the problem of jargon, but rather about a more fundamental matter. The phenomena of nature as they present themselves to us in ordinary experience are understood by the scientist in terms of abstract concepts such as molecules and genes, and ultimately in terms of mathematical formulae. This is to the layman a new cabala, but a cabala with a difference, a cabala that can create new life in test tubes and can send men to the moon. Small wonder that the scientists and technologists have become for many people the new priesthood.[16]

Because this new priesthood has promised its rewards here on earth, it faces perhaps a heavier responsibility—especially when it fails to deliver. The public has acquired high expectations for technology, from which impatience and frustration are easily bred. This point has been surprisingly overlooked, but I think this is what is meant every time someone starts to complain, "Well, if they can put a man on the moon, why can't they . . .?" Given this general disposition with regard to science and technology, is it not likely that the expectations of an infertile couple will be much higher for any baby given them by the rationalized, disinfected procedures of the laboratory, than for a baby born in the usual way and obtained via adoption? Even with adequate warnings, it will be hard to get the science-worshipping patient to face the reality of possible disastrous outcome. Imagine the heartache, and then the outcry, if the child conceived *in vitro* turns out to be hemophiliac or retarded (even for reasons unrelated to the use of the procedure). On this ground alone, prudence dictates caution.

The problem of the specific abuses of specific powers cannot be overlooked. However, because it is more widely appreciated, I will

---

[16] At a recent meeting, scientists were summarizing for a group of educated laymen the current state of knowledge in human genetics and the promising fields for the future. At the conclusion of a summary of studies on mutation, a woman rose in the audience to ask about the meaning of one of the findings for the chance of her having an abnormal child. The answer came back that the matter was complicated, involving some function of "one over the square root of the mean." The woman seated herself with a look of bewilderment on her face, but shaking her head affirmatively, "Amen."

There is of course a revolt in progress against this new priesthood, but primarily because it has been selling indulgences to the Pentagon or because its blessings are not biodegradable. Very few are questioning the intellectual foundations of the modern scientific conception of the world.

spend little time on it. It is sufficient merely to mention the prospects of involuntary breeding programs, the cloning of tyrants, and the production of whole cadres of "gammas" and "deltas" to handle the onerous tasks of an advanced civilization, or the more modest abuses of cloning quintuplets for the circus, for five complete sets of spare organs, or for partners in crime who can always have an alibi. Nevertheless, while events which have occurred in our lifetime should warn us not to dismiss these possibilities, I think we have greater reason to be concerned about the private, well-intentioned, and voluntary use of the new technologies. The major problem to be feared is not tyranny but voluntary dehumanization.

Questions of dehumanization

Human procreation not only issues new human beings; it is itself a human activity (an activity of embodied men and women). The new forms of baby making discussed earlier represent in themselves a radical change in human procreation as a human activity. As already noted, the new beginnings occur in a new locus, the laboratory, and involve a new partner, the scientist. Moreover, the techniques which at first serve merely to provide a child to a childless couple will soon be used to exert control over the quality of the child. A new image of human procreation has been conceived, and a new "scientific" obstetrics will usher it into existence. As one obstetrician put it at a recent conference, "The business of obstetrics is to produce *optimum* babies." The price to be paid for this optimum baby is the transfer of procreation from the home to the laboratory, and its coincident transformation into manufacture. Increasing control over the product is purchased by the increasing depersonalization of the process. (In this continuum, artificial insemination represented the first step.) Perhaps for some techniques used for some purposes, e.g., artificial insemination (husband) to circumvent infertility, the benefits outweigh the costs. But let us not say that there are not costs. The complete depersonalization of procreation (possible with the development of an artificial placenta), and its surrender to the demands of the calculating will, would be in themselves seriously dehumanizing no matter how "optimum" the product.

Human procreation is not simply an activity of our rational wills. Men and women are embodied as well as desiring and calculating creatures. It is for the gods to create in thought and by fiat ("Let the earth bring forth . . ."). And some future race of demigods (or demimen) may obtain its survivors from the local fertilization and decanting station. But *human* procreation is begetting. It is a more complete human activity precisely because it engages us bodily and spiritually, as well as rationally. Is there possibly some wisdom in that mystery of nature which joins the pleasure of sex, the communication of love, and the desire for children in the very activity by which we continue the chain of human existence? Is biological parenthood a built-in "device" selected to promote the adequate caring for posterity? Before we embark on new modes of reproduction, we should consider the meaning of the union between sex, love, and procreation, and the meaning and consequences of its cleavage.

My point is almost certain to be misunderstood. I am not suggesting that one can be truly human only by engaging in procreation. I think there is a clear need for curtailing procreation, and have no objections to the use of any and all contraceptive devices. I am not suggesting that there is something inhuman about adopting children instead of getting them through the pelvis, nor do I think that the most distinctively human activities center in the groin. My point is simply this: There are more and less human ways of bringing a child into the world. I am arguing that the laboratory production of human beings is no longer *human* procreation, that making babies in laboratories—even "perfect" babies—means a degradation of parenthood.

There will be some who object to my calling the new technologies forms of manufacture. I mean "manufacture" in quite a literal sense —*hand made*. It matters not whether we are talking about small- or large-scale manufacture. With *in vitro* fertilization, the natural process of generating becomes the artificial process of making. In the case of cloning, the artistry is taken one step further. Not only is the process *in hand,* but the total genetic blueprint of the cloned offspring is selected and determined by the human artisan. To be sure, the subsequent development is still according to natural processes, and no so-called laws of nature have been or can be violated. What has been violated, even if only slightly, is the distinction between the natural and the artificial, and at its very root, the nature of man himself. For man is the watershed that divides the world of the familiar into those things which belong to nature and those things which are made by men. *To lay one's hands on human generation is to take a major step toward making man himself simply another one of the man-made things.* Thus, human nature becomes simply the last part of nature which is to succumb to the modern technological project, a project which has already turned the rest of nature into raw material at human disposal, to be homogenized by our rationalized technique according to the artistic conventions of the day.

The family: a final solution?

If the depersonalization of the process of reproduction and its separation from human sexuality dehumanize the activity which brings new life, and if the manufacture of human life threatens its humanness, both together add up to yet another assault on the existence of marriage and the human family. Sex is now comfortably at home outside of marriage; child rearing is progressively being turned over to other institutions: the state, the schools, the mass media, the child-care centers. Transfer of procreation to the laboratory undermines the justification and support which biological parenthood gives to the monogamous (or even polygamous) marriage. Cloning adds an additional, more specific, and more fundamental threat: The technique renders males obsolete. All it requires are human eggs, nuclei, and (for the time being) uteri. All three can be supplied by women.

Curiously, both those who welcome and those who fear the new technologies for making babies agree that they will pose serious threats to marriage and the family. Indeed, one of the reasons, not always explicitly admitted, that the new technologies are endorsed in some quarters is precisely that they will help lay these institutions to rest. The congregation of deliberate family wreckers includes persons eager to remove all restraints from human sexuality or to render obsolete the biological differences between the sexes, others who see the destruction of marriage as a needed step in limiting population growth, and yet others who find the modern nuclear family a stifling and harmful institution for education and child rearing. I will not deny that the modern nuclear family shows signs of cracking under various pressures. It may have intrinsic limitations which make it seem, even at best, ill-fitted for modern technological society. But perhaps this should be viewed as a problem of modern technological society rather than of the family. We really ought to be less frivolous and journalistic in discussing such matters, and should keep in mind the essential question: Are we to accept as desirable the final solution which eliminates biological kinship from the foundation of social organization? Yes, laboratory and governmental alternatives could be devised for procreation and child bearing. But at what cost? How much stunting of our humanity would result from the totalitarian orientation that these alternatives require and foster?

Some of the important virtues of the family are, nowadays, too often overlooked. The family is rapidly becoming the only institution in an increasingly impersonal world where each person is loved not for what he does or makes, but simply because he is. The family is also the institution where most of us, both as children and as parents, acquire a sense of continuity with the past and a sense of commitment to the future. Without the family, most of us would have little incentive to take an interest in anything after our own deaths. It would be a just irony if programs of cloning or laboratory-controlled reproduction to improve the genetic constitutions of future generations were to undermine the very institution which teaches us concern for the future. These observations suggest to me that the elimination of the family would weaken ties to past and present, and would throw us even more on the mercy of an impersonal, lonely present. The burden of proof should fall upon those believing our humanness could survive even if the biological family does not.

Dehumanizing the scientist

Finally, there may well be a dehumanizing effect on the scientist himself, and through him, on all of us. The men who are at work on new beginnings in life are out to subdue one of the most magnificent mysteries, the mystery of birth and renewal. To some extent, the mystery has already been subdued. Those who do *in vitro* fertilization are in the business of initiating new life. To the extent that they feel that there is nothing unusual or awesome in what they are doing, to that extent they have already lost the appreciation of mystery, the sense of wonder. The same can be said of the heart surgeon who sees the heart simply as a pump, the brain surgeon who sees the brain simply as a computer, or the pathologist who sees the corpse simply as a body containing demonstrable pathology. The sense of mystery and awe I am speaking of is demonstrated by most medical students on their first encounter with a cadaver in the gross anatomy laboratory. Their uncomfortable feeling is more than squeamishness. It is a deep recognition, no matter how inarticulate, that it is the mortal remains of a human being in which they are to be digging; ultimately it is a recognition of the mysterious phenomena of life and death. The loss of this sense of awe occurs in a matter of days or weeks; mastery drives out mystery in all but a very few.

145

There *is*, I admit, no reason in principle why the sense of mystery needs to be lost by the increase of knowledge or power. And, indeed, in the case of the great men of science knowledge served to increase rather than to decrease their sense of wonder and awe. Nevertheless, for most ordinary men of science and technology, and probably for most men in this technological age, once nature is seen as or transformed into material and given over to their manipulation, the mystery and the appreciation are gone. Awed by nothing, freed from all so-called superstitions and so-called atavistic beliefs, they practice their power without even knowing what price they have paid.

Consider in this connection these excerpts from an editorial in *Nature* ("Premature Birth of Test Tube Baby," Vol. 225, p. 886, March 7, 1970) concerning adverse public reaction in Britain to the announcement (apparently erroneous) that Drs. Edwards and Steptoe were at that time about to do the first transfer of a laboratory-grown human embryo into a woman's uterus to circumvent her infertility.

> What has all this to do with the test tube baby? In terms of scientific fact, almost nothing at all. The test tube baby, as this phrase is usually understood, refers to the growing of a human embryo to full term outside the body and the chief obstacle to this feat (not that anybody has proclaimed it as a goal) is the formidable problem of maintaining the embryo after the stage at which it would normally implant in the uterus. The Oldham procedure concerns only the pre-implantation embryo which, except in the trivial sense that fertilization is carried out *in vitro*, can hardly be equated with the test tube baby. Moreover, it is difficult to see that the wastage of embryos occasioned by the procedure raises moral problems any knottier than those to do with IUCD, a device that probably prevents the embryo from implanting in the womb. What then was all the fuss about? . . .
>
> A curious feature of the public debate is that the letter-writing segment of the public, at least, seemed to believe that human life was about to be created from nothing in the test tube. For example, a correspondent in *The Times* voiced the fear that "The ability of scientists to develop the technique of creating life in a test tube is so serious that I feel human beings should be given the opportunity to express their views on whether or not this line of research should be pursued . . . [sic] Personally I find the idea of creating life at man's will terrifying." These are indeed dark and atavistic fears which have been nurtured, perhaps, by the views of Dr. Edmund Leach that scientists have usurped the creative powers and should assume the moral responsibilities formerly attributed to gods. Whatever the merit of Dr. Leach's thesis, those who are engaged in research that is at all liable to be misinterpreted will doubtless take the present episode as a warning of the misunderstandings that can arise,

146

particularly if the true facts are not readily available from authoritative sources. There is always the danger that lack of information or misinformation may convert legitimate public concern about new knowledge into a paranoia that impedes research.

The moral is clear. Research, the supreme value, is to be protected from the dark "atavistic" fears of an ignorant public by the "authoritative" dispensation of (only) the "true facts."

The first paragraph of the excerpt contains the hard, cold, technical facts, presented "scientifically" by the scientist-editor of *Nature*. That editor finds it "curious" that the public had a somewhat different view of what was done; he erroneously attributes this difference to the public's lack of the "true facts." The source of the difference, however, is not the lack of information, but rather a difference in interpretation; the real reason for the difference is that the editor of *Nature* lacks, whereas the correspondent in *The Times* does not, a sense of mystery and awe concerning the initiation of new life. The editor is correct in distinguishing *in vitro* fertilization from full extracorporeal gestation, and partly correct in analogizing the question of disposal of embryos to the question concerning the IUCD (but not in thus disposing of the question). But by calling "trivial" the fact that fertilization has occurred *in vitro*, and embryonic development has been initiated, he displays his own human impoverishment. I do not insist that the embryo created is a human life worthy of protection; but it surely is alive, and it surely is potentially human. To look upon these embryos as anything less than potential human life in human hands is a misperception so gross as to be alarming.

We have paid some high prices for the technological conquest of nature, but none perhaps so high as the intellectual and spiritual costs of seeing nature as mere material for our manipulation, exploitation, and transformation. With the powers for biological engineering now gathering, there will be splendid new opportunities for a similar degradation in our view of man. Indeed, we are already witnessing the erosion of our idea of man as something splendid or divine, as a creature with freedom and dignity. And clearly, if we come to see ourselves as meat, then meat we shall become.

"Humanization"—man as self-creator

Among those who would take strong exception to my remarks on dehumanization are those who argue that the new biomedical technologies, including those which make possible new methods for

making babies, provide the means for human self-modification and
therefore improvement. They see man as imperfect, unfinished, but
endowed with creative powers to complete and perfect himself.
Included in this group are scientists such as Robert Sinsheimer:

> For the first time in all time a living creature understands its origin and
> can undertake to design its future. Even in the ancient myths man was
> constrained by his essence. He could not rise above his nature to chart
> his destiny. Today we can envision that chance—and its dark com-
> panion of awesome choice and responsibility. . . . We are an historic
> innovation. We can be the agent of transition to a wholly new path of
> evolution. This is a cosmic event.

And theologians, such as Karl Rahner:

> Freedom enables man to determine himself irrevocably, to be for all
> eternity what he himself has chosen to make himself.

We note in passing that the theologians have done the scientists
one better. Many scientists take the fatalistic view that "what can
be done, will be done." Those theologians-turned-technocrats sanc-
tify the new freedoms: "What can be done, should be done."

The notion of man as a creature who is free to create himself, as a
"freedom-event," to use one of Rahner's formulations, is problematic,
to say the least. It is an idea that is purely formal, not to say empty.
It provides no boundaries that would indicate when what was subhuman became truly human, or when what was at first human
became less than human. Moreover, the freedom to change one's
nature includes the freedom to destroy (by genetic manipulation
or brain modification) one's nature, and thereby the capacity and
desire for freedom itself. It is, literally, a freedom which can end
all freedom. Nor can it provide standards by which to measure
whether the changes made are in fact improvements. Evolution
simply means change—to measure "progress" requires a standard
which this view cannot in principle supply.

The new technologies for human engineering may well be the
"transition to a wholly new path of evolution." They may, therefore, mark the end of *human* life as we and all other humans have
known it. It is possible that the non-human life that may take our
place will be in some sense superior—though I personally think it
most unlikely, and certainly not demonstrable. In either case, we are
ourselves human beings; therefore, it is proper for us to have a
proprietary interest in our survival, and in our survival *as human
beings.* This is a difficult enough task without having to confront
the prospect of a utopian, constant remaking of our biological nature
with all-powerful means but with no end in view.

A matter of wisdom

I had earlier raised the question of whether we have sufficient wisdom to embark upon new ways for making babies, on an individual scale as well as in the mass. By now it should be clear that I believe the answer must be a resounding "No." To have developed to the point of introduction such massive powers, with so little deliberation over the desirability of their use, can hardly be regarded as evidence of wisdom. And to deny that questions of desirability, of better and worse, can be the subject of rational deliberation, to deny that rationality might dictate that there are some things that we can do which we must never do—in short, to deny the need for wisdom—can only be regarded as the height of folly.

Let us simply look at what we have done in our conquest of non-human nature. We shall find there no grounds for optimism as we now consider offers to turn our technology loose on human nature. In the absence of standards to guide and restrain the use of this awesome power, we can only dehumanize man as we have despoiled our planet. The knowledge of these standards requires a wisdom we do not possess, and what is worse, we do not even seek.

But we have an alternative. In the absence of such wisdom, we can be wise enough to know that we are not wise enough. When we lack sufficient wisdom to do, wisdom consists in not doing. Restraint, caution, abstention, delay are what this second-best (and maybe only) wisdom dictates with respect to baby manufacture, and with respect to various other forms of human engineering made possible by other new biomedical technologies. It remains for another time to discuss how to give practical effect to this conclusion: how to establish reasonable procedures for monitoring, reviewing, and regulating the new technologies; how to deal with the undesirable consequences of their proper use; how to forestall or prevent the introduction of the worst innovations; how to achieve effective international controls so that one nation's folly does not lead the world into degradation.

Fortunately, there are no compelling reasons to proceed rapidly with these new methods for making babies. Though it saddens the life of many couples, infertility is hardly one of our major social problems. Moreover, there are other means of circumventing it that are free of the enormous moral and social problems discussed earlier. At a time when we desperately need to limit population growth, it may be a questionable sentimentality which seeks to provide every couple with its own biological child rather than continue the practice of adoption. But it would be a foolish senti-mentality to unleash baby-making technologies for this purpose, especially before there are means to limit and control their use. The same arguments apply, with equal force, to the use of these tech-nologies for the eradication of genetic disease. We probably do not know enough about the genetics of man, despite our well-meaning desires, to prevent genetic disease by the practice of eugenic abortion (e.g., following amniocentesis). We certainly don't know enough to escalate our tinkering by the eugenic use of the new baby-making techniques.

I am aware that mine is, at least on first glance, not the most com-passionate view (although it may very well turn out to be so in the long run). I am aware that there are some who now suffer who will not get relief, should my view prevail. Nevertheless we must measure the cost—I do not mean the financial cost—of seeking to eradicate that suffering by any and all means. In measuring the cost, we must of course evaluate each technological step in its own terms—but we can ill afford to ignore its place in the longer journey. For, defensible step by defensible step, we could willingly walk to our own degradation. The road to Brave New World is paved with sentimentality—yes, even with love and charity. Have we enough sense to turn back?

150

Letters to the Editor

BIRTH AFTER THE REIMPLANTATION OF A HUMAN EMBRYO

SIR,—We wish to report that one of our patients, a 30-year-old nulliparous married woman, was safely delivered by cæsarean section on July 25, 1978, of a normal healthy infant girl weighing 2700 g. The patient had been referred to one of us (P.C.S.) in 1976 with a history of 9 years' infertility, tubal occlusions, and unsuccessful salpingostomies done in 1970 with excision of the ampullæ of both oviducts followed by persistent tubal blockages. Laparoscopy in February, 1977, revealed grossly distorted tubal remnants with occlusion and peritubal and ovarian adhesions. Laparotomy in August, 1977, was done with excision of the remains of both tubes, adhesolysis, and suspension of the ovaries in good position for oocyte recovery.

Pregnancy was established after laparoscopic recovery of an oocyte on Nov. 10, 1977, in-vitro fertilisation and normal cleavage in culture media, and the reimplantation of the 8-cell embryo into the uterus $2\frac{1}{2}$ days later. Amniocentesis at 16 weeks' pregnancy revealed normal α-fetoprotein levels, with no chromosome abnormalities in a 46 XX fetus. On the day of delivery the mother was 38 weeks and 5 days by dates from her last menstrual period, and she had pre-eclamptic toxæmia. Blood-pressure was fluctuating around 140/95, œdema involved both legs up to knee level together with the abdomen, back, hands, and face; the blood-uric-acid was 390 µmol/l, and albumin 0·5 g/l of urine. Ultrasonic scanning and radiographic appearances showed that the fetus had grown slowly for several weeks from week 30. Blood-œstriols and human placental lactogen levels also dropped below the normal levels during this period. However, the fetus grew considerably during the last 10 days before delivery while placental function improved greatly. On the day of delivery the biparietal diameter had reached 9·6 cm, and 5 ml of amniotic fluid was removed safely under sonic control. The lecithin: sphingomyelin ratio was 3·9:1, indicative of maturity and a low risk of the respiratory-distress syndrome.

We hope to publish further medical and scientific details in your columns at a later date.

Department of
    Obstetrics and Gynæcology,
General Hospital,
Oldham OL1 2JH   P. C. STEPTOE

University Physiology Laboratory,
Cambridge CB2 3EG   R. G. EDWARDS

Unborn: Historical perspective of the fetus as a patient

Michael R. Harrison, M.D.

The author (AΩA, Harvard University, 1969) is assistant professor of surgery (pediatric surgery) at the University of California, San Francisco, and co-director of the fetal treatment program. He has been in the forefront of research in this emerging area of surgery.

The concept that the fetus may be a patient, an individual whose maladies are a proper subject for medical treatment as well as scientific observation, is alarmingly modern. The fetus could not be taken seriously as long as he remained a medical recluse in an opaque womb; and it was not until the last half of this century that the prying eye of the ultrasonogram rendered the once opaque womb transparent, stripping the veil of mystery from the dark inner sanctum, and letting the light of scientific observation fall on the shy and secretive fetus.

Painted and carved door from Dutch New Guinea displays a fetus.
From *Ethnographische Beschrijving van de West-en Noordkust van Nederlansch New Guinea,* by F.S.A. de Clercq and J.D.E. Schmelz, Leiden, 1893. In *A History of Embryology* by Joseph Needham, New York, Abelard-Schuman, 1959, p. 19.

Historically, we approached the fetus, tucked away in the warm, wet womb, with a wonder bordering on mysticism. Enid Bagnold's description in *The Door of Life* captures the awe engendered by a scene no one had actually witnessed.

Hanging head downwards between cliffs of bone, was the baby, its arms all but clasped about its neck, its face aslant upon its arms, hair painted upon its skull, closed, secret eyes, a diver poised in albumen, ancient and epic, shot with delicate spasms, as old as a Pharaoh in its tomb.[1]

Whether this reverence is a reflection of our profound wonder at the "miracle" of differentiation of a fertilized egg into a human infant or a clandestine Darwinian mechanism to ensure survival of the species, it has certainly hindered investigation of fetal pathophysiology and precluded any consideration of fetal therapy. Only in the last few decades have techniques for visualizing, monitoring, measuring, and prodding the fetus begun to temper our centuries-old reverence for the living human fetus. Only now are we beginning to take the fetus seriously — medically, legally, and ethically. Now that our perspective is undergoing radical change, it is instructive to trace the medical history of our approach to this, our most unsolicited patient.

The Fetus as Seed and Homunculus

Animal husbandry is not complicated unless you think about it. So, in biblical times, it was enough to describe the phenomena: someone begot someone begot someone, et cetera. The intricacies of the begetting, the conception and growth of the individual, were matters assigned to the realm of Solomon's wisdom: "And in my mother's womb was fashioned to be flesh in the time of ten months, being compacted in blood, of the seed of man, and the pleasure that came with sleep." (Wisd. of Sol. 7:2) There was general recognition that physical characteristics passed from parent to offspring.

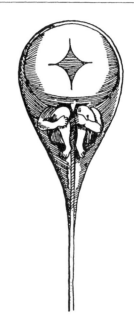

A human spermatazoon is depicted containing a homunculus.
By Hartsoeker in his *Essay de Dioptrique*, Sect. 88, Paris, 1694. In *The Rise of Embryology* by Arthur W. Meyer, Stanford, Stanford University Press, 1939, following p. 78.

spring. This transmission was conceptualized as resulting from the father passing the seed to the mother, where it would grow and later blossom as a child. Even into the second century, Marcus Aurelius could say: "A man passes seed into a womb and goes his way, and anon another cause takes it in hand and works upon it and perfects a babe — what a consummation from what a beginning!"[2]

In attempting to explain how the fetus was related to the child, the Greek and Roman thinkers came up with the idea of the homunculus — a miniature person living and growing within the mother before birth. This concept undoubtedly evolved from practical experience in animal husbandry and from observation of aborted human and animal fetuses. The fetus could be seen to grow in resemblance to the child as gestation progressed. In the absence of modern biologic information, the notion of the homunculus provided a very good explanation for at least the older fetus.

The homunculus was a purely descriptive explanation, and the description could be carried to a wonderfully whimsical extreme. Laurence Sterne, in *The Life and Opinions of Tristram Shandy, Gentleman,* has his hero soliloquize:

The Homunculus, Sir, in however low and ludicrous a light he may appear, in this age of levity, to the eye of folly or prejudice; — to, the eye of reason in scientifick research, he stands confess'd — a being guarded and circumscribed with rights: — The minutest philosophers, who, by the bye, have the most enlarged understandings, (their souls being inversely as their enquiries) shew us incontestably, that the Homunculus is created by the same hand, — engender'd in the same course of nature, — endowed with the same loco-motive powers and faculties with us: — That he consists as we do, of skin, hair, fat, flesh, veins, arteries, ligaments, nerves, cartilages, bones, marrow, brains, glands, genitals, humours, and articulations; — is a Being of as much activity, — and, in all senses of the word, as much and as truly our fellow-creature as my Lord Chancellor of *England.* — He may be benefited, — he may be injured, — he may obtain redress; — in a word, he has all the claims and rights of humanity.[3]

Prior to this century, the development of both obstetrics and pediatrics contributed surprisingly little to our knowledge of the fetus. Although physicians were beginning to take an interest in children as early as 1472, when Paulus Bagellardus published *The Little Book of Diseases of Children,* it was not until 1748 that a

scientific treatise on the care and feeding of infants appeared. William Cadogan of London, in his monograph *An Essay upon Nursing,* denounced the common practice of feeding the infant artificially until the mother's milk came down, and noted that the infant "requires some intermediate time of abstinence and rest to compose and recover the struggle of the birth and the change of circulation (the blood running into new channels),"[4] thus contributing the first significant observation about perinatal circulatory physiology.

The work of Charles Darwin had an indirect but significant impact on our scientific and philosophical attitude toward the fetus. His ideas about the evolution of species provided a macroscopic view of reproduction that required a lawful progression from seed to infant, as well as from generation to generation. Before this view of procreation and speciation, the relation between the seed and the infant was a religious matter, a continuum accepted on faith, which may account for the derivation of the religious doctrine that links copulation directly with procreation and, thus, acribes the sacredness of life to the conceptus or homunculus from its very beginning. Darwin's ideas demystified the role of reproduction. In *The Descent of Man,* he wrote: "Man is developed from an ovule, about 125th of an inch in diameter, which differs in no respect from the ovules of other animals."[5]

By the dawn of the twentieth century, the fetus was shedding its metaphysical trappings in favor of a biological description. Still, the concepts of inheritance and evolution did little to explain exactly how the seed became an infant, and work on the development of the human fetus remained entirely descriptive through the first part of the twentieth century. In the 1920s, Minkowski examined human fetuses obtained at cesarean operations by placing the recently delivered fetus in warm saline solution. Hertig and Rock

studied aborted fetuses at various stages in gestation and were able to piece together the morphology of embryonic and fetal development. This descriptive information was supplemented by study of animal fetuses in various species, an undertaking that blossomed and bore fruit in this century.

Experimental Fetal Observation

Since early observations on the human fetus were limited to an occasional aborted "homunculus," fetal developmental anatomy and physiology had to be derived almost exclusively from observations on animals. This is undoubtedly how Hippocrates arrived at the brilliantly intuitive proposal that the fetus urinates in utero and that amniotic fluid is composed of fetal urine, an observation confirmed only in the last century. Andreas Vesalius must be credited with the first truly analytic observations on the living mammalian fetus. At a time when Galen's dogmatism reigned and before the circulation of blood was discovered, Vesalius described in *De Humani Corporis Fabrica* (1543) the anatomy of the fetal-placental unit and the experimental preparation on which his observations were based:

Quite pleasing is it in the management of the fetus to see how when the fetus touches the surrounding air it tries to breathe. And this dissection is performed opportunely in a dog or pig when the sow will soon be ready to drop her young. . . .
. . . [T]he naked fetus attempts and struggles for respiration and thereupon when the coverings are punctured and broken thou shalt see that then the fetus breathes and that pulsations of the arteries of the fetal membranes and of the umbilicus stop. Up to this moment the arteries of the uterus are beating in unison with the rest of the arteries outside of itself.[6]

This remarkable description, based on direct observation of a dissected living fetus, marked the beginning of

20

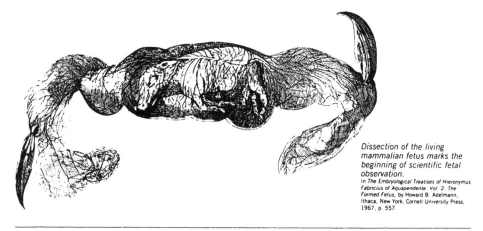

Dissection of the living
mammalian fetus marks the
beginning of scientific fetal
observation.
In The Embryological Treatises of Hieronymus
Fabricius of Aquapendente. Vol. 2. The
Formed Fetus, by Howard B. Adelmann.
Ithaca, New York. Cornell University Press,
1967, p. 557.

scientific fetal observation.

It wasn't until the nineteenth century that experimental animal preparations were used to make physiologic observations on the living mammalian fetus. Bichat in 1803 was the first to study fetal movements. Zuntz (1877) and later Preyer (1885) studied intact fetal guinea pigs suspended in warm saline, noting that the fetus must be kept in warm physiologic salt solution and that a fetus, once allowed to breathe, could not be returned to its mother and survive.

Experimental fetal observation blossomed in the twentieth century, at first reluctantly, and then with a crescendo of enthusiasm. By 1920, the first successful fetal operations had been performed: Mayer removed guinea pig fetuses from the uterus and placed them in the maternal abdominal cavity — a few surviving for several days. Fetal movements had been studied by Graham Brown in the cat, and Lane in the rat. In the 1920s, the first experimental *in utero* manipulation (amputation of the limb of a guinea pig) was reported by Bors. The feasibility of using maternal anesthesia for fetal manipulation was demonstrated by Swenson, and the possibility of normal delivery after *in utero* surgery was established by Nicholas.

In the 1930s and 1940s, experimental fetal observation gained momentum. Barcroft introduced the most productive fetal experimental model when he described operations on the lamb fetus using spinal anesthesia. Surgery was performed through a small uterine incision, without removing the fetus. Hall's work on development of the nervous system in the fetal rat and Barron's work on neurologic development in the fetal lamb extended the techniques for fetal surgery, including the use of purse-string sutures to avoid loss of amniotic fluid.[7]

The first big dividend from experimental fetal manipulation came when Jost demonstrated that removal of the fetal rabbit testes had a profound influence on subsequent sexual development.[8] In the 1950s Louw and Barnard produced intestinal atresia, similar to that seen in human neonates, by interrupting the mesenteric blood supply in fetal puppies.[9] This was an important contribution, although less glamorous than Barnard's later work with cardiac transplantation, because it not only established the ischemic pathogenesis of neonatal intestinal atresia, but also demonstrated the feasibility of simulating human birth defects by appropriate fetal manipulation.

In the 1960s and 1970s, experimental fetal surgery was used to simulate a variety of human congenital anomalies: coarctation of the aorta in the puppy,[10] intestinal atresia in the rabbit,[11] congenital diaphragmatic hernia in the lamb,[12] congenital hydronephrosis in the rabbit[13] and lamb,[14] and congenital heart disease in the fetal lamb.[15] Perhaps more important, the development of a chronically catheterized fetal lamb preparation led to intensive investigation of fetal cardiovascular, pulmonary, and renal physiology.[15, 16] Experimental fetal surgery proved far more difficult in the primate, where uterine contractility and preterm labor were more difficult to control. But in the last two decades advances in surgical and anesthetic techniques and in the pharmacologic control of labor have made experimental manipulation of even the primate fetus feasible.[17]

By the late 1970s a variety of experimental fetal models were widely used to study normal developmental physiology and the pathophysiology of several congenital defects. These models proved to be both descriptive and predictive: for example, removal of a piece of diaphragm not only produced a lesion that mimicked the human analogue (congenital diaphragmatic hernia), but also produced the associated developmental consequence (pulmonary hypoplasia).

Once the pathophysiology of certain congenital anomalies has been clarified, these models may be used to explore treatment. For instance, babies born with congenital diaphragmatic hernia die because their lungs are hypoplastic. It has now been demonstrated experimentally that simulated correction of experimental diaphragmatic hernia *in utero* may prevent development of fatal pulmonary hypoplasia.[18] Fetal experiments designed to explore the pathophysiology of correctable congenital anomalies may soon assume some practical clinical significance. Fetal malformations are now being detected *in utero,* raising the question, So what? What can be done about it anyway?

The Fetus Becomes a Patient

The fetus could not be seriously considered as a patient until the centuries-old accumulated mysteriousness of the biblical seed and the classical homunculus were demystified, until the origin and development of the fetus from embryo to neonate could be explained scientifically. It is only with the development of molecular biology in the last century that we have been able to bridge the conceptional gap between the seed and the fully developed and marvelously complex human infant, between subcellular events like DNA-directed enzyme synthesis and subsequent complex biologic functions like the digestion of food. Modern molecular biology provided the conceptional framework for linking the seed to the infant — the seed contains a microscopic blueprint for the future individual coded in DNA. We can now be confident that there is an explicable order in the differentiation of the seed into the individual, even if we do not know exactly how the microscopic blueprint is acted out in the gorgeously orchestrated ballet of fetal development.

Molecular biology provided a conceptional framework that demystified fetal development. But the fetus could not be taken seriously as a patient until his ailments could be diagnosed. The ability to diagnose fetal disorders evolved slowly. Fetal activity felt by mother or palpated by her physician was the first crude measure of fetal well-being. Then the fetal heartbeat, detected at first by auscultation and later by sophisticated electrical monitors, was found to reflect fetal stress and distress. Then microscopic amounts of gestational hormone were detected in maternal blood and urine and these levels correlated with the condition of the fetus. But the development that had the most profound effect on our approach to the fetus was the introduction of a safe, noninvasive, imaging technique that permitted direct visualization of the living fetus. X rays were recognized as potentially harmful to the developing organism: plain X ray yielded little information, and introduction of radiopaque materials into the amniotic fluid (amniogram) increased the risk without yielding much more diagnostic information. Then came the ultrasonogram. Sonography can accurately delineate normal and abnormal fetal anatomy with astounding detail. It can produce not only static images of the intact fetus, but real-time "live" moving pictures. And, unlike all previous techniques, ultrasonic imaging appears to have no harmful effect on mother or fetus.

The sonographic voyeur, spying on the unwary fetus, finds him or her a surprisingly active little creature, and not at all the passive parasite we had imagined. He sees the fetus kicking and rolling, breathing in peculiar cyclical bursts, swallowing enormous quantities of amniotic fluid, and emptying his or her bladder every few hours. The sonographer can even see the beautiful rhythmic motion of the heart and its valves. Fetal parts can be measured, biparietal diameters recorded to assess fetal growth, and certain anatomic malformations accurately delineated. Sonography can be used to guide needle puncture of the amniotic cavity for amniocentesis, and even needle aspiration of fetal urine, ascites, and cerebrospinal fluid. Analysis of the constituents of amniotic fluid, obtained by amniocentesis, has made possible the prenatal diagnosis of many inherited metabolic and chromosomal disorders, and permitted assessment of fetal pulmonary maturity and the severity of fetal hemolytic reactions. Recently, the development of fetoscopes the size of a large hypodermic needle has allowed endoscopic visualization of the living fetus.

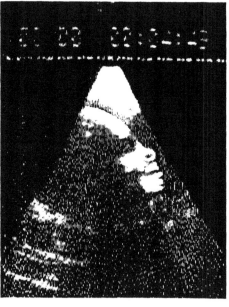

An image of the face of a fetus appears on an ultrasonogram. News and Publications Service, Stanford University, Stanford, California.

Although vision through the fetoscope is limited by the optics, fetal parts can be observed and fetal blood sampled — making possible the diagnosis of fetal hemotologic disorders.

Most of the techniques for prenatal diagnosis evolved to detect serious inherited diseases and malformations that were incompatible with normal postnatal life. The diagnosis of such diseases would usually lead to abortion. An indirect, but more positive, benefit of prenatal testing for potentially fatal lesions is reassurance when the test is negative. Nevertheless, through this early concentration on detecting uncorrectable fetal disorders, prenatal diagnosis has acquired an unsavory connotation: that the only way to prevent birth defects is to eliminate the defective fetus. The possibility of preventing certain birth defects by correcting the defect, rather than eliminating the fetus, has not been seriously entertained. Consequently, the diagnosis of potentially correctable disorders has attracted less attention than the diagnosis of hopeless lesions.

We are only now beginning to take fetal treatment seriously. But the idea is not new. Hydrops fetalis associated with maternal Rh sensitization was the first fetal disorder to be successfully treated. In the early 1960s, treatment of the neonate with severe hydrops fetalis was so discouraging that Liley attempted to transfuse the fetus *in utero*. He demonstrated that intraabdominal infusion of blood ameliorated severe hydrops, thus inaugurating fetal intervention.[19] There is a little known side to this Cinderella story, which marked a rather inauspicious start for more invasive fetal treatment. A logical refinement in the treatment of the erythroblastotic fetus is complete exchange transfusion. This procedure required direct access to the fetal circulation, and thus prompted the first *in utero* fetal operations. In the early 1960s, obstetricians in New York and Puerto Rico exposed several fetuses through

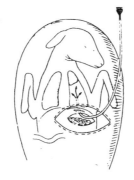

Correction of fetal anatomic defects in utero is being pursued.
From Harrison, M.R., Ross, N.A., and de Lorimier, A.A.: Correction of congenital diaphragmatic hernia in utero. III Development of a successful surgical technique using abdominoplasty to avoid compromise of umbilical blood flow. J. Pediatr. Surg. 16:934-42, 1981. Reproduced with permission of Grune and Stratton, Inc.

uterine incisions in order to cannulate femoral and jugular vessels for exchange transfusion. The overall experience was apparently discouraging: reports are sketchy, and this approach was quickly abandoned. Although surgical exposure of the living fetus would have to await development of better anesthetic agents and surgical techniques, this initial experience at least raised the possibility of fetal surgery.[20]

The next fetal disease to be approached therapeutically was the devastating respiratory distress syndrome of prematurity. Through a combination of clinical experience with severely premature infants and laboratory experiments using fetal lamb and rabbit preparations, surfactant deficiency was established as the physiologic basis for the respiratory distress syndrome. Effective treatment could then be devised. Steroids administered to the mother increase fetal surfactant production, hasten fetal lung maturation, and ameliorate the respiratory distress syndrome. Steroid therapy to induce fetal lung maturation is now common obstetrical practice and, combined with improved methods of respiratory support for tiny premature infants, has greatly reduced the mortality of this dreaded condition.

It is now apparent that there are other fetal disorders that can not only be diagnosed *in utero*, but also treated by prenatal administration of appropriate medications or hormones. The severe developmental consequences of congenital hypothyroidism, for example, might be prevented by giving thyroid hormones to the fetus. The hormone could simply be deposited in the amniotic fluid. The fetus would obediently swallow it along with his usual sizable daily ration of amniotic fluid. The fetus can actually be treated P.O.! And transamniotic fetal treatment has many potential applications. For instance, it is possible that the nutritionally deprived, growth-retarded fetus could be fed by infusing nutrients into the amniotic fluid.

In the last two decades, it has become clear that certain fetal deficiencies can be treated by providing blood cells, hormones, or medications. We are now ready to take the next step, the correction of fetal anatomic defects. Although many fetal anatomic malformations are detectable by sonography, only a few warrant consideration, for only a few have a compelling physiologic rationale for prenatal correction. Congenital hydronephrosis, diaphragmatic hernia, and obstructive hydrocephalus are examples of malformations in which a simple anatomic lesion interferes with fetal organ development and, if the anatomic defect is corrected, fetal development may proceed normally. The physiologic rationale for *in utero* correction of these lesions has been discussed, and the feasibility of *in utero* correction is being pursued.[21, 22]

The Unborn Future

The fetus has come a long way — from biblical "seed" and mystical "homunculus" to an individual with medical problems that can be diagnosed and treated, that is, a patient. Although he cannot make an appointment and seldom even complains, this patient will at times need a physician.

Until recently, the fetus did not need a healer, only an advocate or protector. He has always had his advocates — but the grounds for advocacy were religious, emotional, and philosophical, rather than diagnostic and therapeutic. Even with the advent of prenatal diagnosis of severe uncorrectable lesions, what was needed was a humane executioner, rather than a physician. Now the possibility of treating certain fetal disorders before birth gives an entirely new meaning to prenatal diagnosis. Now correctable disorders can be recognized, and their recognition will directly affect clinical management of the pregnancy. The possibility of fetal therapy raises new questions about the pathophysiology of fetal organ development and the technical feasibility of intervention before birth. It also raises complex ethical questions about risks and benefits, about the rights of mother and fetus as patients. We are only beginning to address these difficult issues.[21]

In 1980, the fetus with a potentially correctable malformation is like the "blue baby" with congenital heart disease in 1940. Although the pathophysiology of the congenital cardiac defects was pretty well understood, precise clinical diagnosis was seldom pursued, because nothing could be done for the patient anyway. Then came the "blue baby" shunt operations and the subsequent explosive development of open-heart surgery. Similarly, the pathophysiology of many fetal defects has been sorted out over the last few decades. Still, the application of powerful new techniques, like sonography, to the diagnosis of potentially correctable fetal malformations has been pursued with alacrity, because there is little point in finding a fetal defect that cannot be repaired and will not alter prenatal management. The possibility of fetal treatment will radically alter this attitude. If fetal treatment becomes feasible, the 1980s may wit-

ness an explosion of interest in fetal diagnosis, fetal pathophysiology, and techniques of fetal intervention.

Treatment of the unborn has had a long and painstaking gestation; the date of confinement is still questionable and viability uncertain. But there is promise that the fetus may become a "born again" patient.

References

1. Bagnold E.: The Door of Life. Quoted in Familiar Medical Quotations, edited by Maurice B. Strauss, M.D., Boston, Little, Brown and Company, 1968, p. 179.
2. Marcus Aurelius: Meditations, X: 26, translated by C. R. Haines. Quoted in Familiar Medical Quotations, edited by Maurice B. Strauss, M.D., Boston, Little, Brown and Company, 1968, p. 64.
3. Sterne, L.: The Life and Opinions of Tristram Shandy, Gentleman. In Selected Works, edited by Douglas Grant, London, Rupert-Hart Davis, 1950, p. 30.
4. Cadogan, W.: An Essay upon Nursing and the Management of Children, From Their Birth to Three Years of Age. Quoted in Source Book of Medical History, compiled with notes by Logan Clendening, M.D., New York, Dover Publications, Inc., 1960, p. 272.
5. Darwin, C.: The Descent of Man. Quoted in Familiar Medical Quotations, edited by Maurice B. Strauss, M.D., Boston, Little Brown and Company, 1968, p. 64.
6. Lambert, S. W.: A reading from Andreae Vesaliis De humani corporis fabrica, Liber VII De vivorum sectione nonulla caput XIX: Proc. Charaka Club 8:3-41, 1935, pp. 11-12.
7. Rosenkrantz, J. G., Simon, R. C., and Carlisle, J. H.: Fetal surgery in the pig with a review of other mammalian fetal techniques. J. Pediatr. Surg. 3.392-97, 1965.
8. Jost, A.: Sur la différenciation sexuelle de l'embryon de lapin. I. Remarques au sujet de certaines operations chirurgicales sur l'embryon. II. Experiences de parabiose. Société de Biologie: Compte Rendus. 140:461-64, 1946.
9. Louw, J. H., and Barnard, C. N.: Congenital intestinal atresia: Observations on its origin. Lancet 2:1065-67, 1955.
10. Jackson, B. T., Piasecki, G. J., and Egdahl, R. H.: Experimental production of coarctation of the aorta in

utero with prolonged postnatal survival. Surg. Forum 14:290-96, 1963.
11. Blanc, W. A., and Silver, L. A.: Intrauterine abdominal surgery in the rabbit fetus: Production of congenital intestinal atresia (Abstr.). Am. J. Dis. Child. 104:548, 1962.
12. de Lorimier, A. A., Tierney, D. F., and Parker, H. R.: Hypoplastic lungs in fetal lambs with surgically produced congenital diaphragmatic hernia. Surgery 62:12-17, 1967.
13. Thomasson, B. H., Esterly, J. R., and Ravitch, M. M.: Morphologic changes in the fetal rabbit kidney after intrauterine ureteral ligation. Invest. Urol. 8:261-72, 1970.
14. Beck, A. D.: The effect of intrauterine urinary obstruction upon the development of the fetal kidney. J. Urol. 105:784-89, 1971.
15. Heymann, M. A., and Rudolph, A. M.: Effects of congenital heart disease on fetal and neonatal circulations. Prog. Cardiovasc. Dis. 15:115-43, 1972.
16. Assali, N. S., (ed.): Biology of Gestation. New York, Academic Press, 1968.
17. Suzuki, K., and Plentl, A. A.: Chronic implantation of instruments in the neck of the primate fetus for physiologic studies and production of hydramnios. Am. J. Obstet. Gynecol. 103:272-81, 1969.
18. Harrison, M. R., Bressack, M. A., Churg, A. M., and de Lorimier, A. A.: Correction of congenital diaphragmatic hernia in utero. II. Simulated correction permits fetal lung growth with survival at birth. Surgery 88:260-68, 1980.
19. Liley, A. W.: Intrauterine transfusion of foetus in haemolytic disease. Br. Med. J. 2:1107-9, 1963.
20. Adamsons, K., Jr.: Fetal surgery. N. Engl. J. Med. 275:204-6, 1966.
21. Harrison, M. R., Filly, R. A., Parer, J. T., Faer, M. J., Jacobson, J. B., and de Lorimier, A. A.: Management of the fetus with a urinary tract malformation. J.A.M.A. 246:635-39, 1981.
22. Harrison, M. R., Golbus, M. S., and Filly, R. A.: Management of the fetus with a correctable congenital defect. J.A.M.A. 246:774-77, 1981.

Send reprint requests to the author at the following address:

Department of Surgery
Room 585 HSE
University of California
San Francisco, California 94143

158

Perspectives on fetal surgery

SHERMAN ELIAS, M.D.

GEORGE J. ANNAS, J.D., M.P.H.

Chicago, Illinois, and Boston, Massachusetts

Recent developments in methods for the antenatal detection and in utero surgical repair of fetal defects have received much attention by both the medical profession and the lay public. We present our perspectives on the technical, legal, and ethical issues surrounding this conceptually dramatic form of biomedical intervention. (AM. J. OBSTET. GYNECOL. 145:807, 1983.)

EXPERIMENTATION with fetal surgery has come of age, and its routine clinical application seems inevitable. Society has a critical stake both in the successful treatment of fetal disorders and in the maintenance of respect for the human dignity of the fetus. Fetal surgery raises not only novel scientific issues but also far-reaching ethical and legal ones, which challenge many of our traditional concepts of the fetus. As the primary health care providers to women and their fetuses, it is time for obstetricians to reflect on the impact of fetal surgery on their own practices and society's responses to them. In this article, we review the major legal and ethical issues which this technology raises.

From the Section of Human Genetics, Department of Obstetrics and Gynecology, Northwestern University Medical School, and Boston University Schools of Medicine and Public Health.

Supported in part by grants from the W. K. Kellogg Foundation Fellowship Program and a Basil O'Connor Research Starter Grant from the March of Dimes–Birth Defects Foundation.

Received for publication August 19, 1982.

Revised October 27, 1982.

Accepted November 2, 1982.

Reprint requests: Sherman Elias, M.D., Prentice Women's Hospital and Maternity Center, Suite 1102, Chicago, Illinois 60611.

Medical aspects of fetal surgery

Ascertainment of abnormal fetuses. Only a small minority of potential candidates for antenatal surgery are currently identified. Sometimes a family history of a heritable condition will lead the obstetrician to monitor a fetus to determine if it is similarly affected. Some cases will be ascertained because of clinical suspicion (for example, oligohydramnios which may be associated with urinary tract obstruction). And most importantly, with the current trend toward liberal use of antenatal ultrasonographic monitoring for a variety of obstetric indications (including perhaps "routine" ultrasonographic studies), the majority of cases in which fetuses are ascertained to be candidates for antenatal surgery are likely to be serendipitously diagnosed.

Prerequisites for fetal surgery. Prior to consideration of a fetus as a candidate for in utero surgery, several prerequisites must be met. The extent of the fetal malformation must be delineated and a careful evaluation must be made for possible coexisting abnormalities. In addition to ultrasonographic studies, ancillary procedures may include genetic amniocentesis for cytogenetic and biochemical testing, fetoscopy, and amniography. Second, the abnormality should be compatible with a reasonable expectation for a healthy infant as a result of the procedure. For example, surgery to repair a urinary tract obstruction would, in general, be contraindicated in a fetus diagnosed as having

April 1, 1983
Am. J. Obstet. Gynecol.

trisomy 18. Third, surgical intervention should be performed only in cases in which the fetus would be better off by performance of surgery before delivery, rather than after birth. In certain situations, the gestational age of the fetus will dictate whether or not preterm delivery followed by neonatal surgery offers an overall safer approach as compared with in utero surgery. The primary rationale is that the fetal condition will progressively deteriorate to a point of irrevocable injury unless surgical intervention is undertaken.

State of the art.

Erythroblastosis fetalis. Surgical treatment of fetal disease is not in itself new. In 1963, Liley[1] reported the first successful intrauterine transfusion for severe erythroblastosis fetalis. Subsequently, this procedure has become an integral part of modern obstetric therapy.[2]

Hydrocephalus. Current methods of ultrasonography permit the antenatal detection of at least some cases of fetal hydrocephalus during the second trimester; however, sensitivity and specificity has not been established. Normally, the outer wall of the lateral ventricle is about two thirds of the distance from the midline to the outer table of the skull at 15 weeks' gestation; at 17 weeks it is less than one half the distance, and at 20 weeks it is approximately one third the distance.[3] In fetal hydrocephalus, the lateral ventricles become abnormally dilated prior to any changes in the outer skull outline (biparietal diameter).

One approach in the treatment of fetal hydrocephalus is a ventriculoamniotic shunt placed under ultrasonographic guidance. Clewell and associates[4] reported the placement of such a silicone rubber shunt with a one-way valve in a 23-week fetus with hydrocephalus caused by X-linked aqueductal stenosis. The function of the shunt was confirmed by an increased cortical mantle thickness, a decreased ventricular-to-hemisphere ratio, and a normal biparietal diameter at 32 weeks' gestation. In addition, this same group has performed two other antenatal ventriculoamniotic shunt procedures. Preliminary follow-up evaluations have suggested that all three infants benefitted from the method of treatment. Other investigators have performed similar operations. Thus far, there have been too few fetuses treated with shunt placement to assess the value of ventricular decompression with respect to ultimate prognosis.

Urinary tract obstruction. Fetal urinary tract obstruction may be ultrasonographically characterized by oligohydramnios, a massively enlarged bladder, and hydronephrotic kidneys.[5] Unrelieved complete urinary tract obstruction as a result of posterior urethral valves or strictures leads to: (1) hydronephrosis, (2) cystic dysplasia of the kidneys, (3) oligohydramnios with sec-

ondary facial deformities (hypertelorism, malformed low-set ears, depressed nasal tip, micrognathia), flexion contractures in the extremities, and pulmonary hypoplasia, and (4) abdominal wall muscle deficiency or "prune-belly." In theory, drainage of urine from the bladder into the amniotic cavity prior to the time of irreversible damage may allow normal kidney development and restore normal amniotic fluid dynamics and thus prevent oligohydramnios and its severe sequelae.

Harrison and co-workers[6] reported a case of an 18-year-old primigravid patient who underwent ultrasonographic evaluation at 20 weeks' gestation because of an inappropriately small uterus. There was oligohydramnios, and it was determined that the male fetus had bilateral hydronephrosis, dilated redundant ureters, and a large thick-walled bladder with a dilated bladderneck. The renal parenchyma did not appear to be cystic. Because of the fetal position and the oligohydramnios, placement of a catheter shunt from the bladder to the amniotic cavity was deemed impossible. Therefore, it was believed that bilateral ureterostomies offered the only hope to save the fetus. At 21 weeks' gestation, the lower part of the fetal body was lifted through a hysterotomy incision; the dilated ureters were exposed through bilateral flank incisions, opened in the midportion, and marsupialized to the skin. The fetus was then replaced and the incision closed. At 35 weeks' gestation a 2,300 gm infant was delivered by cesarean section. The infant had mild facial deformities, limb contractures, a small chest, a slightly protuberant abdomen, and bilateral undescended testes. After 9 hours at maximum supportive measures, the infant was permitted to die at 9 hours of age. At autopsy, the lungs were found to be hypoplastic and the kidneys showed Potter type IV cystic dysplasia.

In a second case, Golbus and colleagues[7] reported a 41-year-old woman who, by ultrasonography prior to genetic amniocentesis at 17 weeks' gestation, was found to have twins, one of which was a male with marked ascites. At 23 weeks' gestation, this twin showed a marked increase in ascites associated with a slight dilation of the left renal pelvis and ureter. At 30 weeks' gestation, the abnormal twin was found to have a significant decrease in ascites, and grossly mild oligohydramnios. An attempt to percutaneously place an indwelling catheter into the fetal bladder under ultrasonographic guidance was unsuccessful. At 32 weeks' gestation, a polyethylene catheter was placed in the fetal bladder under ultrasonographic guidance so that the curled end was in the bladder and the other end drained into the amniotic cavity. At 34 weeks' gestation, the infants were spontaneously delivered vaginally. The affected infant had the features of the "prune-belly" syndrome with bulging flanks and undescended testes. At 1 day of age, the infant underwent bilateral high-loop cutaneous ureterostomies, correction of an associated intestinal malrotation, and excision of a portion of the redundant abdominal wall. Renal biopsies showed mild dysplasia.

From the very limited experience with surgery for fetal urinary tract obstruction, as well as the natural history of such anomalies, it appears chances for fetal salvage are optimized by decompression of the bladder and restoration of amniotic fluid dynamics as early in gestation as possible. However, the success rate measured by survival and ultimate prognosis remains to be established.

Future directions. Several additional approaches to the in utero surgical repair of fetal defects have been proposed.

DIAPHRAGMATIC HERNIA. Many infants with congenital diaphragmatic hernia cannot survive because of pulmonary hypoplasia caused by compression of the herniated viscera into the thoracic cavity. With the use of a fetal lamb model in which diaphragmatic hernias were created at about 100 days' gestation, Harrison and associates[8] have developed a successful in utero surgical technique which involves reduction of the viscera from the thoracic cavity into the peritoneal cavity, repair of the diaphragmatic defect, and enlargement of the abdominal cavity by abdominoplasty by means of an oval silicone rubber patch sutured to the fascial edges followed by closure of the skin over this patch. They have concluded that correction of congenital diaphragmatic hernia in utero appears physiologically sound and technically feasible.

SPINA BIFIDA. With the use of fetal rhesus monkeys induced to develop neural tube defects by administration of synthetic corticosteroids and thalidomide between days 18 and 28 or embryogenesis, Hodgen[9] has developed a technique in which an agar-based medium containing crushed bone particles is used as a "bone paste" for sculpturing antenatal enclosures to overlay and "correct" herniated nerve bundles. While such patching techniques may seal the spina bifida lesion, the effects on neurological development and function is yet to be established.

GASTROSCHISIS. The morbidity and mortality rate in neonates with gastroschisis has been significantly reduced since the mid-1960s by improved surgical techniques; however, serious complications are not infrequent and include respiratory distress, matted viscera, and peel formation. It is conceivable that early in utero repair of gastroschisis may ultimately prove the method of therapeutic choice. Fetal gastroschisis mod-

els have been created in rabbits and lambs which would allow such investigation to be undertaken.

ALLOGENIC BONE TRANSPLANTS. Michejda and coworkers[10] have successfully performed intrauterine allogenic bone transplantations in the rhesus monkey at 120 to 135 days of gestation with the use of either fetus-to-fetus bone transplants or particles of crushed bone mixed with an agar-enriched culture medium. They concluded that the immune surveillance system of fetal rhesus monkeys may be tolerant of such bone allografts, even when performed as late as the second trimester. Moreover, such transplants used in ablative long bone surgery permit normal growth and development as compared with the contralateral unoperated extremity. Of particular interest was the fact that the "bone paste" had strong adhesive properties and could be sculptured into the desired conformation without forfeiting ultimate long bone strength. Accordingly, these investigators suggest that this technique may offer potential in the human fetus for surgical repair of skeletal anomalies in utero.

Legal and ethical issues
Human experimentation.
Federal regulations. Fetal research is one of the most controversial and complex areas in the entire field of experimentation regulation. The National Commission for the Protection of Human Subjects of Biomedical and Behavioral Research, for example, spent the first year of its existence working on it under a Congressional mandate to make recommendations regarding fetal research before working on any other topic. This mandate itself was most influenced by the 1973 decision of the United States Supreme Court (Roe versus Wade, 410 U.S. 113), which provided that the government could not interfere with the decision of a woman and her physician regarding abortion prior to fetal viability. This decision increased the number of fetuses aborted, and thus the amount of fetal material available for research.

As adopted by the United States Department of Health and Human Services, federal regulations currently require that appropriate animal studies be done and researchers have no role in any decision to terminate a pregnancy. The purpose of any in utero experimentation must be ". . . to meet the health needs of the particular fetus and the fetus placed at risk only to the minimum extent necessary to meet such needs" (or, in the case of nontherapeutic research, the risk to the fetus must be "minimal" and the knowledge to be gained "important" and not obtainable by other means). The consent of both the mother and father is required, unless the father's identity is not known, he is not rea-

April 1, 1983
Am. J. Obstet. Gynecol.

sonably available, or the pregnancy resulted from rape.

In addition, the research protocol must be reviewed by an Institutional Review Board (IRB) which, in addition to its normal duties, must take special care to review the subject selection process and the method by which informed consent is obtained. The consent process itself should probably be audited by a representative of the IRB as well.

These federal regulations technically apply only to those who receive federal funds for research or who are affiliated with institutions that have signed an agreement with Health and Human Services that all research in their institutions will be reviewed by an IRB under these federal regulations. But they are so fundamental to the protection of the integrity of the fetus, the potential parents, and the research enterprise itself that they should be followed voluntarily in all institutions doing fetal research. In cases in which surgeons are only doing one procedure and it is contended that treatment is the primary goal, the IRB might consider expediting the review, but routine use of an advocate for the fetus still seems appropriate (since, by definition, it is not known if the risks to the fetus will be outweighed by potential benefit).

State statutes. The states that have legislated in this area have used a much more rigid and punitive approach; fetal research in some states is a criminal activity. At least 17 states have statutes on fetal research; 15 were passed soon after the *Roe versus Wade* decision and in direct response to it. In fact, more state legislation has been enacted regarding fetal research than any other type of research, and this, together with its poor quality, indicates the emotional nature of this issue. Eight states prohibit or restrict both in utero and ex utero research, and the restrictions are generally more stringent than the federal regulations (Illinois, Louisiana, Maine, Massachusetts, Minnesota, Missouri, South Dakota, and North Dakota). In Massachusetts, for example, it is a crime to study the fetus in utero unless the study does not "substantially jeopardize" the life or health of the fetus and the fetus is not the subject of a planned abortion. Thus, therapeutic research, such as that on hydrocephalus, is permissible even in this restrictive state. Utah, the only state to deal exclusively with in utero fetuses, prohibits all research on "live unborn children." California, where much of the research to date has been done, restricts experimentation only on ex utero fetuses, outlawing ". . . any type of scientific or laboratory research or any other kind of experimentation or study, except to protect or preserve the life and health of the fetus."

State regulation is a hodgepodge of restrictions and prohibitions with little consistency over jurisdictions and no clear rationale. Nevertheless, one is bound by the law of the state in which one performs fetal experimentation, and knowledge of its provisions is obviously necessary in states which have such statutes.[11-13]

Therapeutic interventions. The line between therapy and experimentation has never been a completely clear one. Therapy involves procedures done primarily for the benefit of the patient that are considered "good and accepted medical practice," whereas experimentation involves new or innovative procedures (not yet considered standard practice) for the primary purpose of testing a hypothesis to gain new knowledge.[11, 14] Courts have not always been consistent in applying these criterion and even rejected, for example, the notion that the first artificial heart implant was done for other than therapeutic purposes.[11] Nevertheless, it seems fair and accurate to conclude that all of the procedures described in this article must currently be considered experimental and subject to the rules already summarized. Specifically, IRB review, detailed consent, an advocate for the fetus, and a consent auditor are all morally and legally appropriate. Someday, however, it is likely that some types of fetal surgery will become accepted medical practice and thus the basic legal and ethical rules relating to therapy will apply. The most important issues will then be informed consent and resource allocation.

Consent. It is a fundamental premise of Anglo-American law that no one can touch or treat a competent adult without the adult's informed consent. This doctrine is based primarily on the notion of autonomy or self-determination and second on the notion of rational decision making. The first requires that individuals have the ultimate say concerning whether or not their bodies will be "invaded"; the latter requires disclosure of certain material information (a description of the proposed procedure, risks of death and serious disability, alternatives, success rates, problems of recuperation) before one is asked to consent to an "invasion."[11, 14, 15]

All of this is relatively straightforward when dealing with an adult, but what are we to do when the therapy is aimed at the fetus? In the experimental setting, federal regulations call for the consent of both the mother and father prior to any permissible experimentation. In the therapeutic setting, the consent of either one of the parents is generally sufficient consent for beneficial procedures on children. In the case of the fetus, however, if the proposed procedure will place the mother at any risk of death or serious disability at all, only she would have the right to consent (and the corresponding right to withhold consent). Even in the third trimester, *Roe versus Wade* gives the women and her physician the right to abort if her health is endangered. In an analogous case, *Danforth versus Planned Parenthood,* 428

U.S. 52 (1976), the United States Supreme Court ruled that where conflict existed over the issue of an abortion between a potential father and mother, the mother's position should prevail since she has more at stake (that is, her own body and health) than the father. The same logic applies here. Consent of the mother must be a necessary precondition for such surgery. Of course her consent must be informed, and she should be told as clearly as possible about the proposed procedure and its risks to herself and her fetus, as well as the alternatives, success rates, and the problems of recuperation.

Resource allocation. Fletcher[16] has raised this as one of the major ethical issues concerning fetal therapy. Currently, the issue concerns how much funding the federal government and others should allocate to research in this area. This is fundamentally a political question, but it has ethical overtones. Is it acceptable, for example, to continue to place our most heavy emphasis on the extension of life for the elderly rather than on providing fetuses and neonates with the best chance to live a healthy life? In an area in which early treatment can lead to prevention of disease and misery, and perhaps can reduce significantly the cost of a lifetime of care, both research and treatment warrant a high priority. When fetal surgery becomes accepted medical practice, issues of screening, selection, and indications will have to be addressed by insurance payors. A liberal policy of reimbursement seems sensible.

Status of the fetus. Developments in fetal therapy will enhance the status of the fetus and raise new ethical and legal issues.[17] These innovations raise complex ethical questions about the rights of the mother and fetus as patients. If both are really patients, do both need their own physicians? Is the obstetrician to view the fetus or the mother as his patient? Usually there is no need to make a distinction, but what if the physician believes a procedure (such as fetal surgery, if and when it becomes "accepted medical practice") is indicated and the mother refuses to consent? At least two practices are possible: (1) Try to persuade the mother, but if she will not change her mind do what she wants; (2) take legal action to try to compel the woman to submit to the procedure. No court has ever compelled anyone to undergo an experimental treatment, but when fetal surgery becomes accepted medical practice, and if the procedure can be done with minimal invasiveness and risk to the mother and significant benefit to the fetus, there is an argument to be made that the woman should not be permitted to reject it. Such rejection of therapy could be considered "fetal abuse" and, at a late stage in pregnancy, "child abuse," and an appropriate court order sought to force treatment.[18, 19]

Such forcible medical treatment strikes us as brutish and horrible and may even endanger the mother's life if she actively resists the treatment. However, if we are to view the fetus as a patient, someone must be seen as the fetus' advocate in cases in which the pregnant woman refuses standard treatment for her fetus and every attempt has been made to see to it that the fetus gets good care in utero. There are at least two major problems with compulsory fetal therapy: (1) It forces a procedure on a woman for the benefit of another (something we do not now require parents to do even for their live-born children); (2) it requires weighing the indignity and deprival of autonomy involved in invading a woman's body against the potential benefit to life and health of the fetus. There may seem little biological difference between a fetus about to be born and a neonate, but we believe that the fact that the fetus cannot be treated without significantly invading the woman's body makes her consent necessary for lifesaving treatment for her fetus but not her child. We might, however, feel comfortable as a society in requiring women to take certain oral medications or vitamins, or even injections, if the bodily invasion is very slight and the potential benefit to the fetus very great. In such cases the indignity and interference with autonomy is minimal and one can argue that there exists a "duty to rescue" the fetus by the mother under these circumstances.

Fetuses have rights now, but most of them come into being only after birth. Abortion and feticide laws do give some protection during the third trimester against arbitrary pregnancy termination. In most states a suit can be brought for injury while in utero, but only after a live birth. And should property be left to a fetus, its right to it vests only after a live birth.[20] There are also proposals (unlikely to succeed) to amend the United States Constitution to make the fetus a "person" and two courts have ordered women to submit to cesarean sections to save the lives of their fetuses on the basis that the right of the fetus to life outweighed the risk to the mother inherent in the procedure.[21]

The fetus may never be considered a "person" under law, and this is probably the wrong issue on which to concentrate. The functional question is: Whether we call the fetus a person, a patient, or a fetus, what rights will we accord it? When will we leave all decisions regarding its health to its mother, and when, if ever, will we as a society decide to restrict the autonomy of pregnant women for the sake of their fetuses? The more things we can do for the fetus in utero, the more relevant and real this question becomes. We believe that respect for the pregnant woman's dignity and autonomy and the tradition of medicine as a voluntary service require the voluntary and informed consent of the pregnant woman prior to the performance of fetal research or therapy. The right to make a mistake should continue to be the pregnant woman's, not the physician's or judge's. Accordingly, we believe in involvement of not only individual patients but also society as a whole in setting rules and priorities for fetal surgery.

Fetal surgery is a positive development in medi-

cine—one that enhances the status of the fetus and gives mother and child a better chance for a healthy life together. But it is not without vexing ethical and social problems, and it is time for us all to discuss them publicly.

REFERENCES

1. Liley, A. W.: Intrauterine transfusion of fetus in haemolytic disease, Br. Med. J. 2:1107, 1963.
2. Queenan, J. T.: Modern Management of the Rh Problem, ed. 2, Hagerstown, Maryland, 1977, Harper & Row, Publishers.
3. Campbell, S.: Early prenatal diagnosis of neural tube defects by ultrasound, Clin. Obstet. Gynecol. 20:351, 1977.
4. Clewell, W. H., Johnson, M. L., Meier, P. R., Newkirk, J. B., Mendee, R. W., Bowes, W. A., Zide, S. L., Heckt, E., Henry, G., and O'Keeffee, D.: Placement of ventriculoamniotic shunt for hydrocephalus in a fetus, N. Engl. J. Med. 305:944, 1981.
5. Hobbins, J. C., Grannum, P. A. T., Berkowitz, R. L., Silverman, R., and Mahoney, M. J.: Ultrasound in the diagnosis of congenital anomalies, AM. J. OBSTET. GYNECOL. 134:331, 1979.
6. Harrison, M. R., Golbus, M. S., Filly, R. A., Callen, P. W., Katz, M., deLorimier, A. A., Rosen, M., and Jonsen, A. R.: Fetal surgery for congenital hydronephrosis, N. Engl. J. Med. 306:591, 1982.
7. Golbus, M. S., Harrison, M. R., Filly, R. A., Callen, P. W., and Katz, M.: In utero treatment of urinary tract obstruction, AM. J. OBSTET. GYNECOL. 142:383, 1982.
8. Harrison, M. R., Ross, N. A., and deLorimier, A. A.: Correction of congenital diaphragmatic hernia in utero. III. Development of a successful surgical technique using abdominoplasty to avoid compromise of umbilical blood flow, J. Pediatr. Surg. 16:934, 1981.
9. Hodgen, G. D.: Antenatal diagnosis and treatment of fetal skeletal malformations with emphasis on in utero surgery for neural tube defects and limb bud regeneration. JAMA 246:1079, 1981.
10. Michejda, M., Bacher, J., Kuwabara, T., and Hodge, G.:

In utero allogenic bone transplantation in primates. Roentgenographic and histologic observations, Transplantation 32:96, 1981.
11. Annas, G. J., Glantz, L. H., and Katz, B. F.: Informed Consent to Human Experimentation: The Subject's Dilemma, Cambridge, Massachusetts, 1977, Ballinger Publishing Company.
12. Friedman, J. M.: The federal fetal experimentation regulations: An establishment clause analysis, Minn. Law Rev. 61:961, 1977.
13. Brock, E. A.: Fetal research: What price progress? Detroit Coll. Law Rev. 3:403, 1979.
14. Annas, G. J., Glantz, L. H., and Katz, B. F.: The Rights of Doctors, Nurses, and Allied Health Professionals, Cambridge, Massachusetts, 1981, Ballinger Publishing Company.
15. Annas, G. J.: Informed consent, Ann. Rev. Med. 29:9, 1978.
16. Fletcher, J. C.: The fetus as patient: Ethical issues, JAMA 246:772, 1981.
17. Martin, M. M.: Ethical standards for fetal experimentation, Fordham Law Rev. 43:547, 1975.
18. Shaw, M. W., and Damme, C.: Legal status of the fetus, in Milunsky, A., and Annas, G. J., editors: Genetics and the Law, New York, 1976, Plenum Press, pp. 3-19.
19. Shaw, M. W.: The potential plaintiff: Preconception and prenatal torts, in Milunsky, A., and Annas, G. J., editors: Genetics and the Law. II, New York, 1980, Plenum Press, pp. 225-233.
20. Wilson, J. P.: Fetal implications of an ethical conundrum, Denver Law J. 53:581, 1976.
21. Annas, G. J.: Forced cesareans: The unkindest cut of all, Hastings Cent. Rep. 12:16, 1982.

PUBLIC POLICY ON HUMAN REPRODUCTION
AND THE HISTORIAN

In December of 1959, President Dwight Eisenhower made a famous disclaimer. Asked at a press conference for his reaction to the recommendation by his Committee to Study the United States Military Assistance Program that the United States aid foreign governments in their efforts to control population growth, Eisenhower bluntly declared:

> I cannot imagine anything more emphatically a subject that is not a proper political or governmental activity or function or responsibility.... This government will not, as long as I am here, have a positive political doctrine in its program that has to do with the problem of birth control. That's not our business.[1]

The President's belief that public and private life were distinguishable and that government should not meddle in such areas as family life except in extraordinary circumstances was probably shared by most historians. The questions they asked and the subjects they pursued reflected at least tacit agreement with other leaders of opinion that there was a private realm of human activity that should remain closed to the scrutiny of politicians and historians alike.

During the 1960s both politicians and historians began to discard the myth of a private/public dichotomy. In 1963, with a Roman Catholic Democrat serving as President, Eisenhower urged readers of the *Saturday Evening Post* to follow his example in recognizing that population growth posed a threat to world peace and that birth control was a legitimate concern of government.[2] By the end of the 1960s Congress had declared family planning a "special emphasis" program in the War on Poverty and required that at least six percent of all federal maternal and child health funds be spent on family planning services.[3] While Congress discovered the problematic nature of human reproduction, historians found that the family had a history and began the systematic study of the formerly invisible majority – women, children, racial and ethnic minorities, the working poor – who had been neglected in favor of the minority of white males who controlled the government and the economy.

To juxtapose the interests of Congressmen and historians is in some respects deceiving, however, because their concerns often differed. Few federal policy makers were feminists. When Louis M. Hellman, M.D., Deputy Assistant Secretary for Population Affairs in the Department of Health, Education and Welfare, spoke at the dedication of the Harvard Medical School's laboratory of Human Reproduction and Reproductive Biology in May, 1972, he differentiated his interests, as a physician and federal bureaucrat, from those who were most concerned over the rights of women. "Although the opportunity for women to control their own fertility furnishes a high sounding cause," Hellman asserted, "it does not have sufficient impact to influence either the majority of the people or the Congress of the United States, who must vote for the use of tax monies for family planning services."[4] In contrast, much of the vitality of social history in the 1970s sprang from the self-consciousness of women. By 1980 feminist scholars had raised the consciousness of their profession by insisting that the "social relations of production" were so intimately interwoven with the "social relations of reproduction" that they could no longer be separated by the conscientious; the private was political; gender and class, no less than race, were problems for every historian. The doctrine of two spheres – echoed by President Eisenhower in 1959 – was not in 1980 an interesting Victorian fiction but an instrument of oppression, and one could no longer call male elites "the Americans."[5]

In the 1980s more American historians were interested in sexual politics than in the history of political parties. Historians had discovered that public policies on prostitution, smut, contraception, induced abortion, and sterilization were shaped by elites as part of larger struggles between the sexes, between classes, and between ethnic groups. This new scholarship demonstrated the extent to which government had always been involved in reproductive decisions despite the self-serving myth of male politicians and opinion leaders that the family was a private zone into which government ought not to intrude. Unfortunately, some of the historians were so intent on demonstrating the relevance of their work to contemporary audiences that they forgot the concerns of their historical subjects. As a result, some of the new scholarship cannot serve as a guide to the policy making process past or present, nor does it provide a working agenda for those who would influence public policy by constitutional means. Rather, it raises the prospect of a new irrelevance based on a different but still limiting parochialism among historians.

A conscientious reader of such recent work as Linda Gordon's *Woman's Body, Woman's Right: A Social History of Birth Control in America* (1976) or Sheila Rothman's *Woman's Proper Place: A History of Changing Ideals and Practices, 1870 to the Present* (1978) might draw the following "lessons" from the past:

1. A "single-issue" approach to social change, such as Margaret Sanger adopted as the leader of the birth control movement in the United States, is a mistake. Only a holistic strategy aimed at social transformation, rather than pragmatic "liberal" reform, can advance the interests of women.[6]

2. The "medicalization" of birth control, or placing the delivery of any social service in the hands of established professional groups, is not in the interest of women because the professionals are most interested in controlling women for their own purposes.[7]

3. Continuing inequality based on gender and class is the fundamental reality that must be kept in view; incremental changes in the status or condition of women should be downplayed because they detract attention from the main issue of intolerable inequality and exploitation.[8]

None of these "lessons" will bear careful scrutiny. And the lack of tolerance for ambiguity, compromise, coalition-building, or even careful description that they reflect would prevent one informed by these histories from playing a significant role in public affairs. Contrary to the impression left by these works, significant changes in public policy have been engineered by those who were willing to separate issues (contraception from abortion; day care from the equal rights amendment), lobby professional and social elites, and play down larger issues of social justice in the interest of coalition building for short term gains.

It is true that public policies reflected the interests and concerns of elites who wanted to promote social stability and who aggressively defended their own interests. It is *not* true that meaningful change in the interest of women did not occur or that the policy making process was closed to "outsiders" and the opinions of ordinary people. Finally, while the public and private were never separate realms, and both government and private interest groups consistently attempted to shape the decisions made by individuals, these efforts often had unintended and sometimes ironic consequences. A Rockefeller, a President of the United States or of the American Medical Association had immense power, but its use was mediated and compromised in complex ways, and the opportunity to act unilaterally came only when a policy aroused minimal controversy. A review of the history of public policy on issues related to human reproduction illustrates these points.

ii

Changes in the patterns of reproductive behavior among native-born white women inspired the first self-conscious attempts to influence fertility through legislation. Long before the term "Manifest Destiny" was coined in 1845, patriots used rapid population growth as proof of the superiority of American institutions. Benjamin Franklin provided Thomas Malthus with his generalization that under the most favorable conditions humans can double their numbers every 25 years in a 1751 pamphlet intended to show that the vigorous growth of the North American population would lead to a crisis in colonial relations with England.[9] After the first United States census in 1790. President Washington and Secretary of State Jefferson were disappointed that fewer than four million had been counted and feared that this figure would provide ammunition for European critics of the United States.[10] By the 1850s, when Samuel Morton and Josiah Nott, the leaders of the "American School" of ethnographers, provided a scientific rationale for racial caste systems and the extinction of non-Anglo-Saxons who stood in the way of American expansion, the alleged cultural superiority of the republic was increasingly attributed to the biological origins of dominant Caucasian groups.[11] Social facts were soon discovered that mocked this racial interpretation of American destiny.

We now understand the rapid decline of native-born fertility in the nineteenth century as part of the "demographic transition" that occurred during "modernization." As home and workplace were separated, children were less frequently economic assets; as more individuals believed in their ability and right to shape their lives in accordance with the incentives of a rapidly expanding market economy, then ever-larger numbers of men and women perceived an advantage in having fewer children. Parents began seeking ways to control family size, despite guilt for betrayal of traditional values and the wails of social leaders that they were shirking their patriotic duty, sinning against Nature, and committing "race suicide."[12]

While the average native-born white woman bore seven or eight children in the late eighteenth century, by the middle of the nineteenth century she was the mother of five, by the early twentieth century, the mother of three, and by the middle of the Great Depression, the mother of two. One of the remarkable aspects of the American demographic transition is that there were no sustained declines in infant mortality before the end of the nineteenth century. Several generations of American women had fewer children than their mothers despite a murderous infant mortality and vigorous attempts by social leaders to encourage higher fertility through both exhortation and legislation to suppress "Malthusian practices."

Many theories were advanced by contemporaries to account for the increasing prevalence of the small family. Some observers argued that changes in biological capacity were responsible; others blamed the growing hedonism of young adults who sought to maximize their pleasures by avoiding social duties. Francis A. Walker, the economist and director of the Ninth Census (1870), argued that the native-born were being "shocked" into infertility by exposure to and competition with the foreign-born, but he did not explain whether the nativist revulsion worked through biological or psychological changes. No voices were raised in favor of a stable or declining population. Population growth was viewed as an important index of social well-being. The only live issue concerned the circumstances in which individuals had the right to limit their fertility and the legitimate means.

Historians now agree with those Victorians who found the primary cause of the fertility decline to be a combination of practices − contraception, abortion, and abstention from coitus − rather than biological changes or shifts in the percentage of individuals who married or their age at marriage. As might be expected, physicians were prominent in the debate about the causes and consequences of this vital trend. They wrote many of the marriage manuals that provided counsel on the question of family limitation; they received requests for relief from women who were "irregular"; they could not avoid issues of morality in a culture that increasingly looked to science for answers. Doctors, rather than ministers, lawyers, or businessmen, took the lead in campaigns to criminalize abortion.

Prior to 1830, the law on abortion was generally permissive before quickening. This casual attitude reflected the fact that there were no means of determining whether a woman was pregnant or amenorrheic before the fetus began to move in the womb. Those seeking abortions were usually unmarried women, generally viewed as victims of male lust. Between 1840 and 1870 apparent changes in the social status of women seeking relief from pregnancy alarmed many physicians and led them into successful campaigns to outlaw induced abortion at any stage of pregnancy. As historian James Mohr has demonstrated, the medical leaders of the anti-abortion campaigns believed that many married Protestant women had begun to seek abortions, and they seem "to have been deeply afraid of being betrayed by their own women." They reacted with denunciations of feminists and successful lobbying campaigns in state legislatures. By 1880 induced abortion was illegal, many wise-women and irregular practitioners had been driven out of business, and physicians had gained new stature as moral arbiters. [13]

The culmination of the campaigns against abortion in state legislatures coincided with the passage of the Comstock Act (1873), or strengthened national obscenity law, in which no distinctions were made between smut, abortafacients, or contraceptives – all were prohibited. As a result, explicit discussion of contraception was omitted from post-1873 editions of books in which the subject had been given space. [14]

Anthony Comstock was a lobbyist for the New York Society for the Suppression of Vice. Popular accounts of his activities have trivialized his concerns by portraying him as an idiosyncratic fanatic whose success depended on Congressional desire to divert attention away from the Crédit Mobilier scandal. Comstock's concerns were shared, however, by the prominent New York businessmen who paid his salary, the eminent physicians who campaigned for criminalization of abortion, and political leaders at all levels of government. The declining birthrate, the broadly acknowledged dissatisfaction among women, the new visibility of urban vice, the hedonism of popular culture, and the streets teeming with the foreign-born – all seemed to threaten the hegemony of Protestant values and the stability of the middle-class family. [15]

Despite the criminalization of "vice," birth rates continued to decline among the socially ambitious groups of occupationally skilled and property-owning individuals who made up the ever-rising middle class. Both as individuals and as couples, husbands and wives had complex and compelling motives for restrictive behavior. An adequate analysis of them would require a separate monograph, but the declining birthrate is in itself strong evidence of the success of men and women in gaining some measure of control over aspects of their lives that had once been resigned to fate. [16] The birth rate also testifies to the limited capacity of the state to control reproductive decisions.

The conflict between public and individual interests, between eros and civilization, was mediated at great cost, however, for the generation that came of age in the last decade of the nineteenth century. The inablility of many young adults to cope with the demands of "civilized morality" provided the first American specialists in psychosomatic disease with many cases of neurosis. While the new ideal of companionate marriage based on romantic love, first popularized for a mass audience in nineteenth-century marriage manuals, might in retrospect seem to require recognition of erotic bonding and nonprocreative marital sex, most Victorians remained preoccupied with the need to sublimate eroticism to higher ends.[17]

In the long run the "costs" of demographic transitions for individuals would be lowered by the acceptance of an ethos of rational expressiveness. The advertising industry would teach Americans to relax and to enjoy the fruits of an abundant economy. The new values were as problematic as the old, however, and the attempt to articulate new standards of conduct, to work out a compromise between the needs of individuals and of their society, was influenced by the ideas and activities of competing reformers and would-be social engineers.

iii.

At the turn of the twentieth century respectable opinion supported the criminalization of contraception and abortion, large families, and rapid population growth. Those who questioned this consensus may be divided into two groups: civil libertarians, who were most concerned with the right of individuals to manage their sexuality according to personal preference; and quality controllers, those who wanted to redefine "the population problem" from a need for more people to a need for more people of the right kind.

The civil libertarians were a relatively small group that included some advanced feminists, the anarchist Emma Goldman and the medical journalist William Robinson. Margaret Sanger, the charismatic leader of the American birth control movement during its heroic phase (1915-1937), began as a protégé of Goldman. Sanger's chief competitor for leadership of the birth control movement, Mary Ware Dennett, was also a civil libertarian.[18]

The quality controllers were a much larger and socially distinguished group. They included the social housekeepers, academics, and patrician nativists who enlisted in the eugenics movement.[19] Although many eugenicists were rigid hereditarians whose zealous pursuit of Mendelian explanations for crime, prostitution, juvenile delinquency, and poverty alienated geneticists and hurt the effort to redefine the population problem, the most influential of the quality controllers — Raymond Pearl, Frederick Osborn, and Frank Notestein are examples — explicitly dissassociated themselves from naive hereditarianism and were as concerned over the effects of poor environment as poor heredity.[20] A number of prominent physicians — Robert Dickinson, Haven Emerson, Adolpus Knopf, Howard Taylor, Jr., and Alan Guttmacher are examples — were drawn into the effort to influence reproductive behavior by their discovery of the threats posed to stable family life by sexual incompatablility and such diseases as syphilis and tuberculosis.[21]

Quality controllers enjoyed some early legislative victories. In 1907 Indiana was the first of sixteen states that passed eugenic sterilization laws in the decade before World War I. By 1940 eugenic sterilization acts had been on the books in some 30 states, but the advocates of this practice were never able to mobilize a consensus among key medical groups. By 1940 over 36,000 individuals had been sterilizled under the authority of these laws, but this number was insignificant in terms of a scientific program of negative eugenics.[22]

Eugenicists enthusiastically lobbied Congress for the immigration restriction bills passed between 1921 and 1927. These acts mark the end of the open door to mass migration and were intended to discriminate against Eastern and Southern Europe. Recent scholarship has questioned the influence exerted by eugenicists on Congress, however, during a period of strong popular nativist sentiment.[23] By the time the door was shut, the nativists had already lost the war since a century of mass migration guaranteed that the United States would no longer be either a Protestant or a Nordic nation.

Even these hollow victories for negative eugenics were impossible to match on the positive side of the quality control agenda. The birth rate among the middle classes continued to decline. As it became apparent that the "fit" would not breed more, some quality controllers began considering means for "democratizing" birth control practice. A vigorous movement to legitimate contraception had already been organized, however, by civil libertarian feminists whose calls for birth strikes and autonomy for women were not congenial to quality controllers.[24]

By 1925 Margaret Sanger had solidified her position as leader of the birth control movement. Her success depended on a pragmatic resourcefullness that infuriated competitors and baffled opponents. Sanger first gained notoriety in 1912-14 as an organizer for the Industrial Workers of the World and radical journalist with a special interest in the plight of working-class mothers. After some frustrating experiences with male radicals, whose attitudes toward women seemed little different from those of other men, Sanger decided to concentrate on women's issues. Her background as a nurse working in the tenements of Manhattan provided vivid stories of poor women whose lives were destroyed by unwanted pregnancies and septic abortions. As a first step toward liberating women, Sanger set out to remove the stigma of obscenity from contraception through a strategy of flamboyant defiance of the law. Faced with an indictment for publishing obscenity (specifically a defense of political assassination), she fled to England in October of 1914. When she returned a year later, she barnstormed the country urging women to take matters into their own hands by establishing contraceptive advice centers and opened her own in October 1916.[25]

The "clinic" in the Brownsville section of Brooklyn was staffed by Sanger and her sister, also a nurse, who for a fee of ten cents showed how to use pessaries and apparently fitted some women with devices. The police closed the Brownsville center after ten days and 488 clients. The "birth control sisters" went to jail after highly publicized trials, but, on appeal of her case, Sanger got a judicial decision that she used as a mandate for a shift in strategy.

While upholding the constitutionality of the New York Comstock law, the appellate court judge ruled in 1918 that the law would not prevent licensed physicians from prescribing contraceptives when medically necessary. Thus, Sanger's claim that the law was unconstitutional because it compelled women to risk death through unwanted pregnancy was rejected. Characteristically, she turned this apparent defeat into a victory by recruiting physicians to direct the birth control clinics organized under the auspices of her American Birth Control League. Although there was a police raid in New York as late as 1929, the medical community rallied behind the clinic's medical director (Hannah Stone), and the charges against her were dismissed.

Historians who have criticized Sanger's decision to "medicalize" contraception have taken their cue from Mary Ware Dennett, the leader of a series of competing birth control organizations.[26] The differences between Sanger and Dennett began in 1916, when Dennett criticized Sanger's strategy of flamboyant law breaking and argued that a more effective and ethically defensible tactic would be to lobby for amendment of state and national Comstock laws.

In 1917 and 1918 Dennett's group lobbied without success in Albany and in 1919 began work in Washington. A federal bill was introduced in 1923 and 1924 but never got out of committee. Meanwhile, Sanger's American Birth Control League began pushing a "doctors only" bill that simply recognized the right of physicians to give contraceptive advice, in contrast to Dennett's bills, which called for a clean repeal of all prohibitions on contraception. Dennett could not understand how a "radical" like Sanger could campaign for "class and special-privilege legislation" establishing a "medical monopoly" on contraceptive information.

Sanger was willing to compromise on the kind of legislation that she supported despite the fact that she was mailing her own *Family Limitation* to anyone who asked for it. She needed support from at least some medical groups in order to recruit clinicians and to publish case studies from her clinics in medical journals. "Doctor's only" bills aroused less opposition from organized medicine at a time when the profession was gaining status and jealously defending its hard won prerogatives. The "doctors only" approach also mitigated the damaging claim that birth controllers were encouraging immorality and undermining national vitality.

There was another important reason for Sanger's opportunism. In contrast to Dennett, she was pessimistic about the prospects for legislative success. As a detailed diary of lobbying efforts in Washington revealed, most politicians regarded women lobbyists with contempt.[27] The ideal of reproductive autonomy for women was literally a joke among them. Sanger was aware of these attitudes but believed that lobbying campaigns were good publicity and provided a chance to educate the public.

She continued to break the law when necessary to operate a clinic, import diaphragms, or to embarrass the opposition. Breaking the law provided access to the courts, another public forum, but more important, a forum in which the political power of Roman Catholics would be minimized and one could appeal to a relatively well-educated arbiter. It was in the federal courts, in the *One Package* decision of 1936, that Sanger finally won a clarification of the federal obscenity laws and established the right of physicians to receive contraceptive supplies through the mail. Chagrined over Sanger's mounting popularity among women financial backers of the movement, Dennett left the fight for birth control in 1925. Since no bill to amend New York's birth control laws got out of committee until 1965, and contraception was not removed from the federal Comstock Act's prohibitions until 1971, in retrospect Sanger's strategy of combining opportunistic lobbying with selective litigation was the more effective.[28]

Sanger regarded her most important achievement to be the national chain of birth control clinics that she successfully promoted, beginning in New York in 1923. These clinics provided case histories that disproved irresponsible claims by some medical leaders that contraceptive practice did not work and caused disease. The clinics also served as teaching centers where the great majority of physicians who offered contraceptive services in their private practice received training that was not a part of regular medical education. Finally, the clinics made more reliable contraceptive practice possible for thousands of women, a behavioral change documented by the sexual histories gathered by Alred Kinsey and his colleagues.[29] Organized by middle and upper class women who gained new confidence and social skills, as well as access to better birth control, staffed by sympathetic women physicians, and eagerly patronized by working class women who thus "participated" in the movement and demonstrated their desire to gain greater control over their lives, these clinics represented a remarkable effort by women to act on behalf of their gender.

Ironically, this achievement has been subjected to searching criticism and all but dismissed by some historians. Sheila Rothman argued that the clinics represented yet another example of a feminist reform that was coopted by male professionals. Linda Gordon examined the marriage counseling courses that developed as an auxiliary service in some clinics and correctly pointed out that clients were advised that they should enjoy coitus; that "vaginal" orgasm was the mature form of female sexual experience; and that women had a duty to cultivate erotic bonds with their husbands as part of the effort to make marriage work.[30]

Judged by the expectations of the 1970s, both Rothman and Gordon's criticism have some plausibility. As Rothman points out, the mainly urban-based clinics reached only a minority of women. Gordon's complaint about the sexual counseling should be compared with the published studies of client records. They reveal a level of need that places the clinic in a different perspective.[31]

Many of the women who came to the clinics lacked even a rudimentary knowledge of their bodies, did not know if women could or should enjoy coitus, suffered from physical problems ranging from the "subclinical" to gross pathology. To be shown a sexual anatomy and instructed in the mechanics of heterosexual intercourse, to be able to discuss intimate fears and humiliations with sympathetic women who had authoritative and reassuring answers, to have a physical problem recognized, explained, and treated, and finally, to be assured that it was possible to avoid unwanted pregnancies – these were liberating experiences for women of both the working and the middle classes in the 1920s and 1930s. The clinics were a major and at least partially successful effort to promote the feminist goal of reproductive autonomy, whatever the limitations of the Freudian-inspired marriage counseling.

Some quality controllers criticized the birth control clinics because they were not reversing "dysgenic" birth rates. In their view too many able mothers and not enough incompetent women were influenced. The popular appeal of the birth control movement was based, of course, on the desire of a majority of married couples to manage their fertility in the interest of a higher standard of living. The birth control movement was not responsible for this restrictive behavior but simply made it easier than it might have been.

Few policy makers were interested in either feminist or quality control issues. For them "the population problem" remained a declining birthrate. Efforts by birth controllers to obtain public funds were sometimes frustrated by Roman Catholic opposition, but the greatest barrier was a widely-shared fear of the economic and political results of demographic change. For example, the Harvard economist Alvin H. Hansen, one of the most influential exponents of the ideas of Alred Keynes, devoted his 1938 presidential address before the American Economic Association to what became known as the doctrine of "secular stagnation." Hansen argued that the low birth rate was a principal cause of the inadequate demand and the low rate of capital investment associated with the Great Depression. In 1938 Harvard took advantage of the Swedish sociologist Gunnar Myrdal's visit to the United States by inviting him to give the prestigious Godkin Lectures. Myrdal had come under the auspices of the Carnegie Corporation of New York to study the race problem and published his classic study of the conflict between American ideals and its racial caste system in 1944. In 1938, however, he drew attention to *Population: A Problem for Democracy.* Myrdal argued that the Western liberal states would have to develop a state-supported system of family services if they wanted to avoid disastrous declines in population and to compete with Germany and other fascist states.[32]

Although the birth rate began the recovery that led to the 1950s "baby boom" in the late 1930s, scholarly opinion remained bearish on population until there was overwhelming evidence that Americans had regained the will to reproduce themselves. As late as 1950 sociologist David Riesman based his analysis of the country's "character structure" on a presumed "incipient decline" in its population.[33]

Within a decade, however, "the population problem" had been redefined from how to encourage growth, or the right kind of growth, to how to bring about zero population growth with a minimum of social disruption. This remarkable reversal of informed opinion helped set the stage for dramatic changes in public policy regarding human reproduction.

iv.

Around the turn of the twentieth century the United States emerged not only as one of the dominant economic and political powers in the world but also as a center of scientific research. The private foundation rather than the state provided the resources for such institutions as the Rockefeller Institute for Medical Research and a National Research Council with numerous grant making committees, including a Committee for Research in Problems of Sex (Sex Committee). The Sex Committee provides a direct link to the crisis in "civilized morality" since it was funded by the Bureau of Social Hygiene, an organization that quite literally sprang from John D. Rockefeller, Jr.'s concern over the ties between political corruption and commercial vice. In an age that increasingly looked to science for models of social efficiency, the younger Rockefeller believed that effective response to such disturbing phenomena as prostitution, venereal disease, drug abuse, and juvenile delinquency required scientific knowledge. This faith had some unexpected consequences as Sex Committee sponsored research led to the development of the hormone concept, the identification and clinical use of the sex hormones, and eventually to the birth control pill marketed in 1960. The Sex Committee also sponsored the research of Alfred Kinsey, who in effect documented the decline of "civilized morality" through sexual histories of several statistical generations of Americans.[34]

Concern over possibly dysgenic population trends also led American philanthropists to sponsor institutions for the study of vital trends, including the Scripps Foundation for Research in Population (founded 1922), the Research Division of the Milbank Memorial Fund (1928), the Population Association of America (1931), and the Office of Population Research at Princeton (1936).[35]

When Frank Notestein, one of the influentials in population studies, began work at the Milbank in 1928, his first assignment was the analysis of unused data from the 1910 census on the relationship of social class to fertility. His study, published in 1930, was the first extensive empirical investigation of differential fertility, one of a series of distinguished Milbank monographs that provided an empirical basis for analysis of vital trends. Frederick Osborn, the leader of the American eugenics movement, believed that Notestein's work deserved a secure academic setting, and he persuaded Albert Milbank, a fellow trustee of Princeton, to establish an office of population research at Princeton, with Notestein at its head. Notestein went on to develop the now conventional theory of demographic transition which showed how fertility was shaped by socio-economic determinants rather than changes in fecundity.

After World War II, demographers associated with the Office of Population Research and the Milbank Memorial Fund were the prime movers in an effort to focus the attention of world leaders on "the population explosion" and its detrimental impact on Third World economies. Just as differential fertility between classes and regions exacerbated social problems in the United States, they argued that the rapid expansion of population threatened the possibility of engineering rapid economic development in the Third World.

It had proved relatively easy to engineer dramatic decreases in mortality because inexpensive mass procedures could be introduced by small numbers of technicians with little more than passive support from the general public. As Notestein explained in 1947 to a Milbank Round Table on International Approaches to Problems of Underdeveloped Areas:

> Human fertility. . . responds scarcely at all in the initial, and often super-imposed, stages of such changes [in mortality] − changes that too often leave the opportunities, hopes, fears, beliefs, customs, and social organization of the masses of the people relatively

untouched. These latter are the factors that control fertility, and since they are unmodified, fertility remains high while mortality declines. . . .

If gains in production only match those in population growth, 'improvement' may result principally in ever larger masses of humanity living close to the margins of existence and vulnerable to every shock in the world economic and political structure. Such 'progress' may amount to setting the stage for calamity.[36]

Public health campaigns provided examples of the dangers inherent in failure to develop comprehensive strategies for social change that took into account the effects of population growth. Nations newly liberated from colonial status wanted to share the prosperity of he West. Failure to develop their economies would lead to bitter internal divisions and to rejection of Western alliances in favor of communist models of development. Thus, political stability depended on rapid economic development, and, in Notestein's view, that development could only succeed if the rate of population growth did not eat up the necessary capital.

John D. Rockefeller III became the principal sponsor of Notestein's efforts to do something about the world population problem. As the eldest of Rockefeller, Jr.'s five sons, Rockefeller III was expected to assume his father's role as a hard-working philanthropist, but the specific focus of his career, Asian-American relations and population control, was only partly dictated by established family interests. Through John Foster Dulles, a trustee of the Rockefeller Foundation since 1935 and chairman of its board in 1950, Rockefeller III learned that cultural relations were an important aspect of diplomacy and that he could serve his nation by helping to make the Pacific a breakwater against communism. Rockefeller-suppported experts would be mobilized in a comprehensive effort to promote capitalist development in the Third World. In their "transition studies" or models of development, the problems of "less developed nations" would be traced to internal factors, such as high birth rates, while external explanations for national poverty, such as a disadvantageous position in an international division of labor, would be neglected. Population control would be modeled on the enormously successful public health programs that the Rockefeller Foundation had subsidized and shaped.[37]

After attempting to interest the Rockefeller Foundation in population control without success, Rockefeller III decided that there was a need for a small private organization "able to work closely with foreign governments without the publicity about Americans which so often arouses nationalistic feelings."[38] In 1952, after convening a conference of experts to discuss the world population situation, he founded the Population Council with his own funds. The governing body included Rockefeller, Frank Notestein, Frank Boudreau of the Milbank Memorial Fund, and Frederick Osborn as chief executive officer.

The Council began spending half a million dollars a year on population research. Conscious of the need to avoid ideological conflicts, the Council was "anxious to keep the work . . . in the hands of competent scientists, believing that accurate determination of the facts must precede propaganda rather than the other way around, and that when verifiable information was available it would inevitably be used in the guidance of policy."[39]

Throughout the 1950s the "verifiable information" poured in, its collection spurred by Population Council grants that brought students from around the world to the United States to study and subsidized the establishment of demographic studies in foreign universities. The "intolerable pessimism" of demographers that had aroused the contempt of agricultural scientists in the early 1950s became acceptable as study after study showed per capita food production declining, largely, Rockefeller-backed social scientists argued, because of the acceleration of population growth. Fear of famine gained a growing audience for those who insisted that population control was the only alternative to social catastrophe. In 1954, India requested help from the Council in organizing its family planning program and Pakistan followed in 1959.

The Council's growing influence might lead to frustration for both demographers and foreign governments, however, since the available technology and delivery systems, a diaphragm or condom from a physician or drugstore, assumed a developed Western style economy. Notestein, who succeeded Frederick Osborn as President of the Council in 1959, remembered his frustration in believing that something *had* to be done to control rapid population growth but lacking the contraceptive means that would enable the Council to take decisive action. This sense of urgency led the Council to investigate the possibilities of an old and discredited method, the intrauterine device (IUD), in a world where antibiotics had made the risks of infection seem acceptable.[40]

The Council invested more than $2.5 million in the clinical testing, improvement, and statistical evaluation of the IUD, which worked well enough. Armed at last with a method that was relatively inexpensive to deliver and required little motivation from the user, family planning programs began to have an effect on birth rates in South Korea, Taiwan, and Pakistan. By 1967 a review article in *Demography* criticized the overoptimism of the Population Council technocrats about the prospects for controlling world population growth.[41] Other observers argued that population control was getting too much of the development dollar and pointed out that it was no substitute for social justice. Notestein was acutely aware, however, that technology alone could not solve the problems of economic development. Basic social reforms were necessary in many developing countries, but private American agencies could not force others to do anything. "What can a white capitalist do in a very sensitive world?"[42] Notestein's answer was that he could provide high quality technical assistance when asked.

By 1965 the Population Council had managed to establish the need for population control as a relatively noncontroversial part of economic wisdom. The United States, rather than India or China, was the "backward" nation in terms of public policy. Osborn and Notestein were appalled by the growing hysteria over the prospects of ecocatastrophe fanned by such zealots as the Dixie-cup tycoon Hugh Moore, whose Cold War rhetoric and ticking bomb advertisements in *Time* seemed to be parodies of the Council's positions.[43]

Ironically, from 1966 onward the birth rate in the United States was below any recorded for the 1930s, but, instead of a revived pronatalism, the public heard shrill demands for yet fewer children from Zero Population Growth, Inc. According to surveys, the organization's membership was predominantly highly educated, white males, with a median age under 30. These were the boom babies, a generation that from primary grades on had been squeezed into crowded class rooms and was now competing for scarce jobs. From their point of view, the United States was desperately overcrowded.[44]

Baby boom generation women experienced many of the frustrations of their brothers, but they shared grievances with their mothers as well. In the 1960s the voices of women were added to the criticism of population growth and the social values that supported it. The publication of Betty Friedan's *The Feminine Mystique* in 1963 helped to explain the malaise and the rage that found expression in the revival of the feminist movement. Apparently the stage had been set for a mass rebellion among women by structural changes in the economy. With the post-war expansion of the service sector, married women with children were drawn in ever larger numbers into permanent work outside the home. The great majority of American families could not attain the American dream of ever-increasing affluence with a single income. The existential reality of the working mother was ignored, however, as women workers continued to be regarded as transients who should be paid less and promoted with discretion because they were in theory working to supplement someone else's income. The growth of "the universal marketplace," or a society in which ever-more human relationships were commodities, generated immense desires and immense frustration that provided the basis for a renewed feminist movement.[45] Women began to repeat Margaret Sanger's half-forgotten demands that they be allowed to control their lives in their interests. Perhaps the fundamental insight of the new women's liberation movement was the intimate relationship between private and public — women could not achieve equality in the workplace without a rethinkiing of gender roles and a new division of labor within the home.

The redefinition of the population problem meant that for the first time in American history the desire of the majority of married persons to limit the burdens of parenthood was not in conflict with "the public interest." A private vice had become a public virtue. This convergence of private and public need paralleled two other events that helped to shape a revolution in public policy concerning human reproduction: the rapid expansion of the welfare state and of judicially dictated "entitlements" or rights to public services.

In the United States the distinctive role of the judiciary in interpreting a written constitution that has proven extraordinarily flexible and the absence of an effective alternative to the middle-of-the-road two-party system led the dissatisfied to organize outside the political system and to present their demands as claims under the Bill of Rights.[46] As Margaret Sanger's attorney explained after the *One Package* decision, the process of social reform "is a simple one, it is a matter of educating judges to the mores of the day."[47] By the mid-1960s the concerns that had previously inhibited judges from denying police powers to the to the state had eroded. In 1965 the United States Supreme Court, in *Griswold v. Connecticut*, struck down an "uncommonly silly law" that prohibited contraceptive practice. The Court continued to expand the rights of individuals to defy out-dated restrictions in *Eisenstadt v. Baird* (1972) which established that the unmarried had the same right to contraceptives as the married. In 1973 the Court decided that women and their physicians, rather than legislatures, should decide whether or not to abort a fetus during the first three months of pregnancy.[48]

Removing positive barriers to reproductive choice did not, of course, guarantee that all persons would have equal opportunity to control their fertility in an age when the state was expected to assume a paternalistic role on behalf of the "medically indigent." As Lyndon Johnson's War on Poverty emerged from Congress in 1964, a number of planned parenthood groups successfully applied to the Office of Economic Opportunity for funds under the "local option" policy that allowed community groups to initiate welfare programs. Pro-population controllers such as Senator Ernest Greuning orchestrated public hearings that emphasized "freedom of choice" and the "right to equal access" as rationales for including budgets for family planning services in domestic social welfare legislation. Whereas birth controllers had for years been stopped by the problem of justifying use of taxpayers' money for a purpose that many citizens considered immoral, in the context of Lyndon Johnson's War on Hunger abroad and War on Poverty at home, the question became, how can we justify withholding from the poor birth control services that the middle classes already enjoy? With the population problem redefined in such a way that population growth was no longer a national necessity, there was indeed no rationale for denying anyone access to contraceptives.

The Social Security Amendments of 1967 specified that at least six percent of maternal and child health care funds be spent on family planning, and contraceptives were removed from the list of materials that could not be purchased with Agency for

International Development funds. These changes mark the point at which the federal government clearly adopted a pro-family planning policy.[49]

iv.

What lessons are to be drawn from this history? In large measure it confirms the findings of political scientists who have argued that the American political system favors incremental change that does not explicitly challenge the consensus of values held by the voting public.[50] Small victories were possible for such minority interests as civil libertarians or quality controllers, but the major change in public policy, from suppression to active promotion of birth control, depended both upon a sustained effort to organize a revised consensus of public opinion and upon changes in American society that no one anticipated.

Was a single-issue approach a mistake for women's rights activists? No, success depended on limiting the scope of debate. Was "medicalization" of birth control a mistake? No, success for feminists required expanding definitions of constitutional entitlements for individuals and getting women into the medical profession rather than returning the business of health care to the wise-women and popular healers that flourished before the rise of modern medicine. Have women achieved significant gains through liberal reform? Yes, since in the 1980s the debate is over whether the federal government will continue to subsidize certain services for the poor rather than over whether individuals have the right to control their fertility as they see fit.[51]

In considering the relevance of history to public policy, we should not forget the importance of traditional standards of excellence in historical writing. We need accurate descriptions of the groups and individuals involved in particular issues. We need narratives that decribe how conflicts were perceived and mediated. We can maximize our usefullness to other citizens by building upon rather than forsaking the virtues that we profess to respect – a desire to understand the past on its own terms; concern for context, detail, and thick description; the ability to communicate effectively in plain English that is both accessible and pleasing to others. In the case of public policy concerning human reproduction, the past provides ample materials for those who want to mine it for partisan purposes, but the historian's most important role may be to resist the claims of those who imagine their concerns to have unique urgency that would justify rigid solutions for problems arising from deep social changes that are still underway.[52]

Rutgers University   *James Reed*
Dept. of History
New Brunswick, NJ 08903

FOOTNOTES

1. *New York Times*, Dec. 3, 1959, pp. 1, 18; Phyllis T. Piotrow, *World Population Crisis: The United States Response* (New York, 1973), pp. 45-46.

2. "Let's Be Honest with Ourselves," *Saturday Evening Post* (Oct. 26, 1963), p. 27.

3. Piotrow, p. 141; "Anniversaries," editorial in *Family Planning Perspectives* 11 (Jan., 1979): 2-4.

4. "Conception Control as a Health Practice: An Emerging Concept in Government and Medicine," *Perspectives in Biology and Medicine* 16 (Spring, 1973): 367.

5. Mary P. Ryan, "Reproduction in American History," *Journal of Interdisciplinary History* X: 2 (1979): 319-332; "The Explosion of Family History," *Reviews in American History* 10:4 (Dec., 1982): 181-185; Estelle B. Freedman, "Sexuality in Nineteenth Century America: Behavior, Ideology, and Politics," *Ibid.*, pp. 196-215; Elaine Tyler May, "Expanding the Past: Recent Scholarship on Women in Politics and Work," *Ibid.*, pp. 216-233.

6. Gordon, pp. 257, 300-301, 341-344, 404, 410, 414-415; Rothman, pp. 188-209.

7. Gordon, pp. 249-300, 309, 320; Rothman, pp. 142-153, 201, 209-218, 289-290.

8. This assertion is based on the whole of the two works. There are great differences between Gordon and Rothman, but I believe that they are both relentlessly subjective in their treatment of Margaret Sanger and other reformers; they consistently denigrate the activities and motives of reformers and of professionals, and they draw excessively pessimistic, if fashionable, conclusions about woman's fate in modern American society.

9. Robert Wells, *Revolutions in Americans' Lives* (Westport, CT., 1982), pp. 75-76.

10. James H. Cassedy, *Demography in Early America: Beginnings of the Statistical Mind, 1600-1800.* (Cambridge, Mass., 1969), pp. 216-220.

11. William Stanton, *The Leopard's Spots: Scientific Attitudes Toward Race in America, 1815-1859* (Pittsburgh, 1960); Reginald Horsman, *Race and Manifest Destiny: The Origins of American Racial Anglo-Saxonism* (Cambridge, Mass., 1981).

12. James Reed, *From Private Vice to Public Virtue: The Birth Control Movement and American Society since 1830* (New York, 1978), pp. 3-45.

13. James C. Mohr, *Abortion in America: The Origins and Evolution of National Policy* (New York, 1978), p. 168 and *passim*.

14. Heywood Broun and Margaret Leech, *Anthony Comstock: Roundsman of the Lord* (New York, 1927); R. Christian Johnson, "Anthony Comstock: Reform, Vice, and the American Way" (Ph.D. dissertation, University of Wisconsin, 1973); David Pivar, *Purity Crusade: Sexual Morality and Social Control, 1868-1900* (Westport, CT., 1973).

15. Reed, pp. 34-45.

16. Daniel Scott Smith, "Family Limitation, Sexual Control, and Domestic Feminism in Victorian America," *Feminist Studies* 1 (1973): 40-57.

17. Francis P. Gosling, "American Nervousness: A Study in Medicine and Social Values in the Gilded Age, 1870-1900" (Ph.D. dissertation, University of Oklahoma, 1976); Nathan Hale, *Freud and the Americans: The Beginnings of Psychoanalysis in the United States, 1876-1917* (New York, 1971), pp. 116-150; Barbara Sicherman, "The Paradox of Prudence: Mental Health in the Gilded Age," *Journal of American History* 62 (1976): 890-912; "The Uses of Diagnosis: Doctors, Patients, and Neurasthenia," *Journal of the History of Medicine* 32 (1977): 33-54; Reed, pp. 54-63.

18. Reed, 34-45, 97-105.

19. Mark H. Haller, *Eugenics: Hereditarian Attitudes in American Thought* (New Brunswick, NJ, 1963); Kenneth M. Ludmerer, *Genetics and American Society* (Baltimore, 1972).

20. Ludmerer, pp. 121-134; Haller, pp. 180-183; Frank Lorimer and Frederick Osborn, *Dynamics of Population: Social and Biological Significance of the Changing Birth Rates in the United States* (New York, 1934); Frederick Osborn, *Preface to Eugenics* (rev. ed., New York, 1951); Frank W. Notestein, "Demography in the United States: A Partial Account of the Development of the Field," *Population and Development Review* 8:4 (1982): 651-687.

21. Reed, pp. 143-193; John C. Burnham, "Medical Specialists and Movement toward Social Control in the Progressive Era: Three Examples," in *Building the Organizational Society*, ed. Jerry Israel (New York, 1971); Interview with Howard Taylor, Jr., April 27, 1971, New York City; Interview with Mrs. Alan F. Guttmacher, New York City, November 15, 1974.

22. Gerald N. Grob, *Mental Illness and American Society, 1875-1940* (Princeton, 1983), pp. 166-178; Julius Paul, "Population 'Quality' and 'Fitness for Parenthood' in the Light of State Eugenic Sterilization Experience, 1907-1966," *Population Studies* 21 (1967): 295-299.

23. Franz Samelson, "On the Science and Politics of IQ," *Social Research* 42 (1975): 467-488; "World War I Intelligence Testing and the Development of Psychology," *Journal of the History of Behavioral Sciences* 13 (1977): 274-282. Samelson's argument that psychologists had much less impact on public policy than they and their historians would like to believe has in turn been questioned by Gerald Sweeny in an unpublished essay, Alpha-Beta, American Intelligence, and Restriction: Some Objections to a Consensus on Influence, San Diego State University, Spring, 1981.

24. Reed, pp. 211-217.

25. *Ibid.* pp. 67-110; Joan M. Jensen, "The Evolution of Margaret Sanger's *Family Limitation* Pamphlet, 1914-1921," *Signs* 6 (1981): 548-555.

26. Mary Ware Dennett, *Birth Control Laws* (New York, 1926).

27. "Congressional Report: January 1 to May 1, 1926," American Birth Control League Papers, Houghton Library, Cambridge, Mass.; Reed, pp. 100-104.

28. C. Thomas Dienes, *Law, Politics and Birth Control* (Urbana, 1972), pp. 82-83, 898-91, 104-115, 195-196.

29. Reed, pp. 123-128.

30. Rothman, *Woman's Proper Place*, pp. 200-209; Gordon, *Woman's Body, Woman's Right*, pp. 370-390.

31. Marie Kopp, *Birth Control in Practice* (New York, 1934); Regina K. Stix and Frank W. Notestein, *Controlled Fertility: An Evaluation of Clinic Studies* (Baltimore, 1940); Ellen Chesler, "Feminism, Sexuality and Birth Control: A Closer Look at the Birth Control Clinic," paper presented to the Berkshire Conference on the History of Women, Radcliffe College, 1974.

32. Reed, pp. 208-210.

33. *The Lonely Crowd: A Study in the Changing American Character* (New Haven, 1950), pp. 7-31.

34. James F. Gardner, Jr., "Microbes and Morality: The Social Hygiene Crusade in New York City, 1892-1917" (Ph.D. dissertation, Indiana University, 1973); James H. Jones, "Scientists and Progressives: The Development of Scientific Research on Sex in the United States, 1920-1963," a paper presented to the American Historical Association, New Orleans, December, 1972; Diana Long Hall, "Biology, Sex Hormones and Sexism in the 1920s," *Philosophical Forum* 5 (1973-74): 81-96; George W. Corner, *Twenty-Five Years of Sex Research: History of the National Research Council Committee for Research in Problems of Sex, 1922-1947* (Philadelphia, 1953). Jones, now in the Department of History, University of Houston, is working on a full-scale biography of Alfred Kinsey. Hall, now director of the Francis Wood Institute in the College of Physicians, Philadelphia, is working on the history of the Sex Committee.

35. Reed, pp. 202-206.

36. *Milbank Memorial Fund Quarterly* 26 (1948): 250, 252; quoted in Reed, p. 282.

37. Peter Collier and David Horowitz, *The Rockefellers: An American Dynasty* (New York, 1976), pp. 279-285; Dennis Hodgson, "Demography as Social Science and Policy Science," *Population and Development Review* 9 (March 1983): 1-34.

38. Fredereick Osborn to author, May 2, 1973.

39. *Ibid*: The Population Council, *The Population Council: A Chronicle of the First Twenty-Five Years, 1952-1977* (New York, 1978); Notestein, "Demography in the United States."

40. Reed, pp. 303-307.

41. Philip M. Hauser, "Family Planning and Population Programs," *Demography* 4 (1967): 397-414.

42. Interview with Frank Notestein, Princeton, N.J., August 12, 1974; quoted in Reed, p. 307.

43. Reed, pp. 303-304; Notestein Interview.

44. Larry D. Barnett, "Zero Population Growth, Inc." *Bioscience* 21 (1971): 759-765; "Zero Population Growth, Inc.: A Second Study," *Journal of Biosocial Science* 6 (1974): 1-24; Peter Filene, *Himself/Herself: Sex Roles in Modern America* (New York, 1974).

45. Harry Braverman, *Labor and Monopoly Capitalism* (New York, 1974), pp. 271-283; Rothman, pp. 226-253. I do not mean to slight the interpretations of the rebirth of feminism presented in such works as Sara Evans, *Personal Politics: The Roots of Women's Liberation in the Civil Rights Movement and the New Left* (New York, 1979) and Barbara Ehrenreich, *The Hearts of Men: American Dreams and the Flight from Commitment* (New York, 1983). I am simply trying to sketch the context in which feminist demands evoked a large scale response among women from many social strata and political persuasions.

46. Paul Starr, *The Social Transformation of American Medicine* (New York, 1982), pp. 388-393.

47. Morris Ernst quoted in Reed, p. 121.

48. Reed, pp. 377-380.

49. Piotrow, pp. 107-108, 141-142.

50. Joyce Gelb and Marian Palley, *Women and Public Policies* (Princeton, 1982).

51. For informed discussion of the impact of social welfare programs in the 1960s and beyond, see James Patterson, *America's Struggle Against Poverty, 1900-1980* (Cambridge, Mass., 1981).

52. Wells, pp. 280-281. For an example of how careful historical analysis can inform public policy debates over reproductive issues, see Maris Vinovskis, "An 'Epidemic' of Adolescent Pregnancy? Some Historical Considerations," *Journal of Family History* 6 (Summer, 1981): 205-230.

Am J Hum Genet 39:253–264, 1986

Clients' Interpretation of Risks Provided in Genetic Counseling

DOROTHY C. WERTZ,[1] JAMES R. SORENSON,[2] AND TIMOTHY C. HEEREN[3]

SUMMARY

Clients in 544 genetic counseling sessions who were given numeric risks of having a child with a birth defect between 0% and 50% were asked to interpret these numeric risks on a five-point scale, ranging from very low to very high. Whereas clients' modal interpretation varied directly with numeric risks between 0% and 15%, the modal category of client risk interpretation remained "moderate" at risks between 15% and 50%. Uncertainty about normalcy of the next child increased as numeric risk increased, and few clients were willing to indicate that the child would probably or definitely be affected regardless of the numeric risk. Characteristics associated with clients' "pessimistic" interpretations of risk, identified by stepwise linear regression, included increased numeric risk, discussion in depth during the counseling session of whether they would have a child, have a living affected child, discussion of the effects of an affected child on relationships with client's other children, and seriousness of the disorder in question (causes intellectual impairment). Client interpretations are discussed in terms of recent developments in cognitive theory, including heuristics that influence judgments about risks, and implications for genetic counseling.

Received December 4, 1985; revised March 31, 1986.

This study was supported in part by grant G-63 from the March of Dimes Birth Defects Foundation and by Interdisciplinary Incentive Award ISP8114333 from the Program in Ethics and Values in Science and Technology of the National Science Foundation and the Science, Technology, and Society Program of the National Endowment for the Humanities.

[1] Social and Behavioral Sciences Section, Boston University School of Public Health, Boston, Mass.

[2] Department of Health Education, University of North Carolina School of Public Health at Chapel Hill, Chapel Hill, N.C.

[3] Biostatistics and Epidemiology Section, Boston University School of Public Health, Boston, Mass.

INTRODUCTION

Individual variation in interpretation of numeric risks is a topic of concern to genetic counselors, because a client's interpretation of a risk for having a child with a specific birth defect or genetic disorder may affect reproductive behavior. Given the same numeric risk for a disorder, some clients, who may be considered "pessimists," will view their risk as higher than do other clients, who may be called "optimists." In the following analysis, we identify factors that are associated with pessimistic and optimistic client interpretations of risks for having a child with a specific birth defect provided in genetic counseling.

BACKGROUND

Risk interpretation (the judgment that a risk is "high" or "low") is ordinarily associated with acceptability of risk (willingness to take the risk) and with actual decision-making. All three—risk interpretation, acceptability, and decision-making—are linked in a complex interactive process whereby each one affects the other. Most of the literature on risk focuses on the "active" components: acceptibility and decision-making [1–9]. Although we recognize that the process of interpreting a risk is linked with the process of imagining whether one is willing to accept that risk, for purposes of this paper we shall leave aside acceptability and decision-making. We have considered elsewhere the reproductive decision-making of genetically counseled clients under conditions of risk [10–12]. Other researchers have developed models of rational decision-making under conditions of uncertainty in genetic counseling [13–18].

We have chosen instead to focus on the interpretation of a numeric risk as high, moderate, or low, without regard to whether the client considers the risk acceptable or decides to take it. The interpretation of risk is influenced by qualitative features, including, for example, whether the risk is under the individual's control, is reversible or irreversible, visible or invisible (e.g., beating a train approaching a railroad crossing vs. exposure to a possible carcinogen), or familiar or new [3, 6], or whether any cultural aspects of this risk (e.g., new technologies) threaten the person's social or religious values [2].

According to the Bayesian hypothesis, which is a rule for revising probabilistic beliefs on the basis of new information, if new evidence results in a change in a probability, our interpretation of the new probability will be affected by the direction of the change. For example, if a client goes to genetic counseling with a prior belief about her numeric risk of having an affected child, and is then given a lower numeric risk than she thought she had, her interpretation of the new risk should be "optimistic;" if she is given a higher numeric risk than anticipated, her interpretation of the new risk should be "pessimistic" [5, 7].

It is now a well-established fact in cognitive psychology that individuals violate Bayes's rule in systematic ways [19–22]. Tversky and Kahneman [20, 22] describe three heuristics or biasing factors that people frequently apply to expectations about outcomes: representativeness, availability, and anchoring. Representativeness is the degree to which an individual outcome (in the context of genetic counseling, a particular child with a particular birth defect) is regarded as representative or stereotypical of all outcomes. Many birth defects, for example, have widely variable potentials for effective treatment. For example, if the parents of a child at the "poor" end of the treatment spectrum regard that child as "representative" of all children with the disorder in question, they will interpret a given numeric risk as "higher" than they would if they did *not* regard their child as typical of all children with the disorder.

Availability is the ease with which instances of the risked event can be brought to mind (e.g., one already has, or knows someone who has, a child with a particular birth defect). The greater the "availability" of an event, the more likely its occurrence will seem. A numeric risk of its occurrence will tend to be interpreted as higher than will the same numeric risk for an event that is not "available." In genetic counseling, Lippman-Hand and Fraser [23, 24] have speculated on the effects of clients' experiences with previous children with birth defects, including the early death of such a child, on their subsequent interpretations of risks as high.

Anchoring occurs when the person has a starting point or partial computation on which to base an interpretation of risk, for example, a prior belief that a risk is high or low. This prior belief will continue to affect the interpretation of a risk, even after a new numeric risk is provided. Therefore, we would anticipate that, for those clients who have a belief about their numeric risk prior to counseling, these prior beliefs will be reflected in their interpretations of risk after counseling.

Experiments with lotteries have demonstrated that each of these three heuristics influences subjects' interpretations of risk [25]. Cognitive psychologists have further demonstrated that, when subjects are given numeric probabilities, their estimates of the degree of risk do not increase in proportion to increases in numeric risk. Most subjects tend to "overestimate" numerically smaller risks and to "underestimate" numerically larger risks [26]. This occurs because the desire for absolute certainty makes even a high numeric probability, such as 90%, seem insufficiently certain, while at the same time the desire for impossibility of a negative outcome makes even a 5% probability seem too high. Low numeric probabilities are perceived as higher than they are, and higher numeric probabilities are perceived as lower than they are. The result is that a wide range of numeric risks is given the same interpretation.

Clients of genetic counseling face lottery situations somewhat similar to those described by the experimental psychologists. Most clients go to counseling in order to get information that will help them make a decision about whether to have a child [10]. The prize, a normal child, may be won only by taking the risk of conceiving a child with a birth defect. In some cases, clients' interpretations of their numeric risks may affect them for a lifetime, in the form of child with a birth defect living at home. Although the risk of having a child with a birth defect is given by the counselor in terms of a percentage, the outcome is *binary:* the child either will or will not be normal. As Lippman-Hand and Fraser point out, "the one in the numerator never disappears no matter the size of the denominator, and the 'one' could be the counselee's child" [23, 24].

In the following analysis, we examine client characteristics associated with optimistic and pessimistic interpretations of numeric risk. Under real conditions, as opposed to experimental ones, is clients' interpretation of risk affected by the heuristics of representativeness, availability, and anchoring?

METHODS

Genetic counseling cases were ascertained between 1977 and 1979 at 47 genetic counseling clinics located in 25 states and the District of Columbia. The study employed a prospective longitudinal design. Detailed structured self-administered questionnaires were completed by clients immediately before and within 7–10 days following genetic counseling. Data on a large number of topics were collected, including client sociodemographic characteristics, reproductive history, and status, characteristics of the disorder in question, questions and concerns brought to genetic counseling, perceptions of problems associated with having a child with a birth defect, and topics discussed in genetic counseling.

Clients' knowledge of the diagnosis of the birth defect for which their children were at risk and of their numeric risk for having a child with the disorder in question was assessed before and after counseling, as were their views about the burden that a child with this defect would impose upon personal and family life. Detailed medical and genetic information on each case was provided by the genetic counselor through a structured questionnaire completed immediately after each counseling session. Counselors provided the specific diagnosis and the assigned risk of occurrence or recurrence of the birth defect in the client's children.

Clients in 1,369 cases chose to participate. Of those who completed precounseling questionnaires, 1,096 cases (83%) returned a postcounseling questionnaire. In 550 cases where the client reported that the counselor had given a numeric risk, clients were asked to report this risk. Next they were asked, "I think this risk is _____ (choose one) very low, low, moderate, high, very high." The following analysis deals with 544 cases where the numeric risk, as reported by the client, was less than or equal to 50%. The six clients who reported risks greater than 50% were too few for analysis.

For purposes of this paper, the responses of female clients have been used as the basis of analysis. Although 68% of clients were seen as couples, a chi-square analysis of the male clients' interpretations of risk revealed no significant differences from the responses of the (more numerous) female clients. The analysis focuses on identifying factors associated with clients' interpretations of a given numeric risk as low or high, controlling for actual numeric risk. An interpretation that is lower than the modal response for that particular risk in this population of counseled clients is considered optimistic; an interpretation that is higher than the modal response, pessimistic.

RESULTS

The first step in identifying characteristics of clients who were optimistic or pessimistic about a given numeric risk was to identify the modal interpretation category for each numeric risk along a five-point scale ranging from "very low" to "very high." At every numeric risk, up to and including 50%, clients used the full range of responses (very low, low, moderate, high, very high). Changes in the modal interpretation category as numeric risk increased are reported in figure 1. The modal interpretation varied directly with risks between 0% and 15%. The modal category reached "moderate" at 15% and remained there to 50% risk. The clients' modal category of risk interpretation never reached "high" or "very high." The result is a largely "flat" curve for the modal category of interpretation of risks between 10% and 50%. The flatness of this

N=544*

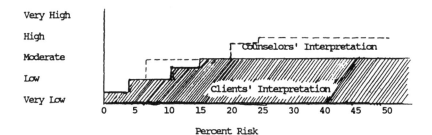

*Genetic Counseling Sessions

FIG. 1.—Modal category of risk interpretation

curve agrees with the empirical lottery research of cognitive psychologists, who have documented both (1) a tendency to give the same risk interpretation to a wide range of numeric probabilities, and (2) the absence of a corresponding increase in interpretation of level of risk as numeric risk becomes very high [25]. Genetic counselors' modal categories of interpretation also produced a flat curve between 25% and 50% risk, with the difference that they interpreted these risks as "high" as opposed to the clients' "moderate."

The percents of clients who chose each category of risk interpretation are reported for selected levels of numeric risk in table 1. The risk levels selected include ≤ 3%, which is sometimes described as the risk faced by the population at large, and the risks of 25% and 50% faced by clients who carry Mendelian genetic disorders. (Too few clients reported risks of 11%–24% or 26%–49% for meaningful inclusion in this table.) Although the modal interpretation category never rose beyond moderate, some clients chose each category of risk interpretation at every level of numeric risk.

TABLE 1

CLIENTS' INTERPRETATIONS OF SELECTED LEVELS OF NUMERIC RISK
(No. = 510)

NUMERIC RISK, AS REPORTED BY CLIENT	RISK INTERPRETATION (%)					No.	(TOTAL %)
	Very low	Low	Moderate	High	Very high		
≤ 3%							
("population risk")....	41.9	35.7	14.5	6.3	1.6	255	(100)
4%–5%	31.3	45.0	18.8	2.5	2.5	80	(100)
6%–10%	16.6	45.8	29.2	6.3	2.1	48	(100)
25%	2.7	19.2	37.0	27.4	13.7	73	(100)
50%	7.4	9.3	38.9	29.6	14.8	54	(100)
Total..						510	

In order to identify characteristics associated with optimistic and pessimistic interpretations of risk, we ran partial correlations between clients' interpretations of risk on the five-category scale of very low to very high and all variables on client demographic background, characteristics of the disorder in question, reasons for coming to counseling, reproductive history, plans, and expectations, present and anticipated problems in caring for an affected child, and topics that clients reported were discussed in the counseling session, a total of 49 variables. Selection of these variables was partially governed by suggestions in the literature on genetic counseling that certain factors could be expected to affect risk interpretations, for example, the existence of a living affected child whose care was burdensome [23, 24]. In the course of the analysis, virtually all data gathered through client pre- and postcounseling questionnaires were examined for strength of association with risk interpretation.

We were interested in examining optimistic and pessimistic interpretations of all numeric risks from 1% to 50%. Therefore, actual numeric risk, as reported by the clients, was controlled for in the partial correlation procedure. Thirty variables with a correlation coefficient of .10 or above ($P < .05$) with client pessimism about risk are listed in table 2, along with the P values and correlation coefficients. Among the characteristics associated with pessimistic interpretations of risk were: having an affected child living at home; the disorder in question causes intellectual (as opposed to physical or neurological) impairment; prenatal diagnosis is not available for the disorder; client is having serious problems in raising an affected child; and client anticipates serious problems raising the child in the future. Clients whose interpretations were pessimistic were more likely than optimists or those whose interpretations followed the mode to have discussed in depth with the counselor (1) whether they would have a child; (2) the effect of the affected child on their relationships with their other (normal) children; (3) the checkup and progress of the affected child; and (4) school programs for the child. They were also more likely to report that the session had not provided all the medical facts they wanted and had not helped with personal concerns. Pessimists were more likely to have believed, before counseling, that their risk was high than were those who were optimistic or whose interpretations followed the mode. Finally, pessimists were less likely to be pregnant at the time of counseling or to have planned, either *before* or *after* counseling, to have a child in the foreseeable future than were those who were optimistic or whose interpretations followed the mode.

Several client characteristics that we thought might be associated with pessimistic interpretations were not significantly associated. These were: education, income, discussion of risk during counseling session, and changes in reported numeric risk from before to after counseling.

The characteristics listed in table 2 were next put into a stepwise linear regression, in order to identify those independently and significantly associated with pessimistic interpretations. To find out whether any of these characteristics had an effect over and above numeric risk, numeric risk was also included in the regression. The results are reported in table 3. Among 427 clients in the regression analysis, the characteristics most strongly associated with pessi-

TABLE 2

Characteristics Associated with Clients' Pessimistic Interpretations of Risk, Controlling
for Numeric Levels of Risk
(No. = 544; Risk ≤ 50%)

	P^*	Correlation coefficient
Background:		
Have affected child living at home	.00	.14
Affected child was firstborn	.00	.12
Genetic disorder:		
Causes intellectual impairment	.00	.16
Prenatal diagnosis not available	.00	.16
Present problems raising an affected child:		
Caring for child at home	.00	.10
Feelings about child, before session	.00	.13
Feelings about child, after session	.05	.10
Effects of child on clients' social life	.02	.12
Anticipated future problems in raising child:		
Feelings about child	.00	.11
Caring for child at home	.01	.11
Medical care	.01	.11
Education	.01	.14
Financial costs	.03	.10
Content of counseling session:		
Discussed *in depth:*		
Whether would have (another) child	.00	.18
Relationship with other (normal) children	.00	.14
Checkup for affected child	.00	.11
School programs for child	.00	.13
Client reported that session		
Did not give all the medical facts	.00	.10
Did not help with personal concerns	.00	.17
Reproductive expectations, before counseling:		
Think risk is high	.00	.28
Not pregnant	.00	.15
Do not intend pregnancy in next 2 years	.00	.10
Do not intend pregnancy after 2 years	.00	.15
Fulfillment as parent not important	.00	.10
Wishes of spouse not important	.00	.10
Ideal family size not important	.00	.00
Reproductive expectations, after counseling:		
Report that counseling session changed 2-year pregnancy plans	.00	.14
Do not intend pregnancy in next 2 years	.00	.21
Do not intend pregnancy after 2 years	.00	.12
Uncertain about ideal number of children	.01	.10

* Partial correlation, controlling for level of numeric risk, was used as the measure of association. Only those variables with a partial correlation coefficient ≥ .10 are reported here.

mistic interpretations, in order of incremental R-squares, were: (1) numeric risk; (2) discussion in depth of the effects of an affected child on relationships with client's other children; (3) discussion in depth in counseling about whether they would have a child; (4) has a living affected child; and (5) the disorder in question causes intellectual impairment.

Another method of ascertaining clients' interpretation of risk is to compare

TABLE 3

RESULTS OF STEPWISE LINEAR REGRESSION ANALYSIS IDENTIFYING CHARACTERISTICS ASSOCIATED
WITH PESSIMISTIC INTERPRETATIONS OF RISK
(No. = 427)

Characteristic	Beta	P value	R-square
Numeric Risk	.47	.00	.218
Discussed in depth, effects of affected child on relationship with other children	.11	.01	.248
Discussed in depth whether to have a child	.14	.00	.264
Has a living affected child	.12	.01	.279
Disorder in question causes intellectual impairment	.09	.03	.287

their reported numeric risk with their estimates of the probability that their next child will be normal or abnormal. In posing this question, we asked clients to interpret their reproductive futures in terms of a zero-sum game [23, 24] in which their child either will or will not have a birth defect, rather than interpreting a numeric risk. Their answers, for clients at selected levels of self-reported numeric risk, are summarized in table 4 (there were 471 clients who answered this question). Their answers parallel the risk interpretations reported in table 1, with the difference that, as numeric risk increases, "pessimism" is expressed in terms of uncertainty about the child's normalcy rather than responses indicating that the child will probably or definitely have a birth defect. Even at population risk (3%), few (12.6%) were prepared to state that the next child would "definitely" be normal. Fewer still were ready to state that the

TABLE 4

CLIENTS' EXPECTATIONS ABOUT NORMALCY OF NEXT CHILD, BY SELECTED LEVELS OF NUMERIC RISK
(No. = 471)

NUMERIC RISK AS REPORTED BY CLIENT	EXPECTATIONS ABOUT NEXT CHILD					TOTAL CLIENTS
	Definitely normal	Probably normal	Probably have defect	Definitely have defect	Not sure	
≤ 3% ("population risk")	33 (12.6)*	182 (69.7)	10 (3.8)	0	36 (13.8)	261
4%–5%	2 (2.4)	53 (63.9)	3 (3.6)	0	25 (30.1)	83
6%–10%	2 (4.0)	36 (72.0)	1 (2.0)	0	11 (22.0)	50
25%	2 (2.7)	25 (33.3)	14 (18.7)	3 (4.0)	31 (41.3)	75
50%	4 (7.0)	19 (33.3)	14 (24.5)	0	20 (35.1)	57
Total clients	43	315	42	3	123	526

* Percentage in parentheses.

child would "definitely" have a defect, even at risks as high as 50%. Instead, clients interpreted their risks in terms of probabilities (as child will "probably" be normal or "probably" have a defect) or uncertainties ("not sure"). The basic uncertainty of their situation is expressed in their modal category of expectation. At risks below 10%, the modal category is "probably normal." At 25%, the modal category of expectation about the next child's normalcy becomes "not sure" rather than "definitely" or "probably" would have a defect.

DISCUSSION

Numeric risk itself, as reported by the client, accounted for most of the explained variance in risk interpretation. Four additional characteristics, over and above numeric risk, emerged from the regression as significantly associated with the interpretation of risk, accounting for 6.9% of the explained variance. The total explained variance of 28.7% compares favorably with the percent of variance explained in other studies of health behavior and outcomes. Part of the unexplained variance may result from interpersonal communication factors, including possible client disagreement with or refusal to accept the numeric risk given by the counselor. Such factors are beyond the scope of our data.

Another factor in the unexplained variance may be that clients who interpret a risk pessimistically may be interpreting the seriousness or burden of the risked disorder rather than the numeric probability of its occurrence. The variable we used to assess seriousness of the disorder (causes intellectual impairment) was a general ranking for each disorder given by a medical geneticist independently of the client's individual case; it is possible that some clients and counselors assessed seriousness and treatment potential in a more pessimistic light than did our medical geneticist. Nevertheless, the emergence of four characteristics, over and above numeric risk, as significant suggests that risk interpretation is affected by two client background and two session characteristics. The two background characteristics ("has living affected child" and "disorder in question causes intellectual impairment") are examples of Tversky and Kahneman's "availability," in this case, a readily visible negative outcome that influences the client's interpretation toward pessimism.

The two session characteristics (discussion of whether client would have a child and discussion of effects of affected child on client's relationships with her other children) suggest that counseling can affect the interpretation of risk, particularly if decision-making in the face of risk and burdens associated with a negative outcome are discussed in depth. Although it is possible that pessimistic clients may push the counselor harder to discuss these topics, it is also possible that counseling could increase clients' pessimistic interpretations of risk if all aspects of the burden of raising an affected child—including financial costs, strain on the marriage, and care of the child as an adult—were to be presented by the counselor. Whether it would be ethical for counselors to attempt to influence client interpretations by these means is questionable. A more ethical approach would be to present *both* positive and negative scenarios—examples of the best and worst outcomes for a child with the disor-

189

der in question, illustrated by the clearest means possible—and to try to avoid purposive influencing of risk interpretations. In any case, counselors should realize that giving or even discussing a numeric risk does not necessarily affect the client's actual interpretation of that risk. Clients are more likely to appreciate the seriousness of a risk if that risk is discussed in terms of the client's actual behavior, in this case, whether the client would have a child. The educational value of giving a risk may be enhanced if that risk is described in behavioral terms.

Counselors should not overlook the effects of "prior beliefs" (Tversky and Kahneman's "anchoring") upon risk interpretation. A regression on a subset of 251 clients who reported an interpretation of their numeric risk prior to counseling documented that, among these particular clients, their precounseling interpretation of risk preceded *all* other variables, *including* numeric risk, in its strength of association with postcounseling interpretation of risk. Those who were pessimists or optimists before counseling tended to retain the same views after counseling.

The effects of the counselor's providing a new numeric risk to a subset of 138 of these clients did not support the Bayesian hypothesis about the revision of probabilistic beliefs on the basis of new information. According to Bayes's theory, learning that your risk is higher or lower than you had previously anticipated it to be should affect your interpretation of the new risk. This was not the case for clients in our study. Changes in clients' reported numeric risks from before to after counseling were not significantly related to optimism or pessimism about risk.

The results of our analysis have implications, not only for genetic counseling, but for health education in general. In genetic counseling, how clients interpret their risk is important because the interpretation of a risk as high or low is associated with subsequent reproductive intentions and behaviors [11, 12]. Clients who were pessimistic about their risk were less likely to plan future pregnancies than were clients who were optimistic. The interpretation of risk was a better predictor of client reproductive intentions than was numeric risk itself.

Client and counselor often approach the counseling situation with different agendas [10]. Clients most frequently give as their major reason for coming to counseling: "to get information that will help me to make a decision about whether to have a child." Counselors are likely to think that clients come to counseling to learn their risk [27]. Some counselors may believe that providing the client with a numeric risk that the counselor considers "high" will act as a deterrent to childbearing. The client, however, is likely to interpret that numeric risk as lower than does the counselor. Clients are not necessarily impressed by what counselors or educators usually consider "high" levels of risk, say 25%–50%. This is perhaps because clients feel that they have control over their own reproductive processes and decisions. Risk analysis research [21] suggests that people are likely to underestimate behavioral risks over which they think they can exert control. For many clients, the effect of higher numeric risk, given the uncertainties inherent in genetic counseling, is to produce more uncertainty about the normalcy of the next child. This leads to reproductive uncertainty [12] rather than reproductive restraint.

Provision of numeric risk in and of itself is not sufficient to reach genetic counseling's goal of providing information on which clients can base informed reproductive decisions. Clients' decisions are based on their own personal interpretation of the risk and their expectations about the normalcy of the next child, rather than on the numeric risk alone. Our results suggest that interpretation of a numeric risk as "high" is more likely to occur if: (1) an example of a risked negative outcome is readily "available" to the client; and (2) the potential burdens of a negative outcome are discussed in depth with the client, along with implications for the client's behavior. Counselors cannot provide clients with the absolute certainty that most people desire. They can facilitate informed decision-making, however, and perhaps reduce reproductive uncertainty, by providing and discussing examples of the risked outcome.

REFERENCES

1. ARROW K: Risk perception psychology and economics. *Econ Inquiry* 20:1–9, 1982
2. DOUGLAS M, WILDAVSKY A: *Risk and Culture*. Berkeley, Univ. of California Press, 1982
3. FISCHHOFF BV: How safe is safe enough? A psychometric study of attitudes towards technological risks and benefits. *Pol Sci* 8:127–152, 1982
4. HAMMOND P: Utilitarianism, uncertainty, and information, in *Utilitarianism and Beyond*, edited by SEN A, WILLIAMS B, Cambridge, England, Cambridge Univ. Press, 1982
5. JEFFREY R: *The Logic of Decision*. New York, McGraw-Hill, 1965
6. MACLEAN D, BROWN PG: *Energy and the Future*. Maryland Studies in Public Philosophy. Totowa, N.J., Rowan and Allanheid, 1983
7. NISBETT R, ROSS L: *Human Influence: Strategies and Shortcomings of Social Judgment*. Englewood Cliffs, N.J., Prentice Hall, 1980
8. RAIFFA H: *Decision Analysis*. Reading, Mass., Addison Wesley, 1968
9. STARR C: Social benefit versus technological risk. *Science* 165:1232–1238, 1969
10. SORENSON JR, SCOTCH NA, SWAZEY JP: *Reproductive Pasts, Reproductive Futures. Genetic Counseling and Its Effectiveness*. New York, Alan R. Liss, 1981
11. SORENSON JR, SCOTCH NA, SWAZEY JP, MUCATEL M, WERTZ DC, HEEREN TC: A prospective study of planned pregnancies among patients counseled for birth defects not diagnosable prenatally. Unpublished manuscript. Boston Univ. School of Public Health, 1985
12. WERTZ DC, SORENSON JR: Genetic counseling and reproductive uncertainty. *Am J Med Genet* 18:79–88, 1984
13. BLACK RB: The effects of diagnostic uncertainty and available options on perceptions of risks. *Birth Defects: Orig Art Ser* XV:SC:341–354, 1979
14. BLACK RB: Genetic counseling and decision-making: implications for social work. Paper presented at the Annual Meeting of the American Public Health Association, Los Angeles, November 1981
15. BRINGLE RG, ANTLEY RM: Elaboration of the definition of genetic counseling into a model for counselee decision-making. *Soc Biol* 27:304–318, 1980
16. LUBS M: Does genetic counseling influence risk attitudes and decision-making? *Birth Defects: Orig Art Ser* 15:355–367, 1979
17. PEARN JH: Patients' subjective interpretations of risks offered in genetic counseling. *J Med Genet* 10:129–134, 1973

18. PAUKER S, PAUKER SG: The amniocentesis decision: an explicit guide for parents. *Birth Defects: Orig Art Ser* 15:289–324, 1979
19. KAHNEMAN D, TVERSKY A: Prospect theory. *Econometrics* 47:263–291, 1979
20. KAHNEMAN D, TVERSKY A: The psychology of preference. *Sci Am* 246:160–171, 1982
21. SLOVIC P, TVERSKY A: Who accepts Savage's axioms? *Behav Sci* 19:368–373, 1974
22. TVERSKY A, KAHNEMAN D: Judgment under uncertainty: heuristics and biases. *Science* 185:1124–1131, 1974
23. LIPPMAN-HAND A, FRASER FC: Genetic counseling: parents' responses to uncertainty. *Birth Defects: Orig Art Ser* 15:325–339, 1979
24. LIPPMAN-HAND A, FRASER FC: Genetic counseling—the post-counseling period: parents' perceptions of uncertainty. *Am J Med Genet* 4:51–71, 1979
25. LOPES LL: Decision-making in the short run. *J Exp Psychol* 7:377–385, 1981
26. LOPES LL: Risk and distributional inequality. *J Exp Psychol* 10:465–475, 1984
27. SORENSON JR, KAVENAUGH CM, MUCATEL M: Client learning of risk and diagnosis in genetic counseling. *Birth Defects: Orig Art Ser* XVII(1):215–228, 1981

SEMINARS IN MEDICINE

OF THE

BETH ISRAEL HOSPITAL, BOSTON

JEFFREY S. FLIER, M.D., *Editor*

LISA H. UNDERHILL, *Assistant Editor*

A NEW ERA IN REPRODUCTIVE TECHNOLOGY

In Vitro Fertilization, Gamete Intrafallopian Transfer, and Donated Gametes and Embryos

MACHELLE M. SEIBEL, M.D.

IT has been estimated that there are at least 2.8 million infertile couples in the United States who want to have children.[1] A decade ago many of these couples remained childless, despite the many advances in reproductive endocrinology and in the treatment of infertility. In 1978, however, the birth of the first baby conceived by in vitro fertilization ushered in a new era of reproductive technology.[2] More than 3000 children have now been born as a result of this technique. Several hundred centers throughout the world perform in vitro fertilization, and there are currently more than 100 such centers in the United States. The pool of patients who might benefit from in vitro fertilization is believed to exceed 1 million, and this population is growing. Two technological spinoffs of in vitro fertilization — gamete intrafallopian trans-

From the Department of Obstetrics and Gynecology, Division of Reproductive Endocrinology and Infertility, Beth Israel Hospital, Harvard Medical School, and the Dana Biomedical Research Institute, Boston. Address reprint requests to Dr. Seibel at the Department of Obstetrics and Gynecology, Beth Israel Hospital, 330 Brookline Ave., Boston, MA 02215.

fer and in vitro fertilization with donated embryos — were successfully introduced in 1984.

This review will present an overview of in vitro fertilization and discuss some of the many issues raised by this advance in reproductive science. Many successful variations on the basic technique have evolved throughout the world; because space constraints do not permit exhaustive discussion of each one, variations in technique are described in a general way, and the salient concepts are highlighted.

NATURAL FERTILIZATION AND IN VITRO FERTILIZATION

In the normal woman of reproductive age, one mature oocyte is released into the fallopian tube each month at midcycle. During intercourse, millions of spermatozoa are deposited in the vagina. Many of them swim through the uterus, but only a few dozen reach the ampullary end of the fallopian tube. When a sperm penetrates the oocyte, fertilization occurs. The resulting embryo remains in the fallopian tube for two to three days and grows to approximately 8 to 16 cells. The embryo then enters the uterus, where it implants itself and grows over the next nine months.

The process of in vitro fertilization bypasses the fallopian tubes. The mature oocyte is removed from the ovary immediately before ovulation and is placed in a petri dish for two or three days with sperm from the woman's husband. The fertilized egg is transferred to the uterus when it has become an embryo with only two to eight cells; it may thus be less developed than embryos that enter the uterus after natural fertilization. After a number of days, this small cluster of cells implants itself and then grows over the next nine months.

Indications for in Vitro Fertilization

When in vitro fertilization was first attempted, the procedure was used only for women whose fallopian tubes were absent or severely damaged. The indications for its use have since been broadened, however, and now include severe endometriosis,[3] immunologic infertility, and cases of unexplained infertility of two or more years' duration. Approximately 50 percent of centers accept women 40 years old or older. The procedure is also performed for certain kinds of male infertility, as when the man's sperm count is severely reduced or his sperm are incapable of living in the woman's cervix.[4]

The process of in vitro fertilization has four principal steps: the induction and timing of ovulation, the retrieval of oocytes, fertilization, and the transfer of the embryo.

Induction and Timing of Ovulation

Before ovulation occurs, the oocyte is in prophase, a resting stage of the first meiotic division. Under the influence of the midcycle surge of luteinizing hormone, the oocyte resumes meiosis and is released into the oviduct at the metaphase II stage, at which time it is receptive to fertilization.[5] Oocytes that have not reached the metaphase II stage cannot be fertilized. Because ovulation normally occurs 36 to 38 hours after the luteinizing hormone level begins to rise at midcycle,[6] it is possible to use the natural cycle in the process of in vitro fertilization and to retrieve the one mature oocyte that was about to be released, by aspiration of the follicle 34 hours after the beginning of the rise in luteinizing hormone.

The natural cycle has two disadvantages, however. First, only a single oocyte can be obtained. If more than one embryo can be transferred to the uterus, the success rate for in vitro fertilization increases considerably.[7] Second, the onset of the surge in luteinizing hormone varies, and unless the serum or urinary level of luteinizing hormone is measured frequently, it is difficult to pinpoint its onset and to schedule the retrieval of the oocyte exactly 34 hours after an increase in serum luteinizing hormone or 28 hours after an increase in urinary luteinizing hormone.[8]

For these reasons, agents that induce ovulation are used both to promote the simultaneous maturation of a number of oocytes and to cause ovulation to occur at a particular time of day, usually in the morning, so that the oocyte can be retrieved at a specific time. Two basic regimens of fertility drugs are used, with many variations. One uses human menopausal gonadotropins (hMG) alone.[9] The other uses both hMG and clomiphene, either sequentially or in combination.[10] More recently, two additional protocols have been used extensively. The first uses urofollitropin (Metrodin), a purified follicle-stimulating hormone administered in combination with hMG. The second uses a gonadotropin-releasing–hormone agonist to reduce the activity of the pituitary gland for approximately two weeks. During this time, the gonadotropin-releasing–hormone agonist is continued, and hMG is administered to stimulate follicular development.

Whichever of these regimens is used, the development of the follicles is monitored by both daily measurement of serum estradiol levels and pelvic ultrasound examinations. Most in vitro fertilization centers measure levels of luteinizing hormone, progesterone, or both before ovulation in order to detect the midcycle rise in luteinizing hormone that indicates impending ovulation.[5,6,11] The goal is a steadily increasing estradiol level combined with a progressive growth of follicles until at least two or, preferably, three or more follicles reach approximately 18 mm in diameter. Before the natural rise in luteinizing hormone can occur, 5000 to 10,000 IU of human chorionic gonadotropin is administered to simulate the surge in luteinizing hormone and to trigger the resumption of meiosis and ovulation. Ovulation occurs 36 to 38 hours after the administration of human chorionic gonadotropin, just as it follows the onset of the rise in luteinizing hormone in the natural cycle. Thus, if the human chorionic gonadotropin is administered at 11:00 p.m., the oocyte can be retrieved 34 hours later, during the morning operating schedule, just before ovulation is anticipated. The development of the follicles is satisfactory in 65 to 75 percent of initial attempts. Lack of success is usually due to poor estradiol response or to an insufficient number of developing follicles. In such cases, the cycle can be halted before the oocyte is retrieved, and different dosages or a different combination of medications can often be used in a subsequent cycle to induce ovulation successfully.

Retrieval of Oocytes

The next step is the aspiration of the oocytes from the follicles. Oocytes are successfully retrieved more than 95 percent of the time. The method most frequently used has been laparoscopy, performed in the routine manner. A grasping instrument is inserted through an incision above the pubic bone to hold and stabilize the ovary. A needle trocar can then be inserted, either through a small third incision 2.5 to 5 cm above the second or through the laparoscope itself. Looking through the laparoscope, the surgeon pushes the needle tip into the follicle, and the oocyte and the surrounding fluid are aspirated by gentle suction. The follicle is then filled to its former volume with culture medium and reaspirated at least once. The collecting tube is taken to the laboratory, which should ideally be adjacent to the operating room. The contents are poured into a petri dish, and the oocyte is identified with a dissecting microscope. The majority of oocytes are found in the initial aspirate and the remainder in the fluid used to irrigate the follicle.

During the past year, an increasing number of centers have begun to retrieve oocytes by means of aspiration guided by ultrasound, as well as by laparoscopy.[12-14] In patients with extensive pelvic adhesions, ultrasonography can obviate the need for a preparatory operative procedure to make the ovaries more accessible to laparoscopic retrieval of oocytes. Ultrasonography is also a less invasive technique and can be performed under local anesthesia augmented by intravenous analgesic agents. Finally, when the need for laparoscopy and the operating room is eliminated, the cost of the procedure is reduced. Although the risks of laparoscopy are well known, the potential hazard of concealed hemorrhage after the ultrasound-guided retrieval of the oocytes should also be mentioned; one

195

patient died after unrecognized intraperitoneal hemorrhage. Optimally, therefore, the procedure should be carried out in a hospital so that emergency backup is available.

Three ultrasound-guided techniques are currently used: transabdominal–transvesical, transvaginal, and transurethral retrieval. For the transabdominal–transvesical approach, the ultrasound transducer is placed on the abdomen and the needle is inserted into the follicle after it passes through the abdomen and the bladder. The transvaginal approach uses a transvaginal ultrasound transducer with a needle guide attached. Alternatively, an ultrasound transducer can be placed on the abdomen and the needle can be inserted transvaginally through the posterior cul-de-sac into the follicle. For the transurethral approach, ultrasound transducer is placed on the abdomen; the needle is passed through the urethra and the posterior wall of the bladder into the follicle (Fig. 1). Success with these methods has resulted in the addition of ultrasonography or an increase in its use in many centers offering in vitro fertilization. Ideally, both laparoscopy and ultrasound-guided oocyte retrieval should be available, so that the anatomical needs of all patients can be accommodated.

Fertilization

After the aspiration of the follicle, the oocytes are identified, either in the liquid initially aspirated from the follicle or in the fluid used later to irrigate the follicle. The stage of maturation of each oocyte is estimated on the basis of the appearance of the granulosa cells that surround the oocyte (cumulus oophorous). A tightly clustered cumulus indicates an immature oocyte, whereas a loosely expanded cumulus indicates

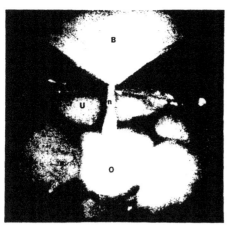

Figure 1. Ultrasonic View of the Distended Bladder.
The needle is seen traversing the bladder (B), and the needle tip (n) can be seen within a follicle. U indicates uterus; O indicates ovary.

maturity.[15] At Beth Israel Hospital, oocytes that appear mature are incubated for 5 to 6 hours and those that are thought to be immature are incubated for 24 to 26 hours before the spermatozoa are added, in order to ensure full maturity.

While the oocytes are being incubated, the man is asked to produce a semen specimen. The ejaculated spermatozoa must then undergo capacitation — that is, specific changes in metabolism and in the plasma membrane must be induced that will allow fertilization to occur.[16] The spermatozoa are washed in a special insemination medium to separate them from the seminal plasma, which is believed to contain several factors that make fertilization less likely, and are capacitated by one of several methods. The specimen may be centrifuged and the sperm that "swim up" from the pellet may be collected, or the specimen may be centrifuged on a percoll gradient and the capacitated sperm may be collected from a defined layer of the gradient. Typically, 50,000 to 100,000 sperm are added to the culture medium that contains the oocyte,[4] although this number varies widely from center to center. After the spermatozoa enter the cytoplasm, the oocyte completes its second meiotic division and the second polar body is extruded. The next morning (16 to 17 hours after insemination), the oocyte is visualized; if fertilization has occurred, two pronuclei are usually seen (Fig. 2). This stage is followed by cleavage. The development of the two-celled stage takes an average of 37 hours (range, 31 to 43 hours); the development of the four-celled stage takes an average of 44 hours (range, 37 to 51). The embryos are usually transferred 48 to 72 hours after fertilization. Studies from our center have demonstrated that a higher percentage of successful pregnancies result from embryos that have attained the four-celled stage by 48 hours after insemination.[17]

Transfer of the Embryo

Technically, the transfer of the embryo — the placement of the embryo in the uterus — is perhaps the simplest and most straightforward part of in vitro fertilization.[18] Yet it is also the most disappointing, since it is the stage at which most of the failures of in vitro fertilization seem to occur. In my opinion, the transfer of the embryo should ideally take place in an operating room adjacent to the culture room, in order to ensure sterile technique; however, a number of centers successfully transfer embryos outside the operating room. In order to benefit from gravity, the woman must be properly positioned. If her uterus is retroverted, she should lie on her back. If it is anteverted, she should be in the knee–chest position. A speculum is placed in the vagina, and the cervix is cleaned with sterile saline. A plastic or metal sheath is advanced into the cervix up to the level of the internal cervical os. The embryo or embryos, in a small volume of culture medium (20 to 30 μl), are loaded into the tip of a thin catheter, which is threaded through the cervix until it is within 1 to 2 cm of the uterine fundus.

Here the embryo or embryos are released. The catheter is held in place for one minute or less and is then slowly removed. Patients remain in the same position while the catheter is examined under the microscope to ensure that all embryos have been released. If any remain in the catheter, the process is repeated.

After the embryologist confirms that the embryos are no longer in the catheter, the speculum is removed. The patient, still on a stretcher, lies flat and is taken back to a recovery area, where she remains at rest for a variable period, usually not more than six hours. She is then discharged from the hospital and asked to remain in bed as much as possible for the next 48 to 72 hours. Most centers, although not all, administer progesterone after the transfer of the embryos for hormonal support of the endometrium. Its efficacy has not been established.

Rates of Pregnancy and Spontaneous Abortion and Incidence of Abnormalities

The first baby conceived by in vitro fertilization in the United States was born at Eastern Virginia Medical School in Norfolk, Virginia, on December 28, 1981. The first baby born in Massachusetts as a result of in vitro fertilization was conceived at Beth Israel Hospital in Boston and was delivered on July 23, 1984.

At the fourth World Conference on In Vitro Fertilization in Melbourne, Australia, in November 1985, Marrs reported the results of a questionnaire distributed to the in vitro fertilization centers in the United States. Of 7057 laparoscopies performed, 891 resulted in clinically confirmed pregnancies (12.6 percent); at the time the questionnaire was administered, 387 live births had resulted from these clinical pregnancies (5.5 percent of laparoscopies), but not all pregnant patients had delivered. Assuming a 23 percent rate of spontaneous abortion among the remaining ongoing pregnancies, one would anticipate a rate of live births of approximately 8 percent per cycle. Only 55 of the 123 centers in the United States (45 percent) had reported their results, however, and many of those that had done so were new programs with few patients and performing few procedures. Fifteen of the 55 centers reported no pregnancies whatsoever. Most of the pregnancies had resulted from procedures in established centers that attempted fertilization during more than 100 cycles per year. The rates of pregnancy among centers that attempt fertilization during more than 100 cycles per year and that have highly selective criteria for admission to their programs may be 16 percent or higher.

Because of variations in the reporting of results and the selection of patients, it is exceedingly difficult to obtain accurate rates of pregnancy. Pregnancies demonstrated by positive pregnancy test only, miscarriages, and pregnancies carried to term are not distinguished in the available data, nor are the number of pregnancies per patient, per cycle started, per oocyte-retrieval procedure, or per embryo transfer easy to determine from the reported pregnancy rates. How-

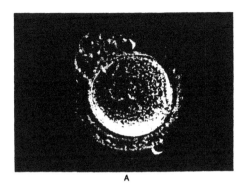

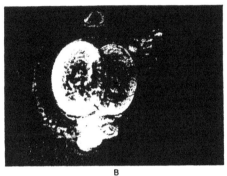

Figure 2. A Fertilized Oocyte.
Panel A shows a fertilized oocyte with two pronuclei. Panel B shows a two-celled embryo.

ever, the rate of ongoing pregnancies at most centers is no more than 10 to 15 percent per oocyte-retrieval procedure, and the rate of pregnancy at many centers is lower.[19] Although relatively few patients go through the procedure repeatedly, a recent study[20] suggests that those who do so have an increasing chance of success. After six attempts, the rate of ongoing pregnancies may approximate 60 percent.

To place this apparently low success rate in perspective, one must recognize that the human reproductive system is very inefficient. Of 100 eggs exposed to potential fertilization among fertile couples, only 31 will actually produce a viable offspring.[21] The other 69 will be lost, usually within the first two to three weeks of pregnancy. Therefore, if we could do as well as nature, we would anticipate only a 31 percent success rate. We can hope, however, that improvements in the techniques of fertilization, embryo growth, embryo transfer, and cryopreservation will increase the success rate of in vitro fertilization over the next few years.

Among the 891 clinical pregnancies reported by U.S. centers in response to Marrs' questionnaire, the incidence of twins was 19.8 percent; of triplets, 3.8

percent; and of ectopic pregnancies, 2.6 percent. Birth defects occurred in 1.7 percent of the babies. In Australia, the incidence of birth defects was 2 percent in singleton pregnancies and 3.4 percent in multiple pregnancies.[22] The broad spectrum of defects reported suggests that they did not result from the in vitro fertilization procedure itself. When one takes into account the fact that many of the patients who undergo in vitro fertilization are in the older reproductive age group, these percentages approximate those found in the general population.

The Australian centers also reported a rate of premature delivery of 18 percent for singleton pregnancies and 27 percent for multiple pregnancies, as well as a 2.4 percent incidence of stillbirth.[22] At Beth Israel Hospital, it appeared at first that babies conceived by in vitro fertilization were more likely than others to be delivered by cesarean section and were more likely to require treatment in the neonatal intensive care unit. When the mothers of these babies were compared with age-matched controls in the same institution, however, the difference was not significant. The only complication experienced significantly more frequently by mothers whose pregnancies were initiated in vitro than the control mothers was the need for agents that arrest premature labor (ritodrine or other beta-sympathomimetic drugs). However, because of the possibility that babies conceived in vitro may be at higher risk of perinatal complications than other babies, it is my opinion that until special precautions are clearly proved unnecessary, women who undergo in vitro fertilization should be considered at high risk and should be cared for by perinatologists at centers for high-risk deliveries. On a brighter note, recent studies of infants conceived in vitro demonstrate that both early psychosocial development[23] and development at the first birthday[24] appear normal.

Financial Considerations

In vitro fertilization is a complex procedure that typically requires, among other things, several outpatient hospital procedures — including at least one operative procedure, numerous laboratory tests, and ultrasound examinations — and the availability of a highly specialized fertilization laboratory and trained personnel. The cost in most centers approaches $6,000 from the start of the menstrual cycle to the outcome of the pregnancy test two or three weeks after the transfer of the embryo. Streamlined protocols and the retrieval of oocytes with the guidance of ultrasonography, using local anesthesia, may reduce costs. A number of private insurance companies cover all or a portion of the costs of the procedure. Recently, under pressure, the Massachusetts Senate and House of Representatives enacted a bill that mandates insurance coverage "for medically necessary expenses and diagnosis and treatment of infertility." This bill went into effect January 6, 1988. However, variable interpretation of the law by insurance companies has resulted in inconsistent and incomplete coverage. Therefore, many couples for whom advances in reproductive technol-

ogy offer the only chance to achieve biologic parenthood still continue to be denied reimbursement for these procedures.

GAMETE INTRAFALLOPIAN TRANSFER

The transfer of gametes into the fallopian tubes is an alternative technique for a subgroup of candidates for in vitro fertilization.[25] This procedure, first performed in 1984, is similar to in vitro fertilization, but with a few important distinctions. The ovulation-inducing medications, pelvic ultrasound examinations, and daily blood tests used for this procedure are similar to those used for in vitro fertilization. The man is asked to produce a semen specimen, and the spermatozoa are capacitated as they are for in vitro fertilization. In gamete intrafallopian transfer, however, the spermatozoa and oocytes are placed in the fimbriated end of the fallopian tube when the laparoscopy is performed. Usually, no more than two oocytes are introduced into each fallopian tube. Candidates for this procedure must have at least one patent fallopian tube that appears to be normal.

Because fertilization occurs in vivo and not in vitro, the Catholic Church has accepted this procedure, with two specific provisions. First, a perforated silastic sheath must be used to collect the semen during intercourse, so that neither masturbation nor contraception is involved. A portion of the collected semen must also be placed in the vagina. Second, the spermatozoa and the oocytes must be separated in the transfer catheter so that fertilization can occur in vivo.

Indications for gamete intrafallopian transfer include unexplained infertility of two or more years' duration, cervical stenosis, immunologic infertility, oligospermia, and endometriosis (when the fallopian tubes are free of disease). Severe pelvic adhesions or distorted tubal anatomy from any cause are contraindications. A previous ectopic pregnancy is a less severe contraindication.

Accurate data from around the world have been difficult to obtain, but estimates are that the pregnancy rates for this procedure are the same as or slightly higher than for in vitro fertilization. Procedures performed because of oligospermia have been particularly disappointing. Ectopic pregnancy occurs 3 to 8 percent of the time. Because this procedure does not require in vitro culture techniques and embryo transfer, it costs approximately 25 percent less than in vitro fertilization. At present, there is no consensus about the method of choice when both gamete intrafallopian transfer and in vitro fertilization are acceptable. My own practice has been to perform in vitro fertilization initially, in order to establish that fertilization can occur, and to perform gamete intrafallopian transfer on subsequent attempts because it is less expensive.

DONATED EMBRYOS, OOCYTES, AND SPERMATOZOA

Initially, in vitro fertilization was intended to provide genetic offspring to women whose fallopian tubes were absent or blocked. Under certain conditions, however, a woman may require the donation of an embryo[26] or oocyte[27] (Table 1). Candidates for these procedures fall into two major groups: those with normal ovarian function and those with ovarian failure.[28] In both instances, the critical component is the synchronization of the recipient's cycle with the donor's. In recipients with normal ovarian function, the surge in luteinizing hormone is identified simply by monitoring daily serum levels of estradiol, luteinizing hormone, and progesterone. A recipient with ovarian failure must have the status of her endometrium made compatible with the donor's by the administration of exogenous estradiol and progesterone, available either as capsules of micronized hormone or as vaginal suppositories or intramuscular injections.[29] The recipients' hormones are usually monitored for one to two months before the procedure so that the appropriate time for transfer can be anticipated. An endometrial biopsy is performed in the luteal phase of a monitored cycle to ensure that the condition of the endometrium is adequate for transplantation. A four-to-eight-cell conceptus must be available for transfer on days 17 to 19 of the recipient's menstrual cycle.[28] After the embryo is transferred, patients with ovarian failure must continue to receive exogenous hormones for 12 to 18 weeks, until it is certain that the placenta will provide adequate endogenous support.[28]

Embryos are obtained from paid volunteers who are ovulating normally.[26] Sperm from the recipient's husband is used to inseminate the donor artificially. Several days later, the donor's endometrium is nonsurgically flushed out with a special catheter. If fertilization has occurred, the embryo is then transferred to the uterine cavity of the recipient, whose endometrial development has been approximately synchronized. The disadvantage of this procedure is that the flushing technique occasionally fails to remove the conceptus, and the pregnancy may continue in the donor.

The donation of oocytes is analogous to artificial insemination by donor in cases of oligospermia or azoospermia; this method has the advantage that the donor is not at risk for continuing the pregnancy. Potential donors are usually patients undergoing in vitro fertilization. Some in vitro fertilization centers do not fertilize more than four or five oocytes per patient because the transfer of more than four or five embryos

Table 1. Indications for the Donation of Oocytes or Embryos.

Normal ovarian function
  Abnormal, unfertilized, or degenerate oocytes
  Inadequate response to ovarian stimulation
  Transmissible genetic disease
No ovarian function
  Primary gonadal failure
    Gonadal dysgenesis
    Resistant-ovary (Savage) syndrome
    Autoimmunity
  Secondary ovarian failure
    Premature menopause
    Surgically absent ovaries
    Failure induced by chemotherapy or radiotherapy

does not improve the success rate and increases the risk of multiple births. Therefore, if excess oocytes are retrieved, patients may choose to discard them, to have them fertilized and cryopreserved if possible, or to donate them anonymously. The identities of the donor and recipient are kept confidential, and donors are not informed of the outcome of the fertilization procedure or of any resulting pregnancy. Women undergoing tubal ligations who agree to receive stimulation by gonadotropin before laparoscopy can also donate oocytes. More recently, the technique of ultrasound-guided retrieval of oocytes under local anesthesia has made it possible to pay volunteers or ask family members to donate oocytes. The fertilization procedure is the same as for in vitro fertilization, and the embryo is transferred in the same way.

In certain circumstances, in vitro fertilization may be carried out when both male and female infertility is present. In such instances, the procedure is identical to that for in vitro fertilization, and donated spermatozoa are used to fertilize the donated oocyte.

EMOTIONAL ASPECTS OF REPRODUCTIVE TECHNOLOGY

In vitro fertilization and its variations are emotionally charged procedures. In my opinion, far too little psychological support is provided to the majority of patients who undergo these procedures. Few centers routinely include psychological counseling for couples undergoing in vitro fertilization and, to my knowledge, virtually no centers offer support either for couples who fail to conceive or for those who succeed. In my opinion, psychological support should be provided to all patients and should be continued throughout the pregnancy and perhaps into the neonatal period.[30]

ETHICAL AND LEGAL CONSIDERATIONS

Techniques to unite the sperm and egg of consenting married adults outside of the body and to transfer the fertilized product into the wife's uterus are now widely accepted. Nevertheless, a number of religious, ethical, and legal issues remain.[31-33] Although a comprehensive discussion of these considerations is beyond the scope of this article, one must be mentioned in particular: the cryopreservation of embryos.

The freezing and thawing of animal embryos to increase reproductive potential has been an established practice in agriculture for more than a decade. The freezing and thawing of human embryos, although still experimental, is slowly becoming part of the routine practice of in vitro fertilization around the world. The major advantage of cryopreservation is that retrieved oocytes that are not used during a particular cycle can be fertilized and stored for later use. It is speculated that the use of cryopreserved embryos could increase the rate of pregnancies per oocyte-retrieval procedure by 8 to 12 percent.[34] Because the use of cryopreserved embryos in subsequent cycles does not require additional oocyte-retrieval procedures, the cost of attempting fertilization during these cycles is reduced. Successful pregnancies from frozen and thawed human embryos have been reported in the United States, Australia, and Europe. Because success rates after the freezing and thawing of embryos are similar to those achieved by conventional in vitro fertilization, the demand for cryopreservation is growing.[35]

The main legal issues that surround cryopreservation concern the authority to dispose of the embryo, the length of storage, posthumous use, inheritance rights, and family relations after embryo donation and surrogacy.[36] The primary ethical questions include whether the freezing and thawing of human embryos results in abnormal or defective births (no evidence thus far indicates that it does), whether the embryo itself has rights before implantation, and whether the freezing of embryos is an unacceptable intrusion into the natural process of reproduction.[36] The Ethics Committee of the American Fertility Society has determined that although embryos deserve special respect, they do not themselves have the rights or status of persons and, therefore, cannot be wronged.[33] The literature on this important topic is becoming extensive.[31-40]

A more practical ethical consideration has to do with which physicians and scientists should perform in vitro fertilization. The American Fertility Society has suggested that the minimal requirements for establishing an in vitro fertilization center[41] include the presence of a board-certified reproductive endocrinologist and an adequately trained scientist. Exceptions clearly exist, however, and many pioneers in the field do not have such certification. Nonetheless, at least in the United States, where certification is available, these guidelines will help ensure that patients receive treatment only from qualified practitioners. The society is in the process of compiling a list of all American in vitro fertilization centers that perform 40 or more procedures annually and already have three or more live births. It is hoped that this list will help patients choose the centers that are most likely to help them achieve a successful pregnancy.

REFERENCES

1. Talbert LM. Overview of the diagnostic evaluation. In: Hammond MG, Talbert LM, eds. Infertility: a practical guide for the physician. 2nd ed. Oradell, N.J.: Medical Economics, 1985:1-8.
2. Steptoe PC, Edwards RG. Birth after the reimplantation of a human embryo. Lancet 1978; 2:366.
3. Chillik C, Rosenwaks Z. Endometriosis and in vitro fertilization. Semin Reprod Endocrinol 1985; 3:377-80.
4. Alper MM, Lee GS, Seibel MM, et al. The relationship of semen parameters to fertilization in patients participating in a program of in vitro fertilization. J In Vitro Fert Embryo Transfer 1985; 2:217-23.
5. Seibel MM, Smith DM, Levesque L, Borten M, Taymor ML. The temporal relationship between the luteinizing hormone surge and human oocyte maturation. Am J Obstet Gynecol 1982; 142:568-72.
6. Taymor ML, Seibel MM, Smith D, Levesque L. Ovulation timing by luteinizing hormone assay and follicle puncture. Obstet Gynecol 1983; 62:191-5.
7. Jones HW Jr, Acosta AA, Andrews MC, et al. Three years of in vitro fertilization at Norfolk. Fertil Steril 1984; 42:826-34.
8. Seibel MM, Shine W, Smith DM, Taymor ML. Biological rhythm of the luteinizing hormone surge in women. Fertil Steril 1982; 37:709-11.
9. Jones HW Jr, Jones GS, Andrews MC, et al. The program for in vitro fertilization at Norfolk. Fertil Steril 1982; 38:14-21.
10. Taymor ML, Seibel M, Oskowitz SP, Smith DM, Lee G. In vitro fertilization and embryo transfer: an individualized approach to ovulation induction. J In Vitro Fert Embryo Transfer 1985; 2:162-5.
11. Trounson AO, Calabrese R. Changes in plasma progesterone concentrations around the time of the luteinizing hormone surge in women superovulated for in vitro fertilization. J Clin Endocrinol Metab 1984; 59:1075-80.
12. Wikland M, Nilsson L, Hansson R, Hamberger L, Janson PO. Collection of human oocytes by the use of sonography. Fertil Steril 1983; 39:603-8.
13. Parsons J, Riddle A, Booker M, et al. Oocyte retrieval for in-vitro fertilisation by ultrasonically guided needle aspiration via the urethra. Lancet 1985; 1:1076-7.
14. Dellenbach P, Nisand I, Moreau L, Feger B, Plumere C, Gerlinger P. Transvaginal sonographically controlled follicle puncture for oocyte retrieval. Fertil Steril 1985; 44:656-62.
15. Laufer N, Botero-Ruiz W, DeCherney AH, Haseltine F, Polan ML, Behrman HR. Gonadotropin and prolactin levels in follicular fluid of human ova successfully fertilized in vitro. J Clin Endocrinol Metab 1984; 58:430-4.
16. Chang MC. The meaning of sperm capacitation: a historical perspective. J Androl 1984; 5:45-50.
17. Claman P, Armant DR, Seibel MM, Wang T-A, Oskowitz SP, Taymor ML. The impact of embryo quality and quantity on implantation and the establishment of viable pregnancies. J In Vitro Fert Embryo Transfer 1987; 4:218-22.
18. Jones HW Jr, Acosta AA, Garcia JE, Sandow BA, Veeck L. On the transfer of conceptuses from oocytes fertilized in vitro. Fertil Steril 1983; 39: 241-3.
19. Medical Research International, American Fertility Society Special Interest Group. In vitro fertilization/embryo transfer in the United States: 1985 and 1986 results from the National IVF/ET Registry. Fertil Steril 1988; 49:212-5.
20. Guzick DS, Wilkes C, Jones HW Jr. Cumulative pregnancy rates for in vitro fertilization. Fertil Steril 1986; 46:663-7.
21. Leridon H. Facts and artifacts in the study of intra-uterine mortality: a reconstruction from pregnancy histories. Popul Stud 1976; 30:319-35.
22. Lancaster PL. In vitro fertilization pregnancies, Australia 1980-1983. National Perinatal Statistics Unit, Commonwealth Institute of Health, University of Sidney, New South Wales, 1984. (Publication ISSN 0814-7205.)
23. Mushin DN, Barreda-Hanson MC, Spensley JC. In vitro fertilization children: early psychosocial development. J In Vitro Fert Embryo Transfer 1986; 3:247-52.
24. Yovich JL, Parry TS, French NP, Grauaug AA. Developmental assessment of twenty in vitro fertilization (IVF) infants at their first birthday. J In Vitro Fert Embryo Transfer 1986; 3:253-7.
25. Asch RH, Ellsworth LR, Balmaceda JP, Wong PC. Pregnancy after translaparoscopic gamete intrafallopian tube transfer. Lancet 1984; 2:1034-5.
26. Bustillo M, Buster JE, Cohen SW, et al. Nonsurgical ovum transfer as a treatment in infertile women: preliminary experience. JAMA 1984; 251:1171-3.
27. Lutjen P, Trounson A, Leeton J, Finday J, Wood C, Renou P. The establishment and maintenance of pregnancy using in vitro fertilization and embryo donation in a patient with primary ovarian failure. Nature 1984; 307: 174-5.
28. Rosenwaks Z. Donor eggs: their application in modern reproductive technology. Fertil Steril 1987; 47:895-909.
29. Navot D, Laufer N, Kopolovic J, et al. Artificially induced endometrial cycles and establishment of pregnancies in the absence of ovaries. N Engl J Med 1986; 314:806-11.
30. Seibel MM, Levin S. A new era in reproductive technologies: the emotional stages of in vitro fertilization. J In Vitro Fert Embryo Transfer 1987; 4:135-40.
31. Quigley MM, Andrews LB. Human in vitro fertilization and the law. Fertil Steril 1984; 42:348-355.
32. American Fertility Society. Ethical statement on in vitro fertilization. Fertil Steril 1984; 41:12.
33. The Ethics Committee of the American Fertility Society. Ethical considerations of the new reproductive technologies. Fertil Steril 1986; 46:Suppl 1:1S-94S.
34. Trounson A. Preservation of human eggs and embryos. Fertil Steril 1986; 46:1-12.
35. Testart J, Lassalle B, Belaisch-Allart J, et al. Human embryo viability related to freezing and thawing procedures. Am J Obstet Gynecol 1987; 157:168-71.
36. Robertson JA. Ethical and legal issues in cryopreservation of human embryos. Fertil Steril 1987; 47:371-82.
37. Cohen J, Simons RF, Edwards RG, Fehilly CB, Fishel SB. Pregnancies following the frozen storage of expanding human blastocysts. J In Vitro Fert Embryo Transfer 1985; 2:59-64.
38. Report of the Committee of Inquiry into Human Fertilisation and Embryology. London: Her Majesty's Stationery Office, 1984:53.
39. Chargaff E. Engineering a molecular nightmare. Nature 1987; 327:199-200.
40. May WF. Religious justification for donating body parts. Hastings Cent Rep 1985; 15(1):38-42.
41. American Fertility Society. Minimal standards for programs of in vitro fertilization. Fertil Steril 1984; 41:13.

FERTILITY AND STERILITY
Copyright © 1989 The American Fertility Society

Vol. 51, No. 3, March 1989
Printed in U.S.A.

The new genetics: molecular technology and reproductive biology

William J. Butler, M.D.
Paul G. McDonough, M.D.

Department of Obstetrics and Gynecology, The Albany Medical College, Albany, New York, and Department of Obstetrics and Gynecology, Medical College of Georgia, Augusta, Georgia

Recombinant deoxyribonucleic acid (DNA) technology has had a major impact during the last decade on the field of biology. Advances in molecular biology have provided new tools to investigate poorly understood disorders of reproduction, yielding new insights into the etiology of recurrent pregnancy loss and intersex syndromes. Gene cloning has allowed the characterization and expression of genes coding for various hormones, with a potential for improved diagnosis and treatment of reproductive dysfunctions. Understanding the basis for this new technology will allow the clinician to use the resulting diagnostic and therapeutic techniques to his patient's benefit.

MOLECULAR BIOLOGY

Understanding the new DNA technology is predicated on a basic understanding of molecular genetics. Deoxyribonucleic acid is the genetic material responsible for coding for the diversity of cellular proteins. It is highly conserved and passed to daughter cells and, through gametes, to progeny. Genes are segments of DNA that code for specific proteins, along with flanking and intervening sequences that serve controlling functions for gene transcription. The study of genetics used to be an inferential science, with analysis of phenotype and family studies used to infer the underlying genetic abnormalities. Diagnosis therefore was dependent on expression of the abnormal genes. Direct DNA analysis makes diagnosis independent of gene expression or identification of gene products, potentially enabling the diagnosis to be made preclinically. As DNA is present and identical in all cells

of an individual, direct analysis can be performed on any available tissue. The technology involved will be briefly reviewed.

RECOMBINANT DNA

The first step is to obtain DNA. Deoxyribonucleic acid can be extracted from any tissue by careful lysis of the cells and nuclei, and separation from attached proteins. The proteins are extracted with organic solvents while the water-soluble DNA remains in the aqueous phase, and then is precipitated with ethanol. Total genomic DNA is comprised of long continuous double strands; working with these is difficult. A method therefore was needed to obtain pieces of manageable length for study. In 1970, Smith and Wilcox[1] and Kelly and Smith[2] at Johns Hopkins described an enzyme they isolated from a bacterium that consistently cut bacteriophage (bacterial virus) DNA at a specific recognition site.[1,2] This was the first restriction endonuclease. Restriction endonucleases are bacterial enzymes that protect the bacteria by degrading foreign (viral) DNA. Over 200 enzymes have now been isolated from different bacterial strains that recognize many different nucleotide sequences (Table 1).[3] These enzymes can be used to reproducibly cut genomic DNA into fragments of specific sizes. Any change in the nucleotide sequence of the DNA that alters a pre-existing recognition site or creates a new site will alter the fragment sizes obtained. These different length fragments provide a basis for the analysis of individual genes.

The individual DNA fragments then must be

201

Table 1 Restriction Endonucleases

Restriction endonuclease	Digestion site[a]	
	5'	3'
BamHI	GGATCC	
Bgl II	AGATCT	
EcoRI	GAATTC	
Hae III	GGCC	
HindIII	AAGCTT	
HinfI	GANTC	
Hpa I	GTTAAC	
Hpa II[b]	CCGG	
Mbo I	GATC	
Msp I[b]	CCGG	
Mst II	CCTNAGG	
Pst I	CTGCAG	
Sal I	GTCGAC	
Taq I	TCGA	
Xba I	TCTAGA	

[a] A, adenine; C, cytosine; G, guanine; T, thymine; N, any base.

[b] Hpa II and Msp I are isoschizomers (cut at same nucleotide sequence).

separated. Deoxyribonucleic acid is a charged molecule, and will migrate when placed in an electrical field, with the distance of migration inversely proportional to the length of the fragment. Agarose (or polyacrilamide) gel electrophoresis therefore will separate the DNA fragments by size. The gel then can be overlaid with a piece of nitrocellulose paper or other filtration membrane and, using a method described by Southern,[4] a flow of buffer is used to transfer the DNA to the filter, preserving the size-dependent orientation (Fig. 1).[4] This is termed southern blotting, and an analogous technique for ribonucleic acid (RNA) analysis is called northern blotting.[5] While still in the gel, the DNA can be denatured with alkali to separate the strands. This will allow hybridization of complementary probes to the single strands to identify specific fragments (genes).

The most difficult task is to identify specific genes within this group of fragments of total genomic DNA. In order to accomplish this, probes, which are pieces of RNA or DNA that are complementary to the gene being sought, must be isolated. There are several ways to obtain probes. One is to obtain messenger RNA (mRNA) transcripts of a certain gene, which can be abundant in certain specialized cells (for example, the mRNA for hemoglobin in erythrocytes). An enzyme called reverse transcriptase then can synthesize copy DNA (cDNA) complementary to the base sequence of the mRNA, and this cDNA probe can be used to identify the DNA fragment with that gene on the

southern blot. Techniques also have been developed to use less abundant mRNAs to generate cDNA probes.[6]

If the amino acid sequence of a protein is known, the corresponding DNA sequence can be deduced and synthesized. These oligonucleotide probes are short, usually about 20 nucleotides; and therefore specifically bind to only one small segment of the gene. This is useful for detection of small mutations within a gene, and therefore has frequent application in prenatal diagnosis. The third way to obtain a probe is to screen a genomic DNA library with a probe for a closely linked region of DNA, hoping to identify a cloned fragment that contains both the known sequence and part of an attached gene of interest. This new DNA fragment then can be used as a probe to find other cloned overlapping fragments in a process known as "gene walking," allowing analysis of long fragments of continuous DNA.[7]

Cloning is a process by which a DNA fragment is inserted into a vector that replicates in bacterial cells, making that fragment available for study or manipulation.[8] Vectors include plasmids, bacteriophage, and cosmids (artificial constructs of plasmid and phage) that have the ability to continue to replicate after incorporation of additional DNA sequences. The choice of a vector depends on the size of the DNA fragments to be cloned, with plasmids accepting the smallest fragments and cosmids the largest. The combined vector and DNA fragment is termed a recombinant. Cloning is a valuable tool. It allows the production of multiple identical copies of a probe, gene, or unidentified DNA fragment

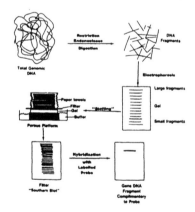

Figure 1 Southern blotting.

Figure 2 Prenatal diagnosis of sickle-cell anemia by direct DNA analysis. The first two lanes show the parental genomic DNA digested with *Mst* II and hybridized to β-globin probe pSS 1.8. Both 1.3 and 1.1 kilobase bands are present, indicating heterozygosity for the sickle gene. The fourth lane is the prenatal sample, also demonstrating both bands. The fetus will not have sickle-cell anemia, but will be a carrier. The third, fifth, and sixth lanes are known homozygous sickle, homozygous hemoglobin A, and heterozygous AS DNA, respectively.

from the original molecule by amplifying the recombinant in bacterial cells.[9] It also can be used to construct a genomic library, which is total genomic DNA cut into fragments and then all fragments inserted into vectors and bacterial cells, under conditions such that each cell gets only one recombinant. This library gives access to all genomic DNA, with probes acting as the index to identify specific genes by screening bacterial colonies for specific cloned inserts.[10]

MOLECULAR DIAGNOSIS

Molecular diagnosis is possible using various applications of these previously described techniques. Direct gene analysis for a specific abnormal gene is possible when the DNA mutation alters a restriction site for cutting the gene with a restriction endonuclease. Adding or subtracting a restriction site will change the size of the DNA fragment carrying the gene, and therefore will alter its position on a southern blot. Hybridization with the probe for the desired gene, usually radioactively labeled to enable visualization via autoradiography, will locate the specific fragment, and its position will reveal whether the normal or abnormal gene is present. The best example of this in clinical practice is sickle-cell anemia. The sickle mutation alters the recognition site for the restriction endonucleases Dde I and Mst II, such that the sickle gene gives a larger fragment than the normal β-globin gene when genomic DNA is cleaved with these enzymes (Fig. 2).[11] A heterozygous carrier of the sickle gene will show both the larger and smaller bands, indicating one normal β-globin gene on one chromo-

some and one sickle gene on the other. Gross insertions or deletions of DNA also can be detected if a probe for the gene in question is available. Either the fragment size will be altered by the excess or deleted DNA, or if the complete gene is deleted, the probe will fail to hybridize with any fragment. Isolated deficiency of human growth hormone is one example.[12]

Oligonucleotide probes are useful when the precise DNA mutation is known, but does not affect a restriction enzyme recognition site. Two short (approximately 20 to 30 nucleotides) DNA probes are synthesized, one complementary to the normal gene and the other complementary to the mutated DNA sequence. These labeled probes are hybridized to the southern blot under high stringency conditions such that only the probe that is perfectly complementary will stably hybridize, allowing detection of a normal or abnormal gene. Some thalassemias and α-1-antitrypsin deficiency can be diagnosed in this fashion.[13,14]

Restriction fragment length polymorphisms (RFLPs) are benign point mutations that alter restriction sites in DNA but are genetically neutral and have no phenotypic effects. They provide a powerful diagnostic tool in cases where the genetic mutation or even the normal gene is unknown. If a RFLP is closely linked to the gene in question, as determined by family studies, the passage of the gene through that family can be followed by following the RFLP.[15,16] The limitations of this process are that the DNA from at least one affected individual must be available to determine if the RFLP is present in a particular family, there must be heterozygosity at both the RFLP locus and gene locus (the family must be informative), and enough family members must be tested to determine the phase of the RFLP; whether it is segregating with the normal or abnormal gene (Fig. 3). False positive or negative results also are possible because of recombination, the exchange of DNA between homologous chromosomes, that may occur between the RFLP marker and the gene. The percentage of recombination is directly correlated to the degree of linkage between the RFLP and the gene.

PRENATAL DIAGNOSIS

Molecular techniques have been applied to the prenatal diagnosis of numerous diseases because they offered several advantages over preexisting methods. Fetal DNA can be easily isolated from amniocytes obtained at amniocentesis or tropho-

203

Figure 3 Prenatal diagnosis by RFLP—Hemophilia A. Blot A: The proband T.N. had an affected brother, R.S., and her mother, E.S., was a carrier for hemophilia A. Genomic DNA digested with *Bcl* I and hybridized to Factor VIII probe 17.18 revealed that T.N. was heterozygous for a RFLP with fragments of 1.16 and 0.87 kilobases. As E.S. was homozygous for the 0.87 kb fragment, T.N. must have received the X chromosome with the 1.16 kb fragment from her father. Blot B: Trophoblast obtained by chorionic villus biopsy revealed the male fetus was hemizygous for the 1.16 kb fragment, and therefore did not inherit T.N.'s maternal X chromosome carrying the hemophilia gene. The fetus was predicted to be unaffected, and this was confirmed at birth.

blast from chorionic villus biopsy with only small fetal risks.[17,18] Diagnosis is independent of gene expression, and therefore is not limited by the fetal tissue available for study or by the developmental stage of the fetus. Deoxyribonucleic acid studies also allow diagnosis of conditions that previously could not be detected in utero because the primary defect is unknown. Direct diagnostic methods have the advantage of detecting the mutation in the fetus, but are limited by the necessity of knowing the genetic defect responsible and being able to identify it by one of the previously described methods. Indirect diagnosis using polymorphisms only enables the prediction of fetal status by following the inheritance pattern of markers (RFLPs) in the pedigree, but can be applied to diseases that result from multiple different mutations or in which the underlying defect is unknown (Table 2).

DIRECT ANALYSIS

Sickle-cell anemia is the prototype of a mutation causing loss of a restriction site, resulting in a larger DNA fragment size after digestion with the appropriate restriction endonuclease.[11] The restriction enzyme *Mst* II cuts DNA at the sequence CCTNAGG (C, cytosine; T, thymine; A, adenine; G, guanine; N, any nucleotide). The sickle mutation substitutes T for A, changing the amino acid coded from glutamine to valine. This prevents cutting of the DNA by *Mst* II at that site, and the sickle globin gene is then detected on a DNA frag-

ment 1.3 kilobases (kb) long, instead of the 1.1 kb fragment seen in hemoglobin A.[11] A patient with sickle-cell anemia will have only the 1.3 kb fragment, whereas a heterozygous carrier (sickle trait) will show both fragments, indicating one normal and one abnormal gene. Several types of thalassemia also are secondary to mutations that alter restriction sites.[19,20]

Gene deletions also allow direct analysis if the differences in fragment size are large enough to be discriminated on the Southern blot. Alpha-thalassemia usually results from homozygous deletion of both α-globin genes.[21] Several restriction enzymes yield different sized fragments after digestion of normal and abnormal genes.[22] Deletions have been found in some cases of β-thalassemia as well.[23]

Many disorders that result from multiple different molecular defects include some affected individuals with gene deletions. The Ilig type IA growth hormone deficiency results from homozygous deletion of the growth hormone gene on chromosome 17. Hybridization with a probe for the growth hormone (GH) gene shows loss of the expected 3.9 kb fragment in affected patients, and decreased intensity of the signal in heterozygotes.[12] Some cases of classical congenital adrenal hyperplasia are secondary to deletion of the structural gene for cytochrome P450-21-hydroxylase, and the deletion may extend into the neighboring *C4A* complement gene.[24,25] Deletions have been detected in over 10% of cases of Duchenne's muscular dys-

Table 2 Molecular Prenatal Diagnosis

Genetic disease	Method of diagnosis
Sickle cell	Loss of restriction site
Thalassemia	Deletion
Thalassemia	Loss of restriction site
	Deletion
	Oligonucleotide probe
	RFLP
Hemophilia A	Deletion
	RFLP
Hemophilia B	Deletion
	RFLP
Duchenne's muscular dystrophy	Deletion
	RFLP
Cystic fibrosis	RFLP
Phenylketonuria	RFLP
Ilig type IA growth hormone deficiency	Deletion
Congenital adrenal hyperplasia	Deletion
	RFLP
α-1-antitrypsin deficiency	Oligonucleotide probe
Myotonic dystrophy	RFLP
Von Willebrand's disease	RFLP
Huntington's chorea	RFLP
Adult polycystic kidney disease	RFLP

trophy.[26,27] Both hemophilia A (Factor VIII deficiency) and hemophilia B (Factor IX deficiency) result from detectable deletions in some patients.[28,29]

Oligonucleotide probes have been useful in some populations at risk for β-thalassemia where a specific mutation is the cause of most cases[30]; however, in populations with mixed mutations, direct analysis is not possible. In contrast, α-1-antitrypsin deficiency is an autosomal recessive condition resulting from a single nucleotide change of G to A and consequent replacement of lysine for glutamine in the peptide chain.[31] Synthetic oligonucleotide probes for the normal and abnormal DNA sequences can distinguish normal, homozygous affected and heterozygous individuals.[13]

INDIRECT ANALYSIS

Restriction fragment length polymorphisms are the basis for indirect analysis of genotype and prediction of disease state. The accuracy of this technique is dependent on the closeness of the marker to the gene of interest (linkage), and its applicability on the criteria previously given that determine if a family is informative. The importance of RFLP analysis is best shown by its usefulness in diagnosing disease states in which the genetic defect is unknown. The classic example is Huntington's disease. Huntington's disease is an autosomal dominant neurodegenerative disorder with late onset of symptoms, usually after reproduction. Study of a large family revealed linkage of the Huntington's gene to a polymorphic DNA marker on chromosome 4, enabling both presymptomatic and prenatal diagnosis.[32] Other diseases resulting from unknown defective genes that are amenable to RFLP diagnosis are Duchenne's muscular dystrophy (DMD) and cystic fibrosis (CF).[33,34] In DMD, multiple linked probes known to hybridize to X chromosome loci have been used to detect RFLPs with multiple restriction endonucleases.[35] The PERT 87 series of probes were cloned from DNA sequences homologous to a deletion in a DMD-affected male, and are very closely linked to, if not within, the DMD locus.[27] Application of multiple RFLPs provides informative pedigrees in over 90% of families studies.[26] Multiple RFLPs also have been described for diagnosis of the autosomal recessive condition CF, using an anonymous DNA probe j3.11 and the Met oncogene probes, which recognize loci on chromosome 7.[36,37] The informative diagnostic rate is 85% to 90% of families. Re-

cently, two new very closely linked probes have been reported with significant linkage disequilibrium for the CF gene (one particular allele of each RFLP usually is present with the abnormal CF gene), and this will significantly expand the diagnostic capability of CF testing.[38]

Restriction fragment length polymorphisms are the primary molecular technique for diagnosis of thalassemia and the hemophilias. Over 35 mutations are known to result in β-thalassemia, and the presence of multiple different alleles within a gene pool necessitates the use of an indirect diagnostic approach.[39] Multiple RFLPs around and within the β-globin gene cluster give informative pedigrees in over 87% of families.[40] Hemophilia A, an X-linked recessive disorder, has two intragenic RFLPs that are informative in 60% of families.[41] With essentially no chance of intragenic recombination, these provide for very accurate diagnosis. Linked probes detect other RFLPs, but with up to 5% recombination rates.[42,43] A new technique involving the enzymatic amplification of the specific DNA sequences in the intragenic polymorphisms allows identification of the RFLP alleles by direct visualization on the southern blot, without the necessity for probe hybridization.[44] Hemophilia B is another X-linked condition in which a cDNA probe for the normal gene (Factor IX) recognizes several intragenic RFLPs.[45–47]

Phenylketonuria (PKU) is an autosomal recessive disease caused by a deficiency of the enzyme phenylalanine hydroxylase.[48] Prenatal diagnosis is not possible using enzyme activity because the enzyme is not expressed in amniocytes or trophoblast. Multiple DNA polymorphisms are coupled to the phenylalanine hydroxylase gene and allow both carrier detections and prenatal diagnosis.[49,50] Ornithine transcarbamylase deficiency is a rare X-linked disease for which several polymorphisms have been identified and used for diagnostic testing.[51] An RFLP also has been identified for each of the autosomal dominant conditions von Willebrand's disease,[52] adult onset polycystic kidney disease,[53] and myotonic dystrophy.[54]

RECURRENT ABORTION

In recent years, the advances made in DNA technology have allowed exploration of the previously speculative role of molecular mutation in embryonic loss. The possibility exists that defects in genetic coding for products critical for embryonic or fetal development play a causative role in euploidic

205

abortion in humans. Suggestive evidence can be found in genetic manipulation experiments in mouse models. Insertional mutagenesis is a process by which foreign DNA is microinjected into an embryo such that on occasion it will be integrated into the native genome, causing a mutation of whatever gene is at the locus of integration.[55] If a mouse embryo is used, the resulting mouse is termed transgenic. A dominant mutation would be expressed in the transgenic animal unless it was lethal, resulting in embryonic death. If the mutation were recessive, crosses between transgenic mice could screen for recessive lethals (No progeny homozygous for the foreign DNA insert) or nonlethals if they express a mutant phenotype. Such an experiment, in which the foreign DNA was integrated into the mouse collagen (I) gene, demonstrated recessive lethality in homozygous offspring.[56] It was shown that the DNA insertion blocked transcription of the collagen genes.[57] The embryos could not produce collagen necessary for formation of a vascular system, and therefore underwent embryonic death. One can hypothesize that mutations in genes critical for human development could have similar outcomes in man. Support for this theory may be derived from the known effect of homozygous deletion of the α-globin genes (α-thalassemia), which causes fetal hydrops and intrauterine death.[21]

Another candidate for critical development genes with a potential for lethal mutation are homeo-box sequences. Homeo-box genes are well studied in drosophila, where they appear to control expression of genes that direct cell differentiation and morphologic development.[58] Homologous human DNA sequences have been identified,[59] and have been shown by mRNA analysis to be expressed in early embryonic development.[60] Over 50 homeo-box mutations have been described in drosophila, with severe conformational consequences. Similar mutations in humans theoretically would yield severely deformed nonviable embryos, or even the massive embryonic disorganization that results in a blighted ovum.

SEXUAL DIFFERENTIATION

Application of the new molecular technologies has significantly advanced understanding of some of the complex processes involved in human reproduction. The genetic control of male and female gametogenesis can now be investigated, and the etiology of some puzzling reproductive anomalies determined.

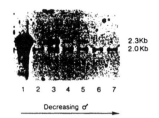

Control of testicular differentiation is vested in the Y chromosome, but the precise genetic mechanism has proved difficult to elucidate. Discovery of a male-specific cell surface transplantation antigen, termed H-Y antigen, resulted in a proposal that this was the product of the testicular determinant genes.[61] Subsequent studies were contradictory,[62,63] possibly because of lack of reproducibility in the assay. Deoxyribonucleic acid studies using specific DNA probes were able to identify specific DNA sequences unique to the Y chromosome. These sequences were potential candidates for the testicular determinant genes and could be detected in very small amounts, making molecular analysis a useful tool for investigating aberrations in male phenotypic development (Fig. 4).[64] Deletion maps of the Y chromosome based on intersex patients have placed the testicular determinant genes on the short (p) arm, either just proximal to the terminus[65] or more proximal and pericentric.[66–68] These genes direct and regulate morphogenesis of the embryonic testis. H-Y antigen is distinct from their gene product, which has yet to be identified, although H-Y may have a role in spermatogenesis.[69,70]

A distal Yp location for the testicular determinant genes is attractive because it has been shown that the short arms of the X and Y chromosomes pair and undergo crossing over during meiosis.[71] This may explain certain intersex disorders. Deoxyribonucleic acid from 46, XX males has been shown to hybridize to cDNA probes specific for Yp sequences.[72,73] This suggests that an exchange of X and Y DNA occurred during meiosis, and that this molecular translocation must have included testicular determinant genes that resulted in a male phenotype for the recipient offspring. Conversely, some patients with XY gonadal dysgenesis (Swyer's syndrome) demonstrate microdeletions of some of these same DNA sequences, with loss of testicular determinants leading to a female phenotype.[67,68]

Ovarian determinants on two intact X chromosomes are necessary for normal ovarian development and functions.[74] Absence of an X chromosome results in follicular atresia and gonadal dysgenesis.[75] Studies of X chromosome deletions have attempted to identify a critical region for ovarian development,[76] but results have been contradictory and limited by the degree of resolution available in chromosome banding. Molecular probes offer a much more precise technique for investigation. Although the exact genetic locus for ovarian development has yet to be defined, studies of a family with dominantly inherited premature ovarian failure have demonstrated a deletion at X (pter–q21.3:–: q27–qter).[77] Molecular analysis of other patients with gonadal failure may enable researchers to locate the actual determinant genes.

REPRODUCTIVE ENDOCRINOLOGY

Recombinant DNA technology has provided a mechanism for the identification, isolation, and characterization of genes coding for important hormones in the reproductive process and reproductive disorders. Congenital adrenal hyperplasia (CAH) is an autosomal recessive disorder in which a deficiency in an enzyme important in adrenal steroidogenesis, most commonly 21-hydroxylase, results in a partial or complete block in cortisol biosynthesis.[78] Failure of cortisol negative feedback leads to increased adrenocorticotropic hormone (ACTH) production from the pituitary and overproduction of androgenic steroid precursors. The classical disorder presents at birth in females with genital ambiguity secondary to high androgen levels, and results from a severe enzyme deficiency.[79] A milder partial deficiency does not present until puberty, with hirsutism and chronic anovulation.[80] Classical genetic studies have shown that the 21-hydroxylase gene is linked to the human leucocyte antigen (HLA) major histocompatibility locus on the short arm of chromosome 6.[81,82] Molecular analysis of this region has shown that the genetic defect involves the structural gene for cytochrome P450-21-hydroxylase[83] (Fig. 5). Two copies of the gene are present on each chromosome, but the A

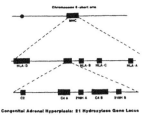

Figure 5   Expanded gene map of the major histocompatibility locus containing HLA, cytochrome p450-21-hydroxylase and complement C4 genes. 21 OHA is a pseudogene; 21 OHB, the active gene.

gene contains sequences that preclude transcription and therefore is a pseudogene.[84] As previously noted, deletions in the active gene have been found in some cases of classical CAH,[24,25] but other classical and nonclassical cases must be secondary to as yet unidentified point mutations that affect gene transcription or mRNA processing. Recently, a RFLP has been reported in the C4A complement gene closely linked with the 21-hydroxylase locus that may be useful for prenatal diagnostic testing.[85]

Molecular analysis of the genes encoding hormones that are involved in the ovulatory process has given new insights into regulation of hypothalamic, pituitary, and gonadal functions. Gonadotropin-releasing hormone (GnRH) has been shown, by isolation and cloning of the GnRH gene on chromosome 8p, to be derived from a large precursor molecule containing both GnRH and a larger peptide, termed gonadotropin-associated peptide (GAP), whose existence was previously unsuspected.[86] Studies of GAP have shown that it is a potent gonadotropin secretagogue and prolactin inhibitor,[87] suggesting that it may be a previously unsuspected mediator of gonadotropin and prolactin release. Although the exact biologic function of GAP has yet to be determined, studies have shown that the peptide is present in the hypothalamus, indicating that the entire GnRH gene is expressed and that GAP may play a role in the regulation of pituitary function.[88]

The gonadotropin genes belong to a family of glycoprotein hormones that share a common α-subunit but have unique β-subunits conferring a specific biologic activity (Table 3). The α-subunit gene has been isolated, sequenced, and localized to chromosome 6.[89,90] The gene for the β-subunit of follicle-stimulating hormone (FSH) is located on chromosome 11p, but has not been fully characterized.[91]

207

Table 3 Gonadotropin Genes

	Alpha	FSH	LH	hCG
Chromosome	6	11p	19q	19q
Copy number	1	?	1	7

The structure of the luteinizing hormone (LH) and human chorionic gonadotropin (hCG) β-subunits are very similar, with 95% homology, showing that the gene structures are highly conserved and derived from a common ancestral gene.[92] They comprise a family of genes on chromosome 19q13. There is a single copy of the LH β gene and seven copies of the hCG β gene, not all of which are functional.[93,94] The structural similarity accounts for the immunologic cross-reactivity and functional overlap of the peptides, but the differences also have functional significance. Different promoter regions allow for different physiologic regulation and tissue transcriptional specificity of the genes, and alteration of the translational stop signal in hCG allows addition of extra glycosylation sites, giving hCG a longer serum halflife.[95]

Gene expression studies use northern blotting to analyze the mRNA within a given tissue to determine if a particular gene is being transcribed. Various putative regulatory factors also can be assessed for their effects on transcription by quantifying the mRNA transcript specific for the gene of interest. For the gonadotropin genes, the α-subunit is expressed at a higher rate than any of the β-subunit genes, and although it does respond to regulatory factors, this constitutive synthesis indicates that the α-subunit probably is not important in regulating the synthesis and biologic activity of these hormones.[96-98] Gonadotropin-releasing hormone has been shown to increase mRNA transcription of both α- and LH β-subunits, indicating that gonadotropin regulation is at least partially based on alterations in gene expression.[99] Estradiol negative feedback decreases mRNA transcripts from both α and β genes, demonstrating an effect at the level of either gene transcription or mRNA processing.[100]

Inhibin is an ovarian follicular hormone whose functional effect in suppressing FSH activity was well documented, but which was difficult to isolate using standard biochemical techniques.[101] Molecular technology was instrumental in its characterization, resulting in sequencing of the gene. There are two forms of inhibin, A and B.[102] Each is a heterodimer of a similar α-subunit and dissimilar but homologous β-subunit, β-A or β-B.[103] Northern blot mRNA analysis demonstrated excess production of α-mRNA, implying that regulation of β gene transcription controls hormonal functional activity. Unexpectedly, FSH releasing activity also was noted, and found to reside in dimers of β-subunits, a homodimer (βA-βA) termed FSH releasing peptide (FRP) and a heterodimer (βA-βB) termed activin.[104,105]

The β-subunit genes for inhibin also were noted to be homologous to the gene for müllerian inhibiting substance (MIS).[106] The MIS gene recently has been isolated and found to be expressed not only in fetal testes but also in granulosa cells.[107] Its biologic role in the ovarian follicle is unknown, but has been theorized to involve regulation of germinal vesicle breakdown and arrest of oocyte maturation.[108]

Recombinant DNA technology not only has provided insight into normal structure and regulation, but is enabling researchers to explore the interactions among and therapeutic uses of these peptides. Molecular cloning can be used to produce multiple copies of genes, and with genetic engineering techniques used to attach appropriate DNA transcriptional control signals, these genes can be introduced into either prokaryotic or eukaryotic cells in a manner that allows gene expression and production of large quantities of pure hormone (Fig. 6). The genes for human insulin,[109,110] GH,[111] and interferon[112] have been cloned and expressed in in vitro systems. Expression of glycoprotein hormones is more complex than expression of these previously cloned hormones because glycosylation

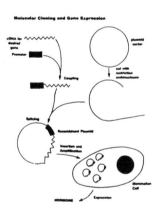

Figure 6 Molecular cloning and gene expression for hormone synthesis.

208

cannot occur in bacterial systems. A mammalian cell line is required for post-translational processing. Biologically active hCG, LH, FSH, and MIS are now being produced, showing that these functions are effectively performed in the mammalian expression system.[113] This will provide large quantities of pure hormones for clinical use as well as for establishing an in vitro system for studying genetic mutations and resultant functional derangements with potential applications in diagnosis and management of clinical disease.

Gene transfection into living organisms offers the possibility for correction instead of replacement therapy for genetic hormone deficiencies. Microinjection of a rat GH gene and linked mouse DNA promoter sequence into the male pronucleus of a fertilized mouse oocyte resulted in integration into the host genome and expression of rat growth hormone in a percentage of the surviving animals.[114] These mice grew to twice the size of their litter mates. The potential application of this technology to Ilig type IA growth hormone deficiency[12] is obvious and exciting.

The new genetic technology has vastly expanded the capability of the reproductive biologist to diagnose and treat reproductive dysfunction. Its impact has already been considerable, but with the rapid advancement of knowledge and techniques seen in recent years, this may only be the beginning of a complete revolution in the clinical practice of reproductive endocrinology. Endocrine dysfunction may be defined totally in molecular terms, without reference to hormone assays or dynamic testing, and treatment may involve manipulation of gene expression or direct gene therapy. The exciting years still lie ahead.

REFERENCES

1. Smith HO, Wilcox KW: A restriction enzyme from Hemophilus influenza, I. Purification and general properties. J Mol Biol 51:379, 1970
2. Kelly TJ Jr, Smith HO: A restriction enzyme from Hemophilus influenza, II. Base sequence of the recognition site. J Mol Biol 51:393, 1970
3. Roberts RJ: Restriction and modification enzymes and their recognition sequences. Nucleic Acids Res 11:135, 1983
4. Southern EM: Detection of specific sequences among DNA fragments separated by gel electrophoresis. J Mol Biol 98:503, 1975
5. Alwine JC, Kemp DJ, Stark GR: Method for detection of specific RNAs in agarose gels by transfer to diazobenzyloxymethyl-paper and hybridization with DNA probes. Proc Natl Acad Sci USA 74:5350, 1977
6. Chan SJ, Noyes BE, Agarwal KL, Steiner DF: Construction and selection of recombinant plasmids containing full-length complementary DNAs corresponding to rat insulins I and II. Proc Natl Acad Sci USA 76:5036, 1979
7. Bender W, Spierer P, Hogness D: Gene isolation by chromosomal walking. (Abstr.) J Supra Molec Struc 10 (Suppl):32, 1979
8. Rougeon F, Kourilsky P, Mach B: Insertion of a rabbit β-globin gene sequence into E. coli plasmid. Nucleic Acids Res 2:2365, 1975
9. Maniatis T, Kee SG, Efstratiadis A, Kafatos FC: Amplification and characterization of a β-globin gene synthesized in vitro. Cell 8:163, 1976
10. Grunstein M, Hogness DS: Colony hybridization: a method for the isolation of cloned DNAs that contain a specific gene. Proc Natl Acad Sci USA 72:3961, 1975
11. Wilson JT, Milner PF, Summer M, Nallaseth FS, Fadel HE, Reindollar RH, McDonough PG, Wilson LB: Restriction endonuclease analysis of HbS gene. Proc Natl Acad Sci USA 79:3268, 1982
12. Phillips JA III, Hjelle BL, Seeburg PH, Zachman M: Molecular basis for familial growth hormone deficiency. Proc Natl Acad Sci USA 78:6372, 1981
13. Kidd JV, Wallace BR, Itakura K, Woo SLC: Prenatal diagnosis of α-1-antitrypsin deficiency by direct analysis of the mutation site in the gene. N Engl J Med 310:639, 1984
14. Boehm CD, Antonarakis SE, Phillips JA III, Stetten G, Kazazian HH Jr: Prenatal diagnosis using DNA polymorphisms: report on 95 pregnancies at risk for sickle cell disease or β-thalassemia. N Engl J Med 308:1054, 1983
15. White R, Schafer M, Barker D, Wyman A, Skolnick M: DNA sequence polymorphism at arbitrary loci. In Human Genetics: The Unfolding Genome, Edited by B Bonne-Tamir, T Cohen, RN Goodman. New York, Alan R. Liss, 1982, p 67
16. Little PFR, Annison G, Darling S, Williamson R, Camba L, Modell B: Model for antenatal diagnosis of β-thalassemia and other monogenic disorders by molecular analysis of linked DNA polymorphisms. Nature 285:144, 1980
17. NICHD National Registry for Amniocentesis Study Group: Midtrimester amniocentesis for prenatal diagnosis: safety and accuracy. JAMA 236:1471, 1975
18. Simoni G, Brambati B, Danesino C, Terzoli GL, Romitt L, Rossella F, Fraccaro M: Diagnostic application of first trimester trophoblast sampling in 100 pregnancies. Hum Genet 66:252, 1984
19. Kazazian HH, Orkin SH, Boehm CD, Sexton JP, Antonarakis SE: β-Thalassemia due to deletion of the nucleotide which is substituted in sickle cell anemia. Am J Hum Genet 35:1028, 1983
20. Thein SL, Wainscoat JS, Lynch JR, Weatherall DJ, Sampietro M, Fiorelli G: Direct detection of β_0 39 thalassemic mutation with MaeI. Lancet 1:1095, 1985
21. Ottolenghi S, Lanyon WG, Paul J, Williamson R, Weatherall DJ, Clegg JB, Pritchard J, Pootrakul S, Wong HB: The severe form of α-thalassemia is caused by a haemoglobin gene deletion. Nature 251:389, 1974
22. Pressley L, Higgs DR, Clegg JB, Weatherall DJ: Gene deletions in β-thalassemia prove that the 5'zeta locus is functional. Proc Natl Acad Sci USA 77:3586, 1980
23. Orkin SH, Old JB, Weatherall DJ, Nathan DJ: Partial deletion of β-globin gene DNA in certain patients with β_0-thalassemia. Proc Natl Acad Sci USA 76:2400, 1979

24. Rumsby G, Carroll MC, Porter RR, Grant DB, Hjelm M: Deletion of the steroid 21-hydroxylase and complement C4 genes in congenital adrenal hyperplasia. J Med Genet 23:204, 1986

25. Werkmeister JW, New MI, Dupont B, White PC: Frequent deletion and duplication of the steroid 21-hydroxylase genes. Am J Hum Genet 39:461, 1986

26. Monaco AP, Bertelson CJ, Middleworth W, Colletti CA, Aldridge J, Fishbeck KH, Bartlett R, Pericak-Vance MA, Roses AD, Kunkel LM: Detection of deletions spanning the Duchenne muscular dystrophy locus using a tightly linked DNA segment. Nature 316:842, 1985

27. Kunkel LM, Monaco AP, Middleworth W, Ochs HD, Lett SA: Specific cloning of DNA fragments absent from the DNA of a male patient with X chromosome deletion. Proc Natl Acad Sci USA 82:4778, 1985

28. Gitschier J, Wood WI, Tuddenham EGD, Shuman MA, Goralka TM, Chen EY, Lawn RM: Detection and sequence of mutations in the factor VIII gene of hemophiliacs. Nature 315:427, 1985

29. Giannelli F, Chook H, Rees DJG, Boyd Y, Rizza CR, Brownlee GG: Gene deletions in patients with haemophilia B and anti-factor IX antibodies. Nature 303:181, 1983

30. Rosatelli C, Tuveri T, DiTucci A, Falchi AM, Scalas MT, Monni G, Cao A: Prenatal diagnosis of β-thalassemia with the synthetic-oligomer technique. Lancet 1:241, 1985

31. Kurachi K, Chandra T, Degen SJF, White TT, Machioro TL, Woo SLC, Davie EW: Cloning and sequence cDNA coding for α-1-antitrypsin. Proc Natl Acad Sci USA 78:7826, 1981

32. Gusella JF, Wexler NS, Conneally PM, Naylor SL, Anderson MA, Tarzi RE, Watkins PC, Ottina K, Wallace MR, Sakaguchi AY, Young AB, Shoulson I, Bonilla E, Martin JB: A polymorphic DNA marker genetically linked to Huntington's disease. Nature 307:234, 1983

33. Bakker E, Goor N, Wrogemann K, Kunkel LM, Fenton WA, Majoor-Krakauer D, Jahoda MGJ, VanOmmen GJB, Hofker MH, Mandel JL, Davies KE, Willard JH, Sandkuyl L, Essen AJV, Sachs ES, Pearson PL: Prenatal diagnosis and carrier detection of Duchenne muscular dystrophy with closely linked RFLPs. Lancet 1:655, 1985

34. White R, Woodward S, Leppert M, O'Connell P, Hoff M, Herbst J, Lalouel J-M, Dean M, Woude GV: A closely linked marker for cystic fibrosis. Nature 318:382, 1985

35. Brown CS, Thomas NST, Sarfarazzi M, Davies KE, Kunkel L, Pearson PL, Kingston HM, Shaw DJ, Harper PS: Genetic linkage relationships of seven DNA probes with Duchenne and Becker muscular dystrophy. Hum Genet 71:62, 1985

36. Beaudet A, Bowcock A, Buchwald M, Cavalli-Sforza L, Farrall M, King MC, Klinger K, Lalouel JM, Lathrop G, Naylor S, Ott J, Tsui LC, Wainwright B, Watkins P, White R, Williamson R: Linkage of cystic fibrosis to two tightly linked DNA markers: joint report from a collaborative study. Am J Hum Genet 39:681, 1986

37. Farrall M, Rodeck CH, Stanier P, Lissens W, Watson E, Law HY, Warren R, Super M, Scambler P, Wainwright B, Williamson R: First trimester prenatal diagnosis of cystic fibrosis with linked DNA probes. Lancet 1:1402, 1986

38. Estivill X, Farrall M, Scambler PJ: A candidate for the cystic fibrosis locus isolated by selection for methylation-free islands. Nature 326:840, 1987

39. Orkin SH, Kazazian HH Jr: The mutation and polymorphisms of the human β-globin gene and its surrounding DNA. Annu Rev Genet 18:131, 1984

40. Orkin SH, Kazazian HH Jr, Antonarakis SE, Goff SC, Boehm CD, Sexton JP, Waber PG, Giardina PJV: Linkage of β-thalassemia mutations and β-globin gene polymorphisms with DNA polymorphisms in the human β-globin gene cluster. Nature 296:627, 1982

41. Antonarakis SE, Waber PG, Kittur SD, Patel AS, Kazazian HH Jr, Mellis MA, Counts RB, Stamatoyannopoulos G, Bowie EJW, Fass DN, Pittman DD, Wozney JM, Toole JJ: Hemophilia A: detection of molecular defects and of carriers by DNA analysis. N Engl J Med 313:842, 1985

42. Oberle I, Camerino G, Heileg R, Grunebaum L, Cazneave JP, Crapanzano C, Manucci PM, Mandel JL: Genetic screening for hemophilia A (classic hemophilia) with a polymorphic DNA probe. N Engl J Med 312:682, 1985

43. Harper K, Pembrey ME, Davies KE, Winter RM, Hartley D, Tuddenham EGD: A clinically useful DNA probe closely linked to haemophilia A. Lancet 11:6, 1984

44. Kogan SC, Doherty M, Gitschier J: An improved method for prenatal diagnosis of genetic diseases by analysis of amplified DNA sequences. N Engl J Med 317:985, 1987

45. Gianelli G, Choo KH, Winship PR, Rizza CR, Anson DS, Rees DJG, Ferrari N, Brownlee GG: Characterization and use of an intragenic polymorphic marker for detection of carriers of haemophilia B (Factor IX deficiencey). Lancet 1:239, 1984

46. Winship PR, Anson DS, Rizza CR, Brownlee GG: Carrier detection in hemophilia B using two further intragenic restriction fragment length polymorphisms. Nucleic Acids Res 12:8861, 1984

47. Camerino G, Oberle I, Drayna D, Mandel JL: A new Msp I restriction fragment length polymorphism in the hemophilia B locus. Hum Genet 71:79, 1985

48. Tourian A, Sidbury JB: Phenylketonuria and hyperphenylalaninemia. In The Metabolic Basis of Inherited Disease, 5th edition, Edited by JB Stanbury, JB Wyngaarden, DS Frederickson, JL Goldstem, MS Brown. New York, McGraw Hill, 1983, p 270

49. Lidsky AS, Ledley FD, DiLella AG, Kwok SCM, Daiger SP, Robson KHJ, Woo SLC: Extensive restriction site polymorphism at the human phenylalanine hydroxylase locus and application in prenatal diagnosis of phenylketonuria. Am J Hum Genet 37:619, 1985

50. Lidsky AS, Gutter F, Woo SLC: Prenatal diagnosis of classical phenylketonuria by DNA analysis. Lancet 1:549, 1985

51. Rosen R, Fox J, Fenton WA, Horwick AL, Rosenberg LE: Gene deletion and restriction fragment length polymorphisms at the human ornithine transcarbamylase locus. Nature 313:815, 1985

52. Verweij CL, Hofker M, Quadt R, Briet E, Panmekoek H: RFLP for a human von Willebrand factor cDNA clone. Nucleic Acids Res 13:8289, 1985

53. Reeders ST, Breuning MH, Davies RE, Nicholls RD, Jarmen AP, Higgs DR, Pearson PL, Weatherall DJ: A highly polymorphic DNA marker linked to adult polycystic kidney disease on chromosome 16. Nature 317:542, 1985

54. Bartlett RJ, Pericak-Vance MA, Yamaoka L, Gilbert J, Herbstreith M, Hung W-Y, Lee JE, Mohandas T, Bruns G, Laberge C, Thibault M-C, Ross D, Roses AD: A new

210

probe for the diagnosis of myotonic muscular dystrophy. Science 235:1648, 1987

55. Jaenisch R, Breindl M, Harbers K: Retroviruses and insertional mutagenesis. Cold Spring Harbor Symp Quant Biol 50:439, 1985

56. Lohler J, Timple R, Jaenisch R: Embryonic lethal mutation in mouse collagen (I) gene causes rupture of blood vessels and is associated with erythropoietic and mesenchymal cell death. Cell 38:597, 1984

57. Hartung S, Jaenisch R, Breindl M: Retrovirus insertion inactivates mouse type (I) collagen gene by blocking initiation of transcription. Nature 320:365, 1986

58. Gehring WJ: Homeotic genes, the homeobox and genetic control of development. Cold Spring Harbor Symp Quant Biol 50:243, 1985

59. Levine M, Rubin GM, Tran R: Human DNA sequences homologous to a protein coding region conserved between homeotic genes of Drosophila. Cell 38:667, 1984

60. Su B, Strand D, McDonald J, McDonough PG: Expression of oncogenes and homeobox genes in early human placenta. (Abstr. 323 P) Presented at the 33rd Annual Meeting of the Society for Gynecologic Investigation, Toronto, Ontario, Canada, March 19–22, 1986 Published by SGI in preliminary program, p 196

61. Wachtell SS: The genetics of intersexuality: clinical and theoretic perspectives. Obstet Gynecol 54:671, 1979

62. Moreira-Filho CA, Toledo SPA, Bagnoli VR, Frota-Pessoa O, Bisi H, Wajntal A: H-Y antigen in Swyer Syndrome and the genetics of XY gonadal dysgenesis. Hum Genet 53:52, 1979

63. Wachtell SS, Koo GC, de la Chapelle A, Kallio H, Heyman JM, Miller OJ: H-Y antigen in 46, XY gonadal dysgenesis. Hum Genet 54:25, 1980

64. McDonough PG, Tho SP, Trill JJ, Byrd JR, Reindollar RH, Tischfield JA: Use of two different deoxyribonucleic acid probes to detect Y chromosome deoxyribonucleic acid in subjects with normal and alterred Y chromosomes. Am J Obstet Gynecol 154:737, 1986

65. Page DC, de la Chapelle A: The prenatal origin of the X chromosome in XX males determined using restriction fragment length polymorphisms. Am J Hum Genet 36:565, 1984

66. Vergnaud G, Page DC, Simmler MC, Brown L, Rouyer F, Noel B, Botstum D, de la Chapelle A, Weissenbach J: A deletion map of the human Y chromosome based on DNA hybridization. Am J Hum Genet 38:109, 1986

67. Muller U, Doulon T, Schmid M, Fitch N, Ritcher CL, Lalande M, Latt SA: Deletion mapping of the testis determining locus with DNA probes in 46, XX males and 46, XY and 46, X, dic(Y) females. Nucleic Acids Res 14:6489, 1986

68. Disteche CM, Casanova M, Saal H, Friedman C, Syberts V, Graham J, Thuline H, Page DC, Fellous M: Small deletions of the short arm of the Y chromosome in 46, XY females. Proc Natl Acad Sci USA 83:7841, 1986

69. Goodfellow PJ, Darling SM, Thomas NS, Goodfellow PN: A pseudoautosomal gene in man. Science 234:740, 1986

70. Burgoyne PS, Levy ER, McLaren A: Spermatogenic failure in male mice lacking HY antigen. Nature 320:170, 1986

71. Ashley T: A re-examination of the case for homology between the X and Y chromosomes of mouse and man. Hum Genet 67:372, 1984

72. de la Chapelle A, Tippett PA, Wetterstrand G, Page D: Genetic evidence of X-Y interchange in a human XX male. Nature 307:172, 1984

73. Guellean G, Casanova M, Bishop C, Geldwerth D, Andre G, Fellous M, Weissenbach J: Human XX males with Y single-copy DNA fragments. Nature 307:172, 1984

74. Gartler SM, Liskay RM, Gant M: Two functional X chromosomes on human fetal oocytes. Exp Cell Res 82:464, 1973

75. Jirasek JE: Principles of reproductive embryology. IV Development of the ovary. In Disorders of Sexual Differentiation, Edited by JL Simpson. New York, Academic Press, 1976, p 75

76. Sarto GE, Therman E, Patau K: X inactivation in man: a woman with t(Xq-;12qt). Am J Hum Genet 25:262, 1973

77. Krauss CM, Turksoy RN, Atkins L, McLaughlin C, Brown LG, Page DC: Familial premature ovarian failure due to an interstitial deletion of the long arm of the X chromosome. N Engl J Med 317:125, 1987

78. White PC, New MI, Dupont B: Congenital adrenal hyperplasia. Part I. N Engl J Med 317:1519, 1987

79. Bongiovanni AM, Root AW: The adrenogenital syndrome. N Engl J Med 268:1283, 1963

80. Kohn B, Levine LS, Pollack MS, Pang S, Lorenzen F, Levy D, Lerner AJ, Rondanini GF, Dupont B, New MI: Late onset steroid 21-hydroxylase deficiency: a variant of classical congenital adrenal hyperplasia. J Clin Endocrinol Metab 55:817, 1982

81. Dupont B, Oberfield, Smitherwick EM, Leli TD, Levine LS: Close genetic linkage between HLA and congenital adrenal hyperplasia (21-hydroxylase deficiency). Lancet 2:1390, 1977

82. Levine LS, Zachman M, New MI, Prader A, Pollack MS, O'Neill GJ, Yang SY, Oberfeld SE, Dupont B: Genetic mapping of the 21-hydroxylase-deficiency gene within the HLA linkage group. N Engl J Med 299:911, 1978

83. White PC, New MI, Dupont B: Congenital adrenal hyperplasia. Part II. N Engl J Med 316:1580, 1987

84. Higoshi Y, Yoshioka H, Yamane M, Gotoh O, Fujii-Kuriyama Y: Complete nucleotide sequence of two steroid 21-hydroxylase genes tandemly arranged in human chromosome: a pseudogene and a genuine gene. Proc Natl Acad Sci USA 83:511, 1986

85. Reindollar RH, Lewis JB, White PC, Fernhoff PM, McDonough PC, Whitney JB: Prenatal diagnosis of 21-hydroxylase deficiency by the complementary deoxyribonucleic acid probe for cytochrome P-450$_{c-21OH}$. Am J Obstet Gynecol 158:545, 1988

86. Seeburg PH, Adelman JP: Characterization of the cDNA for precursor of human luteinizing hormone releasing hormone. Nature 311:666, 1984

87. Nikolics K, Mason AJ, Szonyi E, Ramachandran J, Seeburg PH: A prolactin-inhibiting factor within the precursor for human gonadotropin-releasing hormone. Nature 316:511, 1985

88. Phillips HS, Nikolics K, Branton D, Seeburg PH: Immuno-cytochemical localization in rat brain of a prolactin release-inhibiting sequence of gonadotropin-releasing hormone. Nature 316:542, 1985

89. Fiddes JC, Goodman HM: Isolation, cloning, and sequence analysis of the cDNA for the α-subunit of human chorionic gonadotropin. Nature 281:351, 1979

90. Fiddes JC, Goodman HM: The gene encoding the common

211

alpha subunit of the four human glycoprotein hormones. J Mol Appl Genet 1:3, 1981

91. Fiddes JC, Talmadage K: Structure, expression and evolution of the genes for the human glycoprotein hormones. Recent Prog Horm Res 40:43, 1984

92. Talmadge K, Vamvakopoulos NC, Fiddes JC: Evolution of the genes for the β-subunits of human chorionic gonadotropin and luteinizing hormone. Nature 307:37, 1984

93. Talmadge K, Boorstein WR, Fiddes JC: The human genome contains seven genes for the β-subunit of chorionic gonadotropin but only one gene for the β-subunit of luteinizing hormone. DNA 2:281, 1983

94. Policastro PF, Daniels-McQueen SD, Carle G, Boime I: A map of the hCGβ-hLHβ gene cluster. J Biol Chem 261: 5907, 1986

95. Jameson JL, Lindell CM, Habener JF: Evolution of different transcriptional start sites in the human luteinizing hormone and chorionic gonadotropin beta subunit genes. DNA 5:227, 1986

96. Corbani M, Counis R, Starzec A, Jutisz M: Effect of gonadectomy on pituitary levels of mRNA encoding gonadotropin subunits and secretion of luteinizing hormone. Mol Cell Endocrinol 35:83, 1983

97. Nilson HJ, Nejedlik MT, Virgin JB, Crowder ME, Nett TM: Expression of the α-subunit and luteinizing hormone β genes in the ovine anterior pituitary. Estradiol suppresses accumulation of mRNAs for both α-subunit and luteinizing hormone β. J Biol Chem 258:12087, 1983

98. Laudelfeld T, Kepa J, Karsch F: Estradiol feedback effects on the α-subunit mRNA in the sheep pituitary gland: correlation with serum and pituitary luteinizing hormone concentrations. Proc Natl Acad Sci USA 81:1322, 1984

99. Papavasiliou SS, Zmeili S, Khoury S, Landefeld TD, Chin WW, Marshall JC: Gonadotropin-releasing hormone differentially regulates expression of the genes for luteinizing hormone alpha and beta subunits in male rates. Proc Natl Acad Sci USA 83:4026, 1986

100. Gharib SD, Bowers DM, Need LR, Chin WW: Regulation of rat luteinizing hormone subunit messenger ribonucleic acids by gonadal steroid hormones. J Clin Invest 77:582, 1986

101. Grady RR, Charlesworth MC, Schwartz NB: Characterization of the FSH-suppressing activity in follicular fluid. In Recent Progress in Hormone Research, Vol 38, Edited by RO Greep. (Proceedings of the 1981 Lauretian Hormone Conference). New York, Academic Press, 1981, p 409

102. Forage RG, Ring JM, Brown RW, McInerney BV, Cobon GS, Gregson RP, Robertson DM, Morgan FJ, Hearn MTW, Findlay JK, Wettenhall REH, Burger HG, deKretser DM: Cloning and sequence analysis of cDNA species coding for the two subunits of inhibin from bovine follicular fluid. Proc Natl Acad Sci USA 83:3091, 1986

103. Mason AJ, Niall HD, Seeburg PH: Structure of two human ovarian inhibins. Biochem Biophys Res Comm 135: 957, 1986

104. Ling N, Ying S-Y, Ueno N, Shimasaki S, Esch F, Hotta M, Giullemin R: Pituitary FSH is released by a heterodimer of the beta-subunits from the two forms of inhibin. Nature 321:779, 1986

105. Vale W, Rivier J, Vaughn J, McClintock R, Corrigan A, Woo W, Karr D, Spiess J: Purification and characterization of an FSH releasing protein from porcine ovarian follicular fluid. Nature 321:776, 1986

106. Takahashi M, Hayashi M, Manganaro TF, Donahoe PK: The ontogeny of mullerian inhibiting substance in granulosa cells of the bovine ovarian follicle. Biol Reprod 35: 447, 1986

107. Donahoe PK, Budzik GP, Trelstad R. Müllerian-inhibiting substance: an update. In Recent Progress in Hormone Research, Vol 38, Edited by RO Greep (Proceedings of the 1981 Lauretian Hormone Converence). New York, Academic Press, 1981, p 279

108. Takahashi M, Koide SS, Donahoe PK: Mullerian inhibiting substance as oocyte meiosis inhibitor. Mol Cell Endocrinol 47:225, 1986

109. Villa-Komaroff L, Efstratiadis A, Broome S, Lomedico P, Tizard R, Nabet P, Chick WL, Gilbert W: A bacterial clone synthesizing proinsulin. Proc Natl Acad Sci USA 75: 3727, 1978

110. Goeddel DV, Kleid DG, Bolivard F, Heyneker H, Yansura D, Crea R, Hirose T, Kraszewski A, Itakwra K, Riggs A: Expression in escherichia coli of chemically synthesized genes for human insulin. Proc Natl Acad Sci USA 76:106, 1979

111. Goeddel DV, Heyneker H, Hozumi T, Arentzen R, Itakwark, Yansura D, Ross M, Mizzari G, Crea R, Seeburg P: Direct expression in escherichia coli of a DNA sequence coding for human growth hormone. Nature 381:544, 1979

112. Devos R, Cheroutre H, Taya Y, Degrave W, VanHeuverswyn H, Fiers W: Molecular cloning of human immune interferon cDNA and its expression in eukaryotic cells. Nucleic Acids Res 10:2487, 1982

113. Simon JA, Danforth DR, Hutchison JS, Hodgen GD: Characterization of recombinant DNA derived human luteinizing hormone in vitro and in vivo. JAMA 259:3290, 1988

114. Palmiter RD, Brinster RL, Hammer RE, Trumbauer ME, Rosenfeld MG, Birnberg NC, Evans RM: Dramatic growth of mice that develop from eggs microinjected with metallothionein-growth hormone fusion genes. Nature 300:611, 1982

Received December 28, 1987.

Reprint requests: William J. Butler, M.D., Department of Obstetrics and Gynecology, The Albany Medical College, Albany, New York 12208.

Controversy and the Development of Reproductive Sciences*

ADELE E. CLARKE, *University of California, San Francisco*

Under what conditions does an entire line of scientific work become controversial, and with what consequences? Nuclear physics, environmental studies, and sociobiology are familiar examples of controversial sciences. The case examined here is the development of American reproductive sciences over the past century, shaped by four domains of controversy: (1) association with sexuality and reproduction; (2) association with clinical quackery and hotly debated treatments; (3) association with controversial social movements; and (4) the capacity of reproductive sciences to create "Brave New Worlds." Scientists' strategies for managing controversy are delineated. The framework used is arena analysis, and the paper concludes with rudiments of a conditional theory of status as a controversial "boundary world."

What makes a line of scientific work morally controversial? Latour (1987) has argued that science is politics by other means. Schneider (1984) delineates the profoundly political nature of the construction, maintenance, and change of morality. In the case of reproductive sciences, these two politics intersect: the politics of science confronts the politics of moral change. Here the very doing of certain kinds of scientific work can be construed as demonstrating controversial moral commitments. The outcome of this confrontation is a line of scientific work that has been controversial for over a century. The conditions and consequences of the controversial status and identity of the American reproductive sciences are the focus of this paper.

In what follows I describe my perspective on the emergence and nature of controversial status and identity; offer a brief history of modern reproductive sciences in the United States; examine four, often overlapping, domains of controversy that have plagued reproductive sciences in the past century; consider how scientists have sought to manage these problems, with what effects on funding for their research; and finally, gesture toward a conditional theory of what can make a line of scientific work controversial.

By reproductive sciences, I mean the study of mammalian reproduction through endocrinology, physiology, and anatomy, undertaken in biological, medical, and agricultural settings. I use interview data, archival materials, published research reports and symposia, biographical materials, histories done by reproductive scientists themselves, and secondary sources. In developing the argument here and the rudiments of theory proposed at the close of the paper, I draw on the grounded theory approach developed by Glaser and Strauss (1968; Glaser 1978; Strauss 1987).

* An earlier version of this article was given in the History of Science Lecture Series, Woods Hole Marine Biological Laboratory, August 1987. Excerpts were published under the same title in *MBL Science* 3 (Winter, 1988):36-39. I am indebted to Howard S. Becker, the late Rue Bucher, Merriley Borell, Kathy Charmaz, Nan Chico, Joan H. Fujimura, Elihu Gerson, Kathy Gregory Huddleston, Petra Liljestrand, Marilyn Little, Diana Long, Jane Maienschein, M.C. Shelsnyak, S. Leigh Star, and Anselm Strauss for helpful discussions. Thanks also to the anonymous *Social Problems* reviewers, Kathryn Pyne Addelson, and participants in the Reproduction Study Group organized by Evelyn Fox Keller and sponsored by the Bain Institute at U.C., Berkeley, especially Peter Taylor, and to Gerald Markle for speedy help on scientific controversies. Support for the research on which parts of this paper are based was provided by the University of California, San Francisco, the Rockefeller Foundation, and the NIMH Postdoctoral Program in Organizations and Mental Health, Department of Sociology, Stanford University. Correspondence to: Clarke, Department of Social and Behavioral Sciences, University of California, San Francisco, San Francisco, CA 94143-0612.

214

Social Worlds, Arenas, and Science as Controversial

Controversies happen in an *arena* of interaction, a conceptual location where all of the groups that care about a given phenomena meet (Strauss 1978, 1979, 1982; Becker 1982; Clarke 1990b). Examples include policy arenas such as alcohol, energy, and aging (Estes and Binney1989; Weiner 1971). Participants in an arena can be conceived as social worlds—groups with shared commitments to certain activities, building shared ideologies about what *should* happen in their arena of concern, and gathering resources to achieve their goal. Thus social worlds are collective actors in an arena; they typically have diverse moral and political commitments; and they promote conflicting agendas for action. Controversy derives from these worlds having different perspectives on what should happen, when and how. Controversies occur where different social worlds meet—in arenas (Hughes 1971; Fujimura 1988).

There is, of course, a long-standing arena focused on reproductive issues of all kinds (Clarke 1989). Participating social worlds in the American reproductive policy arena include the reproductive sciences, birth control and population control worlds, religious groups, feminist and consumer groups focused on health, pharmaceutical companies and associations, research funding sources (e.g., foundations and governments), regulatory entities, legislative groups, and professional associations. Each of these social worlds seeks legitimacy, authority, and power to further its own goals; and seeks to limit and repress other worlds with which it is on conflict. Of course, there is often conflict *within* worlds as well as among them, not only about strategies and tactics but also about goals. As the dynamics within worlds alter, so too do the dynamics across their arena. Arenas of interaction are thus quite dynamic over time.

How did the reproductive sciences get into the arena? Recent interactionist, ethnomethodological, constructivist, Marxist, and feminist works in the sociology of science have challenged earlier functionalist claims that sciences are separate and apart from society (e.g., Aronson 1984; Latour 1987; Clarke and Gerson 1990). Ironically and importantly, at least since the turn of this century, most lines of scientific work *cannot* opt out of participation in arenas. For the reproductive sciences, like other modern lines of scientific work, must be "sold" to various audiences in order to obtain the often very expensive resources requisite for continuing scientific work (see Aronson 1984; Gieryn 1987b; Latour 1987; Nelkin 1987). Through these selling or marketing processes, in the arenas where this occurs, scientists must interact with representatives of other social worlds in arenas where opportunities for controversy abound. Lines of scientific work become vulnerable to controversy because scientists must claim "that particular scientific findings are useful, that is, relevant to the concerns of the particular audience[s] addressed" (Aronson 1984:13); to succeed in obtaining resources they must in some sense claim that their science is "socially necessary." For reproductive sciences, the public transparency of the research applications immediately politicizes such claims.

The Rise of Modern Reproductive Sciences

Not a single English-language book on reproductive sciences was published until Marshall's *Physiology of Reproduction* appeared in Britain in 1910. Yet by 1940, reproductive research had emerged and coalesced as a line of scientific work in biology, medicine and agriculture. In the United States, numerous major research centers together formed an established and growing scientific enterprise. The prestigious National Research Council's Committee for Research on Problems of Sex sponsored much of this research with Rockefeller funding (Aberle and Corner 1953). After the First World War, American reproductive scientists were preeminent in the field.

When modern reproductive sciences emerged at the turn of the century, the professions, universities, and the life sciences were being rationalized and organized along more bureaucratic, industrial lines (Beer and Lewis 1974; Light 1983; Clarke 1988). A new, engineering model of the life sciences, often with the goal of controlling life, had begun to emerge (Haraway 1979; Pauly 1987). Biology, by then established in the universities, encouraged specialization, including the study of reproduction. In medicine, practitioners sought jurisdictional monopoly over reproduction, displacing midwives (Leavitt 1986). In gynecology and obstetrics the shift was from surgical anatomy to reproductive physiology—toward a functional understanding of reproduction. In agriculture, farming and ranching became commercial enterprises focused on controlling and increasing animal reproduction through scientific improvements (Rosenberg 1979; Rossiter 1979; Herman 1981).

Reproductive physiology developed comparatively late among the physiological studies of other major organ systems. A common explanation of this late development was that the scientific study of reproduction was not acceptable largely because of its association with human sexuality. But new markets for scientific knowledge about reproduction were emerging. Academic departments of biology, medicine, and agriculture; the medical specialties of obstetrics and gynecology; animal production industries; and the birth control, eugenics, and neo-malthusian movements all became important consumers of this knowledge. The social movements in particular brought reproductive topics "out of the closet" and into the center of public discussion between 1900 and 1920. Their membership included scientists, physicians, and other professional elites, further enhancing the legitimacy of reproductive research. As industrialization expanded from the factory to agricultural and social life, the laws of nature would be replaced by the scientific ingenuity of humans.

Two developments in biology especially enabled the rise of the reproductive sciences. First, since Darwin, life scientists' attention had focused directly on evolution, heredity, and development (Brush 1978; Maienschein 1984). Following the "rediscovery" of Mendel's ratios in 1900, investigators began to separate these problems into distinct subproblems (Allen 1986; Gilbert 1987). From this emerged the separate disciplines of genetics, developmental embryology, and reproductive sciences (Clarke 1990a). The second significant development was the emergence of endocrinology as a lively line of scientific work. For the first time, scientists understood that basic bodily processes were regulated not only by the nervous system, but also by chemical messengers—hormones (Borell 1985; Oudshoorn 1990a,b).

From 1910 to 1925, British and U.S. scientists focused on problems of reproductive physiology and endocrinology. They did so in biological, medical and agricultural settings, constructing a mutually helpful network of centers of reproductive research. The study of reproduction was a new research frontier, contributing further to its legitimacy (Clarke 1990a; Fujimura 1987). The period 1925 to 1940 saw a deepening investment in, and a coalescence and consolidation of, this scientific enterprise. Reproductive endocrinology became the core work. Because endocrinological research of all kinds was at the cutting edge of the life sciences, it brought prestige, new investigators, and more resources to research on reproduction (Corner [1942] 1963; Parkes 1966a,b).

However, after the Second World War, major changes occurred. The cutting edge of endocrinology was dulled as molecular biology began to emerge (Abir-Am 1982). Centers of reproductive research in prestigious biological settings (e.g., Chicago and Harvard), which had carried the basic research "torch," declined after major investigators died or were ousted. Worse, despite the phenomenal rise in federal funding for research, it soon became apparent that neither the National Institutes of Health (NIH) nor the National Science Foundation (NSF) would fund basic research in the reproductive sciences at a level equal to that for research on other major organ systems (Greep, Koblinksi, and Jaffe 1976:367-426). Yet at the same time, the potential for practical applications in medicine and agriculture provided considerable impetus to reproductive research in those domains.[1] The medical work centered on research on infertility and contraception. Both blossomed immediately after the Second World War in an extremely supportive context (Corner 1957; American Foundation 1955). The birth control, eugenics, and population control movements were consolidating under the banner of "family planning and population studies" and were beginning to attract major foundation funding for both reproductive and population research (Clarke 1985; Gordon 1976). By the 1960s, the publicity generated by the population establishment's assertion of a threatening "population bomb" in the Third World was intense (Reed 1984:373). Leading reproductive scientists agreed and were drawn ever more deeply into research that would address "population problems" (e.g., Pincus 1965).

Improved means of animal breeding and reproduction also became a significant focus in the United States and Britain after the Second World War (Cole 1977; Herman 1981). Increased funding of agricultural research by the U.S. Department of Agriculture led to improved production, which gradually included applied reproductive research (Greep, Koblinsky, and Jaffe 1976; Rossiter 1979). In these agricultural settings, various reproductive technologies were refined and applied, including artificial insemination, sex preselection via sperm filtration, superovulation, embryo recovery and transfer, synchronized estrus, and genetic cloning (Biggers 1984; Betteridge 1986). Over the past 20 years, most of these technologies have been transferred from agriculture to medicine and applied to humans.

Four Domains of Controversy

Four domains of controversy shaped the development of American reproductive sciences: (1) association with sexuality and reproduction; (2) association with clinical quackery and hotly debated treatments; (3) association with controversial social movements; and (4) the capacity of reproductive sciences to create "Brave New Worlds."

Sexuality and Reproduction

Because they include study of the sexual organs and sexual functioning, the reproductive sciences from their origins have been in some sense stigmatized, both within and outside science, as "dirty" work that impugns and discredits those who do it (Hughes 1962; Goffman 1963). The National Research Council's (NRC) Committee for Research in Problems of Sex almost foundered on these issues in 1921, when it began, because "It was in fact first necessary to establish and defend the dignity of such studies" before they could proceed (Aberle and Corner 1953:1).

The first barrier confronted by the Committee's sponsors was finding an NRC division to be its administrative "home." The Division of Anthropology and Psychology refused to sponsor it. The proposal was then greeted cooly by the Division of Medical Sciences despite promised external support from Rockefeller funds. Only when a physician familiar with venereal disease became Chairman of the Medical Division did the proposal succeed (Aberle and Corner 1953:11). In ways that anticipated the future, the link to clinical medical work carried the day.

Once established, the Committee existed from 1921 to 1962 (National Academy of Sciences 1979:v). It sponsored publication of a book titled *Sex and Internal Secretions* (Allen 1932) that, through subsequent editions, became the bible of American reproductive sciences for fifty years. In his foreword to the third edition, George Corner (1961:xxxi), a leading reproductive scientist and medical historian, stated:

---

1. Interview with A.V. Nalbandov, 7 April 1984, Urbana, Illinois; see also Nalbandov (1978).

The younger readers of this book will hardly be able to appreciate the full significance of such an alliance between biologists, psychologists, and physicians on one hand, and social philanthropists on the other. It represented a major break from the so-called Victorian attitude, which in the English-speaking countries had long impeded scientific and sociological investigations of sexual matters and placed taboos on open consideration of human mating and childbearing as if these essential activities were intrinsically indecent. To investigate such matters, even in the laboratory with rats and rabbits, required of American scientists . . . a certain degree of moral stamina. A member of the . . . Committee once heard himself introduced by a fellow scientist to a new acquaintance as one of the men who had "made sex respectable."

Many early investigators who studied reproductive problems felt they were placing themselves "beyond the pale" of propriety by doing so. They repeatedly stated to the Committee that reluctant administrators in their universities had finally approved a proposed investigation only because it had the backing of an NRC grant (Aberle and Corner 1953:89).

The uproar in the 1950s surrounding Alfred Kinsey's research demonstrates the continuity of opprobrium adhering to sexual and sex-related topics (Pomeroy 1982). It also cost Kinsey some of his research funding.[2] The association of reproductive sciences with issues of sexuality and sex education remains highly problematic today (Booth 1989; Holden 1989).

Clinical Quackery and Problematic Treatments

The connection between reproductive sciences and charges of quackery began with one of the founders of the modern scientific study of reproduction, Charles-Edouard Brown-Séquard (1817-1894). A renowned French neurologist and physiologist, Brown-Séquard taught in the United States, did major work on the adrenal glands and assumed famed physiologist Claude Bernard's Chair at the Collège de France upon Bernard's death in 1878 (Olmstead 1946). In 1889, he made the following announcement (Medvei 1982:289):

I sent to the Society of Biology a communication, which was followed by several others, showing the remarkable effects produced on myself by subcutaneous injection of a liquid obtained by the maceration on a mortar of the testicle of a dog or of a guinea pig to which one has added a little water.

Many people thought that this could be the start of something big. Testicular extract was used immediately for diverse ailments by physicians, who soon also began using extracts from other major organs in their treatments. Organotherapy was "the therapeutic hope of physicians from Cleveland to Bucharest" (Borell 1976:309). Referring to these events some years later, reproductive endocrinologist Herbert McLean Evans, Chair of Anatomy and later Director of the Institute for Experimental Biology at the University of California, Berkeley, said, "Endocrinology suffered obstetric deformity in its very birth" (Borell 1978:283).

Shortly after Brown-Séquard's work, the discovery that extracts of thyroid alleviated my-exedema (1891), and the identification of adrenalin (1894) and secretin (1902) added fuel to the organotherapy fire between clinicians and laboratory scientists. The "use of organ extracts by practitioners . . .quickly outstripped study of these same preparations by experimentalists" (Borell 1985:3), and was related to the rise of immunology (Borell 1976). Both were viewed as new miracles of scientific medicine.   Such sensationalism endured. Well into the 1920s, Harvey Cushing of Harvard was still concerned about practitioners' prescription of dubious substances to "credulous people" (Medvei 1982:504). Another endocrinologist wrote: "Certain therapeutic products are particularly suited to the game of racketeers. . . . The products of the endocrine glands are also particularly adapted to exploitation" (Pottenger 1942:846). The resulting tensions between practicing physicians and laboratory scientists were characteristic of

2. George Corner (1981:316) reported that when the Rockefeller Foundation was under attack as a left liberal organization by "rabid anti-Communists of the McCarthy type," the head of the Foundation, Dean Rusk, surreptitiously called him in as Chair of the NRC Committee; Foundation support of Kinsey through the NRC was soon terminated.

the early years of endocrinology research; this was especially true with regard to the evaluation of extracts of the testes and ovaries, where "particularly controversial therapeutic claims were made" (Borell 1985:4).

The most outrageous therapeutic claims were made in what came to be called "the monkey gland affair" (Hamilton 1986). Monkey and other testes were transplanted by several practitioners into men, sheep, prize bulls, horses and dogs for sexual and geriatric rejuvenation. The most renowned was Sergé-Samuel Voronoff, Russian emigre to France, a surgeon who held an appointment in physiology at the Collège de France. Voronoff began his testicular transplantation practice in 1917, operating on hundreds of men and animals in Europe and the Middle East through 1930, when the surgery began to lose credibility. Another major practitioner was American John R. Brinkley who advertised his goat gland transplants on his own radio station in rural Kansas and ran for governor. Despite dubious medical credentials, Brinkley became a millionaire. Other practitioners used similar approaches (Hamilton 1986:137).

Why did this surgery seem so reasonable to so many scientists, physicians, and lay persons alike? The promise of clinical endocrinology was at its height. Organotherapy had achieved signal successes with insulin and thyroid extracts. And transplantation techniques had been pioneered by Nobel Prize winner Alexis Carrel of the Rockefeller Institute. Glandular transplantation linked these cutting edge sciences in clinical practice, for it made both common and medical sense that transplanting a gland would have beneficial effects. Moreover, specific testicular therapy to aid rejuvenescence and sexual capacity built upon John Hunter's eighteenth century transplantations and Brown-Séquard's technique of ingesting testicular extracts for these purposes.

It was agricultural rather than medical scientists who first seriously criticized Voronoff's lack of scientific method in his efforts to create "super sheep." They especially criticized the omission of random selection for treatment and control groups. Scientific developments also contributed to the gradual abandonment of the surgery, notably animal studies of the actual effects of testosterone and Mayo clinic research on graft rejection, which essentially repudiated glandular grafting as ineffective (Hamilton 1986:136-37).

But organotherapeutic sensationalism simultaneously cut two ways: on the one hand, according to laboratory scientists, it cast a pall of illegitimacy over endocrinology through continued association with quackery. Yet on the other hand, it put the science of hormone study in the bright light of publicity as offering great clinical promise. When more controlled substances were found specifically effective, prestige was heightened. Sensational or not, endocrinology was "cutting edge" research and reproductive sciences benefitted from the association.

Subsequently, a wide array of drugs and clinical treatments developed through reproductive research and built upon principles of organotherapy have themselves caused serious and extended controversies. In the late 1930s, diethylstilbestrol (DES), a synthetic estrogen, began to be used for a wide array of "female problems" associated with reproduction (Bell 1986). Its carcinogenic risks were routinely ignored by clinicians despite researchers' repeated publication of evidence (Bell 1987; McKinlay and McKinlay 1973). One use was in Estrogen Replacement Therapy (ERT), the sequillae of which have included cancer and other serious problems generating a heated debate as to its necessity and the means used to convince patients of its worth (Olesen 1982; Riessmann 1983; McCrea and Markle 1984).

Even more provocative were the sequillae of the use of DES by pregnant women to prevent miscarriage, begun during the Second World War and approved by the Food and Drug Administration in 1947 (Bell 1986). DES was not patented and therefore could be distributed by any pharmaceutical company. Again, warnings of serious problems were ignored, only to resurface in 1971 with diagnoses of adenocarcinoma of the cervix in DES daughters, various congenital sexual abnormalities in DES sons, and higher rates of infertility in both. Probably about 20,000 to 100,000 DES-exposed babies were born each year between 1960-1970. A strong "victims" movement for education, research and improved clinical treatment extended the controversy through public outreach (Dutton 1988).

Contraceptives such as the pill, intrauterine devices (IUDs) and injectables like Depo-Provera have also generated many heated controversies framed in terms of the general morality of contraception, the ethics of their being tested and distributed in Third World countries and other areas with limited medical follow-up, coercion into use, an array of medical risks to users, and fears of medicalization of women's health (McLaughlin 1982; Pramik 1978; Reed 1984; Ward 1986; Vaughn 1970). The women's health movement in the United States and elsewhere mounted powerful campaigns against the birth control pill (Ruzek 1978; Seaman 1969/1980), against FDA approval of Depo-Provera as a contraceptive (Rakusen 1981; National Women's Health Network 1985), and in support of women maimed by the Dalkon Shield (Mintz 1985). Seemingly all IUDs can cause infections that can lead to infertility, a risk noted vehemently by clinicians in the 1930s (Clarke 1985). Federal regulation of contraceptive research and development combined with the risks of product safety liability are now considered so stringent that pharmaceutical companies do not deem it worthwhile. A movement among reproductive scientists, population control and family planning advocates, and some pharmaceutical companies for federal sponsorship of private pharmaceutical contraceptive research and development has been growing (Greep, Koblinsky, and Jaffe 1976; Djerassi 1981; Lincoln and Kaeser 1988).[3] Thus not only those who oppose birth control have objected to reproductive sciences but also health care consumers seeking safe and reliable contraception and other reproduction-related medications and treatments (Ruzek 1978; Fisher 1986).

Links to Social Movements

Reproductive sciences historically have been associated with the birth control, eugenics, and population control movements. The modern birth control movement began in the United States shortly after the turn of the century as a progressive movement to enhance women's control over their reproductive capacities. The eugenics movement began in Britain in the 1880s and was quickly imported into the United States. Focus was on "better people through better breeding," applying agricultural breeding principles to humans. The neo-malthusian population control movement also began in Britain in the 1870s and was quickly imported into the United States. Focus here was on numbers of people and their distribution in relation to resources. All three shared the goal of enhanced control over human reproduction, though their emphases differed considerably.

Separating sexuality from reproduction is controversial. All three of these science-based social movements sought to do so. During the first half of this century, American reproductive scientists valiantly attempted to distinguish their work from these quite controversial movements (Clarke 1985). But after the Second World War, this became increasingly difficult and, for many reproductive scientists, an undesirable strategy as these movements merged and provided increasing support to reproductive sciences (Borell 1985).

The association of reproductive sciences with these controversial social movements has had both positive and negative consequences. Initially, these movements generated much support and funding. But they also served to mobilize oppositional and/or critical movements around five major issues related to reproductive sciences: contraception, abortion, sterilization, infertility services and reproductive research generally.

Catholic and other religious groups have historically opposed reproductive sciences as enhancing the capacity for intervention in natural processes, including sterilization for eu-

---

3. For example, only one IUD is currently on the U.S. market and a new company that will manufacture and distribute another kind will not carry product liability insurance. The company will be structured so that it can go into bankruptcy quickly while protecting its assets should consumer product liability grow too costly (Klitsch 1988).

genic purposes, "unnatural" methods of contraception, and infertility treatments. A number of publicized cases of sterilization abuse—the misinformed, coerced, or unknowing termination of reproductive capacity of women and men—mobilized civil rights and feminist reproductive rights movements during the 1960s and 1970s (Clarke 1984; Shapiro 1985). Recent anti-abortion movements have increasingly extended their objections to federal and state distribution of contraception and to contraceptive research (Rosoff 1988). Moreover, anti-abortionists' use of fetal imagery in popular appeals has had negative consequences for fetal research on reproductive and other problems (CIBA 1986; Petchesky 1987), though reproductive and other scientists have developed new guidelines for such research (Greeley et al. 1989; Annas and Elias 1989).

"Brave New Worlds"

Since the turn of the century, reproductive sciences have been associated by some constituencies with creating "Brave New Worlds," evoking Huxley's (1932) dystopia. Here reproductive sciences are seen as challenging or threatening the "natural order" of life, substituting a technological order, usually with scientists in charge. Such accusations have taken many forms in response to a variety of investigations over the years.

The first dramatized events in this domain were the artificial parthenogenesis experiments of Jacques Loeb in 1899. Loeb succeeded in artificially producing normal larvae from the unfertilized eggs of the sea urchin by altering the inorganic salt solutions in which they lived. This process had formerly required the sperm of the male urchin. The physical chemistry that Loeb believed should be the basis of modern biology "could be a tool for altering the basic process of reproduction" (Pauly 1987:93).

Loeb was a major figure in turn of the century biology and held prestigious positions at the Universities of Chicago and Berkeley, and the Rockefeller Institute. He served as the model for the modern research scientist in Sinclair Lewis' Arrowsmith, and his work brought him both scientific fame and popular notoriety as a modern Faust via popular accounts of his work. Loeb believed that control over both heredity and reproduction could be achieved. Moreover, as Pauly (1987:94) notes:

> The invention of artificial parthenogenesis represented an attack on the privileged status of natural reproduction. . . . The possibility of human parthenogenesis, the sort of basic social reformation to which science could provide the key, was evident from the start. Artificial parthenogenesis was a vindication of Loeb's hopes and a model for science to come, in which biologists would constantly work to reconstruct the natural order to make it more rational, efficient and responsive to the ongoing development of engineering science.

Gregory Pincus was one of the next generation of biologists to carry Loeb's mantle. Born into a family of agricultural scientists, he brought practical as well as theoretical concerns to bear upon his work, focusing on artificial insemination, in vitro fertilization, and artificial parthenogenesis. In 1934, his work on in vitro fertilization became a hot media item. The New York Times' story on its science page was headlined, "Rabbits Born in Glass: Haldane-Huxley Fantasy made real by Harvard Biologists," referring to Huxley's (1932) work based on a book by geneticist J.B.S. Haldane (Reed 1984:321).[4]

In 1936, Pincus declared he had succeeded in producing artificial parthenogenesis in the rabbit. The media renamed the process "Pincogenesis" (Werthessen 1974). Whether or not Pincus had actually succeeded has been hotly debated (Ingle 1971; Werthessen 1974; Reed 1984:317-33). The media were dubious if not highly negative toward this work, in sharp contrast with the neutral to miraculous portrayals that had greeted Loeb's efforts in an earlier era

4. Pincus had transferred the fertilized eggs to living hosts so the rabbits were not born in a test tube. Moreover, the hosts were sacrificed before the embryos grew to term so they were not "born" either (Reed 1984:321).

when the miracles of modern science were unquestioned. An article on Pincus' work in Collier's magazine in 1937, "No Father to Guide Them," combined antifeminism, anti-Semitism, and criticism of the "tricks" that biologists were playing on nature (Reed 1984:323).

The article caused a sensation in Cambridge, and Pincus was soon denied tenure at Harvard. After piecing his career together with grants and temporary appointments for a few years, Pincus and Hudson Hoagland founded the Worcester Foundation for Experimental Biology in 1944, planning to do enough commercial research to fund their own basic research. It was to the Worcester Institute that Margaret Sanger and biologist, feminist and philanthropist Katherine Dexter McCormick went in 1953 to meet with Pincus and offer serious funding for his birth control research. With John Rock and other Boston-based researchers and clinicians, Pincus became a major developer of the pill (Johnson 1977; McLaughlin 1982; Reed 1984).

The association of reproductive sciences with Brave New Worlds has endured and will continue. The application to humans of artificial insemination, in vitro fertilization, embryo transfer, super-ovulation, sex preselection, surrogacy, and so on have been in the media constantly over the past decade. Consumers both recognize and express appreciation for the daring of reproductive scientists. For example, one unsolicited letter to a developer of sex preselection techniques wrote, "Thank you for being brave enough to develop your sperm separation technique. I'm sure you get threatening mail saying you are tampering with a divine process." Another wrote, "It must take much bravery to even enter such 'sacred ground' as sex selection" (Chico 1989:224).

Yet various constituencies are distraught about this work for different reasons. Catholics and fundamentalists object to interventions in "natural" processes. Some feminists view these technologies alternatively as turning women into breeders, as potentially denying women motherhood, and as raising a host of both medical and ethical problems. Other feminists are concerned about the high and typically uninsured costs of these technologies vis-à-vis both costs of health care in general and the limited access poorer people have to them. Disabled communities have found much to object to in terms of both access and applications of reproductive technologies (Raymond 1989; Rothman 1989; Stallone and Steinberg 1987; Stanworth 1987). Lawyers and ethicists are also involved in these debates (Cohen and Taub 1989; Annas 1987).

Agricultural reproductive scientists are also not immune from controversy over the creation of "Brave New Worlds." Here concerned constituencies currently include animal rights movements, ecological groups concerned about the costs of protein derived from meat and dairy products, and groups concerned about the "pharmaceutical farm," biotechnologically created animals, the survival of small farmers, and so on (Schell 1984; Schneider 1988).

Of all the controversial domains in which reproductive sciences have been embedded, their association with Brave New Worlds—with the capacity to radically manipulate and even create new life—is the most significant in sustaining its controversial status. The capacity to manipulate human and animal reproduction (Austin and Short 1986), conceived as both capital and liability, is truly revolutionary.

The Consequences of Controversy for Reproductive Sciences

The controversial status of reproductive sciences has had important consequences in terms of scientific recognition, access to research funding, and the way these scientists managed their careers and their discipline.

Scientific Recognition

Guy Marrian called the years 1926 to 1940 the "heroic age of reproductive endocrinology" (Parkes 1966b:xx), while Alan Parkes (1966a:72) termed it the "endocrinological gold rush." They did so because during this period, the chief naturally occurring estrogens, androgens, and progesterone were isolated and characterized, and the anterior pituitary, placental, and endometrial gonadotropins were also discovered. Yet there was no Nobel Prize in medicine or physiology for explicitly reproductive research either during this breakthrough period or subsequently. In fact, there seems to have been a pattern of awarding the Prize to reproductive endocrinologists for other work that was somewhat tangential to reproductive endocrinology.[5] Many investigators have argued, on behalf of others if not themselves, that the Nobel Prize was deserved by several who did not receive it.[6] In addition, fewer American investigators were elected to the National Academy of Science than might have been expected (Long 1987:266). And despite many honors, the British founder of modern reproductive sciences, F.H.A. Marshall, was never promoted to professor (Medvei 1982:776).

Funding

Funding of reproductive sciences has been largely private, uneven, and problematic (Greep, Koblinsky, and Jaffe 1976; Clarke 1985). Between the World Wars private agency and industrial support in the United States was largely from Rockefeller funding of the prestigious National Research Council's Committee for Research on Problems of Sex (Greep, Koblinsky, and Jaffe 1976:371). One historian asserts that this Committee "virtually paid for the development in American universities of [reproductive] endocrinology" (Reed 1984:283). Greep and his associates (1976:371) charted funding between 1922 and 1945 at $1,666,800.

Why was all this support forthcoming for a relatively unestablished and controversial line of research during the interwar years when the foundations were just beginning to support scientific research? Reproductive sciences were embedded within and dependent upon industrialization processes that enhanced control over nature and social life. They offered control over the means of reproduction, which many sought. For example, in 1934, Warren Weaver of the Rockefeller Foundation asked:

Can man gain an intelligent control of his own power? Can we develop so sound and extensive a genetics that we can hope to breed, in the future, superior men? Can we obtain enough knowledge of physiology and psychobiology of sex so that men can bring this pervasive, highly important, and dangerous aspect of life under rational control? (Kohler 1976:291).

Reproductive scientists answered loudly in the affirmative. Their successes before the Second World War were based not only on their science, but also on the efforts of entrepreneurial investigators building a new line of work in a society increasingly amenable to the entry of science into the most private of social realms—regardless of controversy.

After the Second World War, the burden of funding basic science in the United States began shifting to the federal government, especially in terms of medical research via the NIH. But all sciences did not benefit equally or immediately. Reproduction was not construed as a chronic disease and thus did not fit within the missions of the NIH. Moreover, the NIH was forbidden from funding birth control research prior to 1959 (Greep, Koblinsky, and Jaffe 1976:367). Nor were reproductive sciences seen to fit within the mission of the NSF. During the period from 1945 to 1960, the USDA was the major federal source of funds for reproductive research that focused on quick and sensitive assays for hormones including development of radioimmunoassay in 1956 (Gruhn and Kazer 1989:98). However, between 1945 and 1960,

---

5. In 1939, German Adolf Butenandt, who had done a considerable amount of reproductive endocrinological research, won the Nobel Prize in chemistry with Leopold Ruzicka for changing cholesterol into a synthetic duplicate of natural testosterone (Reed 1984:314; Maisel 1965:37). And in 1947, Bernardo Houssay, another investigator who had worked on reproductive problems, was awarded a Nobel Prize for studies demonstrating the importance of the anterior pituitary in sugar metabolism (Medvei 1982:505).

6. This issue was often raised in my interviews with senior reproductive scientists and see Corner (1981).

the amount of funding did expand, dipping only during the McCarthy era. The major contributors were certain pharmaceutical companies and various population-focused organizations such as the Ford Foundation. Relative to other areas of medical research, however, funding was not at all significant (Greep, Koblinsky, and Jaffe 1976:378-79).

After 1965, both federal and industry (largely pharmaceutical) support rose dramatically. The National Academy of Sciences Committee on Science and Public Policy selected population problems as its focus in 1961 (National Academy of Sciences 1979:v). In 1963, the National Institute for Child Health and Human Development was established in the NIH and, in the view of the Kennedy administration, its mission included sponsorship of both reproductive sciences and fertility control research. The Rockefeller and Ford Foundations designated population to be a major area of concern, and the World Health Organization began supporting reproductive sciences as well. Funding jumped from under $38 million for 1960-1965 to $332 million for 1969-1974 (Greep, Koblinsky, and Jaffe 1976:382).

The irony here is, of course, that it became quasi-legitimate for the government to fund reproductive sciences only shortly before fiscal austerity hit federal research support in the mid-1970s. Reproductive sciences had only a few expansive years before cutbacks became routine and competition became ever more ferocious. Moreover, private foundation support also began to decrease (Clark 1982).

During the 1980s, the Reagan and Bush administrations, likely due to strong alliances with anti-abortion movements, have been extremely hostile to seemingly all aspects of reproductive sciences, including basic research as well as research on infertility and contraception. Reproductive scientists have felt stymied and abandoned in their quest for research support. For example, it has been claimed that "In real dollars, worldwide spending for contraceptive development by both governments and the pharmaceutical industry has declined by nearly one-quarter since its peak in 1972" (Lincoln and Kaeser 1988:20-1). Whole university-based programs vaunted in the 1960s as signalling a new age are being dismantled, and fewer new scientists are entering the field.[7] Moreover, at the same time both governmental and foundation funding have ebbed, the federal regulatory environment has become increasingly restrictive as agencies are forced to respond to an array of consumer groups, religious groups, and anxious clinicians fearing malpractice claims. Many assert that the United States is losing its preeminence in the field for the first time since 1920 (Lincoln and Kaeser 1988; Rosoff 1988).[8]

Scientists' Management Strategies

How have reproductive scientists managed their controversial status and obtained funding across these highly varying historical conditions? Before the First World War, research itself was relatively cheap and could typically be pieced together through department budgets and minor donations (Clarke 1987). During the interwar years, the NRC Committee for Research on Problems of Sex provided both legitimacy and funding (Aberle and Corner 1953). Reproductive scientists quickly captured control of the Committee and redirected its mission from research on human sexuality to basic research on mammalian reproductive processes. They eschewed the proffered support of birth control advocates and, in fact, typically shunned those reproductive scientists who undertook such applied work (Clarke 1985). In Hall's (1974:92) words:

> Men ambitious for a scientifically respectable study of sex recognized that its basis must be broadly biological and independent of human interest or subjective experience. . . . [T]his was a liberal position, and liberated biology as an autonomous science. . . . [T]he biological scientists of the twenties did not create social policy, but, driven by internal pressures for a respectable science, they

7. Interview with Dr. John Biggers, Harvard Medical School, 1987, and see Lincoln and Kaeser (1988).
8. I draw here on discussions with participants at the Conference on Infertility, University of California, Davis. 1986, and numerous interviews with reproductive scientists.

created an increasingly biochemical science whose voice spoke with . . . the "magic" of esoteric authority.

In fact, one management strategy of reproductive scientists has been to assume positions of authority and speak in the calm, moderate tone of objective science to both popular (Corner 1938, 1939; Hartman 1933) and highly educated audiences (Corner [1942] 1963; Maisel 1965) in order to promote and defend their work.[9] Another strategy, used to protect personal careers, was to simultaneously pursue less controversial but related research. This often involved work on the anterior pituitary gland, which produces both reproductive and other hormones. The discovery of this fact linked reproductive with general endocrinology at the height of endocrinology's heyday in the 1920s and 1930s (Clarke 1985). For example, Houssay worked on the anterior pituitary and sugar metabolism; Herbert McLean Evans worked on growth hormone from the anterior pituitary and on vitamins; and Oscar Riddle worked on problems in genetics (Medvei 1982).

After the Second World War, when the family planning and population control movements merged, reproductive scientists intensified their study of fertility processes that could lead to both infertility and contraceptive treatments and technologies. Considerable emphasis on infertility during the 1940s and 1950s (American Foundation 1955; Corner 1957; McLaughlin 1982) simultaneously aided legitimacy and countered criticism of reproductive sciences for development of contraceptives. However, it is in support of birth control and population control that most non-governmental funding has come since the Second World War. Reproductive scientists have increasingly advocated contraceptive and population control goals (Shelesnyak 1968; Greep, Koblinsky, and Jaffe 1976; Djerassi 1989). In fact, today it is often impossible to distinguish reproductive sciences from organizations focused on those issues.

Like other scientists, reproductive scientists have claimed that their science would solve social problems (cf. Aronson 1984). The problems reproductive scientists claim to address are population and reproduction. In consequence, biomedical reproductive sciences are now inextricably linked to those domains and their status fluctuates according to the status of those problems (Hilgartner and Bosk 1988). For example, some analysts have recently argued that concern about "overpopulation" has abated, decreasing commitment to reproductive research that would help to solve such problems (Djerassi 1989). The American controversy over abortion has led to prohibitions against funding research into any method that may be used to induce abortion or improve abortion techniques, including those that might improve safety and efficiency (Lincoln and Kaeser 1988). A recent House subcommittee report has charged that Center for Disease Control officials have been censored and demoted for publishing abortion related research (Specter 1989).

Since about 1980, one strategic response to funding cuts has been the development of distinctive new medical centers for reproductive sciences focused again on infertility. Such centers typically combine both clinical and basic research with service delivery. In some, research costs are covered in part by service fees. While committed American reproductive scientists in the 1980s may have lacked alternatives, this strategy of refocusing on infertility has again provided them with funding and legitimacy among some constituencies. Ironically, however, these scientists have also confronted extensive controversy through these same "new reproductive technologies." Like the earlier controversy over quackery, the possibly strategic defensive retreat into infertility work to cope with funding cutbacks and hostile political administrations during the 1980s seems to have had both positive and negative outcomes.

In sum, for over a century, reproductive scientists have been managing their controversial status, developing strategies and tactics to handle their divergent audiences and constituencies. Problems are moved from the front to back burner and vice versa depending largely

9. Thanks to Diana Long for clarifying this point.

on the pragmatics of the current situation.[10] The funding, status, and controversy management problems of American reproductive scientists have not abated and are unlikely to do so
in the foreseeable future. Meanwhile, the mantles of Loeb and Pincus have been appropri-
ated by colleagues in Britain, Australia, Sweden, and elsewhere as American preeminence in
reproductive sciences dims and the global scientific economy rises in more "hospitable" envi-
ronments (Atkinson, Lincoln, and Forest 1985). In this historical moment the crisis of scientific authority is intersecting with the fiscal crisis of the state, and a radical restructuring of the
international scientific economy is in process.

Discussion

Little work in the sociology of science directly addresses the question of what makes a
line of scientific work controversial (Nelkin 1984; Mazur 1981; Engelhardt and Caplan 1987;
Chubin and Chu 1989). One exception is Gieryn's (1983, 1987a,b) focus on the boundaries of
science and scientists' use of representations of science-in-culture to legitimate their claims to
authority, patronage, and autonomy. For example, Gieryn (1987a) has contrasted "safe science" and "risky science" as embodied in the work of two individuals competing for a single
academic position. In this situation, he found "safe science" was elitist, pure science done
within established and demarcated disciplines, and offered social roles for scientists that were
distinctly separate from most of human daily life. In contrast, "risky science" was populist,
utilitarian, technocratic science done in situations of blurred disciplinary boundaries, while
the science itself (and scientists' roles) expanded into ethical and moral domains. While this
work breaks new ground, it is left to us to ask, "Safe for whom?" "Risky to whom?" And,
especially, "Who cares?"

It seems a safe bet that there are usually more than two "sides" involved in most controversies. That is, an arena analysis assumes that multiple social worlds are involved and
pushes us to analyze their points of intersection. Thus, drawing upon the case study
presented here, we can more generally ask, "Under what conditions is a line of scientific work
likely to be construed as controversial by at least some other social worlds?" Five major conditions can be seen to obtain, each of which might be more or less salient to the concerns of
different worlds.

First, if a "basic" science has publicly obvious areas of actual or potential application it is
much more vulnerable. This is especially true if the applications directly involve humans, as
in clinical medicine. Second, if the science intrinsically challenges or threatens the traditional
constructions of the "natural order," the relations of humans to nature, controversy looms (but
may or may not occur). Third, if the areas of application are related to politicized social issues
or social problems, the potential is great for some social worlds to designate the science itself
as controversial and the scientists as immoral entrepreneurs. Fourth, if sentiment about those
social issues has been mobilized into social movements, the line of scientific work is even
more vulnerable. Moreover, the social movements may attempt to recruit the scientists, enlist
their support, or even become sponsors of the scientific work. Last, if the line of scientific
work is itself embedded in policy or other arenas, in competition for scarce resources and so
on, it is increasingly likely to be embedded in controversy.

More generally, the line of scientific work (or another social world in the arena) may
itself become the contested terrain, in terms not only of its content and ideology, but also its
very existence. Multiple other worlds may seek to control, direct, or even destroy it. Such
focal worlds can be called "boundary worlds" over and through which other social worlds in
the arena pitch their battles and negotiate their treaties (cf. Star and Greisemer 1989). The

10. Interview with Dr. M.C. Shelesnyak, June 1988.

boundaries are challenged relentlessly in these processes. While the boundary world in an arena may shift over time, the current boundary world is highly vulnerable to controversy. However, status as a controversial boundary world does not necessarily mean that that world is doomed to failure. Rather, it means that an analysis of power and politics within the arena is requisite for participants (as well as sociologists). Such analyses must be ongoing to perceive the shifts and alternative strategies that might lead to survival and/or success despite controversy (Clarke 1985; Fujimura 1987; Star 1986). For there is no consensus, merely successful or unsuccessful negotiations among those involved at any historical moment (Strauss 1979). In fact, the strength of the closure of controversies can be grounded not in consensus, but in the mobilization of effective social networks and/or the construction of "black boxes" around controversial phenomena, rendering them invisible (Rip 1986). The lid on the black box contains the controversy within it—until the box is reopened.

References

Aberle, Sophie D., and George W. Corner
    1953   Twenty-Five Years of Sex Research: History of the National Research Council Committee for Research in Problems of Sex, 1922-1947. Philadelphia: W.B. Saunders.
Abir-Am, Pnina
    1982   "The discourse of physical power and biological knowledge in the 1930s: a reappraisal of the Rockefeller Foundation's 'Policy' in Molecular Biology." Social Studies of Science 12:341-82.
Allen, Edgar, ed.
    1932   Sex and Internal Secretions. 1st ed. Baltimore, Md.: Williams and Wilkins.
Allen, Garland
    1986   "T.H. Morgan and the split between embryology and genetics, 1910-1935." In A History of Embryology: The Eighth Symposium of the British Society for Developmental Biology, ed. T.J. Horder, J.A. Witkowski, and C.C. Wylie,113-46. Cambridge: Cambridge University Press.
American Foundation
    1955   Medical Research: A Midcentury Survey. Vol. 2. Unsolved Clinical Problems in Biological Perspective. Boston: Little, Brown.
Annas, George J., and Sherman Elias
    1989   "The politics of transplantation of human fetal tissue." New England Journal of Medicine 320:1079-82.
Aronson, Naomi
    1984   "Science as claimsmaking: implications for social problems research." In Studies in the Sociology of Social Problems, ed. Joseph W. Schneider and John I. Kitsuse, 1-30. Norwood: Ablex.
Atkinson, L.E., R. Lincoln, and J.D. Forest
    1985   "Worldwide trends in funding for contraceptive research and evaluation." Family Planning Perspectives 17:260-62.
Austin, C.R. and R.V. Short, eds.
    1987   Manipulating Reproduction. Second Edition. New York: Cambridge University Press.
Becker, Howard S.
    1982   Art Worlds. Berkeley: University of California Press.
Beer, John B., and W. Daniel Lewis
    1974   "Aspects of the professionalization of science." In Comparative Studies in Science and Society, ed. Sal P. Restivo and Christopher K. Vanderpool, 764-84. Columbus, Ohio: Merrill.

Bell, Susan E.
    1987   "Changing ideas: the medicalization of menopause." Social Science and Medicine 24:535-
            42.
    1986   "A new model of medical technology development: a case study of DES." In Research in
            the Sociology of Health Care, Vol. 4, ed. Julius Roth and Sheryl Ruzek, 1-32. Greenwich,
            Conn.: JAI Press.
    1986   "Increasing productivity in farm animals." In Reproduction in Mammals: Book 5
            Manipulating Reproduction, 2d ed., ed. C.R. Austin and R.V. Short, 1-47. Cambridge:
            Cambridge University Press.
Biggers, John D.
    1984   "In vitro fertilization and embryo transfer in historical perspective." In In Vitro
            Fertilization and Embryo Transfer, ed. Alan Trouson and Carl Wood, 3-15. London:
            Churchill Livingstone.
Booth, William
    1989   WHO seeks global data on sexual practices. Science 244:418-19.
Borell, Merriley
    1987   "Biologists and the promotion of birth control research, 1918-1938." Journal of the
            History of Biology 20:57-87.
    1985   "Organotherapy and the emergence of reproductive endocrinology." Journal of the
            History of Biology 18:1-30.
    1978   "Setting the standards for a new science: Edward Schafer and endocrinology." Medical
            History 22:282-90.
    1976   "Brown-Séquard's organotherapy and its appearance in America at the end of the
            nineteenth century." Bulletin of the History of Medicine 50:309-20.
Brush, Stephen G.
    1978   "Nettie M. Stevens and the discovery of sex determination by chromosomes." Isis 69:163-
            72.
Chico, Nan Paulsen
    1989   Confronting the Dilemmas of Reproductive Choice: The Process of Sex Preselection.
            Ph.D. dissertation. San Francisco, Calif.: University of California, San Francisco.
Chubin, Daryl and Ellen Chu, eds.
    1989   Science Off the Pedestal: Social Perspectives on Science and Technology. Belmont, Calif.:
            Wadsworth.
CIBA Foundation
    1986   Human Embryo Research: Yes or No? London: Tavistock.
Clark, Anne Harrison
    1982   "Funding support for population research." Women and Health 7:73-81.
Clarke, Adele E.
    1989   "The reproductive policy arena." Department of Social and Behavioral Sciences,
            University of California, San Francisco.
    1990a   "Embryology and the development of American reproductive sciences, 1910-1945." In
            The American Expansion of Biology, ed. Keith Benson, Ronald Rainger, and Jane
            Maienschein. New Brunswick, N.J.: Rutgers University Press.
    1990b   "A social worlds research adventure: the case of reproductive science." In Theories of
            Science in Society, eds. Thomas Gieryn and Susan Cozzens. Bloomington, Ind.: Indiana
            University Press.
    1989   "The industrialization of human reproduction, 1889-1989." Presented at meetings of the
            Society for the Study of Social Problems, Berkeley.
    1988   "Getting down to business: the life sciences c.1890-1940." Presented at meetings of the
            Society for Social Studies of Science, Amsterdam.
    1987   "Research materials and reproductive science in the United States, 1910-1940." In
            Physiology in the American Context, 1850-1940, ed. Gerald L. Geison, 323-50. Bethesda,
            Md.: American Physiological Society/Waverly.
    1985   Emergence of the Reproductive Research Enterprise, c.1910-1940: A Sociology of
            Biological, Medical and Agricultural Science in the United States. Ph.D. dissertation, San
            Francisco, Calif.: University of California, San Francisco.

1984   "Subtle sterilization abuse: a reproductive rights perspective." In Test Tube Women:
          What Future for Motherhood?, ed. Rita Arditti, Renata Duelli Klein, and Shelly Minden,
          188-212. Boston: Pandora/Routledge and Kegan Paul.
Clarke, Adele and Elihu M. Gerson
1990   "Symbolic interactionism in science studies." In Symbolic Interaction and Cultural
          Studies, ed. Howard S. Becker and Michal McCall. Chicago, Ill.: University of Chicago
          Press.
Cohen, Sherrill, and Nadine Taub, eds.
1989   Reproductive Laws for the 1990s. Clifton, N.J.: Humana.
Cole, Harold H.
1977   Adventurer in Animal Science: Harold H. Cole. Davis, Calif.: The Oral History Center,
          Shields Library, University of California at Davis.
Corner, George W.
1981   Seven Ages of a Medical Scientist: An Autobiography. Philadelphia: University of
          Pennsylvania.
[1942]   The Hormones in Human Reproduction. Reprint. New York: Athenium.
1963
1961   "Foreword." In Sex and Internal Secretions, 3rd ed., ed. William C. Young, xxxi-xxxvii.
          Baltimore, Md.: Williams and Wilkins.
1957   "Laboratory and clinic in the study of infertility." Fertility and Sterility 8:494-512.
1939   Attaining Womanhood: A Doctor Talks to Girls About Sex. New York: Harper.
1938   Attaining Manhood: A Doctor Talks to Boys About Sex. New York: Harper.
Djerassi, Carl
1989   "The bitter pill." Science 245:356-61.
1981   The Politics of Contraception: The Present and the Future. San Francisco, Calif.: W.H.
          Freeman.
Dutton, Diana B.
1988   Worse than the Disease: Pitfalls of Medical Progress. Cambridge: Cambridge University
          Press.
Elias, Sherman, and George J. Annas
1987   Reproductive Genetics and the Law. Chicago, Ill.: Yearbook Medical Publishers.
Engelhardt, Tristram, and Arthur L. Caplan, eds.
1987   Scientific Controversies: Case Studies in the Resolution and Closure of Disputes in
          Science and Technology. Cambridge: Cambridge University Press.
Estes, Carroll L., and Elizabeth A. Binney
1989   "The biomedicalization of aging: dangers and dilemmas." The Gerontologist 29:587-96.
Fisher, Sue
1986   In the Patient's Best Interest: Women and the Politics of Medical Decisions. New
          Brunswick, N.J.: Rutgers University Press.
Fujimura, Joan H.
1988   "The molecular biological bandwagon in cancer research: where social worlds meet."
          Social Problems 35:261-83.
1987   Constructing doable problems in cancer research: articulating alignment." Social Studies
          of Science 17:257-93.
Gieryn, Thomas F.
1987a   "Safe science and risky science: competition for the chair of logic and metaphysics at the
          University of Edinburgh, 1836." Presented at the Conference on Argument in Science,
          University of Iowa, October 16.
1987b   "Science and Coca-Cola." Science and Technology Studies 5:12-21.
1983   "Boundary-work and the demarcation of science from non-science: strains and interests
          in professional ideologies of scientists." American Sociological Review 48:781-95.
Gilbert, Scott F.
1987   "In friendly disagreement: Wilson, Morgan, and the embryological origins of the gene
          theory." American Zoologist 27:797-806.
Glaser, Barney G.
1978   Theoretical Sensitivity: Advances in the Methodology of Grounded Theory. Mill Valley,
          Calif.: The Sociology Press.

229

Glaser, Barney G. and Anselm L. Strauss
    1968   The Discovery of Grounded Theory: Strategies for Qualitative Research. New York:
            Aldine.
Goffman, Erving
    1963   Stigma: The Management of Spoiled Identity. Harmondsworth: Penguin.
Gordon, Linda
    1976   Woman's Body, Woman's Right: A Social History of Birth Control in America. New
            York: Penguin.
Greeley, Henry T., Thomas Hamm, Rodney Johnson, Carole R. Price, Randy Weingarten, and Thomas
        Raffin
    1989   "The ethical use of human fetal tissue in medicine." New England Journal of Medicine
            320:1093-96.
Greep, Roy O., Marjorie A. Koblinsky, and Frederich S. Jaffe
    1976   Reproduction and Human Welfare: A Challenge to Research. Boston, Mass.: MIT Press
            for the Ford Foundation.
Gruhn, John G., and Ralph R. Kazer
    1989   Hormonal Regulation of the Menstrual Cycle: the Evolution of Concepts. New York:
            Plenum Medical.
Hall, Diana Long
    1974   "Biology, sex hormones and sexism in the 1920s." Philosophical Forum 5:81-96.
Hamilton, David
    1986   The Monkey Gland Affair. London: Chatto and Windus.
Haraway, Donna
    1979   "The biological enterprise: sex, mind and profit from human engineering to
            sociobiology." Radical History Review 20:206-37.
Hartman, Carl G.
    1933   "Catholic advice on the safe period.' " Birth Control Review 17:117-9.
Herman, Harry A.
    1981   Improving Cattle by the Millions: NAAB and the Development and Worldwide
            Application of Artificial Insemination. Columbia, Mo.: University of Missouri Press.
Hilgartner, Stephen, and Charles Bosk
    1988   "The rise and fall of social problems: a public arenas model." American Journal of
            Sociology 94:53-78.
Holden, Constance, ed.
    1989   "Briefings: sex secrets safe." Science 245:599.
Hughes, Everett C.
    1971   The Sociological Eye. Chicago: Aldine Atherton.
    1962   "Good people and dirty work." Social Problems 10:3-10.
Huxley, Aldous
    1932   Brave New World. London: Chatto and Windus.
Ingle, Dwight J.
    1971   "Gregory Goodwin Pincus, 1903-1967." Biographical Memoirs of the National Academy
            of Science 42:229-47.
Johnson, R. Christian
    1977   "Feminism, philanthropy and science in the development of the oral contraceptive pill."
            Pharmacy in History 19:63-78.
Klitsch, M.
    1988   "The return of the IUD." Family Planning Perspectives 20:19.
Kohler, Robert E.
    1976   "The management of science: the experience of Warren Weaver and the Rockefeller
            Foundation programme in molecular biology." Minerva 14:279-306.
Latour, Bruno
    1987   Science in Action. Cambridge, Mass.: Harvard University Press.
Leavitt, Judith Waltzer
    1986   Brought to Bed: Childbearing in America, 1750-1950. New York: Oxford University
            Press.

Light, Donald W.
  1983   "The development of professional schools in America." In The Transformation of Higher Learning 1860-1930, ed. Konrad Jarausch, 345-65. Chicago, Ill.: University of Chicago Press.
Lincoln, Richard, and Lisa Kaeser
  1988   "Whatever happened to the contraceptive revolution?" Family Planning Perspectives 20:20-24.
Long, Diana E.
  1987   "Physiological identity of American sex researchers between the two World Wars." In Physiology in the American Context, 1850-1940, ed. Gerald L. Geison, 263-78. Bethesda, Md.: American Physiological Society.
Maienschein, Jane
  1984   "What determines sex? A study of converging approaches, 1880-1916." Isis 75:457-80.
Maisel, Albert Q.
  1965   The Hormone Quest. New York: Random House.
Marshall, F.H.A.
  1910   The Physiology of Reproduction. London: Longmans, Green and Co.
Mazur, Allan
  1981   The Dynamics of Technical Controversies. Washington, D.C.: Communications Press.
McCrea, Frances B., and Gerald E. Markle
  1984   "The estrogen replacement controversy in the USA and UK: different answers to the same question?" Social Studies of Science 14:1-26.
McKinlay, Sonja M., and John B. McKinlay
  1973   "Selected studies of the menopause: an annotated bibliography." Biosocial Science 5:533-55.
McLaughlin, Loretta
  1982   The Pill, John Rock, and the Church: The Biography of a Revolution. Boston, Mass.: Little, Brown.
Medvei, Victor C.
  1982   A History of Endocrinology. Boston: MTP Press.
Mintz, Morton
  1985   At Any Cost: Corporate Greed, Women, and the Dalkon Shield. New York: Pantheon/Random House.
Nalbandov, A.V.
  1978   "Retrospects and prospects in reproductive physiology." In Novel Aspects of Reproductive Physiology, ed. Charles H. Spelman and John W. Wilks, 1-30. New York: Halsted/Wiley.
National Academy of Sciences
  1979   Contraception: Science, Technology and Application. Washington, D.C.: National Academy of Sciences.
National Women's Health Network
  1985   "The Depo-Provera debate: A report by the National Women's Health Network." Washington, D.C.: National Women's Health Network.
Nelkin, Dorothy, ed.
  1984   Controversy: Politics of Technical Decisions. 2d ed. Beverly Hills, Calif.: Sage.
  1987   Selling Science: How the Press Covers Science and Technology. New York: W.H. Freeman.
Olesen, Virginia
  1982   "Sociological observations on ethical issues implicated in estrogen replacement therapy at menopause." In Changing Perspectives on Menopause, ed. Ann M. Voda, Myra Dinnerstein, and Sheryl R. O'Donnell, 346-60. Austin, Tex.: University of Texas.
Olmstead, J.M.D.
  1946   Charles-Edouard Brown-Séquard: A Nineteenth Century Neurologist and Endocrinologist. Baltimore, Md.: The Johns Hopkins University Press.
Oudshoorn, Nelly
  1990a   "On the making of sex hormones: research materials and the production of knowledge." Social Studies of Science. Forthcoming.

231

1990b   "Endocrinologists and the conceptualization of sex, 1920-1940." Journal of the History of Biology. Forthcoming.

Parkes, A.S.
1966a   Sex, Science and Society: Addresses, Lectures and Articles. London: Oriel Press.
1966b   "The rise of reproductive physiology, 1926-1940. The Dale Lecture for 1965." Journal of Endocrinology 34:xx-xxxii.

Pauly, Philip J.
1987   Controlling Life: Jacques Loeb and the Engineering Ideal in Biology. New York: Oxford University Press.

Petchesky, Rosalind Pollack.
1987   "Fetal images: the power of visual culture in the politics of reproduction." Feminist Studies 13:263-92.

Pincus, Gregory
1965   The Control of Fertility. New York: Academic.

Pramik, Mary Jean, ed.
1978   Norethindron: The First Three Decades. Palo Alto, Calif.: Syntex Laboratories.

Pomeroy, Wardell B.
1982   Dr. Kinsey and the Institute for Sex Research. New Haven, Conn.: Yale University Press.

Pottenger, F.M.
1942   "The Association for the Study of Internal Secretions: its past: its future." Endocrinology 30:846-52.

Rakusen, Jill
1981   "Depo-Provera: the extent of the problem." In Women, Health and Reproduction, ed. Helen Roberts, 75-108. Boston, Mass.: Routledge and Kegan Paul.

Raymond, Janice G.
1989   "At issue: reproductive technologies, radical feminism and socialist liberalism." Reproductive and Genetic Engineering 2:133-42.

Reed, James
1984   The Birth Control Movement and American Society: From Private Vice to Public Virtue, Second edition. Princeton, N.J.: Princeton University Press.

Riessman, Catherine Kohler
1983   "Women and medicalization: a new perspective." Social Policy 14:3-18.

Rip, Arie
1986   "Controversies as informal technology assessment." Knowledge 8:349-71.

Rosenberg, Charles E.
1979   "Rationalization and reality in shaping American agricultural research, 1875-1914." In The Sciences in the American Context: New Perspectives, ed. Nathan Reingold, 143-63. Washington, D.C.: Smithsonian Institution.

Rosoff, Jeannie I.
1988   "The politics of birth control." Family Planning Perspectives 20:312-20.

Rossiter, Margaret
1979   "The organization of the agricultural sciences." In The Organization of Knowledge in Modern America, 1860-1920, ed. Alexandra Oleson and John Voss, 212-48. Baltimore, Md.: Johns Hopkins University Press.

Rothman, Barbara Katz
1989   Recreating Motherhood: Ideology and Technology in a Patriarchal Society. New York: W. W. Norton.

Ruzek, Sheryl Burt
1978   The Women's Health Movement: Feminist Alternatives to Medical Control. New York: Praeger.

Schell, Orville
1984   Modern Meat: Antibiotics, Hormones, and the Pharmaceutical Farm. New York: Random House.

Schneider, Joseph W.
1984   "Morality, social problems and everyday life." In Studies in the Sociology of Social Problems, ed. Joseph W. Schneider and John I. Kitsuse, 180-205. Norwood: Ablex.

Schneider, Keith
    1988   "Biotechnology's cash cow." New York Times Magazine 12:44-53.
Seaman, Barbara
    1969   The Doctors' Case Against the Pill. New York: Doubleday.
Shapiro, Thomas
    1985   Population Control Politics: Women, Sterilization and Reproductive Choice.
            Philadelphia, Pa.: Temple University Press.
Shelesnyak, M. C.
    1968   "Population problems: medical and public health." In Health Problems in Developing
            States, ed. Moshe Prywes, M.D. and A. Michael Davies, M.D., 197-211. New York:
            Grune and Strattan.
Specter, Michael
    1989   "Panel claims censorship on abortion." San Francisco Chronicle, 13 December.
Stallone, Patricia, and Deborah Lynn Steinberg, eds.
    1987   Made to Order: The Myth of Reproductive and Genetic Progress. New York: Pergamon.
Stanworth, Michelle, ed.
    1987   Reproductive Technologies: Gender, Motherhood and Medicine. Minneapolis, Minn.:
            University of Minnesota Press.
Star, Susan Leigh
    1986   "Triangulating clinical and basic research: British localizationists, 1870-1906." History of
            Science 24:29-48.
Star, Susan Leigh, and James R. Griesemer
    1989   "Institutional ecology, 'translations,' and boundary objects: amateurs and professionals in
            Berleley's Museum of Vertebrate Zoology, 1907-1939." Social Studies of Science 19:387-
            420.
Strauss, Anselm L.
    1987   Qualitative Analysis for Social Scientists. Cambridge: Cambridge University Press.
    1982   "Social worlds and legitimation processes." In Studies in Symbolic Interaction, Vol. 4, ed.
            Norman Denzin, 171-90. Greenwich, Conn.: JAI Press.
    1979   Negotiations: Varieties, Contexts, Processes and Social Order. San Francisco, Calif.:
            Jossey-Bass.
    1978   "A social worlds perspective." In Studies in Symbolic Interaction, Vol. 1, ed. Norman
            Denzin, 119-28. Greenwich, Conn.: JAI Press.
Vaughn, Paul
    1970   The Pill on Trial. New York: Coward-McCann.
Ward, Martha C.
    1986   Poor Women, Powerful Men: America's Great Experiment in Family Planning. Boulder,
            Colo.: Westview.
Werthessen, N.T., and R.C. Johnson
    1974   "Pincogenesis—parthenogenesis in rabbits by Gregory Pincus." Perspectives in Biology
            and Medicine 18:86-93.
Weiner, Carolyn
    1981   The Politics of Alcoholism. New Brunswick, N.J.: Transaction.

233

CURRENT CONCEPTS

JANE F. DESFORGES, M.D., *Editor*

PRENATAL DIAGNOSIS

MARY E. D'ALTON, M.D.,
AND ALAN H. DECHERNEY, M.D.

SERIOUS birth defects, often genetically determined, complicate and threaten the lives of 3 percent of newborn infants.[1] These disorders account for 20 percent of deaths during the newborn period and an even higher percentage of serious morbidity in infancy and childhood.[2] The cost of neonatal intensive care is staggering. Higher still are the costs of rehabilitation programs for the severely handicapped. The family tragedy is immeasurable. With the growing recognition of the frequency and importance of congenital disorders and with current social trends toward smaller families and delays in childbearing, prenatal diagnosis has an important role in the management of many pregnancies.

INDICATIONS FOR PRENATAL DIAGNOSIS

The identification of pregnancies in which there is an increased chance of a diagnosable fetal disorder involves a search for general and specific risk factors. The use of a questionnaire to elicit genetic information is currently recommended by the American College of Obstetricians and Gynecologists.[3] Counseling before prenatal diagnosis is important. The central issue is balancing the risk of the birth of an abnormal child against the risk of an investigative procedure. Prospective parents must understand that a specific diagnosis can be excluded or established with a high degree of reliability but not with complete certainty. One of the most important goals of genetic counseling is to help patients understand the reproductive options available to them. A person's previous experience, ethnic and cultural background, and religious beliefs will affect the acceptability of prenatal diagnosis and the choices to be made if an abnormality is diagnosed. Counseling should be nondirective and should concentrate on the accurate presentation of all the facts and options. Common indications for prenatal counseling and diagnosis are summarized in Table 1.

General Risk Factors

It is standard practice to offer prenatal cytogenetic diagnosis to all women who will be 35 years of age or older at their expected delivery date. Numerical chro-

From the Department of Obstetrics and Gynecology, Tufts University School of Medicine, Boston. Address reprint requests to Dr. D'Alton at the Department of Obstetrics and Gynecology, New England Medical Center, 750 Washington St., Box 324, Boston, MA 02111.

mosomal abnormalities occur with increasing frequency with advancing maternal age (Table 2).[4,5] Tests for biochemical markers in maternal serum identify patients at risk for certain cytogenetic and structural abnormalities. Alpha-fetoprotein, the major circulating protein of early fetal life, is synthesized in the fetal liver and yolk sac. Open neural-tube and ventral-wall defects are associated with exposed fetal-membrane and blood-vessel surfaces that increase the levels of alpha-fetoprotein in both amniotic fluid and maternal serum.[6] Low levels of maternal serum alpha-fetoprotein and unconjugated estriol are associated with trisomy 21 and trisomy 18.[7] The single marker that yields the highest detection rate for Down's syndrome is the level of human chorionic gonadotropin, which is significantly elevated when the fetus is affected by this syndrome. The combined use of the maternal serum human chorionic gonadotropin level, the unconjugated estriol level, the alpha-fetoprotein level, and maternal age can identify approximately 60 percent of cases of Down's syndrome, with a false positive rate of 6.6 percent. The use of ultrasonography to verify gestational age reduces the false positive rate to 3.8 percent.[8]

Screening by measurement of maternal serum alpha-fetoprotein should be offered to women at 16 to 18 weeks of gestation. Careful evaluation of gestational age is critical, because maternal serum alpha-fetoprotein values increase steadily throughout the second trimester. Because of differences among population groups in median maternal serum alpha-fetoprotein values, laboratories should provide interpretations of results that take into account such factors as race, multiple gestation, diabetes mellitus, and maternal weight. Most centers in the United States have chosen a cutoff of 2.0 or 2.5 multiples of the median value for use in screening the general population of pregnant women for fetal neural-tube defects. The use of individual estimates of patients' risks rather than multiples of the median maternal serum alpha-fetoprotein level to define the cutoff value for fetal Down's syndrome is now almost universal. Invasive procedures such as amniocentesis may elevate the maternal alpha-fetoprotein concentration; therefore, blood samples for screening should be obtained before amniocentesis is performed.

Specific Risk Factors

Specific risk factors may be identified in the family history, the history of previous pregnancies, or the mother's medical history. After the birth of one child with trisomy 21, for example, the likelihood that a subsequent child will have a similar chromosomal abnormality is approximately 1 percent.[9] The rate of recurrence for neural-tube defects is 2 to 5 percent, as compared with a rate of 1 to 2 per 1000 births (0.1 to 0.2 percent) in the general population.[10] The rate of recurrence of a cardiac defect is 2 to 4 percent, as compared with the general-population rate of 4 to 8 per 1000 live births (0.4 to 0.8 percent).[11,12] If a

Table 1. Indications for Prenatal Diagnosis.

General risk factors

Maternal age ≥35 years at the time of delivery
Elevated or reduced maternal serum alpha-fetoprotein concentration
Results of triple screening: elevated or reduced maternal serum alpha-fetoprotein,
  human chorionic gonadotropin, and unconjugated estriol concentrations

Specific risk factors

Previous child with a structural defect or chromosomal abnormality
Previous stillbirth or neonatal death
Structural abnormality in the mother or father
Balanced translocation in the mother or father
Inherited disorders: cystic fibrosis, metabolic disorders, sex-linked recessive
  disorders
Medical disease in the mother: diabetes mellitus, phenylketonuria
Exposure to a teratogen: ionizing radiation, anticonvulsant medicines, lithium,
  isotretinoin, alcohol
Infection: rubella, toxoplasmosis, cytomegalovirus

Ethnic risk factors

DISORDER	ETHNIC OR RACIAL GROUP	SCREENING MARKER
Tay–Sachs disease	Ashkenazi Jewish, French Canadian	Decreased serum hexosaminidase A concentration
Sickle cell anemia	Black African, Mediterranean, Arab, Indian and Pakistani	Presence of sickling in hemolysate followed by confirmatory hemoglobin electrophoresis
Alpha- and beta-thalassemia	Mediterranean, Southern and Southeast Asian, Chinese	Mean corpuscular volume <80 μm³, followed by confirmatory hemoglobin electrophoresis

parent has spina bifida, congenital heart disease, or a known chromosomal translocation or inversion, there is an increased chance that a child will have a related defect. Prenatal diagnosis is possible for many inborn errors of metabolism, almost all of which are transmitted in an autosomal recessive fashion. Diabetes mellitus and phenylketonuria in the mother are associated with an increased risk of fetal malformations. Known teratogens include ionizing radiation, therapeutic and illicit drugs, and maternal infections.

Ethnic and Racial Risk Factors

Gene frequencies differ among population groups defined on the basis of geography or ethnic and racial background. Programs exist to detect carrier status for the Tay–Sachs disease, hemoglobin S, and thalassemia genes. The population groups involved and the methods of screening are listed in Table 1.

PROCEDURES

Amniocentesis

Midtrimester amniocentesis with ultrasound guidance is usually performed at 16 weeks of gestation in outpatient facilities. Chromosomal studies are carried out by culturing the few viable cells present in amniotic fluid, and the results are generally available in 10 to 14 days. The safety and accuracy of midtrimester amniocentesis have been established in three prospective collaborative studies.[6,13,14] Amniocentesis carries an estimated risk of fetal loss of 0.5 to 1.0 percent.[15,16] Culture failure occurs in less than 1 percent of cases.

Chromosomal mosaicism is the presence of two or more cell lines with different karyotypes in a single person. This occurs as a result of postzygotic nondis-

junction. The observation of multiple cell lines in a prenatal sample does not necessarily mean that the fetus has mosaicism. The most common type of mosaicism detected by amniocentesis is pseudomosaicism.[17] This phenomenon should be suspected when an abnormality is evident in only one of several cultures of an amniotic-fluid specimen. The abnormal cell lines arise during in vitro cultures; they therefore are not present in the fetus and are not clinically important. Contamination by maternal cells can be minimized by discarding the first few drops of aspirated amniotic fluid. True fetal mosaicism, diagnosed when the same abnormality is present on more than one cover slip, is rare (0.25 percent) but clinically important.[17] Whether true mosaicism is present is best resolved by karyotyping fetal lymphocytes obtained by percutaneous fetal-blood sampling,[18] a method that provides results within 48 hours. Detailed ultrasound examination is also recommended to assess fetal growth and rule out structural anomalies. If both the ultrasound and the fetal-blood-sampling results are normal, the parents can be reassured that the major chromosomal abnormalities have been excluded.[18]

Chorionic-Villus Sampling

Chorionic villi may be obtained by aspiration through a transcervical catheter or transabdominal needle with ultrasound guidance. The choice is based on the location of the placenta and the operator's preference and experience. The main advantage of chorionic-villus sampling over amniocentesis is the earlier availability of results, since the procedure is generally performed between 9 and 12 weeks of gestation. Results based on a direct preparation of spontaneously

Table 2. Relation between Maternal Age and the Estimated Rate of Chromosomal Abnormalities.*

AGE	RISK OF DOWN'S SYNDROME	RISK OF CHROMOSOMAL ABNORMALITY
20	1/1667	1/526
25	1/1250	1/476
30	1/952	1/385
35	1/385	1/202
36	1/295	1/162
37	1/227	1/129
38	1/175	1/102
39	1/137	1/82
40	1/106	1/65
41	1/82	1/51
42	1/64	1/40
43	1/50	1/32
44	1/38	1/25
45	1/30	1/20
46	1/23	1/16
47	1/18	1/13
48	1/14	1/10
49	1/11	1/7

*Ages are at the expected time of delivery. Data have been modified from Hook[4] and Hook et al.[5]

dividing cells are usually available in 24 to 48 hours, and final results from cultured cells in 10 to 14 days. An additional advantage of chorionic-villus sampling is that more tissue is obtained than by amniocentesis. The extra tissue is useful when DNA analysis or enzymatic diagnosis is necessary. Amniocentesis must be used, however, for assays for which amniotic fluid is essential, such as measurement of the alpha-fetoprotein concentration.

Two randomized trials demonstrated that a woman assigned to undergo first-trimester chorionic-villus sampling had a 1.7 to 4.6 percent lower chance of a successful pregnancy outcome than a woman assigned to second-trimester amniocentesis.[19,20] A trial conducted by the National Institute of Child Health and Human Development compared transcervical chorionic-villus sampling in 2278 women with amniocentesis at 16 weeks of gestation in 671 women; the procedure-related rate of fetal loss when chorionic-villus sampling was used exceeded that for amniocentesis by 0.8 percent.[21] A randomized trial by the same group found that the rate of spontaneous fetal loss was 2.5 percent for transcervical sampling and 2.3 percent for transabdominal sampling; these rates were not significantly different,[22] and both represented roughly a 1 percent reduction in the rate of post-sampling loss in the earlier study.[21] These results suggest that the experience of the operator may be crucial in achieving optimal safety.

Firth et al.[23] reported limb-reduction defects in five infants born to a group of 289 women who underwent chorionic-villus sampling 56 to 66 days after the beginning of the last menstrual period. Oromandibular hypogenesis was present in four of the five infants. The proposed mechanism of the defects is a form of vascular insult leading to hypoperfusion of the fetus. Burton et al.[24] also reported transverse limb abnormalities after chorionic-villus sampling; in other reports, however, the incidence of limb defects was not significantly different from the expected rates, and defects were also observed when the procedures were performed after 66 days (Table 3).[25-30] Although the incidence of limb defects may be dependent on the timing of the procedure and the experience of the operator may be a factor, a causal relation between chorionic-villus sampling and limb defects has not been established.

The greater frequency of contamination by mater-

Table 3. Incidence of Limb-Reduction Defects in Groups Undergoing Chorionic-Villus Sampling and Population-Based Studies.

Chorionic-villus–sampling series				
STUDY	56 TO 66 DAYS*	DEFECT	>66 DAYS*	DEFECT
Firth et al.[23]	5/289	1 Transverse limb-reduction defect, 4 combined limb-reduction defects with micrognathia or microglossia	0/250	—
Burton et al.[24]†	—	—	4/394	3 Finger-and-toe abnormalities, 1 finger abnormality
Monni et al.[25]	0/525	—	2/2227	2 Finger abnormalities
Mahoney[26]‡	1/1025	Longitudinal limb-reduction defect	6/8563	1 Longitudinal limb-reduction defect, 5 transverse limb-reduction defects (2 of the 5 were limited to fingers or toes)
Jackson et al.[27]‡	1/2367	Bilateral aplasia of the thumbs	4/10,496	3 Transverse limb-reduction defects, 1 finger abnormality
Schloo et al.[28]	1/636	Microglossia and hypodactyly	3/2200	3 Finger abnormalities (2 of the 3 had a family history of limb defects)

Population-based studies		
STUDY	DEFECT	INCIDENCE
Froster-Iskenius and Baird[29]	Combination of limb-reduction defects and micrognathia or microglossia	1/175,000
	Limb-reduction defects	1/1692
	Terminal longitudinal defects	1/2857
	Terminal transverse defects	1/6250
Froster and Baird[30]	Hand	1/11,035
	Fingers	1/7016
	Foot	1/39,158
	Toes	1/43,354

*The number of days after the beginning of the last menstrual period at the time the procedure was performed.
†The number of procedures performed before 67 days was not reported.
‡There was some overlap between the series studied by Mahoney and Jackson et al.

nal cells and false evidence of mosaicism in chorionic-villus samples contributes to the reduced cytogenetic accuracy of this procedure, as compared with amniocentesis. Maternal-cell contamination is uncommon in experienced cytogenetic laboratories; the reported incidence is approximately 2 percent.[31] Villus specimens need to be meticulously separated from blood and decidual cells before cytogenetic analysis. Evidence of mosaicism occurs in less than 1 percent of cases,[31] is more common with the direct preparations, and is usually due to confined placental mosaicism.[32] This phenomenon occurs as a result of nondisjunction during embryogenesis and leads to the presence of aneuploid cells in the extraembryonic tissues that are not present in the fetus itself. When mosaicism is reported, further testing by amniocentesis is warranted. Amniocentesis is necessary in approximately 1 percent of patients who undergo chorionic-villus sampling. The clinical disadvantages of increased risk and potential diagnostic error with chorionic-villus sampling must be weighed against the disadvantage of the later timing of amniocentesis.

Early Amniocentesis

Experience has been accumulating with early amniocentesis (performed before 15 weeks of gestation).[33] Although early amniocentesis is technically easier to perform than chorionic-villus sampling, the rates of success of the two procedures in obtaining samples are similar.[34] On occasion, aspiration of fluid may be hampered by tenting of the membranes, since the amnion has not completely fused with the chorion early in gestation. Amniocentesis can usually be successfully completed by a more vigorous insertion of the needle, or the procedure may be rescheduled for a later date.

The safety of early amniocentesis cannot be assumed to be the same as for conventional amniocentesis. The volume of fluid removed constitutes a much greater proportion of the total fluid volume, and its withdrawal may increase fetal loss and decrease fetal lung function. An additional disadvantage is that standards are still being established for measuring and interpreting alpha-fetoprotein concentrations in amniotic fluid obtained before 14 to 15 weeks of gestation.

Percutaneous Umbilical Blood Sampling

Fetal blood can be obtained beginning at approximately 18 weeks of gestation with a 20- or 22-gauge spinal needle inserted under ultrasound guidance into the umbilical cord. Percutaneous umbilical blood sampling was developed for the diagnosis of toxoplasmosis.[35] Access to the fetal circulation permits the prenatal evaluation of many fetal hematologic abnormalities, including isoimmunization, hemoglobinopathy, thrombocytopenia, and coagulation-factor abnormalities.[36] Fetal blood can be used for the prenatal diagnosis of some inborn errors of metabolism and permits the assessment of viral, bacterial, and parasitic infections by serologic testing and culture.[36] Fetal-blood sampling may clarify whether chromosomal mosaicism detected by cytogenetic analysis of amniotic-fluid cells or chorionic villi is truly present.[18] The need for rapid karyotyping when congenital abnormalities are suspected is the most common reason for fetal-blood sampling in the United States.

Cytogenetic results from short-term fetal-lymphocyte cultures are usually available within 48 to 72 hours. The rate of fetal loss after percutaneous umbilical blood sampling is about 2 percent more than the background risk to the particular fetus.[35,36] Because many of the fetuses studied have severe congenital malformations, the background loss rate is high in comparison with that for the population undergoing amniocentesis or chorionic-villus sampling. Because percutaneous umbilical blood sampling entails a substantially greater risk of pregnancy loss than amniocentesis, it should be reserved for situations in which rapid diagnosis is essential or in which diagnostic information cannot be obtained by safer means.

Fetal Biopsy

Fetal biopsy was initially performed by fetoscopy but is now performed with ultrasound guidance. Certain genetic skin disorders, such as epidermolysis bullosa, that cannot now be diagnosed by DNA analysis require fetal-skin sampling. Fetal-liver biopsy was used in the past to diagnose ornithine transcarbamy-

lase deficiency.[36] Fetal-muscle biopsy has been used to diagnose Duchenne's muscular dystrophy in a family in which DNA studies were uninformative.[37] It is difficult to assess the safety and accuracy of fetal biopsy because experience with these procedures is limited. Patients should be made aware of their investigational nature. Rapid advances in DNA technology can be expected to elucidate the molecular basis of many diseases that now require fetal biopsy. As such knowledge is accumulated, the need for these procedures will decline.

Ultrasonography

Ultrasonography is an important aid in the assessment of gestational age, the monitoring of fetal growth, the confirmation of the placental site, the detection of multiple gestation, and the diagnosis of major fetal anomalies. Some teratogens and infections produce only structural abnormalities, which are potentially detectable with ultrasound but not with the other prenatal diagnostic approaches. Visualization of the fetal anatomy is essential in diagnosing anatomical defects inherited in polygenic or multifactorial fashion. Individual centers report impressive achievements in the ultrasound diagnosis of renal and bladder anomalies, hydrocephaly, and neural-tube and ventral-wall defects.[38-40] Ultrasound examination is also useful in mendelian disorders characterized by certain anatomical defects, such as skeletal dysplasias.

Opinions differ about the advantages of routine ultrasound screening. In 1984 the Consensus Development Conference sponsored by the National Institute of Child Health and Human Development recommended that ultrasound imaging be used only for a list of 28 specific indications.[41] In contrast, several countries, including Germany, France, and the United Kingdom, have adopted a policy of ultrasound screening in all pregnancies. Most of the randomized trials evaluating the usefulness of routine ultrasound screening have not addressed the question of the diagnostic accuracy of the procedure for fetal anatomical defects. Trials conducted in the early 1980s did not demonstrate any benefit of routine ultrasound; furthermore, screened women had a significantly higher rate of hospital admission.[42-44] Two later trials reported a significant reduction in the rate of induction of labor in the screened group.[45,46] In a randomized trial in Helsinki, Finland, half the serious fetal malformations were detected and the perinatal mortality was significantly lower in the screened group than in the control group (4.6 per 1000 births vs. 9.0 per 1000).[47] This reduction was due mainly to improved early detection of major malformations that led to induced abortion.

The most common congenital abnormalities are cardiovascular malformations, which are among the major malformations most frequently missed in prenatal ultrasound examinations. A four-chamber view of the fetal heart is suggested for ultrasound examination in pregnancy.[48] The use of a four-chamber view in obstetrical sonography has resulted in a substantial increase in referrals to fetal echocardiography units, and the majority of the cardiac anomalies now diagnosed at these centers are in this subgroup of referred cases.[49,50] The sensitivity of the four-chamber view for the detection of congenital heart defects was reported to be 92 percent in a high-risk referred population; the ultrasound examinations were performed by experienced investigators, however.[50] Although it is unlikely that similar results could be achieved by less experienced personnel in an unselected population, Allan has argued that the majority of severe cardiac defects could be detected prenatally if screening with a four-chamber view were universal.[49]

When a fetal anomaly is diagnosed ultrasonographically, echocardiography should be performed, since fetuses with an extracardiac anomaly have a 23 percent risk of a cardiac defect.[51] Conversely, fetuses with cardiac defects have a 25 to 45 percent risk of having another anatomical defect.[51] Karyotype analysis should be offered in most cases when cardiac and extracardiac malformations are identified, because approximately one third of the fetuses will have a chromosomal disorder.[52,53] Amniocentesis is the suggested technique, but fetal-blood sampling or placental biopsy may be performed if a rapid result is required. A new method for rapid chromosomal analysis uses fluorescence in situ hybridization to determine the status of chromosomes 21, 18, 13, X, and Y in uncultured amniotic cells. It may be particularly useful when fetal abnormalities are observed on ultrasound examination.[54] This technique provides results for the five analyzed chromosomes in two days; however, it should be performed only in conjunction with a complete chromosomal analysis. Knowledge of an abnormal karyotype affects perinatal management substantially and may prevent unnecessary cesarean section. Several ultrasonographically identifiable patterns of defects are associated with a higher risk of aneuploidy (Table 4).[55] Karyotyping may not be necessary for all malformations detected with ultrasound, such as congenital adenomatoid degeneration of the lung or isolated pyelectasis.

When fetal anatomical abnormalities are detected, consultation can be sought with a multidisciplinary team of physicians, ethicists, nurse specialists, and psychologists to plan management. The appropriate personnel, location, and mode of delivery should be determined. Occasionally in utero treatment has been undertaken, although it is still investigational. Intrauterine shunts have been placed for bladder-outlet obstruction and isolated pleural effusion. Open fetal surgery has been performed to treat congenital diaphragmatic hernia and complete bladder obstruction.

EFFICACY

As noted earlier, prenatal chromosomal analysis is currently offered to women who will be 35 years of age or older at the time of delivery. Nearly all genetic

Table 4. Sonographic Findings in Cases of Chromosomal Abnormalities.

Trisomy 21
Duodenal atresia, tracheo-esophageal fistula, esophageal atresia; hydramnios is usual if these gastrointestinal lesions are present
Cardiac abnormalities: atrioventricular-canal defects, ventricular septal defects, atrial septal defects
Hypoplasia of the middle quadrant of the fifth digit
Second-trimester findings: thickened nuchal fold (>6 mm), ratio of actual to expected femur length = 0.91

Trisomy 18
Intrauterine growth retardation
Hydramnios
Clenched hands with overlapping digits
Club feet, rocker-bottom feet
Cardiac abnormality: ventricular septal defect
Omphalocele, diaphragmatic hernia
Choroid-plexus cysts

Trisomy 13
Holoprosencephaly (midline facial defect)
Cleft lip and palate
Cardiac abnormality: ventricular septal defect
Polydactyly
Omphalocele
Polycystic kidneys

procedures carried out in the United States are performed in the 5 percent of pregnant women who are in this age group. Targeting such pregnancies detects only 20 percent of cases of Down's syndrome, however. The use of maternal serum alpha-fetoprotein testing in all pregnant women can identify an additional 25 percent of cases.[56] As noted earlier, the use of the maternal serum alpha-fetoprotein level, the human chorionic gonadotropin level, the unconjugated estriol level, and the maternal age identifies approximately 60 percent of cases of Down's syndrome.[8] A screening program based on maternal serum alpha-fetoprotein levels can identify 80 to 90 percent of fetuses with neural-tube defects, almost all cases of gastroschisis, and 70 to 80 percent of cases of omphalocele.[6,57] The use of routine ultrasound screening, including a four-chamber view of the heart, can lead to the diagnosis of approximately 50 percent of major cardiac, kidney, and bladder anomalies that would not be detected by maternal serum alpha-fetoprotein screening.[47,49] When targeted ultrasound examination is performed by skilled ultrasonographers to detect malformations suspected on the basis of the history or the screening ultrasonogram, the sensitivity and specificity of this procedure are greater than 90 percent.[58-60]

THE FUTURE

The mapping of the human genome is expected to be completed in the next 10 to 15 years,[61] and as a result molecular genetic techniques are likely to be available for the detection of all common monogenic disorders. Cost-effective screening is expected to be possible for many disorders, including cystic fibrosis. Diagnosis before implantation is even possible in certain circumstances and may allow fetal treatment before organogenesis.

A promising new technique for isolating fetal cells from maternal blood is under study.[62] This technique is designed to separate fetal cells reliably by identifying unique fetal-cell surface antigens and will use modified molecular genetic techniques such as the polymerase chain reaction to analyze small samples of fetal genetic material. It is likely that continued research into the methods of fetal-cell separation will make noninvasive prenatal diagnosis a reality.

We are indebted to Diana Bianchi, M.D., and Brian Ward, Ph.D., for their helpful comments.

REFERENCES

1. McKusick VA. Mendelian inheritance in man: catalogs of autosomal dominant, autosomal recessive, and X-linked phenotypes. 9th ed. Baltimore: Johns Hopkins University Press, 1990.
2. Contribution of birth defects to infant mortality — United States, 1986. MMWR Morb Mortal Wkly Rep 1989;38:633-5.
3. Antenatal diagnosis of genetic disorders. Washington, D.C.: American College of Obstetricians and Gynecologists, 1987:1-8 (technical bulletin no. 108).
4. Hook EB. Rates of chromosome abnormalities at different maternal ages. Obstet Gynecol 1981;58:282-5.
5. Hook EB, Cross PK, Schreinemachers DM. Chromosomal abnormality rates at amniocentesis and in live-born infants. JAMA 1983;249:2034-8.
6. Maternal serum-alpha-fetoprotein measurement in antenatal screening for anencephaly and spina bifida in early pregnancy: report of the U.K. Collaborative Study on Alpha-fetoprotein in Relation to Neural-tube Defects. Lancet 1977;1:1323-32.
7. Merkatz IR, Nitowsky HM, Macri JN, Johnson WE. An association between low maternal serum alpha-fetoprotein and fetal chromosomal abnormalities. Am J Obstet Gynecol 1984;148:886-94.
8. Haddow JE, Palomaki GE, Knight GJ, et al. Prenatal screening for Down's syndrome with use of maternal serum markers. N Engl J Med 1992;327:588-93.
9. Stene J. Detection of higher recurrence risk for age-dependent chromosome abnormalities with an application to trisomy G1 (Down's syndrome). Hum Hered 1970;20:112-22.
10. Cowchock S, Ainbender E, Prescott G, et al. The recurrence risk for neural tube defects in the United States: a collaborative study. Am J Med Genet 1980;5:309-14.
11. Lin AE, Garver KL. Genetic counseling for congenital heart defects. J Pediatr 1988;113:1105-9.
12. Nora JJ, Nora AH. Update on counseling the family with a first-degree relative with a congenital heart defect. Am J Med Genet 1988;29:137-42.
13. The NICHD National Registry for Amniocentesis Study Group. Midtrimester amniocentesis for prenatal diagnosis: safety and accuracy. JAMA 1976;236:1471-6.
14. Simpson NE, Dallaire L, Miller JR, et al. Prenatal diagnosis of genetic disease in Canada: report of a collaborative study. Can Med Assoc J 1976;115:739-48.
15. Working party on amniocentesis: an assessment of the hazards of amniocentesis. Br J Obstet Gynaecol 1978;85:Suppl:12-6.
16. Tabor A, Philip J, Madsen M, Bang J, Obel EB, Nørgaard-Pedersen B. Randomised controlled trial of genetic amniocentesis in 4606 low-risk women. Lancet 1986;1:1287-92.
17. Hsu LYF, Perlis TE. United States survey on chromosome mosaicism and pseudomosaicism in prenatal diagnosis. Prenat Diagn 1984;4:97-130.
18. Gosden C, Nicolaides KH, Rodeck CH. Fetal blood sampling in investigation of chromosome mosaicism in amniotic fluid cell culture. Lancet 1988;1:613-7.
19. MCR Working Party on the Evaluation of Chorion Villus Sampling. Medical Research Council European Trial of chorion villus sampling. Lancet 1991;337:1491-9.
20. Canadian Collaborative CVS-Amniocentesis Clinical Trial Group. Multicentre randomised clinical trial of chorion villus sampling and amniocentesis: first report. Lancet 1989;1:1-6.
21. Rhoads GG, Jackson LG, Schlesselman SE, et al. The safety and efficacy of chorionic villus sampling for early prenatal diagnosis of cytogenetic abnormalities. N Engl J Med 1989;320:609-17.
22. Jackson LG, Zachary JM, Fowler SE, et al. A randomized comparison of transcervical and transabdominal chorionic-villus sampling. N Engl J Med 1992;327:594-8.
23. Firth HV, Boyd PA, Chamberlain P, MacKenzie IZ, Lindenbaum RH, Huson SM. Severe limb abnormalities after chorion villus sampling at 56–66 days' gestation. Lancet 1991;337:762-3.
24. Burton BK, Schulz CJ, Burd LI. Limb anomalies associated with chorionic villus sampling. Obstet Gynecol 1992;79:726-30.
25. Monni G, Ibba RM, Lai R, Olla G, Cao A. Limb-reduction defects and chorion villus sampling. Lancet 1991;337:1091.
26. Mahoney MJ. Limb abnormalities and chorionic villus sampling. Lancet 1991;337:1422-3.
27. Jackson LG, Wapner RJ, Brambati B. Limb abnormalities and chorionic villus sampling. Lancet 1991;337:1423.
28. Schloo R, Miny P, Holzgreve W, Horst J, Lenz W. Distal limb deficiency following chorionic villus sampling? Am J Med Genet 1992;42:404-13.
29. Froster-Iskenius UG, Baird PA. Limb reduction defects in over one million consecutive livebirths. Teratology 1989;39:127-35.
30. Froster UG, Baird PA. Limb-reduction defects and chorionic villus sampling. Lancet 1992;339:66.
31. Ledbetter DH, Martin AO, Verlinsky Y, et al. Cytogenetic results of chorionic villus sampling: high success rate and diagnostic accuracy in the United States collaborative study. Am J Obstet Gynecol 1990;162:495-501.
32. Kalousek DK, Dill FJ, Pantzar T, McGillivray BC, Yong SL, Wilson RD. Confined chorionic mosaicism in prenatal diagnosis. Hum Genet 1987;77:163-7.
33. Choo V. Early amniocentesis. Lancet 1991;338:750-1.
34. Byrne D, Marks K, Azar G, Nicolaides K. Randomized study of early amniocentesis versus chorionic villus sampling: a technical and cytogenetic comparison of 650 patients. Ultrasound Obstet Gynecol 1991;1:235-40.
35. Daffos F, Capella-Pavlovsky M, Forestier F. Fetal blood sampling during pregnancy with use of a needle guided by ultrasound: a study of 606 consecutive cases. Am J Obstet Gynecol 1985;153:655-60.
36. Shulman LP, Elias S. Percutaneous umbilical blood sampling, fetal skin sampling, and fetal liver biopsy. Semin Perinatol 1990;14:456-64.
37. Evans MI, Greb A, Kunkel LM, et al. In utero fetal muscle biopsy for the diagnosis of Duchenne muscular dystrophy. Am J Obstet Gynecol 1991;165:728-32.
38. D'Alton ME, Romero R, Grannum PM, DePalma L, Jeanty P, Hobbins JC. Antenatal diagnosis of renal anomalies with ultrasound. IV. Bilateral multicystic kidney disease. Am J Obstet Gynecol 1986;154:532-7.
39. Mercer S, Mercer B, D'Alton ME, Soucy P. Gastroschisis: ultrasonographic diagnosis, perinatal embryology, surgical and obstetric treatment and outcomes. Can J Surg 1988;31:25-6.
40. Van den Hof MC, Nicolaides KH, Campbell J, Campbell S. Evaluation of the lemon and banana signs in one hundred thirty fetuses with open spina bifida. Am J Obstet Gynecol 1990;162:322-7.
41. National Institutes of Health Consensus Development Conference. The use of diagnostic ultrasound imaging in pregnancy. Washington, D.C.: Government Printing Office, 1984.
42. Bennett MJ, Little G, Dewhurst J, Chamberlain G. Predictive value of ultrasound measurement in early pregnancy: a randomized controlled trial. Br J Obstet Gynaecol 1982;89:338-41.
43. Neilson JP, Munjanja SP, Whitfield CR. Screening for small dates fetuses: a controlled trial. BMJ 1984;289:1179-82.
44. Bakketeig LS, Eik-Nes SH, Jacobsen G, et al. Randomised controlled trial of ultrasonographic screening in pregnancy. Lancet 1984;2:207-11.
45. Eik-Nes SH, Økland O, Aure JC. Ultrasound screening in pregnancy: a randomised controlled trial. Lancet 1984;1:1347.
46. Waldenström U, Axelsson O, Nilsson S, et al. Effects of routine one-stage ultrasound screening in pregnancy: a randomised controlled trial. Lancet 1988;2:585-8.
47. Saari-Kemppainen A, Karjalainen O, Ylöstalo P, Heinonen OP. Ultrasound screening and perinatal mortality: controlled trial of systematic one-stage screening in pregnancy: the Helsinki Ultrasound Trial. Lancet 1990;336:387-91.
48. Ultrasound in pregnancy. Washington, D.C.: American College of Obstetricians and Gynecologists, 1988:1-3 (technical bulletin no. 116).
49. Allan LD. Fetal echocardiography. Clin Obstet Gynecol 1988;31:61-79.
50. Copel JA, Pilu G, Green J, Hobbins JC, Kleinman CS. Fetal echocardiographic screening for congenital heart disease: the importance of the four-chamber view. Am J Obstet Gynecol 1987;157:648-55.
51. Copel JA, Pilu G, Kleinman CS. Congenital heart disease and extracardiac anomalies: associations and indications for fetal echocardiography. Am J Obstet Gynecol 1986;154:1121-32.
52. Copel JA, Cullen M, Green JJ, Mahoney MJ, Hobbins JC, Kleinman CS. The frequency of aneuploidy in prenatally diagnosed congenital heart disease: an indication for fetal karyotyping. Am J Obstet Gynecol 1988;158:409-13.
53. Platt LD, DeVore GR, Lopez E, Herbert W, Falk R, Alfi O. Role of amniocentesis in ultrasound-detected fetal malformations. Obstet Gynecol 1986;68:153-5.
54. Klinger K, Landes G, Shook D, et al. Rapid detection of chromosome aneuploidies in uncultured amniocytes by using fluorescence in situ hybridization (FISH). Am J Hum Genet 1992;51:55-65.
55. Benacerraf BR. The use of sonography for the antenatal detection of aneuploidy. Clin Diagn Ultrasound 1989;25:21-54.

56. Schoenfeld DiMaio M, Baumgarten A, Greenstein RM, Saal HM, Mahoney MJ. Screening for fetal Down's syndrome in pregnancy by measuring maternal serum alpha-fetoprotein levels. N Engl J Med 1987;317:342-6.

57. Palomaki GE, Hill LE, Knight GJ, Haddow JE, Carpenter M. Second-trimester maternal serum alpha-fetoprotein levels in pregnancies associated with gastroschisis and omphalocele. Obstet Gynecol 1988;71:906-9.

58. Manchester DK, Pretorius DH, Avery C, et al. Accuracy of ultrasound diagnoses in pregnancies complicated by suspected fetal anomalies. Prenat Diagn 1988;8:109-17.

59. Sabbagha RE, Sheikh Z, Tamura RK, et al. Predictive value, sensitivity, and specificity of ultrasonic targeted imaging for fetal anomalies in gravid women at high risk for birth defects. Am J Obstet Gynecol 1985;152:822-7.

60. Hill LM, Breckle R, Gehrking WC. Prenatal detection of congenital malformations by ultrasonography: Mayo Clinic experience. Am J Obstet Gynecol 1985;151:44-50.

61. Watson JD. The Human Genome Project: past, present, and future. Science 1990;248:44-9.

62. Bianchi DW, Flint AF, Pizzimenti MF, Knoll JH, Latt SA. Isolation of fetal DNA from nucleated erythrocytes in maternal blood. Proc Natl Acad Sci U S A 1990;87:3279-83.

SPECIAL ARTICLES

THE COST OF A SUCCESSFUL DELIVERY WITH IN VITRO FERTILIZATION

PETER J. NEUMANN, SC.D., SOHEYLA D. GHARIB, M.D., AND MILTON C. WEINSTEIN, PH.D.

Abstract *Background.* The use of in vitro fertilization has engendered considerable debate about who should have the procedure, whether health insurance should cover the cost, and if so, to what extent. We investigated the cost of a successful delivery with in vitro fertilization.

Methods. We calculated the cost per successful delivery with in vitro fertilization (defined as at least one live birth) for a general population of couples undergoing in vitro fertilization and for two subgroups: couples with a diagnosis of tubal disease (who have a better chance of success), and couples in which the woman is over the age of 40 years and the man has a low sperm count (who have a lower chance of success). Information on charges per cycle of in vitro fertilization was obtained from six facilities across the country; delivery rates with this procedure were estimated from the literature.

Results. On average, the cost incurred per successful delivery with in vitro fertilization increases from $66,667

for the first cycle of in vitro fertilization to $114,286 by the sixth cycle. The cost increases because with each cycle in which fertilization fails, the probability that a subsequent effort will be successful declines. Sensitivity analyses indicated that the cost per delivery ranges from $44,000 to $211,940. For couples with a better chance of successful in vitro fertilization (i.e., those with a diagnosis of tubal disease), it costs $50,000 per delivery for the first cycle and $72,727 for the sixth. For couples in which the woman is older and there is a diagnosis of male-factor infertility, the cost rises from $160,000 for the first cycle to $800,000 for the sixth.

Conclusions. The debate about insurance coverage for in vitro fertilization must take into account ethical judgments and social values. But analyses of costs and cost effectiveness help elucidate the economic implications of using in vitro fertilization and thus inform the policy discussion. (N Engl J Med 1994;331:239-43.)

T HE use of in vitro fertilization has engendered considerable debate about who should receive treatment, whether health insurance should cover the costs, and if so, to what extent. The debate has been fueled by conflicting and sometimes misleading claims. Conflicting statistics on in vitro fertilization reflect differences in how the rate of success associated with the procedure is defined and measured.[1] Moreover, success rates differ widely among programs.[2] When data on costs have been presented, they have sometimes been used inappropriately. For example, one group has lobbied for coverage of in vitro fertilization on the grounds that it represents an exceedingly small fraction of health care costs, accounting for only 0.03 percent of the total.[3]

The key question, however, is not how much we spend on in vitro fertilization but what is gained from the investment. In other words, what is the cost of a successful outcome of in vitro fertilization? Measurement of this cost was the goal of our study. In particular, we considered the cost incurred per cycle of in vitro fertilization and how it varied according to the number of cycles in which the procedure was used. We calculated overall cost estimates as well as estimates for two subgroups defined by the woman's age and the cause of infertility. Because of the uncertainty surrounding these estimates, we used sensitivity analyses to determine how the results

would change when we used high-range and low-range assumptions about costs.

METHODS

At the outset, we defined three terms: a cycle of in vitro fertilization, delivery after in vitro fertilization, and the marginal cost per delivery for each successive cycle of in vitro fertilization. A complete cycle of in vitro fertilization involves four steps: first, a woman takes fertility drugs intended to stimulate her eggs to develop; second, these eggs are retrieved from the woman's ovaries by an outpatient transcervical procedure; third, the eggs are fertilized in a laboratory with her husband's or partner's sperm; finally, the fertilized eggs are reimplanted in her uterus for completion of the pregnancy. Couples often proceed through several cycles of in vitro fertilization before an infant is conceived and carried to term or the effort is abandoned.

We defined delivery after in vitro fertilization as the birth of at least one live baby as a result of a cycle of in vitro fertilization. We focused on deliveries instead of births in order to avoid the potential confusion involved in considering multiple births separately. For example, two couples, each having a single baby through in vitro fertilization, should not be treated the same as one couple having twins and the other having no infants at all. We did, however, consider the important implications of the fact that many cycles of in vitro fertilization result in multiple births.

Finally, we defined the marginal cost per delivery for each successive cycle of in vitro fertilization as the cost incurred for a given cycle divided by the probability of achieving a delivery as a result of that cycle.

Cost per Cycle

In estimating the cost of a cycle of in vitro fertilization, we relied on the 1992 listings of charges from published brochures obtained from large in vitro fertilization facilities at six sites in the United States: Brigham and Women's Hospital, Boston; Cornell University Medical Center, New York; Genetics and IVF Institute, Fairfax, Virginia; IVF America, Waltham, Massachusetts; Jones Institute for Reproductive Medicine, Norfolk, Virginia; and the Vanderbilt University Center for Fertility and Reproductive Research, Nashville.

The charge for a single cycle of in vitro fertilization at these centers in 1992 ranged from $7,000 to $11,000, including the costs

From the Project HOPE Center for Health Affairs, Bethesda, Md. (P.J.N.); the Division of General Medicine and Primary Care, Brigham and Women's Hospital and Harvard Medical School, Boston (S.D.G.); and the Department of Health Policy and Management, Harvard School of Public Health, Boston (M.C.W.). Address reprint requests to Dr. Neumann at the Project HOPE Center for Health Affairs, Suite 600, 7500 Old Georgetown Rd., Bethesda, MD 20814.

for initial consultations, laboratory tests, medications, ultrasonography, egg retrieval, gamete culturing, embryo transfer, and physician and nursing services. On the basis of these published charges, we assumed a base case with an average cost of $8,000 per cycle. We used a figure at the low end of the range because we relied on charges that probably overstate the economic costs. Our base case takes into account only the direct costs associated with the procedure, since in general, the measurement of other factors is more uncertain, and their inclusion more controversial among cost analysts. However, in sensitivity analyses we adjusted the cost for the base case to take into account several other factors (Table 1).

In vitro fertilization may involve costs associated with time and complications. Since the time that couples devote to in vitro fertilization could be spent productively in other endeavors, it should be considered a component of economic costs. The cost associated with time varies considerably among couples, from a few hours or days to extended leaves of absence from work in some cases.[4] In the sensitivity analyses we estimate that time away from work adds $880 to the cost of an average cycle, assuming that both partners miss one week of work per cycle and we value time at $11 per hour (average hourly earnings in 1992: 2 × (5 days/cycle) × ($11/hour) × (8 hours/day) = $880 per cycle.

Complications associated with the procedure may also add to its cost. Maternal risks from in vitro fertilization include ovarian hyperstimulation syndrome, bleeding, infection, cysts, anesthesia-related complications, and possibly an increased risk of thromboembolism, stroke, myocardial infarction, and ovarian cancer.[5-13] In general, as compared with in vivo fertilization, in vitro fertilization is associated with higher rates of difficult pregnancies and deliveries.[13-16]

The incidence of adverse outcomes is higher among babies born after in vitro fertilization, though the difference appears to be attributable to the higher incidence of multiple births and not to the procedure itself.[6,14,17] It is well documented that in vitro fertilization, because of the methods of ovarian stimulation used and the practice of implanting multiple embryos in a woman's uterus to improve the chance of pregnancy, increases the incidence of multiple births.[18,19] Twins naturally occur in about 7 deliveries per 1000, and triplets, quadruplets, and quintuplets occur in 1 in 9520, 1 in 600,000, and 1 in 15,000,000, respectively.[15,20] On the other hand, some 20 to 25 percent of pregnancies with in vitro fertilization result in multiple births, including triplets and higher-order multiple births in 2 to 3 percent of such pregnancies.[21]

Multiple births are associated with higher rates of neonatal complications and stays in neonatal intensive care units than are births of singleton infants,[14,16,22-24] as well as higher costs after discharge due to chronic health problems and developmental disabilities.[23,25] The average hospital charge for an infant in the neonatal intensive care unit ranges from $10,000 to $100,000 (in 1992 dollars), depending on the birth-weight category,[22,23,26] and lifetime costs can add another $100,000 or more.[25]

Most reports have noted that the incidence of serious maternal complications is low — 0.1 to 0.2 percent of all cases.[6] On the basis of a comprehensive review of the literature, Schenker and Ezra estimated that complications — major and minor — occur in 3 to 6 percent of cases.[13] If we assume that serious maternal complications occur in 0.2 percent of cases and cost $20,000 per case (for a hospital stay of several weeks), and minor complications occur in 5 percent of cases and cost $2,500 per case (for approximately a

four-day hospital stay), the additional cost is (0.002 × $20,000) + (0.05 × $2,500) = $165. We further assume that 20 percent (roughly the proportion of multiple births) of pregnancies resulting from in vitro fertilization are problematic, requiring an average of four extra weeks away from work. Again, if time is valued at $11 per hour, this adds $11/hour × 8 hours × 5 days/week × 4 weeks = $1,760. But since only about 20 percent of initiated cycles result in a pregnancy (see below), this adds only another 0.20 × $1,760 = $352.

If multiple births occur in 3 percent of cases and cost as much as $200,000 per case ($66,667 per baby), the additional cost is $6,000 per case, but since the probability of a delivery is only 12 percent to begin with (see below), the added cost is $720 per case. If twins are delivered in 20 percent of cases and the cost of delivery is $20,000 per case, the additional cost is 0.20 × $20,000 × 0.12 = $480 per case. Adding these amounts together, we assume that the additional cost of all complications is $165 + $352 + $720 + $480 = $1,717 per case. These are probably generous assumptions, since a cost per case of this magnitude will occur only in the lowest birth-weight categories.

On the other hand, our base-case cost of $8,000 per cycle may overstate the actual cost for several reasons. First, our assumption of constant marginal costs may be questioned. The first cycle may cost somewhat more than subsequent cycles because of the initial consultation with a social worker and other steps of the workup that are not repeated for subsequent cycles. However, most components of the cost of in vitro fertilization are variable, including the costs of laboratory tests, medications, and physicians' and nurses' time. In sensitivity analyses we assumed that the initial cycle costs $2,000 more than subsequent cycles.

Second, only about 86 percent of cycles in which treatment is initiated result in successful egg retrieval,[21] though most of the costs are still incurred, including the cost of medications and ultrasonography and much of the required physician and nursing services. The costs associated with embryo transplantation, however, are averted. If we assume that the cost is 25 percent lower (roughly equivalent to the reduction in charges in the facilities noted above) in the 14 percent of cases with unsuccessful egg retrieval, our estimate is reduced by (0.14 × $2,000) = $280.

Finally, the cost may also be lower because some cycles use embryos that have been cryopreserved from previous cycles. Cycles in which frozen embryos are used may cost less, because the steps of egg retrieval and gamete preparation are bypassed and the frozen embryos are thawed and transferred to the woman's uterus. However, any overall cost reduction is probably small. Frozen embryos are used in only a small fraction of cycles — about 11 percent in 1990.[21] Also, the process involves additional costs (ranging from $1,500 to $4,000 at the facilities noted above) for storage and thawing. In addition, the use of frozen embryos may result in a lower success rate, so the cost per delivery with frozen embryos may actually be higher.[27] For these reasons, we have not made a separate adjustment for this factor.

Delivery Rate with in Vitro Fertilization

We estimated the probability of a delivery with in vitro fertilization from reports in the literature. This probability — which we refer to as the delivery rate with in vitro fertilization — has a long and somewhat controversial history in the literature, in part because definitions vary among studies. Some studies report pregnancy rates, for example, whereas others report delivery rates; some report the outcome per initiated cycle, whereas others report the outcome per successful retrieval. Studies have generally reported that between 10 and 15 percent of initiated cycles result in at least one live birth.[5,21,28-32] In general, the rates are higher among couples in which the woman is relatively young or has tubal disease and lower among couples in which the woman is older or there is an indication of a low sperm count, severe endometriosis, or unexplained infertility.

Researchers have disputed whether and, if so, to what extent the probability of a delivery varies with the number of cycles (the cycle rank). This issue lies at the heart of a key question: What does a failed cycle reveal about a couple's ultimate chance of a successful

Table 1. Cost Assumptions for the Base Case of in Vitro Fertilization.

VARIABLE	COST ($)
Base case	8,000
Adjustment for lost time	+880
Adjustment for complications	+1,717
Adjustment for incomplete cycles	−280
Adjustment for marginal costs*	−2,000

*Does not apply to the first cycle.

delivery with in vitro fertilization? Put another way, if a couple is told that they have a 15 percent chance of having a baby on the first cycle, what is the revised probability on the second try, if the first one fails? What is the probability on the third attempt, given the failure of the first two?

Table 2 shows data from five studies on the rate of successful in vitro fertilization according to the cycle rank. Note that the rates in these studies differ because of differences in definitions. Some studies report a constant probability per cycle,[33,34] whereas others report that the probability declines with the number of cycles attempted.[30,32,35] But a declining rate is to be expected in a heterogeneous population, because as Hershlag et al. note, for a randomly chosen woman, failure to achieve a pregnancy in each successive cycle constitutes increasing evidence of an inherently low potential for fertility.[30] That is, the women with the higher potential for fertility are more likely to become pregnant early, leaving those with a lower potential to continue with subsequent cycles. Thus, we should expect to observe a declining probability of success as the number of cycles increases. The fact that studies do not always bear this out is probably explained by selection bias at the point of entry or between cycles.[30] In other words, some programs weed out patients with a lower potential for fertility, encouraging only those candidates considered to have a higher potential to pursue additional cycles.

In our analyses, we assumed a gradually declining rate of success in all cases. For simplicity, we used round numbers. Our base-case assumption was that the probability declines one percentage point per cycle, from 12 percent on the first cycle to 7 percent on the sixth. We also considered how probabilities vary for two subgroups: couples with a diagnosis of tubal disease and couples in which the woman is 40 years old or older and there is a diagnosis of male-factor infertility (defined as a sperm concentration of less than 20 million per milliliter or motility below 40 percent).[2] Assumptions for these subgroups are based on those reported in the literature.[32,35]

In the sensitivity analyses, we varied the probabilities by two percentage points in each direction to take into consideration several areas of uncertainty. In general, it is difficult to know the true probability of success per cycle, because after every failed attempt, many couples — on the order of 50 percent — discontinue their pursuit of in vitro fertilization.[30,32,35] All published data on the rate of successful in vitro fertilization according to the cycle rank reflect this bias. But in the hypothetical case of a sample with no dropouts, we would expect an even greater decline in the success rate, because more lower-risk couples would initiate each cycle. Thus, the assumptions used here are probably conservative.

Two other points about success rates are also important. First, among some couples seeking in vitro fertilization, pregnancy occurs without the intervention, either before its inception or after its discontinuation.[5] Second, since some of the embryos frozen in earlier cycles will ultimately result in live births, the true success rate per retrieval is somewhat higher than that stated above.[36]

RESULTS

Table 3 shows the marginal cost per cycle of in vitro fertilization. Under the assumptions for the base case, it costs $66,667 per delivery for couples undergoing their first cycle of treatment and rises to $114,286 per delivery for couples attempting their sixth cycle. The marginal cost increases because, with each failed cycle, we revised downward the probability that the next cycle will be successful.

Table 3 also shows the results of analyses with low-range (optimistic) and high-range (pessimistic) assumptions for both costs and success rates. Under the low-range assumptions, the marginal cost is $55,143 for the first cycle and $44,000 for the second, rising gradually to $63,556 for the sixth. Under the high-range assumptions, the cost per deliv-

Table 2. Rate of Successful in Vitro Fertilization per Cycle in Selected Studies.

STUDY	SUCCESS RATE (%)					
	CYCLE 1	CYCLE 2	CYCLE 3	CYCLE 4	CYCLE 5	CYCLE 6
Guzick et al.[33]*	13.2	12.5	17.1	16.1	7.7	15.4
Padilla and Garcia[34]†	24.5	28.8	28.1	33.3	34.6	30.0
Hershlag et al.[30]‡	13.1	10.7	6.9	4.3	4.9	0.0
Tan et al.[32]§	10.6	10.3	9.1	9.2	6.1	7.0
French in Vitro National Study[35]¶	18.3	17.1	17.8	16.8	16.5	16.6

*Pregnancy rate per egg retrieval (530 couples).
†Pregnancy rate per embryo transfer (512 couples).
‡Delivery rate per initiated cycle (571 couples).
§Delivery rate per initiated cycle (2735 couples).
¶Pregnancy rate per egg retrieval (25,666 couples).

ery increases from $105,970 for the first cycle to $211,940 for the sixth.

Table 4 shows the substantial difference in cost per delivery for two groups of infertile couples. For couples in which the woman has tubal disease, the cost per delivery is $50,000 for the first cycle and $72,727 for the sixth, whereas for couples in which the woman is older and there is male-factor infertility, the cost rises from $160,000 for the first cycle to $800,000 for the sixth.

DISCUSSION

In vitro fertilization forces us to confront difficult questions: Who should have access to the procedure? Everyone? Only childless couples? Only couples with a presumably good chance of success? And who should pay for the procedure?

To be sure, some of these questions can be answered not by analyses of costs and outcomes, but

Table 3. Marginal Cost per Delivery with in Vitro Fertilization for the Base Case and Low- and High-Range Assumptions.

ASSUMPTION	DELIVERY RATE PER INITIATED CYCLE (%)	CUMULATIVE DELIVERY RATE (%)*	COST PER INITIATED CYCLE ($)	MARGINAL COST PER DELIVERY ($)†
Base case				
Cycle 1	12	12.0	8,000	66,667
Cycle 2	11	21.7	8,000	72,727
Cycle 3	10	29.5	8,000	80,000
Cycle 4	9	32.9	8,000	88,889
Cycle 5	8	41.0	8,000	100,000
Cycle 6	7	45.1	8,000	114,286
Low-range assumptions				
Cycle 1	14	14.0	7,720	55,143
Cycle 2	13	25.2	5,720	44,000
Cycle 3	12	34.2	5,720	47,667
Cycle 4	11	41.4	5,720	52,000
Cycle 5	10	47.3	5,720	57,200
Cycle 6	9	52.0	5,720	63,556
High-range assumptions				
Cycle 1	10	10.0	10,597	105,970
Cycle 2	9	18.1	10,597	117,744
Cycle 3	8	26.4	10,597	132,463
Cycle 4	7	29.9	10,597	151,386
Cycle 5	6	34.1	10,597	176,662
Cycle 6	5	37.4	10,597	211,940

*Sample calculation of the cumulative delivery rate: for cycle 3 in the base case, the rate is $1 - (0.88 \times 0.89 \times 0.90) = 0.295$.

†Marginal cost per delivery $= \dfrac{\text{cost incurred in cycle 1}}{\text{probability of delivery in cycle 1}}$

Table 4. Marginal Cost per Delivery with in Vitro Fertilization for Couples with Selected Fertility Problems.

FERTILITY PROBLEM	DELIVERY RATE PER INITIATED CYCLE (%)	CUMULATIVE DELIVERY RATE (%)*	COST PER INITIATED CYCLE ($)	MARGINAL COST PER DELIVERY ($)†
Tubal disease				
Cycle 1	16	16.0	8,000	50,000
Cycle 2	15	28.6	8,000	53,333
Cycle 3	14	38.6	8,000	57,143
Cycle 4	13	46.6	8,000	61,538
Cycle 5	12	53.0	8,000	66,667
Cycle 6	11	58.2	8,000	72,727
Woman ≥40 yr and male-factor infertility				
Cycle 1	5	5.0	8,000	160,000
Cycle 2	4	10.7	8,000	200,000
Cycle 3	3	11.5	8,000	266,667
Cycle 4	2	13.3	8,000	400,000
Cycle 5	1	14.2	8,000	800,000
Cycle 6	1	15.0	8,000	800,000

*Sample calculation of the cumulative delivery rate: for couples with tubal disease, the rate for cycle 3 is $1 - (0.84 \times 0.85 \times 0.86) = 0.386$.

†Marginal cost per delivery $= \dfrac{\text{cost incurred in cycle 1}}{\text{probability of delivery in cycle 1}}$

only by consideration of ethical judgments and social values. However, our analyses can help elucidate the economic implications of in vitro fertilization and thus inform the debate. Even those who believe that procreation is a fundamental right would not provide unlimited resources for a couple's pursuit of a child. Inevitably, we are forced to confront trade-offs. The results here suggest that, on average, it costs approximately $67,000 to $114,000 per successful delivery with in vitro fertilization. For older couples with more difficult problems of infertility, the cost is approximately $800,000 per delivery.

An implication of these results is that more deliveries will result per dollar expended if we give priority to the couples with the best chance of success — if, for example, couples in which the woman has tubal disease are allowed more cycles than couples in which the woman is older and there is male-factor infertility.

Of course, efficiency is not the only goal of public policy. Ideally, we would like to know not only the cost per delivery but the associated benefit or value as well. In a community survey, Neumann and Johannesson found that the amount the respondents were willing to pay for a delivery with in vitro fertilization ranged from $170,000 to $1.7 million and depended considerably on how the questions were framed.[37] In weighing public-policy options, it is important to consider public attitudes. Notwithstanding individual preferences, for example, the public may want to guarantee that every infertile couple has at least some access to in vitro fertilization, instead of simply maximizing the total number of deliveries.

Our analyses did not take into account several factors that will be important to explore in further research. First, our base case was calculated from charges, not economic costs, and we assumed that costs are constant among groups of couples. Second, in vitro fertilization may result in an additional benefit to the extent that it raises the economic productivity of

infertile couples who are despondent over not being able to have a child. Third, we did not explore fully the social effects of multiple births.

It is also important to investigate the cost effectiveness of in vitro fertilization as compared with other interventions to correct problems of infertility, such as tubal surgery. Haan, for example, reported that the average cost per ongoing pregnancy for tubal surgery was roughly comparable to the cost per ongoing pregnancy for three attempts at in vitro fertilization.[38] Finally, it is important to compare the cost of in vitro fertilization with alternative uses of public funds, including current expenditures for costly life-saving technology. Such comparisons will be difficult, since they will require a comparison of the life-years of living persons and the life-years of those not yet born or even conceived.

Those who oppose the inclusion of in vitro fertilization as a health insurance benefit tend to argue that there are more important uses for society's scarce resources. Advocates counter that the benefits outweigh the costs and that treatment for infertility should be regarded in the same way as treatment for other diseases. Over the years, some insurers have covered the procedure, though many have not. Eight states have required private insurers to cover the procedure.[39] Recently, in vitro fertilization was one of the few procedures explicitly excluded from the standard benefit package in the Clinton administration's health plan.[40]

In the absence of government intervention, a private market for in vitro fertilization would still exist. The procedure would be available to those who could afford it. How the procedure is treated in the changing U.S. health care system will thus reveal something about Americans' ability to tolerate inequities in access to expensive procedures, even those that make a considerable contribution to the quality of life. As we debate this issue, the least we can do is to improve the discourse over policy by a better understanding of the costs and results involved.

REFERENCES

1. Saltus R. Interpreting IVF statistics is no easy task. Boston Globe. November 27, 1989:26.
2. American Fertility Society, Society for Assisted Reproductive Technology. Annual clinic — specific report. Birmingham, Ala.: American Fertility Society, 1990.
3. Infertility and national health care reform fact sheet. Washington, D.C.: American Fertility Society, Office of Government Relations, 1993.
4. Bonnicksen AL. In vitro fertilization: building policy from laboratories to legislatures. New York: Columbia University Press, 1989.
5. Wagner MG, St Clair PA. Are in-vitro fertilisation and embryo transfer of benefit to all? Lancet 1989;2:1027-30.
6. Edwards RG, Brinsden P, Elder K, Lewis F, Macnamee M, Rainsbury P. Benefits of in-vitro fertilisation. Lancet 1989;2:1328.
7. Howe RS, Wheeler C, Mastroianni L Jr, Blasco L, Tureck R. Pelvic infection after transvaginal ultrasound-guided ovum retrieval. Fertil Steril 1988; 49:726-8.
8. Smith BH, Cooke ID. Ovarian hyperstimulation: actual and theoretical risks. BMJ 1991;302:127-8.
9. Bromwich P, Walker A. Benefits of in-vitro fertilisation. Lancet 1989;2: 1327.
10. Travers B. Risks associated with assisted conception. BMJ 1992;305:50-1.
11. Whittemore AS, Harris R, Itnyre J, et al. Characteristics relating to ovarian cancer risk: collaborative analysis of 12 US case-control studies. Am J Epidemiol 1992;136:1175-220.

12. Spirtas R, Kaufman SC, Alexander NJ. Fertility drugs and ovarian cancer: red alert or red herring? Fertil Steril 1993;59:291-3.
13. Schenker JG, Ezra Y. Complications of assisted reproductive techniques. Fertil Steril 1994;61:411-22.
14. Seoud MA, Toner JP, Kruithoff C, et al. Outcome of twin, triplet, and quadruplet in vitro fertilization pregnancies: the Norfolk experience. Fertil Steril 1992;57:825-34.
15. Petrikovsky BM, Vintzileos AM. Management and outcome of multiple pregnancy of high fetal order: literature review. Obstet Gynecol Surv 1989;44:578-84.
16. Kingsland CR, Steer CV, Pampiglione JS, Mason BA, Edwards RG, Campbell S. Outcome of triplet pregnancies resulting from IVF at Bourn Hallam 1984-1987. Eur J Obstet Gynecol Reprod Biol 1990;34:197-203.
17. Brandes JM, Scher A, Itzkovits J, Thaler I, Sarid M, Gershoni-Baruch R. Growth and development of children conceived by in vitro fertilization. Pediatrics 1992;90:424-9.
18. Botting BJ, Davies IM, Macfarlane AJ. Recent trends in the incidence of multiple births and associated mortality. Arch Dis Child 1987;62:941-50.
19. Levene MI, Wild J, Steer P. Higher multiple births and the modern management of infertility in Britain. Br J Obstet Gynaecol 1992;99:607-13.
20. Kelly TE. Clinical genetics and genetic counseling. 2nd ed. Chicago: Year Book Medical, 1986:218.
21. Medical Research International, Society for Assisted Reproductive Technology (SART), American Fertility Society. In vitro fertilization-embryo transfer (IVF-ET) in the United States: 1990 results from the IVF-ET Registry. Fertil Steril 1992;57:15-24. [Erratum, Fertil Steril 1993;59:250.]
22. Neonatal intensive care for low birthweight infants: costs and effectiveness. Health technology case study 38. Washington, D.C.: Congress of the United States, Office of Technology Assessment, 1987.
23. Blackman JA. Neonatal intensive care: is it worth it? Pediatr Clin North Am 1991;38:1497-511.
24. Sassoon DA, Castro LC, Davis JL, Hobel CJ. Perinatal outcome in triplet versus twin gestations. Obstet Gynecol 1990;75:817-20.

25. Boyle MH, Torrance GW, Sinclair JC, Horwood SP. Economic evaluation of neonatal intensive care of very-low-birth-weight infants. N Engl J Med 1983;308:1330-7.
26. Ewald U. What is the actual cost of neonatal intensive care? Int J Technol Assess Health Care 1991;7:Suppl 1:155-61.
27. Levran D, Dor J, Rudak E, et al. Pregnancy potential of human oocytes — the effect of cryopreservation. N Engl J Med 1990;323:1153-6.
28. Page H. Economic appraisal of in vitro fertilization: discussion paper. J R Soc Med 1989;82:99-102.
29. In vitro fertilization. In: Speroff L, Glass RH, Kase NG. Clinical gynecologic endocrinology and infertility. 4th ed. Baltimore: Williams & Wilkins, 1989:611-20.
30. Hershlag A, Kaplan EH, Loy RA, DeCherney AH, Lavy G. Heterogeneity in patient populations explains differences in in vitro fertilization programs. Fertil Steril 1991;56:913-7.
31. Hull MGR, Eddowes HA, Fahy U, et al. Expectations of assisted conception for infertility. BMJ 1992;304:1465-9.
32. Tan SL, Royston P, Campbell S, et al. Cumulative conception and livebirth rates after in-vitro fertilisation. Lancet 1992;339:1390-4.
33. Guzick DS, Wilkes C, Jones HW Jr. Cumulative pregnancy rates for in vitro fertilization. Fertil Steril 1986;46:663-7.
34. Padilla SL, Garcia JE. Effect of maternal age and number of in vitro fertilization procedures on pregnancy outcome. Fertil Steril 1989;52:270-3.
35. FIVNAT (French in Vitro National). French national IVF registry: analysis of 1986 to 1990 data. Fertil Steril 1993;59:587-95.
36. Dalton M, Lilford RJ. Benefits of in-vitro fertilisation. Lancet 1989;2:1327.
37. Neumann PJ, Johannesson M. Willingness to pay for in vitro fertilization: a pilot test using contingent valuation. Med Care 1994;32:686-99.
38. Haan G. Effects and costs of in-vitro fertilization: again, let's be honest. Int J Technol Assess Health Care 1991;7:585-93.
39. State laws on infertility insurance coverage. Washington, D.C.: American Fertility Society, Office of Government Relations, 1991.
40. Health Security Act section-by-section analysis. Washington, D.C.: Executive Office of the President, 1993:21.

Birth Control

MOTHERS!

Can you afford to have a large family?
Do you want any more children?
If not, why do you have them?

DO NOT KILL, DO NOT TAKE LIFE, BUT PREVENT

Safe, Harmless Information can be obtained of trained

Nurses at

46 AMBOY STREET
NEAR PITKIN AVE. — BROOKLYN.

Tell Your Friends and Neighbors.   All Mothers Welcome
A registration fee of 10 cents entitles any mother to this information.

מוטערס!

זייט איהר פערמעגליך צו האבען א גרויסע פאמיליע?
ווילט איהר האבען נאך קינדער?
אויב ניט, ווארום האט איהר זיי?

מערדערט ניט, נעהמט ניט קיין לעבען, נור פערהים זיד.

זיכערע, אונשעדליכע אויסקינפטען קענט איהר בעקומען פון טרפארענע נוירסעס אין

46 אמבאי סטרים נוער פיטקין עוועניו ברוקלין

זאגט דאס בעקאנט צו אייערע פריינד און שכנות.   יעדער מוטער איז ווילקאמען.

פיר 10 סענט איינשרייב־געלד זיינט איהר בערעכטיגט צו דיעזע אינפאר־מיישאן.

MADRI!

Potete permettervi il lusso d'avere altri bambini?
Ne volete ancora?
Se non ne volete piu', perche' continuate a metterli al mondo?

NON UCCIDETE MA PREVENITE!

Informazioni sicure ed innocue saranno fornite da infermiere autorizzate a

46 AMBOY STREET   Near Pitkin Ave. Brooklyn

a cominciare dal 12 Ottobre. Avvertite le vostre amiche e vicine.

Tutte le madri sono ben accette. La tassa d'iscrizione di 10 cents da diritto a qualunque madre di ricevere consigli ed informazioni gratis.

Margaret H. Sanger.

OF FEMINISM AND BIRTH CONTROL PROPAGANDA
(1790–1840)

NELLA FERMI WEINER

ABSTRACT

The growth of feminism in the 19th century United States was paralleled by a decline in the birth rate. Connections between these two developments are suggested. The work of some other authors is reviewed and the historical context is delineated. The activities and motivations of major birth control propagandists in England and America from 1790 to 1840 are considered in detail. Previously, connections between birth control and feminism in this period have been neglected.

INTRODUCTION

Wollstonecraft and Malthus

Two books of great import for women (as well as other people) were published by Joseph Johnson, bookseller of London, in the last decade of the 18th century. The first was Mary Wollstonecraft's *A Vindication of the Rights of Women* (1792). This book is often thought of as the beginning of the 19th century movement for the emacipation of women.

Six years later came the Reverend Thomas Robert Malthus' *Essay on Population*. Malthus warned that people could increase far faster than the means of feeding them. As a result, population would be limited by such natural checks as war, famine and disease. Extreme poverty of some portion of the population was inevitable. The possibility of limiting population by "moral restraint" (late marriage and abstinence until marriage) was added in mitigation of this rather harsh doctrine in the second edition of the *Essay* (1803). Malthus was aware of the existence of contraceptives, but did not advocate their use. Like Wollstonecraft, Malthus had his precursors, but his work is often thought of as the beginning of the 19th century movement to limit population. (1)

Both books arose from the ideas of the Enlightenment and the French Revolution. The one was a plea for the inclusion of the rights of women in the expanding rights of men. The other was a caveat, a warning of the limits to the perfectibility of human society. There is no reason to suppose that Malthus was concerned with the rights of women, and it is ironic that Malthus' essay was written in reply to William Godwin's *Political Justice* (2) — Godwin was Wollstonecraft's husband. But it could hardly be coincidence that these two movements — for the rights of women and for limitation of population — began to stir at the same time, nor that they grew up side by side.

The economic and demographic burdens of excess fertility (fertility refers to actual numbers of children born as distinct from fecundity which is the biological potential for

411

reproduction) are shared by women and men, but women are the exclusive child bearers, and in most cultures have major responsibility for child rearing. A decline in fertility would most immediately affect women.

Purpose

It is the purpose of this paper to show that feminism partly motivated early birth control propaganda in England and America. The term feminism is here used to indicate interest in improving the condition of women both within and outside the family and increasing their freedom and opportunities, but is not tied to a specific ideology.

Before 1840, four major writers wrote in English on contraception, detailing methods and also developing the social, economic and medical theory of the subject. These were Francis Place and Richard Carlile in England, and Robert Dale Owen and Charles Knowlton in America. In addition, the influence of Jeremy Bentham appears to have been crucial. (3) It will be shown that all of these men, except Knowlton, were feminists or had some feminist leanings. (I have found no evidence that Knowlton had feminist tendencies, but neither have I found evidence which excludes the possibility.) Their feminism was not necessarily a part of their writings on birth control; both birth control and feminism went against the accepted values of their societies. The combination of the two subjects, touching as it did on feminine purity and the sanctity of the home, would have been even more explosive — for this reason some of these authors may have found it prudent to separate the two subjects in their writing.

Nor is it surprising that it was men rather than women who wrote on birth control — the subject was difficult enough for men to broach, much more so for women. Nonetheless, of the very few women who were active outside the home in the early 19th century, one, Frances Wright, spoke (but did not write) about birth control and is likely to have influenced Robert Dale Owen's writing. Another woman, Harriet Martineau, wrote on population limitation, but followed Malthus in advocating "moral restraint" rather than contraceptives. Even so, she feared that her writing on population would be ruinous to her reputation and said that as she wrote "the perspiration many times streamed down my face, though I knew there was not a line in it which might not be read aloud in any family. The misery arose from my seeing how the simplest statements and meanings might . . . be perverted;" (4) Even though birth control was of great importance for women, they could not act for themselves, but needed the help of sympathetic men in making knowledge and techniques available.

It is necessary to consider the historical context and to review the work of some previous authors before a more detailed discussion of the writings and motivations of the early birth control propagandists.

HISTORICAL CONTEXT AND REVIEW

In the United States, the rise of feminism in the 19th century was paralleled by a fall in the birth rate. The first United States census was taken in 1790. Before that time, data

412

are sketchy. Eighteenth century America was a young country and despite a low immigration rate from 1690 to 1790, growth was rapid. (5) Benjamin Franklin noted the speed of population growth in these circumstances and Malthus probably took his cue from Franklin when he used the United States as an example of how population could double every 25 years when unchecked. Where Malthus was pessimistic, Franklin was optimistic, believing that it would take "many ages" to fill the country. (6)

The ink had not dried on Malthus' pen when the birth rate in the United States began a decline which was to continue until the mid-1930s. The total fertility rate (average number of live births per woman living to age 45) in 1800 was 7. By mid-century it had dropped to 5.4 and in 1900 it was 3.6. It reached a low of 2.1 in 1934 and climbed back to 3.6 by 1957. Since then it has again declined.

In the course of the 19th century, the total fertility rate was approximately halved. Such a change must have revolutionized the lives of women. At the beginning of the century women who had children would have spent most of their adult lives in child-bearing and rearing with a child born every two or three years during their fertile years and with the burdens of child care continuing well past menopause for those women who survived. By the end of the century, the situation for many women was quite different – now the burdens of motherhood were lessened and there might be energies to spare for other pursuits.

And indeed, between 1800 and 1900, women gained new rights and freedoms and became active in a number of areas previously closed to them. Feminism became organized, legal and economic battles were won and by the end of the century the woman's suffrage movement was well launched with the vote having been won in at least a few states.

According to Nancy F. Cott, for many women new rights began at home. Starting in the late 17th century, with the industrial revolution, men's work was increasingly removed from the home. This created a separation between man's world and woman's sphere. Women became the guardians of religious and secular morality, angels of the home, providing a shelter in which children could be safely reared and to which men could retreat from the harshness of the outside world. Though productivity was increasingly removed from the home "a cultural halo ringing the significance of home and family . . . reconnected woman's 'separate' sphere with the well-being of society." Women thus gained power within the family. Their primacy in the domestic sphere gave them a sense of their own vocation and a solidarity with their own sex. The cult of domesticity also opened possibilities for women's education (different from men's) as training for wifehood and motherhood. Cott considers this stage a necessary preliminary to organized feminism, though the cult of domesticity – at least for a time – barred other alternatives, especially for married women. (8)

It is likely that this new power within the home gave women some leverage in decisions about family size. But the ideal of feminine purity and of women as guardians of morality

413

would make it very difficult for women to express themselves openly about birth control or indeed any matters related to sex. (In fact, very few women, in the late 18th and early 19th century were expressing themselves publicly at all.)

The beginnings of industrialization may also have provided some economic motive for limiting families. James Reed argues that this was not in itself sufficient to explain the 19th century decline in fertility which occurred in rural as well as industrial areas. However, even in rural areas attitudes were changing; people were better educated and increasingly confident of their abilities to control nature and their own destinies. These aspects of modernization would facilitate the spread of birth control. (9)

Feminism is another aspect of modernization which, both in its domestic and its public manifestations, might be expected to be intimately related to the spread of birth control and declining fertility.

A woman whose desires and aspirations are limited to marriage and motherhood might want to limit births in order to give better care to each child, or to preserve her own health. Lessened burdens of motherhood might free energies to create other options for women. A woman who saw possibilities for herself outside of home and family, would be likely to want fewer children than one whose most important role was that of mother. As new roles opened for women, they might be expected to be more desirous of limiting births. Thus freedoms for women and birth control have the potential to reinforce each other.

Most histories of 19th century American feminism and American women have not dealt exhaustively with the subject of the relation between the diffusion of contraceptives and the rise of the feminist movement. Recently, Linda Gordon has written a detailed history of the birth control movement in the United States and its relations to feminism. However, the main body of Gordon's work begins with the last third of the 19th century. (10)

Connections between feminism and family limitation from 1740 to 1840 have been largely overlooked. This period is important because it contains the genesis of widespread family limitation (as indicated by declining fertility in the United States) and the first stirrings of 19th century feminism.

Early Effective Family Limitation

Louis Henry and his associates pioneered demographic studies of small towns by painstaking family reconstruction from examination of parish records of births, marriages and deaths. By studying ages at marriage, birth intervals and other family patterns it is possible to infer the causes of a decline in fertility. Thus Gautier and Henry conclude that a slow decline of births in the Norman Parish of Crulai after 1750 was probably due to late marriages and perhaps fewer marriages, but a much sharper decline during and after the French Revolution can be attributed to the use of birth control within marriage. Such studies have been done of a number of French towns. (11) Wrigley, using similar methods, found evidence of deliberate family limitation in the English town of Colyton as early as the mid-seventeenth century. (12)

414

Wells studied a group of Quaker families in New York, New Jersey and Pennsylvania in the late 18th and early 19th century. He found that fertility in this group declined well in advance of the United States population as a whole. An analysis of the patterns of child-bearing of these Quakers provides evidence that at least part of the decline was due to deliberate family limitation. (13) Wells does not speculate as to why this particular group was so far ahead of the rest of the nation. However, the position of women among the Quakers has long been a relatively favorable one. The early lowering of the birth rate in this group may be an expression of the strong position of its women.

Methods

Norman E. Himes discusses several effective methods of contraception which are quite old. These methods are perhaps not effective by modern standards — we now expect near perfect control — but they could significantly lower fertility.

Among male methods, withdrawal or *coitus interruptus* is mentioned in the Bible and has probably been a widespread method. It requires no equipment and could be rediscovered at different times and places. Condoms made of skins are of uncertain origin but there are published reports as far back as the 16th century. They were probably used more often to prevent syphilitic infection, but were used also as contraceptives.

Female methods include douches and a variety of intravaginal methods. As far back as 1550 B.C. Egyptian women used pessaries made of lint tampons moistened with honey and a slightly acid substance. This combination could have served both as a physical barrier and a chemical spermicide. Vaginal sponges were also used alone or in combination with vinegar or other spermicides. (14)

Abstinence cannot be dismissed as a method of contraception — it was widely recommended in the 19th century: less frequent intercourse could have reduced fertility. (15) Rhythm was certainly attempted, but the ovulatory cycle was poorly understood, so it is unlikely to have been effective. (16) Methods involving partial or total abstinence (though no doubt abstinence frequently meant purity for the wife and prostitutes for the husband) were consistent with 19th century morality, but methods using contraceptive devices were abhorrent to many people.

Abortion and infanticide have also been widely used as methods of family limitation. Abortion was not illegal in the United States until the latter part of the 19th century and there is evidence that it was widely practiced. Infanticide was, of course, criminal, but was also practiced. (17)

All the methods discussed above have been used since ancient times. The diffusion of birth control and the decline in birth rate at the beginning of the 19th century were not due to improved technology, but to the desire to limit births, and perhaps to the diffusion of knowledge of effective methods.

There have been two major breakthroughs in technology. The discovery of the vulcanization of rubber in the mid-nineteenth century made possible a variety of vaginal caps and the modern diaphragm, as well as the development of inexpensive, lightweight rubber

415

condoms. (18) The introduction of the pill and the IUD about 1960 marked another major step forward. (A primitive form of the IUD has been known since ancient times and was used in the 1920s, but being quite dangerous it fell in disrepute. [19] It is interesting to note that these advances in technology occurred in a context of declining birth rate — the birth rate in the United States had been declining for half a century when rubber was vulcanized. The baby boom peaked in 1957 and the birth rate was declining again when the pill was introduced in 1960. Technology *followed* the desire to limit family size.

It is widely believed that withdrawal was by far the most prevalent method of birth control in the 19th century and earlier. This belief apparently rests on reports of methods used before coming to clinics for contraceptive advice in the early part of this century. (19) Not only is this inferential for the 19th century, but a likely bias is that women (and it was mostly women) going to clinics for advice would have been those who found the methods they were using unsatisfactory. Other evidence for withdrawal appears in the objections of some feminists and others to the use of contraceptive devices.

No doubt withdrawal was widespread; however, female methods such as douching and the sponge may also have been prevalent. Certainly advertisement for these methods, for "rubber goods" and for "regulation of the monthlies" were common in the latter part of the century. All of the early birth control propagandists, including Knowlton but excepting Owen, favored female methods. (This, of course does not tell us what methods were actually used.) In particular, the vaginal sponge may have been quite effective. It was still recommended in the 1920s by Marie Stopes (and various clinics) in some cases because of its simplicity or for women who could not be fitted satisfactorily with diaphragms or caps. Stopes recommends its use with alum, soap powder, quinine ointment or vinegar and water, though she thinks it often succeeds by itself. As late as 1962, *The Consumer Union Report on Family Planning* lists the sponge, in connection with a modern spermicide, as a method with high to medium effectiveness. (20)

JEREMY BENTHAM (1748-1842)

Jeremy Bentham has been credited with some influence on practically every English law reform since his time. He interested himself in prison reform, universal education, cheap postage and taking of censuses. He believed in the greatest good for the greatest number, and his Utilitarianism needed the collection of data for a kind of hedonistic calculus. He was a nuts and bolts philosopher, took nothing for granted, and figured things out from the beginning. Yet Sir John Bowring, his literary executor says "Many of his writings I have not deemed it safe to give to the world, even after his death, so bold and adventurous were some of his speculations." (21) His views on contraceptives are perhaps illustrative. He apparently preceded Malthus in his concern for overpopulation, but he understood human nature too well to put his faith in "moral restraint." In birth control, he found a more feasible solution. We see at once what was surely in its time "adventurous speculation" and also a sense of the practical, the utilitarian.

416

254

Bentham's Riddle

The American and French revolutions made a democrat of the previously rather conservative Bentham. Perhaps he also learned something else from France where contraceptives were better known than in England. (22) Bentham's "The Situation and Relief of the Poor" was published in Arthur Young's *Annals of Agriculture* in 1797. Much of the article argues the need for better enumeration of paupers. In the midst of this discourse is the following passage:

> In some such ways as these we begin, all of us; and if we did not begin a little at random, how would any thing ever be done?
> Come, my Oedipus, here is another riddle for you: solve it, or by Apollo! — You remember the penalty for not solving riddles — *Rates* are encroaching things. You, as well as another illustrious friend of mine, are, I think, for *limiting* them. — Limit them? — Agreed. — But how? — Not by a prohibitory act — a remedy which would neither be applied, nor if applied, be effectual — not by a *dead letter*, but by a *living body:* a body which, *to stay the plague,* would, like *Phineas,* throw itself into the *gap;* yet not, like *Curtius,* be swallowed up in it.
> When I speak of *limitation,* do not suppose that limitation would content me. My *reverend* friend, hurried away by the torrent of his own eloquence, drove beyond *you,* and let drop something about a *spunge.* I too have my spunge; but that a slow one, and not quite so rough a one [Italics Bentham's]. (23)

Norman Himes, who is responsible for finding and decoding this passage believes that Bentham is recommending the vaginal sponge as a means of contraception. The poor rates were a tax for the relief of the poor. Bentham is punning on the rate of reproduction. By limiting the rates of reproduction, the taxes for relief of the poor would also be lowered. He suggests that this could be done not by a "dead letter" — possibly a condom which was also known as a "French letter" (the French sometimes called it a "capote anglais") and also an unenforcible law, or a "prohibitory act" but by a "living body" that is, a sponge. The "reverend friend" is probably the Reverend Joseph Townsend who had visited France in 1789. The "gap" is a common 18th century word for the vagina. Bentham seems to be recognizing that his remedy would be slow, but thinks it "not so tough" that is, not such a strain on human nature. (24)

This may well be the first time that contraception was tied to an economic motive. There are some earlier allusions in English to the sponge, but this may have been a method well known only to prostitutes. Prostitutes and houses of prostitution also sold, and advertised, condoms in England in the 17th century. (25) Thus Bentham's riddle may also be a very early "respectable" mention of contraception in English — an important point for women since women have traditionally been divided (as men have not) between the "respectable" and the "not respectable" with little if any communication between the two. Birth control had to become respectable before respectable women could have access to it.

417

We cannot know if Arthur Young and his readers managed to decode Bentham's riddle, but this evidence of Bentham's advocacy is important because he influenced, directly or indirectly, Place, Carlile, Wright and Owen as well as both James and John Stuart Mill.

Bentham's Feminism

In her biography of Bentham, Mary P. Mack tells us (as several of his other biographers do not) that "though Bentham usually kept within the limits of accepted institutions, there were moments when his imagination soared beyond them. It invariably did whenever he considered the subjection of women, for few social facts angered him more. . . . In the area of women's rights, 18th century England was still barbarian." Like his ideas about contraception and limitation of population, women's rights also went beyond the accepted values of his society.

"Open and ardent feminism," says Mack, "was a luxury that Bentham could not afford. All the same he did what he could to further the equality of women." Bentham was in favor of universal suffrage. (26)

Bentham advocated two unpopular causes — birth control and feminism. Even separately he could not express them openly. To put the two together would have been harder. For while birth control might be cautiously advocated to limit the pauper population and to reduce poor rates — a motive which would have some appeal for conservatives — to place birth control is an example of this, for his motivation is to improve the condition of ery and sin. It is not certain that Bentham linked the two subjects in his own mind, but surely he had good reason for not putting them together in his writings.

FRANCIS PLACE (1771-1854)

Francis Place loved his wife and fathered 15 children (nine survived to adulthood). (27) After that he began his propaganda for birth control. Place could not accept Malthus' prescription of "moral restraint" as a curb to population. He had married Elizabeth Chadd before he was 20 and this had steadied his character. (28) Late marriage he believed would lead to debauchery (i.e., the resort of young men to prostitutes) just as surely as early marriage led to many children. "Moral restraint," he wrote ruefully to James Ensor, "which has served so well in the instances of you and I — and Mill, and Wakefield — mustering among us no less than 36 children . . . rare fellows we to teach moral restraint." (29)

Place was a tailor. In 1793 he and his wife underwent a period of exreme poverty and sometimes near starvation during which their first child died of smallpox. Place was among the organizers of a strike of leather britches makers. Employers retaliated and for a time he could not find work. (30) Looking back on this period it is his wife's miseries which struck him most forcibly:

> My temper was bad, and instead of doing every thing in my
> power to sooth and comfort and support my wife in her
> miserable condition . . . I used at times to give way to passion

418

and increase her and my own misery . . . Nothing conduces so much to the degradation of a man and a woman in the opinion of each other, and of themselves in all respects; but most especially of the woman; than her having to eat and drink and cook and wash and iron and transact all her domestic concerns in the room in which her husband works, and in which they sleep. (31)

But eventually he adapted to his situation, "and at length I obtained such a perfect command of myself that excepting commiseration for my wife, and actual hunger I suffered but little." (32) During this period of unemployment he read widely "in history, voyages, and travels, politics, law and Philosophy." He attributes his later success to the education thus gained. (33)

Matters improved for Place and his wife. As soon as he found sewing to do, both of them worked at it for sixteen or eighteen hours a day and hired a servant to do the housework. He saved enough to go into business for himself and by 1817 was sufficiently wealthy to put his tailoring business in his son's hands in order to devote himself to politics.

But the road to wealth was long and hard and "to some extent destroyed my wife's cheerful disposition . . . Her situation was necessarily worse than mine on account of the two children." (34) Beginning with his wife, Place shows considerable empathy with the situation of women and awareness of the burdens placed on them by children.

His own experience with severe poverty and his lower-class backgound gave him an empathy with the working people and a clarity of purpose. His work on population and birth control. There had been a scandal at Nashoba — word was that free love was being practiced there. Wright's advocacy of birth control only added fuel to the fire. (55) This practicalities. He would not settle for impractical solutions or impose on others what had not worked for himself.

Place's biographer, Wallas, gives an able account of Place's political activities, always in radical causes. Here I will only remark on Place's interest in birth control and in the education of women (part of larger interest in education particularly for the working people).

Place's Book on Population
In 1808 Place became acquainted with James Mill and they became close friends as did their wives. Through Mill, Place also met Bentham. In 1817 he made a two-month visit to Bentham's Ford Abbey where James Mill and his family were living at that time. In 1819, James Mill was appointed to India House and Bentham lost much of Mill's companionship. Bentham turned more and more to Place as a replacement. (35)

"He loved quiet power for the purpose of promoting good ends," said the *Spectator* on Place's death. (36) But though usually working behind the scenes, he did one thing publicly where none had dared before. In 1822 he published *Illustrations and Proofs*

419

of the Principle of Population. This was the first book in English to advocate the use of contraceptives, and the only book length work by Place published in his lifetime.

If Place and Bentham discussed the matter, and very likely they did, they no doubt decided that there was need for a public and respectable work on population advocating contraceptives. They might have concluded that it was safer if Place rather than Bentham wrote the work. Bentham had an established reputation and many followers. The tarnishing of his name might have damaged Utilitarianism as a whole. Place was little known publicly, and having quit tailoring, was no longer dependent on the good will of his customers. Himes states that "the genealogy of the idea of [English] Neo-Malthusianism is . . . now clearer: Bentham . . . was the fountain-head of the idea. He . . . inspired James and J.S. Mill and Place . . . Place was the real leader of the social agitation." (37) (The elder Mill never publicly advocated birth control, though he did write on population. His son, John Stuart Mill, did advocate birth control, but much later than the period discussed in this paper.)

In his book Place stops short of detailing methods of birth contol but says:

> If, above all, it were once clearly understood, that it was not disreputable for married persons to avail themselves of such precautionary means as would, without being injurious to health, or destructive of female delicacy, prevent conception, a sufficient check might at once be given to the increase of population beyond the means of subsistence. (38)

Place advises educating the "commonest mechanics and labourers" in the economics and means of checking population:

> If without airs of superiority and dictation; if without figure and metaphor, means were adopted to show them how the market came to be overstocked with labour . . . if a hundredth, perhaps a thousandth part of the pains, were taken to teach these truths that are taken to teach dogmas, a great change for the better might . . . be expected. (39)

Throughout the book, the main thrust of Place's argument for population limitation is economic. That he was aware of the special difficulties for women of unchecked reproduction is clear from comments about his wife's problems and from other writings to be discussed below. But arguing for contraception on economic grounds was no doubt safer and less controversial than arguing for it on grounds of easing the burdens of women. I do not mean to suggest that the economic argument was unimportant to Place, only that he left out the feminist argument. A feminist rationale for contraception would likely be interpreted as intended to undermine female chastity. It was feared by many that separating sex from reproduction would lead to promiscuous intercourse for both sexes but particularly for women − such fears would be muted in the case of an economic argument, but exacerbated by a feminist one. Place himself saw the matter of chastity

420

and contraception in another light — he believed that, freed from the fear of too many children, all would marry young and this would reduce prostitution and illicit intercourse for both sexes.

Place's book did not have a large circulation and though quoted by scholars, it probably had no great impact at the time. (40)

Place's Handbills on Birth Control
A year later, in 1823, Place took his own advice on educating the people and circulated a number of handbills around London and the industrial districts of the North. In these he recommended the use of the vaginal sponge as a method of contraception and in one version withdrawal and lint tampons also. Young John Stuart Mill, then seventeen, was arrested and kept in jail overnight for helping to distribute them. Only much later were these handbills definitely attributed to Place. (41)

The three versions of the handbills were titled: "To the Married of Both Sexes," "To the Married of Both Sexes in Genteel Life," and "To the Married of Both Sexes of the Working People." All three were clear, concise and specific. Much of the argument for the use of contraceptives is economic: "It is a great truth . . . that when there are too many working people . . . they are worse paid . . . and are compelled to work more hours than they ought." But Place was also mindful of the special applicability to women. In the second version he speaks of the miseries of women with "constitutional peculiarities" and "malconformation of the bones of the pelvis" which "has caused thousands of respectable women to linger on in pain and apprehension." He is also sensitive to the con-nection between prostitution and the absence of birth control:

> At present every respectable mother trembles for the fate of her daughters as they grow up. Debauchery is always feared . . . And why is there so much debauchery? . . . because many young men, who fear the consequences which a large family produces, turn to debauchery . . . Other young men . . . marry early and produce large families . . . These are the causes of the wretched-ness which afflicts you.

And as for the remedy:

> The anwer is short and plain: the means are easy. Do as other people do, to avoid having more children than they wish to have, and can easily maintain.
> What is done by other people is this. A piece of soft sponge is tied by a bobbin or penny ribbon, and inserted just before sexual intercourse takes place, and is withdrawn again as soon as it has taken place. Many tie a piece of sponge to each end of the ribbon, and they take care not to use the same sponge again until it has been washed. (42)

421

259

Place returns to the question of the ill effects of late marriages particularly on women in a letter to Harriet Martineau in which he tries to persuade her of the impracticality of "moral restraint":

> Young men of every rank and station pride themselves in the want of chastity. And I see no chance that this can ever be otherwise until the time shall come when all are married when young. The consequences of delayed marriages are dissolute practices . . . and, towards women, injustice of the greatest possible extent, and most fatal in its consequences, inflicted without the least remorse . . . You can form nothing like a correct opinion of the evils; no respectable woman can do so, since all they can ever have on which to form opinions are a few innuendos and occasional displays of neglect and barbarity. (43)

Place's Concern for Women's Education
In addition to the above concerns about the welfare of women, Place was also interested in women's education. When he visited Bentham and the Mill family at Ford Abbey in 1817 he was much struck by the education James Mill was giving not only John Stuart but also John's younger sisters, Wilhelmina and Clara. He comments extensively and approvingly on the Mill children's education in letters to his wife: "Now I could not be so severe; but the learning these children have acquired is not equalled by any children in the whole world." (44) Place regretted the lack of education of his wife, "I mean her school education was very narrow reading and writing and common sewing constituted the whole," and was sorry that she herself felt inadequate because of this deficiency. (45) He saw to it that his daughters got a good education while at the same time helping and encouraging a young female teacher to found her own school where "my two eldest daughters were well grounded in Languages, and then I employed the best masters . . . They were well instructed in French and Italian, Arithmetic, Geography — some Astronomy, Algebra, Mathematics and History, Needlework of all kinds and the making of their own cloaths." (45)

Place's feminist convictions are perhaps not as clear as Bentham's, but there is evidence in his handbills and other writings that he connected contraception with the welfare of women. His concern for women's education is also unmistakable.

RICHARD CARLILE (1790-1843)

It was difficult for Francis Place to win his younger friend, Richard Carlile, to the cause of birth control, but once converted there was no stopping him.

Carlile began as a tinsmith and became a radical writer and publisher. In 1817 he became agent and then owner of the *Political Register* (later renamed *The Republican*). In a few months he was in jail for publishing parodies of the Lord's Prayer. He was released after several months, but was soon in jail again for publishing more parodies and the works of

422

Tom Paine. While he was in jail, his wife Jane ran the business. She had "talents for business, which were of the greatest value to her husband." Jane also wrote for the paper. While Carlile was publishing seditious writings from jail his wife and then his sister took over the printing business and were successively arrested. Carlile spent nearly a third of his adult life in jail. (47)

Carlile's marriage did not prosper. He and Jane were first separated in 1819, but the separation did not become final until 1832. In 1831 he met Eliza Sharples who had previously been attracted by his writings. When he was in jail she offered to help organize his defense. Later she agreed to live with him. She was one for the first English women to lecture publically and she also published a journal *Isis* in which she defended her relationship with Carlile. (48)

Contraception and Sexual Equality

In 1825, Carlile published "What is Love?" in *The Republican*. This was reprinted in 1826 as a separate booklet under the title *Every Woman's Book: or What is Love?* In it Carlile advocated the sponge primarily, but also withdrawal and the sheath as means of contraception. Carlile also advocated equality for women in the sexual sphere:

> It is a barbarous custom that forbids the maid to make an advance in love, or that confines that advance to the eye, the fingers, the gesture . . . It is ridiculous. Why should not the female state her passion to the male, as well as the male to the female? . . . All the affairs of genuine love are claimed for the female . . . Young women! assume an equality, plead your passions. (49)

Nor is he a believer in the double standard: "In the old maid, the passions of love, like an overflowing gall-bladder, tinges every other sensation with bitterness . . . to make ripeness wholesome it must be enjoyed." He goes on to make the same prescription for the "real old bachelors," (50) and adds:

> Let it not be understood that this work advocates indiscriminate intercourse, such as exists among some animals . . . Where there is an equal number of males and females, each should be contented with one of the other sex . . . Equality between the sexes is the source of virtue. If there were two women to a man, a plurality of wives would be prudent. If two men to a woman a plurality of husbands. None unmatched that desire to be matched. (51)

I thought the book charming, but it is understandable that it was not so received. Place himself made an effort to dissaude Carlile from publishing "What is Love?" (52) But Carlile was not a man easy to muzzle — he is chiefly remembered for his struggle for freedom of the press.

423

Carlile was the first to make a clear connection between equality of the sexes and the use of contraceptives. He was also associated with capable, independent women — his first wife and his sister — and his second wife was an avowed feminist.

ROBERT DALE OWEN (1801-1877) AND FRANCES WRIGHT (1795-1892)

Owen's Feminism

Robert Dale Owen (son of Robert Owen) was a consistent advocate of women's rights. He favored legal equality and equal education and urged women to enter trades and professions in order to become economically independent. When he married Mary Jane Robinson in 1832, it was with a marriage contract in which he divested himself of "the unjust rights which, in virtue of this ceremony, an iniquitous law tacitly gives me over the person and property of another." (53)

Robert Dale Owen was born in Scotland and came to America to join his father's Utopian community at New Harmony, Indiana. It was here that he and Frances Wright became acquainted.

Frances Wright was also born in Scotland. She and her younger sister Camilla were orphaned at an early age and reared by a series of relatives. In 1818, Frances and Camilla Wright traveled alone to America. Fanny Wright's travel memoirs, *Views of Society and Manners in America* (1821) were well received and so much liked by Jeremy Bentham that he invited her to visit. "My hermitage" he wrote, "is at your service . . . I am a single man, turned seventy, but as far from melancholy as a man needs be. Hour of dinner, six; tea between nine and ten; bed quarter before eleven. Dinner and tea in company. Breakfast my guests . . . have in their rooms and at their own hour . . . " The visit was a success and Wright returned often. She had already been impressed by Bentham's writings and the two spent many hours in earnest discussion. (54)

In 1825, Wright returned to America, purchased a tract of land in Tennessee where, with her sister and a few others, she founded Nashoba. She intended this as a model community in which slaves could work to earn their freedom. The project was a failure. During the years Nashoba lasted, Wright made frequent trips to New Harmony for inspiration. She and Robert Dale Owen became friends and then associates in editing and publishing the *Free Enquirer* (originally the *New Harmony Gazette*). In 1829, Wright, Owen and the *Free Enquirer* moved to New York, where Wright felt opportunities would be better.

Wright's Advocacy of Birth Control

Beginning in 1828, Wright made a number of lecture tours. She was an impressive speaker, and though a woman lecturing shocked some, she also developed a devoted following. She was outspoken — she condemned organized religion and capital punishment, demanded equal education and legal rights for women and advocated divorce reform and birth control. There had been a scandal at Nashoba — word was that free love was being practiced there. Wright's advocacy of birth control only added fuel to the fire. (55) This

424

262

may have discouraged her for apparently she stopped lecturing on birth control. Later, when Owen wrote his own treatise on the subject, Wright wrote him an open letter, published in the *Free Enquirer*, telling him not to expect too much good from his birth control treatise since ignorance and hypocrisy would be against it. She urged him to publish it anyway for the good of those who could accept its advice. (56) It is likely that Wright stopped lecturing on birth control because of the pressure of public opinion. The argument was probably more shocking coming from a woman and Owen might hope to meet with less opposition than Wright.

Owen's Advocacy of Birth Control

Owen had a copy of Carlile's *Every Woman's Book,* which he was asked to reprint. He declined to do so, in part because he felt the book was not in good taste, though he paid tribute to the author's honesty and good intentions. Owen believed that "there was too much . . . prejudice . . . to render such discussion . . . generally useful." He changed his mind on this last score and *Moral Physiology: Or a Brief and Plain Treatise on the Population Question* (57) was the result. Owen was the first American writer on birth control — Wright preceded him in speaking on the subject but she never wrote about it.

Himes believes that Robert Dale Owen's work was independent of the British movement except for his reading of *Every Woman's Book.* Himes says:

> There is no evidence that Robert Dale Owen had any direct contact with that Benthamite circle of 'philosophical radicals' who, between 1818 and 1822, were contemplating and discussing among themselves the advisability of a practical neo-Malthusian propaganda among the working classes. These discussions seem to have been for the most part secret, open only to a select group enjoying the confidence of the leaders. (58)

Fryer cites evidence that Owen had dinner at Bentham's sometime after 1824 — apparently only once and in a large company. (59) But Frances Wright had extensive contact and a number of private conversations with Bentham. It is more likely that Bentham discussed birth control with Wright and Wright with Owen, particularly as Wright's lecturing preceded Owen's writing. That Owen was probably aware of Bentham's interest in birth control is shown by the inscription on the title page of *Moral Physiology:* " 'The Principle of utility is the foundation of the present work.' Bentham *On Morals and Legislation.*"

Owen's Feminism Connected to Birth Control Advocacy

Owen is the only one of the early propagandists to advocate a male method of birth control — withdrawal. (Both Knowlton and Owen mention the "baudruche" or condom, but neither believes that this method would ever be widely adopted. Presumably it was the making of cheap, lightweight rubber condoms later in the century that nullified this prediction.) To the objection that withdrawal gave control to the male, Owen replied that the "only effectual defence for women is to refuse connection with any man devoid of honour," and this would encourage more regard for the rights of women in sexual matters. (60)

425

In Owen's work, the connection between feminism and family planning is evident. In part, because he "abandoned his earlier restraint and discussed the subject of contraceptives," says his biographer, Leopold, his ideas on the rights of women were misunderstood:

> In the eyes of his contemporaries he was planning to destroy
> the institution of marriage, abolish all ideals of chastity, and
> make of their wives and sisters common prostitutes. Yet Owen
> did no more than advocate logically, consistently, and indefati-
> gably the liberating measures he had first broached at New
> Harmony. (61)

The advocacy of birth control for feminist reasons was explosive no matter who the writer.

CONCLUSION AND SUMMARY

In 19th century America, birth control and feminism grew side by side. As the birth rate declined, women gained new rights. Between 1800 and 1900, the birth rate was halved. Such a development must have made radical changes in the lives of women. Not only would a decrease in births have eased the lives of women within the family, but possibilities for women outside of family and motherhood might have expanded as their responsibility for child rearing lessened. To the extent that women could exert — directly or through men — control of their fertility, they gained power over their destinies.

It might be expected that feminists, foreseeing these possibilities, would be among those favoring birth control before it became generally accepted. The subject of the connections between the development of feminism and the diffusion of contraceptives is only beginning to be explored. The present work has concentrated on the motivations of early birth control propagandists in England and America.

At the end of the 18th century, Jeremy Bentham wrote, in a veiled manner, recommending the vaginal sponge as a method of contraception. Bentham was a feminist, but did not associate feminism and birth control in his writing. This may have been because he feared that the combination of the two subjects would have been even less acceptable than either alone, rather than because he did not make the connection in his own mind.

Bentham's disciple, Francis Place, was the first "respectable" author to write openly and extensively on contraceptives, though most of his writing on this subject was, like Bentham's, tied to an economic motive for population limitation. Nonetheless, Place was concerned with the special needs and interests of women and the burdens that frequent child bearing placed on them. He was sensitive to the connection between prostitution and late marriage (sometimes advocated as a means of limiting births). He considered prostitution a great injustice to women. Place also advocated improved education of women.

426

Carlile connected birth control with equality between women and men, particularly in the sexual sphere. His book was too radical to be well received in its time. Carlile was also connected with strong, independent women. In America, Frances Wright, probably having been influenced directly by Bentham, spoke but did not write about birth control. She may have been discouraged by pressure of public opinion. Robert Dale Owen, Wright's friend and associate, wrote the first American work on birth control. The influence of Bentham on Owen is probably via Wright. Both Owen and Wright were feminists.

These were most of the major, "respectable" writers on birth control from 1790 to 1840 in England and America (the only other was Charles Knowlton who wrote shortly after Owen). Bentham's influence appears to have been crucial in motivating the writing of the others. All these men were feminists or had leanings in that direction. At least some of them explicitly connected birth control to feminist aims. That it was men, rather than women, who propagandized for birth control, may be explained by the greater difficulty women would have had in writing on a tabooed subject.

The connection between feminism and birth control propaganda in this period has been previously overlooked in part because writers advocating birth control were all men, in part because some of these writers were at pains to omit or downplay feminism in their arguments for birth control and in part because the history of the connections between feminism and birth control is just beginning to be explored.

Feminists, then as now, had an interest in making birth control available. In the late 18th and early 19th century, women may have been able to act privately, within their families, to limit births. But they needed the help of sympathetic men to publicize the cause and the methods of birth control. The men who advocated birth control in this period were men who were sensitive to the needs of women, who were avowed feminists or at least sympathetic to feminist causes.

Nella Fermi Weiner, Senior Teacher,
University of Chicago, Laboratory Schools, Chicago, Ill 60637

NOTES

(1) Thomas Robert Malthus, *An Essay on the Principle of Population as it Affects the Future Improvement of Society, With Remarks on the Speculations of Mr. Godwin, M. Condorcet, and other Writers.* Also D.V. Glass, Ed., *Introduction to Malthus* (London: Watts & Co., 1953); and Norman E. Himes, "Introduction and Appendix A," Francis Place, *Illustrations and Proofs of the Principle of Population* (New York: Augustus M. Kelly, 1967) pp. 7-63, 283-298.
(2) William Godwin, *An Enquiry Concerning Political Justice, and its Influence on General Virtue and Happiness* (London, 1973).
(3) Norman E. Himes, *Medical History of Contraception,* (1936; rpt. New York: Schocken Books, 1970), p. 211; and Norman E. Himes, "Jeremy Bentham and the Genesis of English Neo-Malthusianism,"

427

Economic History, 3, 11 (1936), 268-270. The *Medical History of Contraception*, originally published in 1936, is still the basic work in its field. It is impressive in its scholarship, the more so because sources are hidden and difficult to find. Himes complains of "the almost total lack of reliable secondary sources" and of the locked cases and uncatalogued material on sexual matters which were common in libraries in the thirties. In spite of these difficulties, Himes was able to collect an enormous amount of information on the history of contraception from earliest times to his own. He includes contraceptive practices in Eastern cultures and in twentieth century pre-literate societies. Approximately half the book in devoted to developments in England and America since 1800.

(4) Harriet Martineau, *Autobiography*, ed. Maria Weston Chapman (Boston: James R. Osgood, 1877), pp. 151-2.

(5) Donald J. Bogue, *Principles of Demography* (New York: Wiley, 1969), pp. 129-33.

(6) Benjamin Franklin, "Observations Concerning the Increase of Mankind and the Peopling of Countries" (1751), *Works of Benjamin Franklin*, ed. Jared Sparks (Boston, 1836), 2, pp. 311-321; and Norman E. Himes, "Benjamin Franklin on Population: A Re-Examination with Special Reference to the Influence of Franklin on Francis Place," *Economic History*, 3, 12 (1937), 438-40.

(7) Ansley J. Coale and Melvin Zelnik, *New Estimates of Fertility and Population in the United States* (Princeton: Princeton University Press, 1963), p. 36. Figures are for whites only.

(8) Nancy F. Cott, *The Bonds of Womanhood: "Woman's Sphere" in New England, 1780-1835* (New Haven: Yale University Press, 1977). See especially the concluding chapter. The quote is from page 199. My brief summary does not do justice to Cott's sophisticated argument. See also Daniel Scott Smith, "Family Limitation, Sexual Control, and Domestic Feminism in Victoria America.. in *Clio's Consciousness Raised, New Perspectives on the History of Women*, ed. Mary S. Hartman and Lois Banner (New York: Harper Torchbooks, 1974), pp. 119-136 for more on birth control and women's power within the family. See Carroll Smith-Rosenberg, "The Female World of Love and Ritual: Relations Between Women in Nineteenth-Century America," *Signs*, 1 (Autumn 1975), 1-30 for a discussion on friendships between women.

(9) James Reed, *From Private Vice to Public Virtue: The Birth Control Movement and American Society Since 1830* (New York: Basic Books, 1978), pp. 4-5.

(10) Linda Gordon, *Woman's Body, Woman's Right: A Social History of Birth Control in America* (New York: Grossman Publishers, 1976).

(11) Etienne Gautier and Louis Henry, "The Population of Crulai, a Norman Parish" *Popular Attitudes Toward Birth Control in Pre-Industrial France and England*, ed. Orest Ranum and Patricia Ranum (New York: Harper Torchbooks, 1972), pp. 45-52. Also Etienne van de Walle, *The Female Population of France in the Nineteenth Century* (Princeton: Princeton University Press, 1974), p. 7.

(12) E.A. Wrigley, "Family Limitation in Pre-Industrial England" in Ranum and Ranum, pp. 53-99.

(13) Robert V. Wells, "Family Size and Fertility Control in Eighteenth-Century America: A Study of Quaker Families," *Population Studies* (1971), 73-82.

(14) Himes, *Medical History of Contraception*, pp. 63-4, 72-3, 183-4, 187-206.

(15) Reed, pp. 5-6.

(16) Gordon, p. 45.

(17) Ibid., pp. 49-60.

(18) Himes, *Medical History of Contraception*, p. 187.

(19) Ibid., pp. 342-6.

(20) Marie Carmichael Stopes, *Contraception: Its Theory History and Practice* (1923; rpt. London: John Bale, Sons & Danielsson, Ltd., 1928), pp. 138-143; and Alan F. Guttmacher and the Editors of Consumer Reports, *The Consumers Union Report on Family Planning*, pp. 36-7, 68-9.

(21) C.K. Ogden, *Jeremy Bentham* (London: Kegan Paul, Trench, Trubner & Co. Ltd., 1932), pp. 18-20; and *Autobiographical Recollections of Sr. John Bowring* (London: Henry S. King & Co., 1877), p. 339.

428

(22) Mary P. Mack, *Jeremy Bentham: An Odyssey of Ideas* (New York: Columbia University Press, 1963), pp. 408-10.

(23) Arthur Young, *Annals of Agriculture* (1797), 29, 167 and (1797), 393-426. The quote is from pp. 422-23.

(24) Himes, "Jeremy Bentham"; and *Medical History of Contraception*, p. 194. See also Peter Fryer, *The Birth Controllers* (London: Secker and Warburg, 1965), pp. 33, 67-72, 90.

(25) Himes, *Medical History of Contraception*, pp. 194, 211; Fryer, p. 32, 69.

(26) Mack, pp. 323, 416, 451-452.

(27) Francis Place, *Autobiography of Francis Place*, ed. Mary Thale (Cambridge: Cambridge University Press, 1972, not previously published), p. 298.

(28) Graham Wallas, *The Life of Francis Place, 1771-1854*, (1898, Fourth Edition 1925; rpt. London: George Allen & Unwin Ltd., 1951), p. 5.

(29) Quoted in James Aldred Field, *Essays on Population and Other Papers*, ed. Helen Fisher. Hohman (Chicago: University of Chicago Press, 1931), p. 111.

(30) Wallas, pp. 10-18.

(31) Place, *Autobiography*, pp. 115-6.

(32) Ibid., p. 117.

(33) Ibid., p. 119.

(34) Wallas, p. 31.

(35) Ibid., pp. 65-79.

(36) Ibid., quoted on p. 1.

(37) Himes, "Jeremy Bentham," p. 275.

(38) Francis Place, *Illustrations and Proofs of the Principle of Population*, ed. Norman E. Himes (1822, First Himes edition 1930; rpt. New York: August M. Kelley Publishers, 1967), p. 165.

(39) Ibid., pp. 165-6.

(40) Place, *Population*, Himes' "Introduction," pp. 42-3.

(41) Fryer, pp. 43-46. See also *Medical History of Contraception*; and Field.

(42) *Medical History of Contraception*, the version quoted is pp. 216-7. All three handbills are reproduced and discussed, pp. 212-20.

(43) Place, *Population*, Appendix B, pp. 324-5.

(44) Quoted in Wallas, pp. 74-76.

(45) Place, *Autobiography*, pp. 100, 255.

(46) Ibid., pp. 266-267.

(47) G.D.H. Cole, *Richard Carlile: 1790-1843* (London: Victor Gollancz Ltd. and the Fabian Society, 1943), pp. 4-7; George Jacob Holyoke, *The Life and Character of Richard Carlile* (London: 1949), p. 9-13; and Fryer, p. 66.

(48) Holyoke, pp. 9-10; and Guy A. Aldred, *Richard Carlile, Agitator: His Life and Times* (London: The Pioneer Press, 1923), pp. 152-155.

(49) Richard Carlile, *Every Woman's Book: or What is Love?* (4th ed. London, R. Carlile, 1826), p. 7.

(50) Ibid., p. 9.

(51) Ibid., p. 15.

(52) Himes, *Medical History of Contraception*, p. 222.

(53) Richard William Leopold, *Robert Dale Owen, A Biography* (Cambridge: Harvard University Press, 1940), pp. 59, 76, 110-11.

(54) E.T. James, J.W. James and P.S. Boyer, ed., *Notable American Women 1607-1950* (Cambridge: The Belknap Press, 1971), III, 675-676; Margaret Lane, *Frances Wright and the 'Great Experiment'* (Manchester: Manchester University Press, 1972), pp. 5, 14-15; and A.J.G. Perkins and Teresa Wolfson, *Frances Wright: Free Enquirer* (New York: Harper & Brothers, 1939), pp. 58-60.

(55) *Notable American Women*, III, 677-678.

429

(56) Norman E. Himes, "Robert Dale Owen, the Pioneer of American Neo-Malthusianism," *American Journal of Sociology*, 35 (1930), 529-547, see p. 539n.

(57) Robert Dale Owen, *Moral Physiology: Or a Brief and Plain Treatise on the Population Question* (1830; rpt. New York: B. Vale, 1858).

(58) Himes, "Robert Dale Owen," p. 531.

(59) Robert Dale Owen, "A German Baron and English Reformers," *Atlantic Monthly*, 31, 188 (1873), 740-741, as reported by Fryer, p. 90.

(60) Owen, *Moral Physiology*; Charles Knowlton, *The Fruits of Philosophy, or the Private Companion of Young Married People* (Original "By a Physician," New York, 1832; rpt. London: Freethought Publishing Co., 1876); and *Medical History of Contraception*, pp. 201-202, 225.

(61) Leopold, p. 76.

Contraception and the Working Classes: The Social Ideology of the English Birth Control Movement in its Early Years

ANGUS McLAREN
Oxford University

A host of social movements which had as their goal the improvement of the living conditions of the working classes emerged in England in the 1820s and 1830s. Owenism and Chartism come first to mind, but historians have recently acknowledged the social significance of a number of less well-known groups that proclaimed the benefits of temperance or mechanics' institutes or phrenology or infidel missions. The birth control movement in its early years has as yet received little attention from the historians of the English working classes.[1] A possible reason is that the opposition of the 'pauper press' to the movement has led later observers to adopt the view that it was simply a middle-class Malthusian crusade which set out to convince the poor that the only escape from poverty lay in individual self-help. In what follows I shall sketch out the general lines of argument advanced by the advocates of birth control and their antagonists in the working-class movement. The purpose of the paper is not to provide yet another history of the first neo-Malthusians, but to use the arguments their activities elicited to gain a better understanding of nineteenth-century working-class culture.[2]

Much of the hostility of nineteenth-century reformers to any discussion of the population question stemmed from the fact that it was Thomas Malthus who first informed the Anglo-Saxon world of the dangers of

---

[1] For a brief account by a sociologist see John Peel, 'Birth Control and the Working Class Movement', *Society for the Study of Labour History Bulletin*, 7 (1963), 16–22.

[2] For the history of birth control, see Norman E. Himes, *A Medical History of Contraception* (Baltimore, 1936); James Alfred Field, *Essays on Population* (Chicago, 1931); J. A. Banks, *Prosperity and Parenthood: A Study of Family Planning Among the Victorian Middle Classes* (London, 1954); F. H. Amphlett Micklewright, 'The Rise and Fall of English Neo-Malthusianism', *Population Studies*, 15 (1961–1962), 32–51.

over-population. His name became inextricably associated in Englishmen's minds with the advocacy of every form of family planning, and to appreciate the originality of the early proponents of birth control it is essential that his position be clearly understood.[3] Malthus undertook his investigations with the purpose of destroying William Godwin's argument in favor of a sharing out of property as advanced in *An Enquiry Concerning Political Justice* (London, 1793). The result of such a utopian scheme, declared Malthus, would be that the population would soon overshoot the food supply and be overwhelmed by vice and misery. For his contemporaries his most striking revelation, however, was not that food supplies were increasing in arithmetic and population in geometric progressions, but that the major cause of existing poverty was the 'reckless over-breeding of the poor'. The upper classes, slowly having to relinquish traditional arguments for the maintenance of their privileged status, could now fall back on a modern, scientific justification. Malthus had proven to their satisfaction that reform was futile.

A man who is born into a world already possessed if he cannot get subsistence from his parents on whom he has a just demand, and if the society do not want his labour, has no claim of right to the smallest portion of food, and, in fact, has no business to be where he is. At Nature's mighty feast there is no vacant cover for him. She tells him to be gone, and will quickly execute her orders.[4]

The poor could save their children from such a fate only by practising 'moral restraint'. Now it is important to recall that Malthus' famous plea for 'moral restraint' consisted of convincing the poor to postpone marriage as long as possible. Many of his followers went a step further and called on the lower classes to show as much foresight *after* marriage as before by practising sexual abstinence. The main point, however, is that Malthus was completely opposed to birth control defined as the use of techniques to prevent conception during intercourse. In the first edition of *An Essay on the Principle of Population* he specifically attacked Condorcet for alluding to such stratagems.

He then proceeds to remove the difficulty of over-population in a manner, which I profess not to understand . . . he alludes, either to a promiscuous concubinage, which would prevent breeding or to something else as unnatural. To remove the difficulty in this way, will, surely, in the opinion of most men, be, to destroy that virtue, and purity of manners, which the advocates of equality, and of the perfectibility of man, profess to be the end and object of their views.[5]

---

[3] For the Malthusian debate, see D. V. Glass, ed., *Introduction to Malthus* (London, 1953); H. A. Boner, *Hungry Generations: The Nineteenth Century Case Against Malthusianism* (New York: 1955); D. E. C. Eversley, *Social Theories of Fertility and the Malthusian Debate* (Oxford, 1959); G. T. Griffith, *Population Problems in the Age of Malthus* (London, 1967).

[4] Thomas Malthus, *An Essay on the Principle of Population* (London, 1803), p. 531.

[5] Malthus, *An Essay*, p. 154.

All Malthus could offer the poor were counsels of prudence.

Malthus could not accept contraception because he considered it unnatural and unchristian. The utilitarians adopted his theory of population growth but were not prevented by either religious qualms or by any faith in the capability of the married poor to control family size by 'natural' means from accepting the logical necessity of contraception. Their apologies for the practice were restricted, however, to allusive asides. Jeremy Bentham referred in 'Situation and Relief of the Poor' (1797) to the use of 'sponges' to keep down the poor rate.[6] James Mill, writing on the topic 'Colony' in the *Supplement to the Encyclopedia Britannica* (1824) hinted that other means than emigration could be found to maintain a balanced population if 'the superstitions of the nursery were discarded'.[7] Middle-class radicals were not prepared to sacrifice their reputations by speaking more openly. Had they done so it is still unlikely that they would have won over many members of the working class, suspicious as they were of the utilitarians' motives. In the person of Francis Place, the 'radical tailor of Charing Cross', the Benthamites did have, however, a go-between who could carry out a campaign of propaganda in favor of contraception designed to win the attention of the artisan.[8]

In 1823 Place had distributed in working-class districts in London and the midlands unsigned handbills which explained in detail the use of the 'sponge' and withdrawal methods of contraception.[9] At the same time he sought the support of the well-known champion of the 'pauper press', Richard Carlile. Carlile was the first man in England to put his name to a work devoted to the subject of birth control: *Every Woman's Book; or What is Love? Containing Most Important Instructions for the Prudent Regulation of the Principle of Love and the Number of the Family* (1826).[10] Carlile described three methods of birth prevention: the woman's use of the 'sponge', the man's use of the *baudruche* or 'glove', and partial or complete withdrawal. The cautious approval of Carlile's book which was voiced by Robert Dale Owen in the American paper *New Harmony Gazette* brought down on Owen the charge of aiding in

---

[6] *Annals of Agriculture and Other Useful Arts*, 29 (1797), 422.

[7] *Supplement to the Encyclopedia Britannica* (Edinburgh, 1824), III, 261.

[8] On Place's role see the notes and introduction of Norman E. Himes to Francis Place, *Illustrations and Proofs of the Principle of Population* (London, 1830).

[9] G. J. Holyoake, *Sixty Years of an Agitator's Life* (London, 1900), I, pp. 126ff; Norman E. Himes, 'The Birth Control Handbills of 1823', *Lancet*, N. S. 3 (August 6, 1927), 313–7; see also the documents conserved at the British Museum in the Place Collection, 68.

[10] *What is Love?* first appeared in Carlile's *Republican* (May 6, 1825) and was brought out as a pamphlet in February 1826. Eight editions were produced by 1828. Carlile's imprisoned shopmen supported his argument in the *Newgate Monthly Magazine*. See Carlile to Place, (August 8, 1822), Place to Carlile, (August 17, 1822, September 1, 1824), Place Collection 68; Place to Carlile, *Republican*, 10 (November 12, 1824), 581.

the propagation of immoral doctrines.[11] Owen was in fact distressed by the style, if not the content, of Carlile's work and by his unfounded assertion that Robert Owen had introduced French birth control techniques to Britain. To make his own position clear, the younger Owen published *Moral Physiology: or a Brief and Plain Treatise on the Population Question* (New York, 1830; London, 1832). After devoting sixty-five of the pamphlet's seventy-two pages to a justification of the morality of contraception, Owen gave an account of the three methods recommended by Carlile but warned against the dangers of incomplete withdrawal. Charles Knowlton's *Fruits of Philosophy or, the Private Companion of Young Married People* which appeared in New York in 1832 and was reprinted in London in 1834 approached the question in a more clinical fashion than the earlier works. Knowlton's main contribution to the discussion was to cast doubts on the effectiveness of the sponge if it were not accompanied with a saline douching solution.[12]

With the utilitarian paternity of the birth control movement having been noted, the question inevitably posed is, was it, as its detractors in the popular press contended, simply a refined version of Malthusianism, still another means of forcing the poor to sacrifice themselves for the maintenance of the rich? What was the general social ideology propounded by the advocates of birth control and how was it received by the spokesmen of the working-class movement?

The first impression that one receives from a perusal of the early birth control literature is that all its disseminators were very much aware of the need to respond to the challenge that they were not the true friends of the working class, that by seeking to limit the laboring population they were serving the interests of the propertied. They saw as their first task the necessity of disassociating themselves from Malthus. Their arguments on population growth were, they admitted, based on his calculations but they held that their opinion of the ability of the working class to improve its situation was diametrically opposed to Malthus' pessimistic prophecies. He had no understanding, stated Place, of the workers' ceaseless attempts at self-improvement.

He can know but little of the shifts continually made to preserve a decent appearance. Of the privations endured, of the pains and sorrows which the working people suffer in private, of the truly wonderful efforts long continued, even in the most hopeless circumstances, which vast numbers of them make "to keep their heads above water."[13]

Malthus had sentenced the poor to a life of abstinence, at best, or of vice and misery, at worst. His call for late marriages, argued the birth

[11] *The Free Enquirer* (August 7, October 16, October 23, 1830).
[12] Norman E. Himes, 'Charles Knowlton's Revolutionary Influence on the English Birth Rate', *New England Journal of Medicine*, 199 (1928), 461–5.
[13] Place, *Illustrations and Proofs*, p. 155.

controllers, would, if acted upon, result in increased prostitution and im-morality.[14] His plea for the ill-fed and uneducated to practice a 'moral restraint' not demonstrated by their 'betters' was strangely hypocritical.

The proponents of birth control asserted that they were responding to a situation to which Malthus had turned a blind eye. The working class was already using injurious means to attempt to regulate births and was thereby making known its desire for reliable methods. To supply such knowledge would result in the elimination of abortion and infanticide.[15] It would spare women repeated unwanted pregnancies, attendant ill-nesses, and the dangers of miscarriage. Marriages would no longer have to be postponed for financial reasons, and as a result masturba-tion, prostitution, and similar substitutes for legitimate intercourse would disappear. The married would be more faithful and youth more chaste. Women would no longer be viewed as simple propagators of the species, but as men's helpmates. The plight of orphans, bastards, and women dying in labor or during induced abortions could, claimed the birth controllers, all be avoided.[16]

The advocates of birth control thus offered the lower classes a means of avoiding Malthus' options of either abstinence or misery. But did their adoption of a new solution to the population problem mean that they nevertheless agreed with liberal economists that the poor were the authors of their own distress? The proponents of birth control were obviously sensitive to the charge that they offered only short term palliatives. They responded by asserting that in favoring immediate improvements they were not seeking to postpone more fundamental forms of social change but providing for its advancement. 'Let us do the good we can do', wrote Carlile, 'and still pursue, and thus strengthen ourselves in the pursuit of that which is remote'.[17] A contributor to the *Newgate Monthly Magazine*, published by Carlile's shopmen, declared, 'That man cannot be a friend of the working classes who knowing an evil which affects them fails to point it out . . . .'[18] It was over-population, contended the birth control-lers, that led to extremes of wealth and poverty, of social and political inequality. If the working-class family was of moderate size it could win not only a more respectable standard of living but more time for self-improvement: '. . . all might be comfortably circumstanced and leisure and means for acquiring knowledge brought within their reach'.[19] The

[14] Knowlton, *Fruits of Philosophy* (London, 1841), pp. 7–8; Owen, *Moral Physi-ology* (New York, 1831), p. 25.

[15] Carlile, *Every Woman's Book* (London, 1838), p. 19; Owen, *Moral Physiology*, p. 36; Knowlton, *Fruits of Philosophy*, p. 40.

[16] Carlile, *Every Woman's Book*, pp. 23–24; Knowlton, *Fruits of Philosophy*, p. 37; Owen, *Moral Physiology*, p. 35.

[17] *The Lion* (October 3, 1828).

[18] *Newgate Monthly Magazine* (June 1, 1826).

[19] *Newgate Monthly Magazine* (January 1, 1826).

healthy, educated worker would prove more successful in struggling for social reforms. No debased or ignorant people, wrote Place in reference to Ireland, had ever overthrown a tyranny.[20] The advocates of birth control were, in short, sketching out a theory that the rising expectations of the lower classes would be a force for, not against, social and political change. This was a remarkably modern view of politics, but one that owed its elaboration chiefly to the need to respond to the discrediting assertion that the birth controllers were defenders of the *status quo*.

That reformers should have so viewed the believers in birth control was due to the fact that the latter, after having declared at length their dissatisfaction with existing society, tended to fall back on some variant of the wage fund theory. Accepting the premise that the amount of money available for wages was limited, they argued that workers could avoid exploitation only by limiting the supply of labor. Owen was the most cautious in spelling out the economic benefits of birth control; he described it as an 'alleviation', not a 'cure'.

It is true, and ought to be remembered, that the check I propose, by diminishing the number of laborers, will render labor more scarce and consequently of higher value in the market; and in this view, its political importance is considerable: but it may also be doubted whether our present overgrown system of commercial competition be not hurrying the laborer towards the lowest rate of wages, capable of sustaining life, too rapidly to be overtaken, except in individual cases, even by a prudential check to population.[21]

Place and Carlile were more sanguine. The effectiveness of unions, they contended, was restricted to opposing repressive legislation that interferred with the free movement of labor; they could not drive wages up beyond a level determined by the forces of supply and demand.[22] If the lower classes restricted the numbers entering the labor pool, however, they could command higher compensation. Working men would win sufficient wages for the support and comfort of their families; no longer would their women be forced into prostitution and their children into factory work.

Contraception was, of course, not to be forced on the worker, but the knowledge of such an option had to be available. No truth was dangerous, declared Owen; one was free to act only after having been informed of the choice.[23] Moreover it was frequently stated that only a limited degree

[20] Place Collection, 68.

[21] Owen, *Moral Physiology* (London, 1832, 8th edition), p. 25. This section on the wage fund was not included in earlier editions.

[22] On Place's laissez-faire views see Place to George Rogers (January 15, 1832), Place Collection, 68; Place to the *London Dispatch* (January 29, 1837); W. E. S. Thomas, 'Francis Place and Working Class History', *Historical Journal*, 5 (1962), 61–70; Brian Harrison, 'Two Roads to Social Reform: Francis Place and the Drunken Committee of 1834', *Historical Journal*, 11 (1968), 272–300.

[23] Owen, *Moral Physiology*, 19.

of family limitation was necessary to attain the economic advantages posited by the birth controllers. William Longson, a correspondent of Place, estimated that the prevention of only one or two births per hundred would make all the difference in the living standards of the poor.[24] A letter by 'Amicus' in the *Republican* included the assurance that Carlile did not seek to have the population 'unnecessarily' restricted.[25] Knowlton suggested that contraception would be used, not so much to restrict population growth *per se* as to permit it to take place in an orderly fashion. 'In my opinion, the effect would be a good many more families (and on the whole as many births), but not so many overgrown and poverty stricken ones'.[26]

To appease the suspicions of those artisans who persisted in regarding any discussion of the population question as a ploy of the propertied classes, several of the proponents of birth control claimed Robert Owen as a comrade-in-arms. Place stated in a letter to the *Labourer's Friend* of August 5, 1823 that Owen was preoccupied by the danger over-population could pose to his socialist society and had visited France for the purpose of informing himself on methods of contraception. The claim that Owen had brought back to England such information was repeated by Carlile.[27] The *Newgate Monthly Magazine* went so far as to assert that Owen's 'Society of Harmonists' in America was utilizing certain contraceptive techniques.[28] Robert Owen was to deny these claims, but the fact that his son produced *Moral Physiology* helped to buttress the argument that there was no necessary contradiction between defending contraception and struggling for a reformed society. Those arguing for birth control, of course, insisted that the former should enjoy top priority. Robert Dale Owen stated that even the most perfect society could not function if over-populated. The *Newgate Monthly Magazine* followed the same line in declaring the knowledge of birth control as essential to good government: 'With this knowledge the co-operative communities may be prosperous, but without it they will not only devour themselves but augment an evil the consequences of which are even now grievous and lamentable'.[29]

[24] William Longson, 'Letter I: On Population and Their Wages; Addressed to the Labouring Classes by an Operative Weaver' (September 15, 1824), Place Collection, 68.

[25] *Republican* (May 5, 1826).

[26] Knowlton, *Fruits of Philosophy*, p. 40; see also Knowlton to the *Boston Investigator* (January 11, 1833).

[27] Place to the *Labourer's Friend* (August 5, 1823).

[28] *Newgate Monthly Magazine* (January 1, 1826), and for the continuation of the myth, see the *Black Dwarf* (September 24, 1823; October 1, 1823); Holyoake, *Sixty Years*, I, pp. 126ff; J. F. C. Harrison, *Robert Owen and the Owenites in Britain and America: The Quest for the New Moral World* (London, 1969), p. 61. Owen denied the charge in a letter to the *Morning Chronicle* (October 8, 1827).

[29] *Newgate Monthly Magazine* (June 1, 1826). William Thompson foresaw the possible use of some 'preventive check' in his reformed community but his sugges-

The advocates of birth control concluded their argument that their activities were not part of a veiled attack upon the working class by the agents of the propertied by pointing to the hostility with which their pronouncements were met by the wealthy. Though they emphasized that an immediate improvement in the living conditions of the poor depended more on self-help than on class struggle, they did suggest that the knowledge of contraception had been kept secret by some sort of class conspiracy. The theme that the lower classes should take advantage of methods heretofore monopolized by the wealthy was repeatedly sounded. 'The remedy has long been known in this country', wrote Carlile, 'and to the aristocracy in particular, who are always in search of benefits which they can particularly hold, and be distinct from the body of the labouring people'.[30] 'Do as other people do', wrote Place, 'to avoid having more children than they wish to have and cannot easily maintain. What is done is this . . . .'[31] Similarly, it is not unlikely that Place's handbill, 'To the Married of Both Sexes in the Genteel Life', was so entitled to attract the curious artisan's eye.

The proponents of birth control were appealing to suspicion of one's 'betters', not to class solidarity, but many of their pronouncements, though lacking in political sophistication, were obviously gauged to play on traditional hostilities. A defender of Knowlton attributed his persecution to the upper classes' fear of an enlightened work force: 'He has made the knowledge too cheap and that is not the worse of it, he has permitted common people, people who can be benefitted by the knowledge to have access to it'.[32] Carlile, in response to attacks on his work, retorted: 'The only natural enemies that my book should find, are among the Royal families, the Aristocracy, and the Priests, and that only because it is calculated to elevate mankind above the injuries of their political and religious machinations'.[33] Even their opposition was mere cant; they already used the methods revealed to the poor in *Every Woman's Book:* '. . . they would hold it as a luxury too good for the poor and too dangerous to themselves politically in its enlightenment'.[34]

In parading his animosity towards the 'priests', Carlile was touching

tion, because of its hypothetical nature, seems to have stimulated little interest: *An Enquiry Into the Distribution of Wealth* (London, 1824), pp. 536, 547–9; *Practical Directions for the Speedy and Economical Establishment of Communities* (London, 1830), pp. 229–48.

[30] Carlile, *Every Woman's Book*, pp. 25–26; cf. Owen, *Moral Physiology*, pp. 37–38.

[31] Place, 'To the Married of Both Sexes of the Working People', Place Collection, 68.

[32] *Boston Investigator* (February 15, 1833); Knowlton stated that this was why he had been jailed '. . . while other works of like purpose, as well as that dirty, useless thing called "Aristotle," are publically sold with impunity', *Boston Investigator* (January 11, 1833).

[33] *The Lion* (October 3, 1828).

[34] *Ibid.*

on a theme in the birth controllers' ideology that has been often ignored, the attempt to turn popular anticlericalism to their own purposes. They not only adopted a utilitarian, materialistic approach to morality, they waged an active campaign against the forces of revealed religion. Robert Dale Owen was relatively cautious in his dealings with the clergy; Carlile and Knowlton went out of their way to attack the Church, the former interspersing his examination of contraceptive practices with suggestive asides on the sexual proclivities of nuns and the phallic symbolism of the cross. Those favoring birth control were in turn lumped together as propagators of '. . . a system combining blasphemy, atheism, infidelity, adultery, lewdness, removing all moral and religious and legal checks upon human depravity, and leading to a community of property and striking directly at the foundation of civil society'.[35] Carlile's main pre-occupation in the 1820s was in fact his crusade against priestcraft. He was attracted to birth control at least in part because it was the most direct means of attacking the moral hegemony of the Church.[36] In dedicating Volume eleven to the *Republican* 'To Women' Carlile wrote,

. . . you cannot be dutiful wives and mothers while you are religious. . . . The religious mind must be in a certain ratio destitute of social and domestic affections, destitute of a degree of useful knowledge that might have been otherwise acquired, destitute of health that might have otherwise been studied and acquired, and estranged from all that is really good and solid in human happiness.[37]

Evil, he declared in the *Lion* of October 10, 1828, was caused by '. . . the foul pretence of future-life-pleasures' which permitted sorrows such as unwanted children to be stoically accepted as part of God's plan.[38] Knowlton was of similar opinions and insisted that it was his infidelity which provided the basis for his birth control arguments that was the real source of his persecution.[39]

If one were to advocate contraception in the nineteenth century, there was no alternative but to appeal to a morality other than that offered by the Christian church. Freethinkers were thus attracted to the movement because they recognized that to teach the artisan class that

---

[35] S. D. Parker, *Report on the Argument of the Attorney of the Commonwealth at the Trial of Abner Kneeland for Blasphemy* (Boston, 1834), p. 89.

[36] G. Alfred, *Richard Carlile, Agitator: His Life and Times* (London, 1923). No adequate biography of Carlile has yet been written.

[37] *Republican* (January 1, 1825), preface. See also the writings of Eliza Sharples Carlile in the *Isis* of 1832.

[38] *The Lion* (October 10, 1828). Place spoke of the need for overcoming the 'squeamishness' induced by traditional religions; Place to *Morning Chronicle* (August 30, 1825).

[39] Charles Knowlton, *A History of Recent Excitement in Ashfield* (Boston, 1833); *Two Remarkable Lectures* (Boston, 1833); *Elements of Modern Materialism* (Adams, Mass., 1829); for Knowlton's lectures in Frances Wright's Hall of Science, see the *Correspondent* (April 11, June 6, 1829) and the *Free Enquirer* (May 6, June 10, July 1, July 22, 1829).

it could control the very creation of life was the most dramatic and intimate manner in which to undermine Christian faith. Supporters of birth control who on the other hand sought to refute the charge that they were defenders of the *status quo* could flaunt their infidelity as a token of their revolutionary fervor.

What was the reaction of the working class movement to the arguments made by those favoring birth control? The claims of the latter that their only enemies were the reactionary elements of English society did not dissuade the working-class press from expressing its hostility to the diffusion of contraceptive literature. Despite the efforts of the birth controllers to disassociate themselves from Malthus, they were invariably portrayed as his creatures. Cobbett described Carlile as an

... instrument in the hands of others. ... He is a tool, a poor half-mad tool, of *the enemies of reform.* He wants no reform, for the end of his abominable book, is, to show, that the sufferings of the people do not *arise from the want of reform;* but from the *'indiscreet breeding' of the women!*[40]

'A retailer of the wisdom of Francis Place' and 'a bought tool of the Malthusian party' were the labels used by the *London Mercury* to damn writers sympathetic to birth control.[41]

The main reason for the opposition of the 'pauper press' to the discussion of contraception was its fear that it would be seen to be tacitly accepting the utilitarian argument that unemployment was a 'natural' problem to be overcome by restricting the labor pool. Malthusians and advocates of birth control were tarred with the same brush as formulators of this anti-working class argument.

If Messrs Malthus, M'Culloch, Place & Co are to be believed, the working classes have only to consider how they can most effectively restrict their numbers, in order to arrive at a complete solution of all their difficulties . . . . Malthus & Co . . . would reduce the whole matter to a question between Mechanics and their sweethearts and wives rather than a question between the employed and their employers —between the Mechanic and the corn-grower and monopolist—between the taxpayer and the tax inflictor.[42]

The *Trades Newspaper* declared that the Malthusians' intention was to make marriage among the poor harmless to the rich.[43] The Brighton *Co-*

---

[40] *Weekly Register* (April 15, 1826); see also the issues of April 22 and September 30, 1826 and November 3, 1827. Carlile replied that he had not even read Malthus; *Republican* (August 21, 1826). See also 'Marriages and No Mothers; or the Stage Coach Battle of Cobbett and Carlile', *Paul Pry* (May 6, 1826), Place Collection, 61.

[41] *London Mercury* (April 2, May 28, 1837); Dr. Robert Black and Henry Hetherington were the men under attack.

[42] *Trades Newspaper* (July 31, 1825), cited by E. P. Thompson, *The Making of the English Working Class* (London, 1966), p. 777.

[43] *Trades Newspaper* (September 11, 1825). As early as 1805 Charles Hall had

operator noted that the spectre of over-population had been conjured up by the propertied classes; laborers knew instinctively that no problem would be posed once the 'consuming non-producers' were eliminated.[44] The *Labourer's Friend and Handicraft's Chronicle* described Malthus' propositions as 'monstrous';[45] Bronterre O'Brien thundered, '. . . in spite of the devil and Malthus the work people are resolved to live and breed'.[46]

The opponents of birth control did not say that population pressure would never pose a problem to society but they insisted that one served the forces of reaction in suggesting that it should be dealt with before social reform. Carlile had been of the same opinion when he first broached the question in 1822. 'I maintain . . . *that bad government and a priest-hood constitute the evil which at present degrade* [sic] *the people of this country. . . . I will never complain of too many human beings, whilst all these removable evils exist'.*[47] Carlile of course abandoned this position but the bulk of the working-class press continued to argue that the discussion of contraception only hindered the advancement of social reform; if there was a problem it could only be resolved in an equitable fashion after the establishment of good government. *The Black Dwarf* declared that one could not talk of lack of food as a sign of population pressure as long as one man expended that which could keep another hundred men alive.[48] *The Northern Star* stated that after the upper classes had eliminated every social inequality the regulation of population could be discussed.[49] Four hundred years would pass, *The London Dispatch* predicted, before it was a serious issue.[50]

Turning to the political ramifications of contraception *The Black Dwarf* asked what would be the consequences of a restricted labor pool.[51] The answer was that workers would still be forced to accept subsistence wages; to contend that they were able to bargain freely and so push up salaries was ludicrous. But would birth control be nevertheless a good thing if it in fact improved the health of the people? The editor of *The*

warned of plans abroad to use 'the preventive method' to restrict the poor—like cattle—to the numbers needed by the rich; *Effects of Civilization* (London, 1850 2nd edition), pp. 246–7.

[44] *The Co-operator* (September 1, 1829, October 1, 1829).

[45] *The Labourer's Friend and Handicraft's Chronicle* (January 1, 1821).

[46] Cited by Patricia Hollis, *The Pauper Press* (Oxford, 1970), 231. For O'Brien's attacks on the Malthusians, see the *London Mercury* (April 2, 9, 1837). For the argument that the employment of labor-saving machinery had caused the 'population problem', see the *Pioneer* (January 25, February 1, 1834) and the London *Co-operative Magazine* (April 1, May 1, 1827).

[47] *Republican* (November 12, 1824).

[48] *The Black Dwarf* (December 3, 1823).

[49] *The Northern Star* (March 31, 1838).

[50] *The London Dispatch* (January 29, 1838).

[51] The debate over birth control that appeared in the columns of *The Black Dwarf* from November 1823 to January 1824 was between its editor, Wooler, Francis Place writing as 'A. Z.' and the young John Stuart Mill writing as 'A. M.'

Black Dwarf replied that it would not if the effect of improved conditions was to postpone reforms.

The natural remedy for such a corrupt state of things, is the INCREASE of population, even to the extreme of pressure against the means of subsistence, for it is the nature of the multitude to bear with oppression and want, as long as their animal necessities will permit them; and it is only by reducing them to a state bordering on despair, that they will ever be induced to avenge their wrongs, or to claim their rights.[52]

This idea that the misery that attended population growth in an unreformed society was a harbinger of revolution had been earlier enunciated by Piercy Ravenstone in his attack on Malthus: 'The wretchedness of a people is a sure forerunner of revolt'.[53] In other words, the 'rising expectations' argument of the birth controllers was rejected and a *politique du pire* maintained; the revolutionary was to regard a deterioration of social conditions as the surest form for social change.

What of the suggestion made by those arguing for birth control that short term improvements could be attained by the artisan following the lead of his 'betters' in adopting contraception? This ran counter to both the political analysis of the reformers and the traditional view held by the working class of the degeneracy of the aristocracy. Outraged articles on the sexual habits of the wealthy were familiar features in the popular press.[54] Indications that contraceptive practices were percolating down the social structure to the lower classes were construed as a spread, not of enlightenment, but of debasement. 'Thus we may trace to our artificial right of property, by neither a long nor a circuitous route, that vanity,—that excessive love of expense, in all classes, which makes prostitutes of our women and fraudulent knaves of our men, and plunges all classes in vices and crimes'.[55]

Underlying the socioeconomic opposition of the working-class movement to propaganda favoring birth control was a hostility born of a traditional moralism. It was its reluctance to interfere with the workings of God, providence or nature that made the suggestions of the birth controllers appear even more shocking than those of Malthus. A strong element of anticlericalism ran through much of the writings of the Owenites and Chartists, but this manifested itself primarily in attacks on the Church, not on Christian morality. Indeed it was frequently stated by

[52] *The Black Dwarf* (September 17, 1823).
[53] Piercy Ravenstone, *A Few Doubts as to the Correctness of Some Opinions Generally Entertained on the Subject of Population and Political Economy* (London, 1821), p. 428.
[54] E.g., the *Working Man's Friend and Political Magazine* (January 5, 19, 1833); *The Examiner* (September 11, 1825).
[55] Thomas Hodgskin, *The Natural and Artificial Rights of Property Contrasted* (London, 1832), pp. 154–5.

the popular press that the object of overthrowing the tutelage of the clergy was to permit the artisan to return to the simple teachings of the primitive Church. On those infrequent occasions when the working-class press distinguished between the Malthusians and the proponents of birth control, the message of the latter was always characterized as the more immoral.

The supply and demand philosophers again divide themselves into two parties; one which is named after its leader, Mr. Malthus, and the other, which is without a name, because nobody has as yet been found so regardless of public opinion, as openly to couple his name with its almost nameless doctrines. The Malthusians content themselves with simply expounding the abstract doctrine, that when the working classes cannot subsist by what the employing classes choose to give them for their labor, they are necessarily too numerous, and have themselves to blame for being so. But the nameless take a wide stride beyond this; they do not choose, like the Malthusians, *to leave matters where they found them, and trust to Providence for a remedy;* they think it wise to go a shorter way to work, and try their own hand at mending the order of things; they have no idea of allowing Madame Nature to go on as they say she has been doing, in overpeopling the domains of capital, and are determined to place such checks upon her prolific propensities, that she shall produce just as many labourers, and no more, as are wanted to do the work of those only lords of creation (in the eyes of the Nameless), the holders of a thousand pounds a year and upwards.[56]

This defense of the workings of 'Madame Nature' and attacks on 'unnatural' acts found a place in all the arguments opposing birth control. *The Examiner* declared simply, 'It is unnatural to check the increase of population'.[57] *The Bulldog* likened those who followed such practices to catamites and prostitutes.[58] For the *Trades Newspaper* these acts were a 'damning impiety', unnatural manuevres analogous to, '. . . a mill going constantly, yet grinding nothing, except the tithe boll of the parson; a gardner setting out potatoes but first cutting all the eyes out of them; a hen hatching marbles; what a foil is to a sword, or glove to the hands of a boxer'.[59] *The Black Dwarf* insisted that it was not worried about 'the ordinary cant about violating the laws of nature', but then repeated the argument that contraception was immoral, both in itself and inasmuch as one would be led inevitably from 'prevention to destruction' of children.[60]

Godwin was responding to Malthus in *On Population* (London, 1820), but his portrayal of a beneficent nature was to be taken over by the opponents of birth control. 'She has not left it to the caprice of the human

[56] *Trades Newpaper* (September 11, 1825); cf. the *Bolton Express* (October 27, 1827).
[57] *The Examiner* (August 21, 1825).
[58] *The Bulldog* (September 9, 1826).
[59] *Trades Newspaper* (September 11, 1825).
[60] *The Black Dwarf* (December 3, 1823).

will, whether the noblest species of beings that She has planted on this earth, shall continue or not. She does not ask our aid to keep down the excess of human population'.[61]

As the spokesmen for a movement which they believed to be ordained by providence or nature as the natural development of human society, the writers of the working-class press treated as unacceptable the suggestion that the lower classes utilize 'artificial' means to regulate family size. Such a plan, in their eyes, implied that the birth controllers did not even have the Malthusians' confidence in the poor's prudence and self control. So intent were the writers of the 'pauper press' to respond to such supposed slurs that they were at times led to argue in favor of what was in everything but name Malthus' 'moral restraint'.

To marry, with no other prospect than want before you is to do a very wicked thing, for the same nature which dictates to you to marry dictates to you the duty of providing for your offspring. But beyond the necessity of making this provision we know of no restriction to which the state of marriage can or ought to be subjected.[62]

Birth control offered and symbolized a new individualism. It is not surprising that it should have been viewed with suspicion by men who stressed the citizen's responsibilities to nature and the community. The controversy over contraception revealed the extent to which the working-class movement shared with traditional moralists the belief that the family was the source of all public morality and the sacrifice of individual pleasures the basis of social cohesion.

. . . it is one of the clearest duties of a citizen to give birth to his like, and bring offspring to the state. Without this he is hardly a citizen: his children and his wife are pledges he gives to the public for his good behavior; they are his securities, that he will truly enter into the feeling of a common interest, and be desirous of perpetuating and increasing the immunities and prosperity of his country from generation to generation.[63]

Opponents of birth control concluded that the question was not whether the individual had a right to contraception, and in particular whether the woman had a right to avoid unwanted pregnancies; the question was whether self-interest was to prevail over loyalty to class and community.

The question still unanswered is what effect the opposition of the working-class press to birth control had on its readership. The present article has dealt, of course, only with the prescriptive literature produced by the advocates and opponents of contraception. It is far more difficult to determine what the artisan thought of the debate, but the small amount

[61] William Godwin, *On Population* (London, 1820), p. 219.
[62] *Trades Newspaper* (July 17, 1825).
[63] Godwin, *On Population*, pp. 585-6.

of evidence at hand suggests that the press was successful in fixing in working men's minds the idea that contraception was a highly individualistic act prompted by self-interest, an act unworthy of the artisan who had faith in either God or the forces of reform. A letter addressed to Francis Place provides a touching indication of the quandry in which some artisans found themselves.

Sir having read in the Republican your Advice for Regulating Family according to their Income By Means of Preventing Corception and as I Am situated as a Journeyman Shoe Maker I find it moor than I Can Do to Support My Present Family which Is myself wif & 4 Children under seven years of Age and I'm likely to have as many moor Should Life Remain, So I have taken the liberty of Applying to you in hopes you will Be So Kind as to Inform Me of all the means I Am to Apply to Prevent me from Being Launched further into Poverty By having a Larger Family and I should wish to git through without coming to a Workhouse Though I believe that Need not be the case under a Better system than what we live under. So By Such Information you will much oblige your Humble Servant

Benjamin Base.[64]

The case could also be made that the fact that there was no serious discussion of contraception following the appearance of the first birth control works in the late 1820s and early 1830s until Drysdale's in the 1850s was yet another indication of the success of the movement's opponents. But such a claim would be misleading. Additional books were not needed. The early works, as crude as they were, provided the nineteenth-century Englishman with all the information necessary for relatively effective contraception. Moreover, these works continued to circulate in large numbers. Witnesses before the Committee on the Factories Bill of 1832 and Dr. Loudon's Medical Committee of 1833 testified that 'certain books, the disgrace of the age' were circulating in working class areas.[65] In 1837 Dr. Michael Ryan asked his readers, 'Are not the most revolting vices now unblushingly recommended as checks to population? and are not the most immoral works circulated in almost every bye-street through which we pass'?[66] It has been estimated that from 700 to 1,000 copies of Knowlton's pamphlet were sold each year from 1834 to 1876 and that by 1877, 75,000 copies of Owen's were in print. The publicity of the Charles Bradlaugh—Annie Besant trial of 1877 resulted in the additional sale of 125–185,000 copies of Knowlton's *Fruits of Philosophy*.[67] Radical freethinkers such as James Watson (one

---

[64] Base to Place (June 21, 1825), Place Collection, 68; and see also 'P. H. G.' to the *Poor Man's Guardian* (March 9, 1833).

[65] 'Committee on the Factories' Bill', *Parliamentary Papers*, 15 (1831–32), 132, 545; 'Dr. Loudon's Medical Report', *Parliamentary Papers*, 21 (1833), 2, 15, 18.

[66] Dr. Michael Ryan, *Philosophy of Marriage* (London, 1837), p. 6.

[67] On circulation figures, see D. V. Glass, *Population Policies and Movements in Europe* (London, 1967, 2nd edition), pp. 31ff, and Himes, 'Charles Knowlton', 461–5.

of Carlile's shopmen), Henry Hetherington, John Cleave, Edward True-love, and Austin Holyoake distributed birth control tracts as part of their anticlerical campaigns.[68] In addition advertisements for these works were occasionally carried in papers ostensibly opposed to their doctrines. I have found, for example, publicity for Owen's *Moral Physiology* in the *Examiner*, the *Crisis*, the *Poor Man's Guardian*, the *Destructive*, and the *Working Man's Friend and Political Magazine*.

A glance at the advertisements for the birth control tracts that appeared in the working-class press serves to put the movement in context. These appeared on the same page along with ads for a vast variety of quack medicines: Dr. Henry's French Meroine Pills, Solomon's Guide to Health, Dr. Hallet's Pills Napolitaines, Dr. H.'s Golden Anti-Veneral Pills, Dr. Lambert's Cordial Balm of Life. In the same columns were announcements of the meetings of phrenological and temperance societies. Expanded publishing facilities were allowing any practitioner a hearing; new advertising schemes popularized revolutionary medicaments. The desire for a common man's medicine, preferably with a monistic message, resulted in an upsurge of health movements combatting alcoholism, gluttony, 'folly in dress', and a host of other ills.[69] The campaign for birth control was in many ways just one more self-help medical movement. Like the others, its motto was 'Know thyself'. The fact that the literature it produced provided the artisan with a new understanding of his body's functions and an accompanying sense of independence and self-control was as important as the actual practice of contraception.

Why did the working-class movement deny the birth controllers the sort of sympathetic hearing offered the sponsors of a host of health fads and self-help medical movements? It was not simply because the population theories of the birth controllers seemed, in the eyes of the writers of the popular press, to undermine their social theories. Equally important was the fact that they had not cast aside the moral preoccupations of their conservative contemporaries. I have said very little in this paper of the early working-class movement's views on women and the family, a subject which I will discuss in a later paper. Hence it can only be noted that few writers recognized that sexual and political repression were related; an even smaller number accepted the notion that the freedom of women to control their own bodies was an essential liberty.

[68] F. B. Smith, 'The Atheist Mission, 1840–1900', in Robert Robson, ed., *Ideas and Institutions of Victorian Britain: Essays in Honour of George Kitson Clark* (London, 1967), p. 220.
[69] On health fads, see Richard H. Shryock, *The Development of Modern Medicine* (London, 1948), 205ff; J. F. C. Harrison, *The Early Victorians, 1832–1851* (London, 1971), pp. 171ff; Brian Harrison, *Drink and the Victorians* (London, 1971).

DEMOGRAPHY◦ Volume 22, Number 2 May 1985

BIRTH SPACING AND FERTILITY LIMITATION: A BEHAVIORAL ANALYSIS OF A NINETEENTH CENTURY FRONTIER POPULATION

Douglas L. Anderton
  Department of Sociology, University of Chicago, Chicago, Illinois 60637
Lee L. Bean
  Department of Sociology, University of Utah, Salt Lake City, Utah 84112

Historical demographic research focusing on the transition from natural fertility to controlled fertility suggests two conclusions.[1] First, prior to the onset of the fertility transition populations are uniformly marked by natural fertility regimes. That is, populations may have varied markedly in terms of levels of fertility, but the pattern of marital-age-specific fertility schedules (Henry, 1961) and derivative indices (Coale and Trussell, 1974, 1978a, 1978b) confirm the absence of fertility limitation. Second, the initial adoption of fertility limitation occurs only after the selected number of children are born. Fertility limitation is thus indexed by declining age of last birth—truncation of the child bearing experience—and increasing last and, in some cases, next to last closed birth intervals. The adoption of fertility limitation among natural fertility populations is, therefore, parity dependent.

Although these conclusions are supported by a large body of literature, contrary evidence exists and questions persist. Carlsson (1966) has questioned whether natural fertility populations are homogeneous. He argues that prior to the onset of the fertility transition in Sweden, populations in some areas (counties) appear to have been limiting fertility. Extensive studies by Knodel (1977; 1979) and Wrigley (1966) using more refined measures and smaller units of analysis in Europe as well as data from Asia find no support for Carlsson's argument. Yet the suspicion remains that

subpopulations may have been limiting fertility during periods marked by natural fertility at an aggregate level. Wilson's detailed analysis of 14 parishes in pre-industrial England, 1600–1799, concludes: ". . . while the existence of family limitation in pre-industrial England cannot be ruled out, it is highly unlikely that it was of any significance in determining the overall patterns of marital fertility" (Wilson, 1984:240). Nevertheless he notes that failure to identify subpopulations which limited fertility may have been due to the fact that the data set had "few parishes . . . large enough for any disaggregation by period . . ." (Wilson, 1984:240).

Dupâquier and Lachiver (1969) recognize the possibility of rational spacing of childbearing among natural fertility and transitional populations and propose a contraceptive typology based upon birth intervals to test their proposition. Knodel (1981) evaluates the typology and rejects inferences drawn from it (see also Dupâquier and Lachiver, 1981). Tolnay and Guest (1984) investigate the possibility of birth spacing during early periods of increasing fertility limitation by focusing on U.S. data using the 1900 public use sample constructed from the manuscript census. They conclude that at the turn of the century in the United States there was no evidence of birth spacing. Yet the limitations of their data require the use of a methodology which is inadequate to identify spacing explicitly. Using a restricted sample and examining

DEMOGRAPHY, volume 22, number 2, May 1985

only the first three closed birth intervals (including the interval from marriage to first birth which is often short), they examine the proportion of women having births within 24 months of marriage or the previous birth (first or second). In contrast to the conclusions which they draw, the data presented by Tolnay and Guest demonstrate that after the short first interval (marriage to first birth), the majority of the couples studied do not have short birth intervals and that birth intervals increase progressively across all successive parities.[2] Moreover, David and Sanderson (1984), using in part the same data set but a different methodology, suggest couples may have used "precautionary spacing."

Although the available data and the methodologies used have not confirmed variations in spacing patterns among natural fertility populations and among populations adopting fertility limitation, they also have not laid the discussion to rest. As Knodel (1983:63) states in his analysis of contemporary and historical populations, "Although empirical situations of this type (spacing) have yet to be convincingly identified, they do remain a logical possibility."

Using nineteenth century birth cohorts in Utah, we have found that the adoption of family limitation practices among our population is evident in the dramatic decline in completed fertility levels from 8.8 children ever born in the 1845–1849 birth cohort to 5.5 in the 1890–1894 birth cohort (Bean et al., 1983). In this paper we focus upon three propositions elaborating the relative significance of spacing and stopping behavior during this fertility decline. First, within both natural fertility birth cohorts and cohorts following the onset of fertility limitation, we hypothesize that substantial groups of women with long interbirth intervals across the entire childbearing career can be identified. Specifically, we propose that birth intervals at all parities beyond the first will be inversely related to completed family size in both natural fertility and transition cohorts.

Truncation of childbearing is traditionally indicated by an increase of the last closed birth interval, presumably reflecting failed attempts to stop childbearing. Our second hypothesis is that rational spacing of births also plays a significant role in the onset of fertility limitation in the population, and this is reflected in a compositional shift of the population to patterns of birth spacing associated with smaller family sizes. Specifically, we hypothesize that patterns of birth intervals associated with a given completed family size do not substantially change across the transition, whereas the proportion of women adopting a career long pattern of birth intervals associated with smaller family sizes will increase across the transition.

Finally, we hypothesize that aggregate distributions of the population by length of birth interval changed during the fertility transition not only because of the shift of women to spacing schedules associated with smaller completed family sizes, but also, to a lesser extent, because of changing ages of marriage within the population.

The data base used to address these propositions consists of individual level data with sufficient numbers to allow disaggregation based on theoretical grounds rather than on traditional and arbitrary political or geographical units. Second, the data cover all childbearing across all years of exposure, and are detailed enough, therefore, to allow the calculation of refined measures of spacing rather than approximate indices (Dupâquier and Lachiver, 1969; Tolnay and Guest, 1984). The methodology follows Hollingsworth's (1969:22) dictum that ". . . to study the intervals between births, tabulations are, accordingly, required of births according to parity."

In summary, the issues addressed in this paper are not new, but they remain unresolved. Drawing upon a set of genealogies developed for the Mormon Historical Demography Project (Bean et al., 1978), we evaluate the argument that the transition from a natural fertility regime to controlled fertility is marked not only by parity dependent truncation, with the recognized impact on ultimate and penultimate birth intervals, but also by an increasing proportion of subcohorts adopting distinctive interbirth spacing patterns distributed across all parities.

DESCRIPTION OF THE DATA

In this section of the paper two issues are discussed. First, we describe the nature of the data employed and the constraints introduced in order to eliminate confounding influences for the analysis of birth spacing. Second, because we are dealing with a data base representing a specific U.S. historical population marked by distinctive fertility patterns, we compare data from our study population with other data sets to demonstrate similarities, and thereby to suggest that the conclusions drawn from this analysis may be generalizable to a wider set of populations.

The data used in this paper are derived from approximately 180,000 computerized Utah genealogies. Initially this project, which is part of a larger medical genetics research program, selected family group sheets if at least one individual in the family had experienced a birth or death on the Mormon pioneer trail or in Utah. The recorded data include birth date and place, marriage date and place, and death date and place for each spouse, together with the names of parents of each spouse. The family records also include the birth date and place, marriage date, spouse's name(s), and death date for each child of the family. The data repository from which the family group sheets were selected, the nature of the data set, its utilization, and strengths and weaknesses have been described in a number of other publications; the details published elsewhere will not be repeated here (Bean et al., 1978, 1980; Skolnick et al., 1978, 1979; Mineau et al., 1979, 1981; and Mineau, 1980).

Recognizing the problems inherent in the analysis of birth intervals (Sheps and

Menken, 1972; Sheps and Perrin, 1964; Sheps et al., 1970; Menken and Sheps, 1972; Wolfers, 1968; Whelpton, 1964), we have attempted to eliminate a number of confounding problems through the selection of a particular subset for analysis. First, to eliminate effects arising as a consequence of variations in divorce, separation, and remarriage, we have restricted our analysis to couples who married only once. Second, to eliminate confounding influences arising as a consequence of variations in mortality, only couples in which both husband and wife survived until the wife reached the end of the normal childbearing years of life, age 49, were included. Third, to avoid problems which might arise in the analysis of birth intervals among women whose low parity was due primarily to variations in sterility, we eliminated zero and one parity women from our sample. Fourth, although the number of children with missing birth dates is relatively small, the necessity of precisely measuring all intervals for a particular woman required us to drop from our analysis women for whom any birth was not reported by date.

Our data set has the advantage of allowing us to examine birth intervals for a series of cohorts of women who have all completed their reproductive period, but the constraints cited above reduce the number of couples in our file to a total of 49,451 once-married couples where the wife was born between 1800 and 1895. A total of 24,144 women were eliminated from our analysis, with over 70 percent of these exclusions resulting from our theoretical specifications. Of these, 2,767 were women who died before reaching age 49, and 1,083 were zero or one child families. To clarify the geographic unit of analysis, 2,767 women (coincidentally the same number dying before age 49) whose last birth was outside the territory of settlement were dropped from the analysis. To restrict our analysis to legitimate marital fertility, 4,347 women whose first birth was illegitimate or within the first 8 months of

DEMOGRAPHY, volume 22, number 2, May 1985

marriage were also eliminated from the study. A total of 286 women were eliminated who married after age 49 or were reported having a child after the age of 55. An additional 3,600 women born between 1800 and 1839 were excluded because they represent a pretransition group with rising fertility.

A number of once married women (9,294 or approximately 18.8 percent of the total) were also excluded because of missing items of information. This group included 3,155, or 6.3 percent of the total, for whom no date of marriage was recorded. Death dates were missing for the husband or wife in 3,408 cases (6.9 percent); birthdates were missing for one or more births in 1,262 cases (2.5 percent of the total); place of birth information was missing in 1,393 cases (2.8 percent); and another 76 cases were excluded for obvious recording errors. In summary, of the 49,451 women in our files born during the nineteenth century, 82.1 percent had the information necessary to analyze all birth intervals within marriage. However, to eliminate potential sources of bias, we selected a group of 25,307 women born 1840 to 1895, married only once, remaining in an intact marriage through age 49, and resident within the territory of Utah throughout the childbearing career.

To demonstrate the representativeness of our U.S. frontier population, we compare our data with data from selected European populations. The comparisons are limited by the lack of detailed data in many European family reconstitution studies. In his analysis of birth intervals, Henry (1976) used data from a number of sources. Henry's patterns of parity-specific birth intervals controlling for children ever born (CEB) are among the most detailed European data published and are plotted alongside similar Utah data in Figure 1. Henry's data represent only one aggregated natural fertility period (families from marriages contracted between 1675 and 1780) and

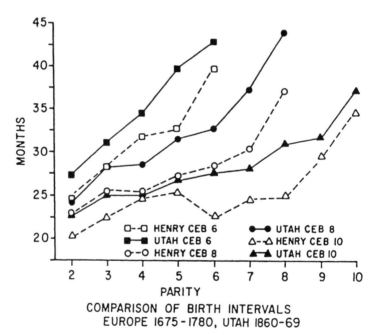

COMPARISON OF BIRTH INTERVALS
EUROPE 1675-1780, UTAH 1860-69

Figure 1.

do not provide sufficient detail to compare longitudinal trends within the Utah and European populations. However, the similarity in birth interval patterns controlling for current parity and children ever born is clear. Only at parity six for women ultimately having ten children (a group with only 41 cases in Henry's data) is there any substantial discrepancy in patterns.

Knodel (1979) has provided a remarkable body of data on birth intervals from his studies of German village genealogies. Using these data he tabulates both mean birth intervals and mean last closed birth intervals for 25 to 50 year marriage cohorts ranging over the years 1750–1799 through 1875–1899. For comparison, we calculate the mean birth intervals and mean last closed intervals for ten year birth cohorts of Utah women from 1840–1849 through 1880–1889 drawn from the current sample analyzed in this paper. Knodel's data for Grafen-

hausen and four Waldeck villages are plotted alongside our data in Figure 2. Over the course of the study periods, last birth intervals in Utah increase somewhat more than in Waldeck and less than in Grafenhausen. Average birth intervals in Utah are approximately one month longer than those in Grafenhausen in the first cohort, and approximately two months longer in the final cohort. Given the widely different definitions of cohorts in the studies, these variations are negligible compared to the overwhelming similarity of birth intervals across the two populations.

The results reported in Figures 1 and 2 suggest that Utah and European populations show differences in lengths of birth intervals no greater than variations within European populations themselves. We believe that an analysis of birth intervals during the Utah fertility transition, utilizing the extremely detailed data available, might provide interesting elaborations.

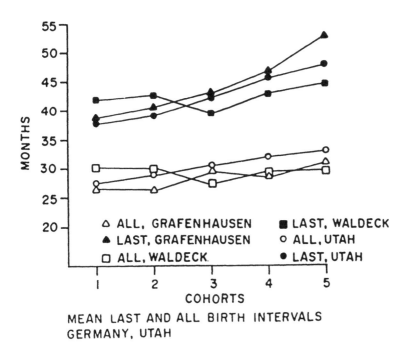

MEAN LAST AND ALL BIRTH INTERVALS
GERMANY, UTAH

Figure 2.

This simple comparison with historical European data suggests such an analysis would contribute to sociological and demographic evidence within the same theoretical framework as European studies.

ANALYSES

In Table 1 and Figure 3, we present data on parity-specific mean birth intervals for five-year birth cohorts of women over the periods of 1840–1844 through 1890–1894. These cohorts represent the transition from natural to controlled fertility (Willigan et al., 1983; Mineau et al., 1979). In Table 1 and subsequent tables, a "minimum interbirth interval" of 17.5 months is subtracted from all computations so that the intervals reported represent an excess over this minimum, largely physiologically determined, interval.[3]

From the data presented in Table 1, we can trace the pattern of increasing

Table 1.—Mean birth interval (−17.5 months) and number of contributing cases by wife's birth cohort and parity for once-married women surviving to age 49 or older

Parity		Cohort										
		1840	1845	1850	1855	1860	1865	1870	1975	1880	1885	1890
2	m	5.92	6.06	6.49	6.77	7.30	7.59	9.48	10.08	10.29	11.76	12.36
	N	566	763	1182	1825	2274	2615	2923	3216	3442	3430	3071
3	m	9.49	9.39	9.22	9.65	10.64	11.55	12.78	14.05	14.88	16.30	16.99
	N	562	754	1162	1787	2223	2552	2824	3086	3257	3187	2805
4	m	9.86	9.98	10.42	10.99	12.63	13.58	15.12	16.35	17.40	18.96	20.50
	N	551	743	1131	1743	2158	2438	2655	2832	2966	2813	2397
5	m	10.42	11.22	12.01	12.10	13.91	15.04	15.18	17.15	18.32	18.82	21.10
	N	532	723	1094	1680	2031	2266	2405	2487	2543	2341	1854
6	m	10.84	11.43	12.19	12.97	14.92	14.86	16.34	17.02	18.00	18.30	18.67
	N	513	696	1043	1570	1845	2012	2086	2082	2079	1816	1348
7	m	11.91	12.61	12.63	13.52	14.37	15.75	15.32	16.67	16.52	17.64	19.31
	N	480	636	956	1431	1631	1746	1716	1676	1591	1330	935
8	m	11.25	12.62	12.32	13.19	14.81	14.25	14.38	16.42	14.85	15.89	16.50
	N	427	566	858	1245	1358	1415	1336	1228	1165	923	639
9	m	11.94	12.16	12.78	13.26	14.76	14.43	13.35	13.85	14.26	15.24	16.43
	N	363	477	718	1008	1054	1083	967	874	810	612	410
10	m	11.38	11.56	13.83	13.10	14.22	13.91	13.40	14.41	13.01	13.82	13.62
	N	276	357	554	740	743	740	686	581	529	380	264
11	m	13.71	13.29	12.86	14.83	13.99	12.55	12.53	12.61	12.87	13.75	13.46
	N	200	249	387	509	478	472	434	343	304	225	152
12	m	9.26	9.40	12.51	12.76	10.66	14.62	12.76	13.55	12.51	10.50	12.01
	N	121	145	217	285	255	253	244	180	161	112	(83)
13	m	13.94	11.17	13.07	12.56	11.79	10.85	11.10	11.53	11.44	11.85	12.82
	N	(65)	(78)	117	141	140	116	115	(81)	(79)	(52)	(36)
14	m	10.29	8.87	10.72	12.95	10.67	16.37	11.92	12.52	10.78	14.03	4.66
	N	(26)	(41)	(45)	(61)	(75)	(44)	(52)	(33)	(33)	(22)	(19)

() indicates less than 100 cases and must be viewed with caution.

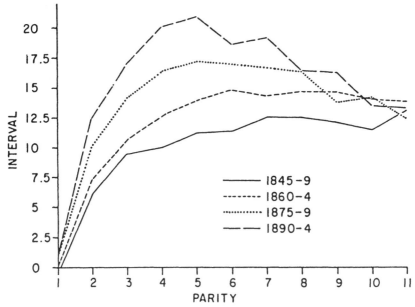

MEAN BIRTH INTERVAL (-17.5 MONTHS) BY WOMENS BIRTH COHORTS AND PARITY

Figure 3.

birth intervals from natural fertility birth cohorts to cohorts exhibiting a substantial fertility control. Note that the average birth intervals increase dramatically at all except the highest parities.[4] One, of course, would not anticipate any significant change in birth intervals among those subcohorts who had ten or more births within the normal childbearing years of life. However, from the data presented in Table 1, it is not possible to determine whether increases in the length of birth intervals over the course of the fertility decline are due to interbirth delay or to failed truncation attempts because each mean interval includes intermediate, as well as ultimate and penultimate intervals.

As indicated in Figure 2 discussed above, there is a slight increase in the average of all birth intervals presumably due to the increase in ultimate birth intervals. In contrast, Figure 3 which presents plots for selected birth cohorts suggests substantial lengthening of birth intervals at lower parities. The length of the early birth intervals clearly increased across the subcohorts representing the transition from natural to controlled fertility.[5]

To address the first hypothesis that distinctive patterns of birth spacing among all birth cohorts are dependent on completed family size, in Table 2 and Figure 4 we present mean birth intervals by parity and children ever born (CEB) for four selected birth cohorts spanning the fertility transition.[6] Comparing birth intervals at all parities across women with different completed family sizes, we find that birth intervals at lower parities are longer for women with smaller family sizes. For natural fertility cohorts as well as controlling cohorts, birth intervals are inversely related to completed fertility.

Controlling for both children ever born and parity allows us to examine the different effects of stopping and spacing within the population. For example, suppose one compares the birth intervals of eight-child families from a natural fertility cohort with six-child families from a transition cohort. If only stopping behavior were involved in this change, we would expect that intervals up to the fourth child would be similar while the intervals to the fifth and sixth child would be lengthened in the transition

Table 2.—Mean birth interval (−17.5 months), standard deviation, number of contributing cases by wife's birth cohort, parity, and children ever born

Cohort	N	% total cohort		Parity								
				2	3	4	5	6	7	8	9	10
Children ever born equals 4												
1845	20	2.6	m	8.53	17.36	35.55						
			s	11.79	19.00	39.69						
1860	125	5.5	m	12.04	16.23	33.96						
			s	17.89	18.35	31.18						
1875	343	10.7	m	13.97	23.41	39.71						
			s	15.84	23.86	38.05						
1890	543	17.7	m	13.97	23.16	38.97						
			s	15.73	21.23	35.20						
Children ever born equals 6												
1845	60	7.9	m	8.51	12.80	16.12	18.47	24.86				
			s	12.89	15.91	15.49	18.27	24.79				
1860	214	9.4	m	9.82	14.58	17.03	22.24	35.40*				
			s	9.82	13.18	16.78	20.46	30.23				
1875	407	12.7	m	9.23	13.23	16.68	22.15	32.61				
			s	11.68	13.67	16.10	21.01	28.48				
1890	413	13.5	m	6.83*	11.07*	16.85	23.97	31.03				
			s	10.85	13.24	16.21	22.86	28.97				
Children ever born equals 8												
1845	89	11.7	m	6.87	11.34	10.51	13.38	12.81	16.14	23.38		
			s	8.07	8.55	8.84	9.33	12.43	14.17	19.83		
1860	304	13.4	m	6.59	10.69	11.01	14.01	15.22	19.79*	26.52		
			s	8.21	9.09	10.50	12.82	13.40	17.74	22.59		
1875	355	11.0	m	6.49	8.34*	10.67	11.93*	15.25	18.82	27.63		
			s	9.02	9.39	10.51	11.42	14.12	17.28	27.26		
1890	230	7.5	m	5.28	7.35	10.56	10.68	13.01	20.57	27.24		
			s	9.67	8.94	11.31	11.42	14.37	21.50	25.69		
Children ever born equals 10												
1845	108	14.2	m	5.71	7.99	7.75	9.91	10.78	11.73	12.25	12.65	16.67
			s	7.02	7.00	7.49	8.49	8.82	10.77	12.06	10.23	14.48
1860	266	11.7	m	5.13	7.50	7.52	9.23	10.19	10.64	13.40	14.19	19.74
			s	6.50	6.49	7.72	9.14	13.54	10.09	12.15	13.69	16.86
1875	237	7.4	m	5.08	7.20	7.26	9.38	9.44	10.03	11.63	12.17	21.60
			s	7.30	7.66	7.67	10.10	8.61	10.12	10.78	13.44	20.53
1890	111	3.6	m	3.04*	5.19*	5.80	7.18*	7.08*	8.41	10.20	16.09*	19.69
			s	7.10	7.04	6.63	7.08	10.35	10.49	10.95	18.52	20.40

* Significantly different from preceding cohort (p=.05).

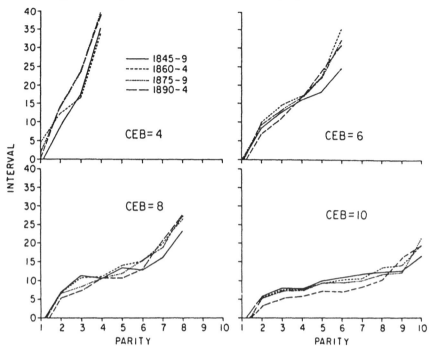

MEAN BIRTH INTERVAL (−17.5 MONTHS) BY WOMENS BIRTH COHORTS AND PARITY

Figure 4.

cohort. This is consistent with the logic that people did not change their behavior until they reached or approached the number of children they desired. On the other hand if all intervals are lengthened in the transition cohorts, we would have evidence that spacing was adopted, i.e. there was a change in behavior early in the childbearing experience.

Within each birth cohort and CEB schedule, there is an increase in ultimate birth intervals. However, comparisons of each birth interval across cohorts shows only a slight increase of ultimate birth intervals. The curves of interval length displayed remarkable stability over the course of dramatic fertility reductions within the population. Given these birth interval schedules, it is also apparent that a shift of increasing proportions of the mothers to subcohorts

with long spacing schedules which produced smaller family sizes across cohorts would increase aggregate ultimate birth intervals without any necessary increase in attempts to truncate childbearing earlier.

The data presented numerically (Table 2) and graphically (Figure 4) lend strong support to our first proposition that distinctive birth spacing patterns are associated at all parities with completed family size. In addition, they provide support for the proposition that birth spacing patterns were relatively constant over the transition when controlling for children ever born. Early cohorts display significant groups of women limiting family size. Among the 1845–1849 birth cohort, 2.6 percent of all women in the study group stopped with four children and 7.9 stopped with six children while

DEMOGRAPHY, volume 22, number 2, May 1985

the mean number of children ever born is 8.8 for the cohort. The relatively low parity and long intervals of these subcohorts are unlikely to reflect subfecundity. By sample definition we have eliminated infecund and highly subfecund women with the exclusion of zero and one parity women. The evidence that low parity is not due to subfecundity is thus indirect, but supportive. The consistency of the birth intervals of low parity women across the transition (upper level panel) of Figure 4 provides further supporting evidence. The birth interval distributions are remarkably consistent whether one is dealing with 2.6 percent of the women in 1845 with only four births or 17.7 percent of the women among the 1890–1894 birth cohort with only four births. The shifts in the proportion of women falling in small family size groups confirms our second proposition that changes in fertility represent a systematic compositional shift in the population described by parity and birth interval schedules.

Our discussion thus far has focused upon the mean length of birth intervals. We are also interested in examining the changing variability of individual behavior over the childbearing career. For example, do failed attempts to truncate increase the dispersion of terminal birth intervals? The standard deviations of mean CEB-parity specific intervals are also presented in Table 2. The standard deviations of CEB and parity-specific birth intervals, show an apparent consistency over the entire transition. Further analysis suggests that all of the standard deviations increase as the length of the birth intervals increases, again controlling for parity and children ever born. We find that, even at low parities such as the interval between second and third birth, variation among those with smaller CEB is greater. For example, in the birth cohort of 1890–1894, the standard deviations for these birth intervals decrease steadily with children ever born; 21.23, 13.24, 8.94, and 7.04 for CEB of four,

six, eight, and ten respectively. Thus increasing variability of behavior would appear in aggregate measures, due to compositional effects of those moving to more variable birth interval schedules associated with smaller family sizes.

Consistent with our third proposition, a number of studies have suggested that age of marriage may affect subsequent spacing of births (Bumpass et al., 1973). Thus, the patterns of interbirth intervals in the population may also be affected by changes in age at marriage over the cohorts analyzed. Trussell and Menken (1978) state that it is well known that an early age at first marriage results in both a high level and rapid pace of subsequent fertility in contemporary U.S. populations (see also Finnas and Hoem, 1980). In contrast, Marini and Hodsdon (1981) suggest that while age at first marriage affects the first birth interval, it is not causally related to the second birth interval. Any relationship between age at marriage and the second birth interval is an indirect association due to the causal relationship between the length of the first and second birth intervals.

To address the effects of changing nuptiality behavior on birth intervals within the populations, Table 3 presents the mean birth intervals for our population stratified by age of marriage. These data suggest substantial shifts in age of marriage distributions over the cohorts tabulated. For example, the percentage of women marrying at less than 21 years of age decreases from 60.89 in the 1845 birth cohort to 34.24 in the 1890 cohort; while the percentage marrying between age 21 and 25 rises from 31.10 to 51.16 percent.[7] Among those married at ages 21 to 25 there appears to be a general lengthening of birth intervals over the course of the fertility transition at all but the highest parities. Because birth intervals are, in agreement with Finnas and Hoem (1980), generally shorter at younger ages of marriage, we can suggest that a part of the onset of longer interbirth intervals in our population is at least indirectly associated with increasing ages of marriage. The small group of those marrying after age 25 was originally separated to control for possible age related subfecundity; however, birth intervals are actually shorter among this group than those marrying at ages 21 to 25. There is some evidence to indicate that later marrying women maintain higher age-specific marital fertility rates

Table 3.—Mean birth interval (−17.5 months), standard deviation, number of contributing cases by wife's birth cohort, parity, and age of marriage

Cohort	% total cohort		Parity								
			2	3	4	5	6	7	8	9	10
Age at marriage less than 21											
1845	60.89	m	6.74	8.96	10.68	11.85	11.73	12.87	13.44	12.97	11.42
		s	9.40	9.56	13.78	13.25	13.22	14.91	14.24	13.94	13.01
		N	464	459	453	446	431	402	371	329	262
1860	52.42	m	6.89	9.66	12.38	14.14	15.07	13.81	15.08	14.87	14.52
		s	10.05	11.74	16.19	16.51	18.34	15.73	16.88	17.48	14.95
		N	1190	1176	1157	1116	1047	951	848	687	540
1875	35.46	m	9.14	12.75	15.34	17.40	16.62	16.91	16.68	13.91	13.83
		s	17.91	18.72	22.18	21.96	19.89	19.85	19.30	17.14	15.46
		N	1135	1113	1074	999	898	772	620	493	349
1890	34.24	m	11.14	16.93	18.84	20.46	19.33	19.36	17.18	16.86	14.13
		s	16.94	22.76	24.49	25.29	25.10	24.59	21.44	20.11	17.82
		N	1050	1002	899	756	595	450	324	221	153
Age at marriage 21 through 25											
1845	31.30	m	5.12	10.19	8.26	10.32	10.88	12.25	11.09	10.16	12.32
		s	8.31	11.46	9.64	11.71	12.34	10.19	11.70	13.30	13.55
		N	237	233	230	226	221	207	179	135	90
1860	38.94	m	7.51	11.58	12.33	13.61	15.16	15.37	14.42	14.60	13.37
		s	12.55	15.55	15.28	16.20	17.68	16.41	15.54	14.78	13.67
		N	884	863	836	786	706	616	475	346	198
1875	48.58	m	11.19	14.73	16.58	17.01	17.55	16.80	16.38	13.88	15.63
		s	16.73	19.86	22.18	19.74	20.22	17.49	18.72	16.40	18.40
		N	1555	1490	1350	1184	993	789	550	357	224
1890	51.16	m	13.37	17.62	21.85	21.99	18.79	19.48	15.64	16.75	13.10
		s	21.22	21.73	25.52	25.07	20.99	21.80	19.08	20.72	17.12
		N	1569	1443	1224	927	649	430	286	175	105
Age at marriage greater than 25											
1845	8.01	m	4.46	9.48	10.87	9.19	11.74	11.41	10.89	12.47	5.10
		s	11.22	12.35	13.72	11.85	13.05	14.50	14.42	13.96	13.09
		N	61	61	59	50	43	26	16	13	5
1860	8.64	m	8.89	12.61	15.76	13.84	11.76	12.21	12.75	13.75	18.90
		s	12.10	19.40	19.02	17.05	16.61	12.60	15.19	9.41	19.15
		N	196	180	161	127	91	63	37	23	6
1875	15.96	m	8.83	14.64	17.60	16.65	16.25	14.04	15.05	11.37	8.55
		s	13.35	16.82	20.20	19.11	17.08	17.19	26.87	14.03	19.28
		N	511	468	395	293	185	109	54	19	5
1890	14.60	m	11.63	14.43	20.34	19.09	14.03	18.10	17.12	8.14	3.82
		s	17.99	19.58	22.18	24.24	20.03	19.36	20.97	9.64	7.87
		N	448	356	270	167	100	52	26	12	5

(Mineau and Trussell, 1982). It is possible that this group reflects an attempt by later marrying women to compensate for late marriage and achieve a desired family size within less time. Since older marrying women (after age 25) have shorter birth intervals than those marrying at 21–25 years of age, we can suggest that the birth intervals of the latter reflect intentional behavior rather than subfecundity.

Summary

Our analysis of changing birth interval distributions over the course of a fertility transition from natural to controlled fertility has examined three closely related propositions. First, within both natural fertility populations (identified at the aggregate level) and cohorts following the onset of fertility limitation, we hypothesized that substantial groups of women with long birth intervals across the individually specified childbearing careers could be identified. That is, even during periods when fertility behavior at the aggregate level is consistent with a natural fertility regime, birth intervals at all parities are inversely related to completed family size. Our tabular analysis enables us to conclude that birth spacing patterns are parity dependent; there is stability in CEB-parity specific mean and birth interval variance over the entire transition. Our evidence does not suggest that the early group of women limiting and spacing births was marked by infecundity. Secondly, the transition appears to be associated with an increasingly larger proportion of women shifting to the same spacing schedules associated with smaller families in earlier cohorts. Thirdly, variations in birth spacing by age of marriage indicate that changes in birth intervals over time are at least indirectly associated with age of marriage, indicating an additional compositional effect.

The evidence we have presented on spacing behavior does not negate the argument that parity-dependent stopping

behavior was a powerful factor in the fertility transition. Our data also provide evidence of attempts to truncate childbearing. Specifically, the smaller the completed family size, the longer the ultimate birth interval; and ultimate birth intervals increase across cohorts controlling CEB and parity. But spacing appears to represent an additional strategy of fertility limitation. Thus, it may be necessary to distinguish spacing and stopping behavior if one wishes to clarify behavioral patterns within a population (Edlefsen, 1981; Friedlander et al., 1980; Rodriguez and Hobcraft, 1980).

Because fertility transition theories imply increased attempts to limit family sizes, it is important to examine differential behavior within subgroups achieving different family sizes. It is this level of analysis which we have attempted to achieve in utilizing parity-specific birth intervals controlled by children ever born. First, our analysis of all birth intervals for a series of cohorts, across the fertility transition, controlling for CEB, suggests that even during the periods of natural fertility, there were, at least among our population, a group of women who both spaced and limited the number of children ever born. Over the course of the transition, because increasing numbers of women modify their behavior to terminate childbearing at lower levels of children ever born, they appear to adopt a pattern of birth spacing already evident among a small group of women in earlier, seemingly natural fertility birth cohorts. Our discussion has also identified several difficulties or compositional effects which may arise in computing simple averages of birth intervals as indicators of the relevant extent of spacing and stopping behaviors.

This study was designed also to determine whether the transition within our study population was principally the result of women adopting birth spacing schedules already extant within subgroups of the population or the product of fertility limitation late in the childbearing career. Our data indicate that birth intervals for the population as a whole increase following the early period of natural fertility. These conclusions are consistent with earlier macrosimulation studies (Anderton et al., 1982; Willigan et al., 1982) in which it was found that a hypothetical simulation model fits observed data only when the model specified a set of increasing birth intervals spread across a considerable range of parities for women of any given completed family size. Accompanying, and abetting, this change in fertility behavior was a shift to older ages of marriage characterized by longer birth intervals.

There are, of course, practical issues to be addressed if one is to accept the argument that spacing was involved in limiting fertility, and that the fertility transition involved a compositional shift from subcohorts with short intervals to subcohorts with long birth intervals. First, there is substantial evidence from historical demographic research that the fertility transition is marked primarily by parity dependent control, reduced age at last birth and lengthened penultimate and ultimate birth intervals. Such conclusions are also supported by evidence from a number of family planning programs in developing societies with high fertility. Specifically, adoption of contraception is most frequent among high parity, older women. We do not dispute these findings. We do, however, suggest they may be incomplete. Large samples of historical populations which make it possible to disaggregate populations over the course of the entire transition by CEB and parity provide additional information.

A second practical issue is whether the means to space systematically and thereby to limit family size were available prior to the introduction of modern contraceptive technology. Prior to the time studied in this paper, and in the absence of modern contraceptive technology, women in France reduced family size substantially. Fertility had also declined

significantly among the populations of the Eastern sections of the United States from which the original members of the LDS Church, heavily represented in our sample, had been recruited. It seems, therefore, that modern contraceptive technology is not a necessary condition for family limitation.

The results of this analysis have implications for both historical demographic research and contemporary family planning policies. We have clearly supported previous authors' suggestions that birth interval analyses provide valuable insights to behavioral change and supplement other fertility indices in historical demographic research. The combined evidence of increasingly long birth intervals over cohorts and the presence of individuals demonstrating such behavior during the "natural" fertility period suggest adaptive fertility behavior rather than innovative behavior underlying the fertility transition within this population. In turn, our analysis suggests that further study of birth intervals and the spread of fertility limitation may be relevant to contemporary settings.

NOTES

[1] Conclusions drawn from historical demographic analysis of the fertility transition in Western Europe are more numerous than the two cited in this paper. These have been summarized and policy implications specified by Knodel and van de Walle (1979), Freedman (1979) and others.

[2] For example, reference to Table 1 of the Tolnay and Guest article indicates, reading down the last column to create synthetic cohorts across parities, that 71.1 percent of women married between the ages of 20–24, did not close the third interval within 24 months. That leaves the possibility of a substantial proportion of the population adopting longer birth intervals.

[3] Potter (1963) suggests that birth intervals of 30–33 months are possible in the absence of birth control with a minimal 18 month interval, assuming no lactation or pregnancy loss. In our tabular analyses the subtraction of this minimum interval (17.5 for rounding) highlights changes in intervals which may well be due to behavioral variations related to control.

[4] The increasing trend among the oldest (natural fertility) cohorts is consistent with the birth intervals observed among other North American natural fertility populations, although the length of the

intervals are consistently higher. See Charbonneau (1979).

[5] The difficulty in identifying possible changes in spacing from data of the type plotted in Figure 2, representing average birth intervals without parity controls, is easily elaborated. A woman having only four births may contribute somewhat longer birth intervals to the calculation of average intervals than would a woman with eight births. The weight placed on the woman with only four births in calculating the average also declines because she contributes only three intervals to the average, while her contemporaries with eight continue to contribute seven shorter closed intervals to the average (Wolfers, 1968.) In addition, even over the period of increasing fertility control, these smaller family size groups also constitute a minority again contributing fewer intervals to the average than larger families. This suggests that the averages of intervals across CEB (children ever born) and parity may be misleading and less than adequately sensitive to shifts of women to smaller family sizes with longer interbirth intervals.

[6] We have limited the presentation to only four cohorts in the interest of space. The patterns are consistent across all cohorts. Tables containing data for all cohorts are available from the authors.

[7] The age groups used in the analysis were selected on the basis of the distribution of cases.

ACKNOWLEDGMENTS

This research was supported by NIH Grant HD-15455 and is a revision of an earlier paper by the same title presented at the Population Association of America April 1983 Meetings in Pittsburgh, Pennsylvania. We wish to acknowledge the substantial contributions of the anonymous reviewers and members of the Editorial Board.

REFERENCES

Anderton, D. L., L. L. Bean, J. D. Willigan, and G. P. Mineau. 1982. A macrosimulation approach to the investigation of the fertility transition. Paper presented at the American Sociological Association Meetings, San Francisco, California, (forthcoming in Social Biology, 1984).

Bean, L. L., G. P. Mineau, K. A. Lynch, and J. D. Willigan. 1980. The Genealogical Society of Utah as a data resource for historical demography. Population Index 46:6–19.

Bean, L. L., D. L. May, and M. Skolnick. 1978. The Mormon Historical Demography Project. Historical Methods 11:45–53.

Bean, L. L., G. Mineau, and D. Anderton. 1983. Residence and religious effects on declining family size: an historical analysis of the Utah population. Review of Religious Research 25:91–101.

Bumpass, L., R. Rindfuss, and R. Janosik. 1973. Age and marital status at first birth and the pace of subsequent fertility. Demography 15:74–86.

Carlsson, G. 1966. The decline of fertility: innovation or adjustment process. Population Studies 20:149–74.

Charbonneau, H. 1979. Les Régimes de Fécondité Naturelle en Amerique du Nord. In H. Leridon and J. Menken (eds.), Natural Fertility. Liege: Ordina Publications.

Coale, A. J. and T. J. Trussell. 1974. Model fertility schedules: variations in the age structure of childbearing in human populations. Population Index 40:185–201.

———. 1978a. Technical note: finding the two parameters that specify model schedules of marital fertility. Population Index 40:185–201.

———. 1978b. Erratum. Population Index 41:572.

David, P. A. and W. C. Sanderson. 1984. Spacing versus stopping in the past: martial duration-specific patterns of fertility control among U. S. native white women, 1880–1910. Abstract appearing in Population Index 50:390.

Dupâquier, J. and M. Lachiver. 1969. Sur les Débuts de la Contraception en France ou les Deux Malthusianisms. Annales Economies, Sociétés, Civilisations 24:1391–1406.

———. 1981. Du Contresens à l'Illusion Technique. Annales Economies Sociétés, Civilisations 36:478–492.

Edlefsen, L. E. 1981. An investigation of the timing pattern of childbearing. Population Studies 35:375–386.

Finnas, F. and J. M. Hoem. 1980. Starting age and subsequent birth intervals in cohabitational unions in current Danish cohorts, 1975. Demography 17:275–295.

Freedman, R. 1979. Theories of fertility decline: a reappraisal. Social Forces 58:1–17.

Friedlander, D., Z. Eisenbach and C. Goldschieder. 1980. Family-size limitation and birth spacing. The fertility transition of African and Asian immigrants to Israel. Population and Development Review 6:581–593.

Henry, L. 1961. Some data on natural fertility. Eugenics Quarterly 8:81–91.

———. 1976. Population Analysis and Models. London: Edward Arnold.

Hollingsworth, T. H. 1969. Historical Demography. Ithaca: Cornell University Press.

Knodel, J. 1977. Family limitation and the fertility transition: evidence from the age patterns of fertility in Europe and Asia. Population Studies 31:219–249.

———. 1979. From natural fertility to family limitation: the onset of fertility transition in a sample of German villages. Demography 16:493–521.

———. 1981. Espacement des Naissances et Planification Familiale: Une Critique de la Méthode Dupâquier-Lachiver. Annales Economies, Sociétés, Civilisations 36:473–488 (and Réponse de John Knodel à Jacques Dupâquier, 493–494.

———. 1983. Natural Fertility: Age Patterns, Levels, and Trends. In R. A. Bulatao and R. Lee (eds.) Determinants of Fertility in Developing Countries, Vol. I, Supply and Demand for Children. New York: Academic Press.

Knodel, J. and E. van de Walle. 1979. Lessons from the past: policy implications of historical fertility studies. Population and Development Review 5:217-246.

Marini, M. M. and P. J. Hodsdon. 1981. Effects of the timing of marriage and first birth on the spacing of subsequent births. Demography 18:529-548.

Menken, J. A. and M. C. Sheps. 1972. The Sample Frame as a Determinant of Observed Distributions of Duration Variables. Pp. 57-87 in T. N. E. Greville (ed.), Population Dynamics. New York: Academic Press.

Mineau, G. P. 1980. Fertility on the Frontier: An Analysis of the Nineteenth Century Utah Population. Unpublished Ph.D. Dissertation, University of Utah.

Mineau, G. P., L. L. Bean, and M. Skolnick. 1979. Mormon demographic history II: the family life cycle and natural fertility. Population Studies 33:429-446.

Mineau, G. P., D. L. Anderton, L. L. Bean, and J. D. Willigan. 1981. A log-linear approach to heterogeneity during the fertility transition in a nineteenth century american population. Paper presented at the Pacific Sociological Association meetings, Portland, Oregon, (translated and reprinted in Annales D. H., 1984:219-236).

Mineau, G. P. and J. Trussell. 1982. A specification of marital fertility by parents' age, age at marriage and marital duration. Demography 19:335-350.

Potter, R. G. 1963. Birth intervals: structure and change. Population Studies 17:155-166.

Rodriguez, G. and J. N. Hobcraft. 1980. Illustrative Analysis: Lifetable Analysis of Birth Intervals in Colombia. World Survey: Scientific Reports, No. 16, ISI, The Hague, Netherlands.

Sheps, M. D. and E. B. Perrin. 1964. The distribution of birth intervals under a class of stochastic fertility models. Population Studies 17:321-331.

Sheps, M. C., J. A. Menken, J. C. Ridley, and J. W. Lingner. 1970. Truncation effect in birth interval data. Journal of the American Statistical Association 65:678-693.

Sheps, M. C. and J. A. Menken. 1972. Distributions of birth intervals according to the sampling frame. Theoretical Population Biology 1:1-26.

Skolnick, M., L. L. Bean, V. Arbon, K. de Nevers, and P. Cartwright. 1978. Mormon demographic history I: nuptiality and fertility of once-married couples. Population Studies 32:5-19.

Skolnick, M., L. L. Bean, S. M. Dintelman, and G. Mineau. 1979. A computerized family history data base system. Sociology and Social Research 601-619.

Tolnay, S. E. and A. M. Guest. 1984. American family building strategies in 1900: stopping or spacing. Demography 21:9-18.

Trussell, J. and J. A. Menken. 1978. Early childbearing and subsequent fertility. Family Planning Perspectives 10:209-218.

Willigan, J. D., G. P. Mineau, D. L. Anderton, and L. L. Bean. 1982. A macrosimulation approach to the investigation of natural fertility. Demography 19:161-176.

Wilson, C. 1984. Natural fertility in pre-industrial England, 1600-1799. Population Studies 38:225-240.

Wolfers, D. 1968. The determinants of birth intervals and their means. Population Studies 22:253-262.

Whelpton, P. K. 1964. Trends and differentials in the spacing of births. Demography 1:83-93.

Wrigley, E. A. 1966. Family limitation in pre-industrial England. Economic History Review 19:82-109.

VOLUNTARY MOTHERHOOD; THE BEGINNINGS OF FEMINIST BIRTH CONTROL IDEAS IN THE UNITED STATES

Linda Gordon

Voluntary motherhood was the first general name for a feminist birth control demand in the United States in the late nineteenth century.* It represented an initial response of feminists to their understanding that involuntary motherhood and child-raising were important parts of woman's oppression. In this paper, I would like to trace the content and general development of "voluntary-motherhood" ideas and to situate them in the development of the American birth-control movement.

The feminists who advocated voluntary motherhood were of three general types: suffragists; people active in such moral reform movements as temperance and social purity, in church auxiliaries, and in women's professional and service organizations (such as Sorosis); and members of small, usually anarchist, Free Love groups. The Free Lovers played a classically vanguard role in the development of birth-control ideas. Free Love groups were always small and sectarian, and they were usually male-dominated, despite their extreme ideological feminism. They never coalesced into a movement. On the contrary, they were the remnants of a dying tradition of utopian socialist and radical protestant religious dissent. The Free Lovers, whose very self-definition was built around a commitment to iconoclasm and to isolation from the masses, were precisely the group that could offer intellectual leadership in formulating the shocking arguments that birth control in the nineteenth century required.[1]

---

*The word "feminist" must be underscored. Since the early nineteenth century, there had been developing a body of population-control writings, which recommended the use of birth-control techniques to curb nationwide or worldwide populations; usually called neo-Malthusians, these writers were not concerned with the control of births as a means by which women could gain control over their own lives, except, very occasionally, as an auxiliary argument. And of course birth control practices date back to the most ancient societies on record.

5

The suffragists and moral reformers, concerned to win mass support, were increasingly committed to social respectability. As a result, they did not generally advance very far beyond prevalent standards of propriety in discussing sexual matters publicly. Indeed, as the century progressed the social gap between them and the Free Lovers grew, for the second and third generations of suffragists were more concerned with respectability than the first. In the 1860s and 1870s the great feminist theoreticians had been much closer to the Free Lovers, and at least one of these early giants, Victoria Woodhull, was for several years a member of both the suffrage and the Free Love camps. But even respectability did not completely stifle the mental processes of the feminists, and many of them said in private writings—in letters and diaries—what they were unwilling to utter in public.

The similar views of Free Lovers and suffragists on the question of voluntary motherhood did not bridge the considerable political distance between the groups, but did show that their analyses of the social meaning of reproduction for the women were converging. The sources of that convergence, the common grounds of their feminism, were their similar experiences in the changing conditions of nineteenth-century America. Both groups were composed of educated, middle-class Yankees responding to severe threats to the stability, if not dominance, of their class position. Both groups were disturbed by the consequences of rapid industrialization—the emergence of great capitalists in a clearly defined financial oligarchy, and the increased immigration which threatened the dignity and economic security of the middle-class Yankee. Free Lovers and suffragists, as feminists, looked forward to a decline in patriarchal power within the family, but worried, too, about the possible desintegration of the family and the loosening of sexual morality. They saw reproduction in the context of these larger social changes, and in a movement for women's emancipation; and they saw that movement as an answer to some of these large social problems. They hoped that giving political power to women would help to reinforce the family, to make the government more just and the economy less monopolistic. In all these attitudes there was something traditional as well as something progressive; the concept of voluntary motherhood reflected this duality.

Since we all bring a twentieth-century understanding to our concept of birth control, it may be best to make it clear at once that neither Free Lovers nor suffragists approved of contraceptive devices. Ezra Heywood, patriarch and martyr, thought "artificial" methods "unnatural, injurious, or offensive."[2] Tennessee Claflin wrote that the "washes, teas, tonics and various sorts of appliances known to the initiated" were a "standing reproach upon, and a permanent indictment against, American women. . . . No woman should ever hold sexual relations with any man from the possible consequences of which she might desire to escape."[3] *Woodhull and Claflin's Weekly* editorialized: "The means they (women) resort to for . . . prevention is sufficient to disgust every natural man. . . ."[4]

6

On a rhetorical level, the main objection to contraception * was that it was "unnatural", and the arguments reflected a romantic yearning for the "natural," rather pastorally conceived, that was typical of many nineteenth-century reform movements. More basic, however, particularly in women's arguments against contraception, was an underlying fear of the promiscuity that it could permit. The fear of promiscuity was associated less with fear for one's virtue than with fear of other women—the perhaps mythical "fallen" women—who might threaten a husband's fidelity.

To our twentieth-century minds a principle of voluntary motherhood that rejects the practice of contraception seems so theoretical as to have little real impact. What gave the concept substance was that it was accompanied by another, potentially explosive, conceptual change: the reacceptance of female sexuality. As with birth control, the most open advocates of female sexuality were the Free Lovers, not the suffragists; nevertheless both groups based their ideas on the traditional grounds of the "natural." Free Lovers argued, for example, that celibacy was unnatural and dangerous—for men and women alike. "Pen cannot record, nor lips express, the enervating, debauching effect of celibate life upon young men and women. . . ."[5] Asserting the existence, legitimacy and worthiness of female sexual drive was one of the Free Lovers' most important contributions to sexual reform; it was a logical correlate of their argument from the "natural" and of their appeal for the integration of body and soul.

Women's rights advocates, too, began to demand recognition of female sexuality. Isabella Beecher Hooker wrote to her daughter: "Multitudes of women in all the ages who have scarce known what sexual desire is—being wholly absorbed in the passion of maternity, have sacrificed themselves to the beloved husbands as unto God—and yet these men, full of their human passion and defending it as righteous & God-sent lose all confidence in womanhood when a woman here and there betrays her similar nature & gives herself soul & body to the man she adores."[6] Alice Stockham, a Spiritualist Free Lover and feminist physician, lauded sexual desire in men and women as "the prophecy of attainment." She urged that couples avoid reaching sexual "satiety" with each other, in order to keep their sexual desire constantly alive, for she considered desire pleasant and healthful.[7] Elizabeth Cady Stanton, commenting in her diary in 1883 on the Whitman poem, "There is a Woman Waiting for Me," wrote: "he speaks as if the female must be forced to the creative act, apparently ignorant of the fact that a healthy woman has as much passion as a man, that she needs nothing stronger than the law of attraction to draw her to the male."[8] Still, she loved Whitman, and largely because of that openness about sex that made him the Free Lovers' favorite poet.

According to the system of ideas then dominant, women, lacking sexual drives, submitted to sexual intercourse (and notice how Beecher Hooker

---

*Contraception will be used to refer to artificial devices used to prohibit conception during intercourse, while birth control will be used to mean anything, including abstinence, which limits pregnancy.

7

continued the image of a woman "giving herself", never taking) in order to please their husbands and to conceive children.The ambivalence underlying this view was expressed in the equally prevalent notion that women must be protected from exposure to sexuality lest they "fall" and become depraved, lustful monsters. This ambivalence perhaps came from a subconscious lack of certainty about the reality of the sex-less woman, a construct laid only thinly on top of the conception of woman as highly sexed, even insatiably so, that prevailed up to the eighteenth century. Victorian ambivalence on this question is nowhere more tellingly set forth than in the writings of physicians, who viewed woman's sexual organs as the source of her being, physical and psychological, and blamed most mental derangements on disorders of the reproductive organs.[9] Indeed, they saw it as part of the nature of things, as Rousseau had written, that men were male only part of the time, but women were female always.[10] In a system that deprived women of the opportunity to make extra-familial contributions to culture, it was inevitable that they should be more strongly identified with sex than men were. Indeed, females were frequently called "the sex" in the nineteenth century.

The concept of maternal instinct helped to smooth the contradictory attitudes about woman's sexuality. In many nineteenth-century writings we find the idea that the maternal instinct was the female analog of the male sex instinct; it was as if the two instincts were seated in analogous parts of the brain, or soul. Thus to suggest, as feminists did, that women might have the capacity for sexual impulses of their own automatically tended to weaken the theory of the maternal instinct. In the fearful imaginations of self-appointed protectors of the family and of womanly innocence, the possibility that women might desire sexual contact not for the sake of pregnancy—that they might even desire it at a time when they positively did not want pregnancy—was a wedge in the door to denying that women had any special maternal instinct at all.

Most of the feminists did not want to open that door either. Indeed, it was common for nineteenth-century women's-rights advocates to use the presumed "special motherly nature" and "sexual purity" of women as arguments for increasing their freedom and status. It is no wonder that many of them chose to speak their subversive thoughts about the sexual nature of women privately, or at least softly. Even among the more outspoken Free Lovers, there was a certain amount of hedging. Lois Waisbrooker and Dora Forster, writing for a Free Love journal in the 1890s, argued that while men and women both had an "amative" instinct, it was much stronger in men; and that women—only women—also had a reproductive, or "generative" instinct. "I suppose it must be universally conceded that men make the better lovers," Forster wrote. She thought that it might be possible that "the jealousy and tyranny of men have operated to suppress amativeness in women, by constantly sweeping strongly sexual women from the paths of life into infamy and sterility and death," but she thought also that the suppression, if it existed, had been permanently inculcated in woman's character.[11]

Modern birth control ideas rest on a full acceptance, at least quantitatively, of female sexuality. Modern contraception is designed to permit sexual intercourse

8

as often as desired without the risk of pregnancy. Despite the protestations of sex counsellors that there are no norms for the frequency of intercourse, in the popular view there are such norms. Most people in the mid-twentieth century think that "normal" couples have intercourse several times a week. By twentieth-century standards, then, the Free Lovers' rejection of artificial contraception and "unnatural" sex seems to preclude the possibility of birth control at all. Nineteenth-century sexual reformers, however, had different sexual norms. They did not seek to make an infinite number of sterile sexual encounters possible. They wanted to make it possible for women to avoid pregnancy if they badly needed to do so for physical or psychological reasons, but they did not believe that it was essential for such women to engage freely in sexual intercourse.

In short, for birth control, they recommended periodic or permanent abstinence. The proponents of voluntary motherhood had in mind two distinct contexts for abstinence. One was the mutual decision of a couple. This could mean continued celibacy, or it could mean following a form of the rhythm method. Unfortunately all the nineteenth-century writers miscalculated women's fertility cycle. (It was not until the 1920s that the ovulation cycle was correctly plotted, and until the 1930s it was not widely understood among American doctors.)[12] Ezra Heywood, for example, recommended avoiding intercourse from 6 to 8 days before menstruation until 10 to 12 days after it. Careful use of the calendar could also provide control over the sex of a child, Heywood believed: conception in the first half of the menstrual cycle would produce girls, in the second half, boys.[13] These misconceptions functioned, conveniently, to make practicable Heywood's and others' ideas that celibacy and contraceptive devices should *both* be avoided.

Some of the Free Lovers also endorsed male continence, a system practiced and advocated by the Oneida community, in which the male avoids climax entirely.[14] (There were other aspects of the Oneida system that antagonized the Free Lovers, notably the authoritarian quality of John Humphrey Noyes's leadership.)[15] Dr. Stockham developed her own theory of continence called "Karezza," in which the female as well as the male was to avoid climax. Karezza and male continence were whole sexual systems, not just methods of birth control. Their advocates expected the self-control involved to build character and spiritual qualities, while honoring, refining and dignifying the sexual functions; and Karezza was reputed to be a cure for sterility as well, since its continued use was thought to build up the resources of fertility in the body.[16]

Idealizing sexual self-control was characteristic of the Free Love point of view. It was derived mainly from the thought of the utopian communitarians of the early nineteenth century,[17] but Ezra Heywood elaborated the theory. Beginning with the assumption that people's "natural" instincts, left untrammeled, would automatically create a harmonious, peaceful society—an assumption certainly derived from liberal philosophical faith in the innate goodness of man—Heywood applied it to sexuality, arguing that the natural sexual instinct was innately moderated, self-regulating. He did not imagine, as did Freud, a powerful, simple libido that could be checked only by an equally

9

powerful moral and rational will. Heywood's theory implicitly contradicted Freud's description of inner struggle and constant tension between the drives of the id and the goals of the super-ego; Heywood denied the social necessity of sublimation.

On one level Heywood's theory may seem inadequate as a psychology, since it cannot explain such phenomena as repression and the strengthening of self-control with maturity. It may, however, have a deeper accuracy. It argues that society and its attendant repressions have distorted the animal's natural self-regulating mechanism, and have thereby created excessive and obsessive sexual drives. It offers a social explanation for the phenomena that Freud described in psychological terms, and thus holds out the hope that they can be changed.

Essentially similar to Wilhelm Reich's theory of "sex-economy," the Heywood theory of self-regulation went beyond Reich's in providing a weapon against one of the ideological bastions of male supremacy. Self-regulation as a goal was directed against the prevalent attitude that male lust was an uncontrollable urge, an attitude that functioned as a justification for rape specifically and for male sexual irresponsibility generally. We have to get away from the tradition of "man's necessities and woman's obedience to them," Stockham wrote.[18] The idea that men's desires are irrepressible is merely the other face of the idea that women's desires are non-existent. Together, the two created a circle that enclosed woman, making it her exclusive responsibility to say No, and making pregnancy her God-imposed burden if she didn't, while denying her both artificial contraception and the personal and social strength to rebel against male sexual demands.

Heywood developed his theory of natural sexual self-regulation in answer to the common anti-Free Love argument that the removal of social regulation of sexuality would lead to unhealthy promiscuity: ". . . in the distorted popular view, Free Love tends to unrestrained licentiousness, to open the flood gates of passion and remove all barriers in its desolating course; but it means just the opposite; it means the *utilization of animalism,* and the triumph of Reason, Knowledge, and Continence."[19] He applied the theory of self-regulation to the problem of birth control only as an afterthought, perhaps when women's concerns with that problem reached him. Ideally, he trusted, the amount of sexual intercourse that men and women desired would be exactly commensurate with the number of children that were wanted. Since sexual repression had had the boomerang effect of intensifying our sexual drives far beyond "natural" levels, effective birth control now would require the development of the inner self-control to contain and repress sexual urges.   But in time he expected that sexual moderation would come about naturally.

Heywood's analysis, published in the mid-1870s, was concerned primarily with excessive sex drives in men. Charlotte Perkins Gilman, one of the leading theoreticians of the suffrage movement, reinterpreted that analysis two decades later to emphasize its effects on women. The economic dependence of woman on man, in Gilman's analysis, made her sexual attractiveness necessary not only for winning a mate, but as a means of getting a livelihood too. This is the case

10

308

with no other animal. In the human female it had produced "excessive modification to sex," emphasizing weak qualities characterized by humans as "feminine." She made an analogy to the milk cow, bred to produce far more milk than she would need for her calves. But Gilman agreed completely with Heywood about the effects of exaggerated sex distinction on the male; it produced excessive sex energy and excessive indulgence to an extent debilitating to the whole species. Like Heywood she also belived that the path of progressive social evolution moved toward monogamy and toward reducing the promiscuous sex instinct.[20]

A second context for abstinence, in addition to mutual self-regulation by a couple, was the right of the wife unilaterally to refuse her husband. This idea is at the heart of voluntary motherhood. It was a key substantive demand in the mid-nineteenth century when both law and practice made sexual submission to her husband a woman's duty.[21] A woman's right to refuse is clearly the fundamental condition of birth control—and of her independence and personal integrity.

In their crusade for this right of refusal the voices of Free Lovers and suffragists were in unison. Ezra Heywood demanded "Woman's Natural Right to ownership and control over her own body-self—a right inseparable from Woman's intelligent existence. . . ."[22] Paulina Wright Davis, at the National Woman Suffrage Association in 1871, attacked the law "which makes obligatory the rendering of marital rights and compulsory maternity." When, as a result of her statement she was accused of being a Free Lover, she responded by accepting the description.[23] Isabella Beecher Hooker wrote her daughter in 1869 advising her to avoid pregnancy until "you are prepared in body and soul to receive and cherish the little one. . . ."[24] In 1873 she gave similar advice to women generally, in her book *Womanhood*.[25] Elizabeth Cady Stanton had characteristically used the same phrase as Heywood: woman owning her own body. Once asked by a magazine what she meant by it, she replied: ". . . womanhood is the primal fact, wifehood and motherhood its incidents . . . must the heyday of her existence be wholly devoted to the one animal function of bearing children? Shall there be no limit to this but woman's capacity to endure the fearful strain on her life?"[26]

The insistence on women's right to refuse often took the form of attacks on men for their lusts and their violence in attempting to satisfy them. In their complaints against the unequal marriage laws, chief or at least loudest among them was the charge that they legalized rape.[27] Victoria Woodhull raged, "I will tell the world, so long as I have a tongue and the strength to move it, of all the infernal misery hidden behind this horrible thing called marriage, though the Young Men's Christian Association sentence me to prison a year for every word. I have seen horrors beside which stone walls and iron bars are heaven. . . ."[28] Angela Heywood attacked men incessantly and bitterly; if one were to ignore the accuracy of her charges, she could well seem ill-tempered. "Man so lost to himself and woman as to invoke legal *violence* in these sacred nearings, *should*

11

have solemn meeting with, and look serious at his own penis until he is able to be lord and master of it, rather than it should longer rule, lord and master, of him and of the victims he deflowers."[29] Suffragists spoke more delicately, but not less bitterly. Feminists organized social purity organizations and campaigns, their attacks on prostitution based on a critique of the double standard, for which their proposed remedy was that men conform to the standards required of women.[30]

A variant of this concern was a campaign against "sexual abuses"—a Victorian euphemism for deviant sexual practices, or simply excessive sexual demands, not necessarily violence or prostitution. The Free Lovers, particularly, turned to this cause, because it gave them an opportunity to attack marriage. The "sexual abuses" question was one of the most frequent subjects of correspondence in Free Love periodicals. For example, a letter from Mrs. Theresa Hughes of Pittsburgh described:

> . . . a girl of sixteen, full of life and health when she became a wife . . . . She was a slave in every sense of the word, mentally and sexually, never was she free from his brutal · outrages, morning, noon and night, up almost to the very hour her baby was born, and before she was again strong enough to move about . . . Often did her experience last an hour or two, and one night she will never forget, the outrage lasted exactly four hours.[31]

Or from Lucinda Chandler, well-known moral reformer:

> This useless sense gratification has demoralized generation after generation, till monstrosities of disorder are common. Moral education, and healthful training will be requisite for some generations, even after we have equitable economics, and free access to Nature's gifts. The young man of whom I knew who threatened his bride of a week with a sharp knife in his hand, to compel her to perform the office of 'sucker,' would no doubt have had the same disposition though no soul on the planet had a want unsatisfied or lacked a natural right.[32]

From an anonymous woman in Los Angeles:

> I am nearly wrecked and ruined by . . . nightly intercourse, which is often repeated in the morning. This and nothing else was the cause of my miscarriage . . . he went to work like a man a-mowing, and instead of a pleasure as it might have been, it was most intense torture. . . .[33]

Clearly these remarks reflect a level of hostility toward sex. The observation that many feminists hated sex has been made by several historians,[34] but they have usually failed to perceive that feminists' hostility and fear of it came from the fact that they were women, not that they were feminists. Women in the nineteenth century were, of course, trained to repress their own sexual feelings, to view sex as a duty. But they also resented what they experienced, which was not an abstraction, but a particular, historical kind of sexual encounter—intercourse dominated by and defined by the male in conformity with his desires and in disregard of what might bring pleasure to a woman. (That this might have resulted more from male ignorance than malevolence could not change women's experiences.) Furthermore, sexual intercourse brought physical danger. Pregnancy, child-birth and abortions were risky, painful and isolating ex-

12

periences in the nineteenth century; venereal diseases were frequently communicated to women by their husbands. Elmina Slenker, a Free Lover and novelist, wrote, "I'm getting a host of stories (truths) about women so starved sexually as to use their dogs for relief, and finally I have come to the belief that a CLEAN dog is better than a drinking, tobacco-smelling, venerally diseased man!"[35]

"Sex-hating" women were not just misinformed, or priggish, or neurotic. They were often responding rationally to their material reality. Denied the possibility of recognizing and expressing their own sexual needs, denied even the knowledge of sexual possibilities other than those dictated by the rhythms of male orgasm, they had only two choices: passive and usually pleasureless submission, with high risk of undesirable consequences, or rebellious refusal. In that context abstinence to ensure voluntary motherhood was a most significant feminist demand.

What is remarkable is that some women recognized that it was not sex per se, but only their husbands' style of making love, that repelled them. One of the women noted above who complained about her treatment went on to say: "I am undeveloped sexually, never having desires in that direction; still, with a husband who had any love or kind feelings for me and one less selfish it *might* have been different, but he cared nothing for the torture to *me* as long as *he* was gratified."[36]

Elmina Slenker herself, the toughest and crustiest of all these "sex-haters," dared to explore and take seriously her own longings, thereby revealing herself to be a sex-lover in disguise. As the editor of the *Water-Cure Journal,* and a regular contributor to *Free Love Journal,*[37] she expounded a theory of "Dianaism, or Non-procreative Love," sometimes called "Diana-love and Alpha-abstinence." It meant free sexual contact of all sorts except intercourse.

> We want the sexes to love more than they do; we want them to love openly, frankly, earnestly; to enjoy the caress, the embrace, the glance, the voice, the presence & the very step of the beloved. We oppose no form or act of love between any man & woman. Fill the world as full of genuine sex love as you can ... but forbear to rush in where generations yet unborn may suffer for your unthinking, uncaring, unheeding actions.[38]

Comparing this to the more usual physical means of avoiding conception—*coitus interruptus* and male continence—reveals how radical it was. In modern history, awareness of the possibilities of nongenital sex, and of forms of genital sex beyond standard "missionary-position" intercourse has been a recent, post-Freudian, even post-Masters and Johnson phenomenon. The definition of sex as heterosexual intercourse has been one of the oldest and most universal cultural norms. Slenker's alienation from existing sexual possibilities led her to explore alternatives with a bravery and a freedom from religious and psychological taboos extraordinary for a nineteenth-century Quaker reformer.

In the nineteenth century, neither Free Lovers nor suffragists ever relinquished their hostility to contraception. But among the Free Lovers, free speech was always an overriding concern, and for that reason Ezra Heywood

13

agreed to publish some advertisements for a vaginal syringe, an instrument the use of which for contraception he personally deplored, or so he continued to assure his readers. Those advertisements led to Heywood's prosecution for obscenity, and he defended himself with characteristic flair by making his position more radical than ever before. Contraception was moral, he argued, when it was used by women as the only means of defending their rights, including the right to voluntary motherhood. Although "artificial means of preventing conception are not generally patronized by Free Lovers," he wrote, reserving for his own followers the highest moral ground, still he recognized that not all women were lucky enough to have Free Lovers for their sex partners.[39]

> Since Comstockism makes male will, passion and power absolute to *impose* conception, I stand with women to resent it. The man who would legislate to choke a woman's vagina with semen, who would force a woman to retain his seed, bear children when her own reason and conscience oppose it, would waylay her, seize her by the throat and rape her person.[40]

Angela Heywood enthusiastically pushed this new political line.

> Is it "proper", "polite", for men, real *he* men, to go to Washington to say, by penal law, fines and imprisonment, whether woman may continue her natural right to wash, rinse, or wipe out her own vaginal body opening—as well legislate when she may blow her nose, dry her eyes, or nurse her babe. . . .Whatever she may have been pleased to receive, from man's own, is his gift and her property. Women do not like rape, and have a right to resist its results.[41]

Her outspokenness, vulgarity in the ears of most of her contemporaries, came from a substantive, not merely a stylistic, sexual radicalism. Not even the heavy taboos and revulsion against abortion stopped her: "To cut a child up in woman, procure abortion. is a most fearful, tragic deed; but *even that* does not call for man's arbitrary jurisdiction over woman's womb."[42]

It is unclear whether Heywood, in this passage, was actually arguing for legalized abortion; if she was, she was alone among all nineteenth-century sexual reformers in saying it. Other feminists and Free Lovers condemned abortion, and argued that the necessity of stopping its widespread practice was a key reason for instituting voluntary motherhood by other means. The difference on the abortion question between sexual radicals and sexual conservatives was in their analysis of its causes and remedies. While doctors and preachers were sermonizing on the sinfulness of women who had abortions,[43] the radicals pronounced abortion itself an undeserved punishment, and a woman who had one a helpless victim. Woodhull and Claflin wrote about Madame Restell's notorious abortion "factory" in New York City without moralism, arguing that only voluntary conception would put it out of business.[44] Elizabeth Cady Stanton also sympathized with women who had abortions, and used the abortion problem as an example of women's victimization by laws made without their consent.[45]

Despite stylistic differences, which stemmed from differences in goals, nineteenth-century American Free Love and women's rights advocates shared the same basic attitudes toward birth control: they opposed contraception and

14

abortion, but endorsed voluntary motherhood achieved through periodic abstinence; they believed that women should always have the right to decide when to bear a child: and they believed that women and men both had natural sex drives and that it was not wrong to indulge those drives without the intention of conceiving children. The two groups also shared the same appraisal of the social and political significance of birth control. Most of them were favorably inclined toward neo-Malthusian reasoning (at least until the 1890s, when the prevailing concern shifted to the problem of under-population rather than over-population).[46] They were also interested, increasingly, in controlling conception for eugenic purposes.[47] They were hostile to the hypocrisy of the sexual double standard and, beyond that, shared a general sense that men had become become over-sexed and that sex had been transformed into something disagreeably violent.

But above all their commitment to voluntary motherhood expressed their larger commitment to women's rights. Elizabeth Cady Stanton thought voluntary motherhood so central that on her lecture tours in 1871 she held separate afternoon meetings for *women only* (a completely unfamiliar practice at the time) and talked about "the gospel of fewer children & a healthy, happy maternity."[48] "What radical thoughts I then and there put into their heads & as they feel untrammelled, these thoughts are permanently lodged there! That is all I ask."[49] Only Ezra Heywood had gone so far as to defend a particular contraceptive device—the syringe. But the principle of woman's right to choose the number of children she would bear and when was accepted in the most conservative sections of the women's rights movement. At the First Congress of the Association for the Advancement of Women in 1873, a whole session was devoted to the theme "Enlightened Motherhood," which had voluntary motherhood as part of its meaning.[50]

The general conviction of the feminist community that women had a right to choose when to conceive a child was so strong by the end of the nineteenth century that it seems odd that they were unable to overcome their scruples against artificial contraception. The basis for the reluctance lies in their awareness that a consequence of effective contraception would be the separation of sexuality from reproduction. A state of things that permitted sexual intercourse to take place normally, even frequently, without the risk of pregnancy, inevitably seemed to nineteenth-century middle-class women as an attack on the family, as they understood the family. In the mid-Victorian sexual system, men normally conducted their sexual philandering with prostitutes; accordingly prostitution, far from being a threat to the family system, was a part of it and an important support of it. This was the common view of the time, paralleled by the belief that prostitutes knew of effective birth-control techniques. This seemed only fitting, for contraception in the 1870s was associated with sexual immorality. It did not seem, even to the most sexually liberal, that contraception could be legitimized to any extent, even for the purposes of family planning for married couples, without licensing extra-marital sex. The fact that contraception was not morally acceptable to respectable women was, from a woman's point of view, a guarantee that those women would not be a threat to her own marriage.

15

The fact that sexual intercourse often leads to conception was also a guarantee that men would marry in the first place. In the nineteenth century women needed marriage far more than men. Lacking economic independence, women needed husbands to support them, or at least to free them from a usually more humiliating economic dependence on fathers. Especially in the cities, where women were often isolated from communities, deprived of the economic and psychological support of networks of relatives, friends and neighbors, the prospect of dissolving the cement of nuclear families was frightening. In many cases children, and the prospect of children, provided that cement. Man's responsibilities for children were an important pressure for marital stability. Women, especially middle-class women, were also dependent on their children to provide them with meaningful work. The belief that motherhood was a woman's fulfillment had a material basis: parenthood was often the only creative and challenging activity in a woman's life, a key part of her self-esteem.

Legal, efficient birth control would have increased men's freedom to indulge in extra-marital sex without greatly increasing women's freedom to do so. The pressures enforcing chastity and marital fidelity on middle-class women were not only fear of illegitimate conception but a powerful combination of economic, social and psychological factors, including economic dependence, fear of rejection by husband and social support networks, internalized taboos and, hardly the least important, a socially conditioned lack of interest in sex that may have approached functional frigidity. The double standard of the Victorian sexual and family system, which had made men's sexual freedom irresponsible and oppressive to women, left most feminists convinced that increasing, rather than releasing, the taboos against extra-marital sex was in their interest, and they threw their support behind social-purity campaigns.

In short, we must forget the twentieth-century association of birth control with a trend toward sexual freedom. The voluntary motherhood propaganda of the 1870s was associated with a push toward a more restrictive, or at least a more rigidly enforced, sexual morality. Achieving voluntary motherhood by a method that would have encouraged sexual license was absolutely contrary to the felt interests of the very group that formed the main social basis for the cause—middle-class women. Separating these women from the early-twentieth-century feminists, with their interest in sexual freedom, were nearly four decades of significant social and economic changes and a general weakening of the ideology of the Lady. The ideal of the Free Lovers—responsible, open sexual encounters between equal partners—was impossible in the 1870s because men and women were not equal. A man was a man whether faithful to his wife or not. But women's sexual activities divided them into two categories—wife or prostitute. These categories were not mere ideas, but were enforced in reality by severe social and economic sanctions. The fact that so many, indeed most, Free Lovers in practice led faithful, monogamous, legally-married lives is not insignificant in this regard. It suggests that they instinctively understood that Free Love was an ideal not be realized in that time.

As voluntary motherhood was an ideology intended to encourage sexual purity, so it was also a pro-motherhood ideology. Far from debunking

16

motherhood, the voluntary motherhood advocates consistently continued the traditional Victorian mystification and sentimentalization of the mother. It is true that at the end of the nineteenth century an increasing number of feminists and elite women—that is, still a relatively small group—were choosing not to marry or become mothers. That was primarily because of their increasing interest in professional work, and the difficulty of doing such work as a wife and mother, given the normal uncooperativeness of husbands and the lack of social provisions for child care. Voluntary motherhood advocates shared the general belief that mothers of young children ought not to work outside their homes but should make mothering their full-time occupation. Suffragists argued both to make professions open to women and to ennoble the task of mothering; they argued for increased rights and opportunities for women *because* they were mothers.

The Free Lovers were equally pro-motherhood; they only wanted to separate motherhood from legal marriage.[51] They devised pro-motherhood arguments to bolster their case against marriage. Mismated couples, held together by marriage laws, made bad parents and produced inferior offspring, Free Lovers said.[52] In 1870 *Woodhull and Claflin's Weekly* editorialized, "Our marital system is the greatest obstacle to the regeneration of the race."[53]

This concern with eugenics was characteristic of nearly all feminists of the late nineteenth century. At the time eugenics was mainly seen as an implication of evolutionary theory and was picked up by many social reformers to buttress their arguments that improvement of the human condition was possible. Eugenics had not yet become a movement in itself. Feminists used eugenics arguments as if they instinctively felt that arguments based solely on women's rights had not enough power to conquer conservative and religious scruples about reproduction. So they combined eugenics and feminism to produce evocative, romantic visions of perfect motherhood. "Where boundless love prevails. . .," *Woodhull and Claflin's Weekly* wrote, "the mother who produces an inferior child will be dishonored and unhappy . . . and she who produces superior children will feel proportionately pleased. When woman attains this position, she will consider superior offspring a necessity and be apt to procreate only with superior men."[54] Free Lovers and suffragists alike used the cult of motherhood to argue for making motherhood voluntary. Involuntary motherhood, wrote Harriet Stanton Blatch, daughter of Elizabeth Cady Stanton and a prominent suffragist, is a prostitution of the maternal instinct.[55] Free Lover Rachel Campbell cried out that motherhood was being "ground to dust under the misrule of masculine ignorance and superstition."[56]

Not only was motherhood considered an exalted, sacred profession, and a profession exclusively woman's reponsibility, but for a woman to avoid it was to choose a distinctly less noble path. In arguing for the enlargement of woman's sphere, feminists envisaged combining motherhood with other activities, not rejecting motherhood. Victoria Woodhull and Tennessee Claflin wrote:

> Tis true that the special and distinctive feature of woman is that of bearing children, and that upon the exercise of her function in this regard the perpetuity of race depends. It is also true that those who pass through life failing in this special feature of

17

their mission cannot be said to have lived to the best purposes of woman's life. But while maternity should always be considered the most holy of all the functions woman is capable of, it should not be lost sight of in devotion to this, that there are as various spheres of usefulness outside of this for woman as there are for man outside of the marriage relation.[57]

Birth control was not intended to open the possibility of childlessness, but merely to give women leverage to win more recognition and dignity. Dora Forster, a Free Lover, saw in the fears of underpopulation a weapon of blackmail for women:

I hope the scarcity of children will go on until maternity is honored at least as much as the trials and hardships of soldiers campaigning in wartime. It will then be worth while to supply the nation with a sufficiency of children ... every civilized nation, having lost the power to enslave woman as mother, will be compelled to recognize her voluntary exercise of that function as by far the most important service of any class of citizens.[58]

"Oh, women of the world, arise in your strength and demand that all which stands in the path of true motherhood shall be removed from your path," wrote Lois Waisbrooker, a Free Love novelist and moral reformer.[59] Helen Gardener based a plea for women's education entirely on the argument that society needed educated mothers to produce able sons (not children, sons).

Harvard and Yale, not to mention Columbia, may continue to put a protective tariff on the brains of young men: but so long as they must get those brains from the proscribed sex, just so long will male brains remain an 'infant industry' and continue to need this protection. Stupid mothers never did and stupid mothers never will, furnish this world with brilliant sons.[60]

Clinging to the cult of motherhood was part of a broader conservatism shared by Free Lovers and suffragists—acceptance of traditional sex roles. Even the Free Lovers rejected only one factor—legal marriage—of the many that defined woman's place in the family. They did not challenge conventional conceptions of woman's passivity and limited sphere of concern.[61] In their struggles for equality the women's-rights advocates never suggested that men should share responsibility for child-raising, housekeeping, nursing, cooking. When Victoria Woodhull in the 1870s and Charlotte Perkins Gilman in the early 1900s suggested socialized child care, they assumed that only women would do the work.[62] Most feminists wanted economic independence for women, but most, too, were reluctant to recommend achieving this by turning women loose and helpless into the economic world to compete with men.[63] This attitude was conditioned by an attitude hostile to the egoistic spirit of capitalism; but the attitude was not transformed into a political position and usually appeared as a description of women's weakness, rather than an attack on the system. Failing to distinguish, or even to indicate awareness of a possible distinction between women's conditioned passivity and their equally conditioned distaste for competition and open aggression, these feminists also followed the standard Victorian rationalization of sex roles, the idea that women were morally superior. Thus the timidity and self-effacement that were the marks of women's powerlessness were made into innate virtues. Angela Heywood, for example,

18

praised women's greater ability for self-control, and, in an attribution no doubt intended to jar and titillate the reader, branded men inferior on account of their lack of sexual temperance.[64] Men's refusal to accept women as human beings she identified, similarly, as a mark of men's incapacity: ". . . man has not yet achieved himself to realize and meet a PERSON in woman. . . ."[65] In idealistic, abstract terms, no doubt such male behavior is an incapacity. Yet that conceit failed to remark on the power and privilege over women that the supposed "incapacity" gave men.

This omission is characteristic of the cult of motherhood. Indeed, what made it a cult was its one-sided failure to recognize the privileges men received from women's exclusive responsibility for parenthood. The "motherhood" of the feminists' writings was not merely the biological process of gestation and birth, but a package of social, economic and cultural functions. Although many of the nineteenth-century feminists had done substantial analysis of the historical and anthropological origins of woman's social role, they nevertheless agreed with the biological-determinist point of view that women's parental capacities had become implanted at the level of instinct, the famous "maternal instinct." That concept rested on the assumption that the qualities that parenthood requires— capacities for tenderness, self-control and patience, tolerance for tedium and detail, emotional supportiveness, dependability and warmth—were not only instinctive but sex-linked. The concept of the maternal instinct thus also in- volved a definition of the normal instinctual structure of the male that excluded these capacities, or included them only to an inferior degree; it also carried the implication that women who did not exercise these capacities, presumably through motherhood, remained unfulfilled, untrue to their destinies.

Belief in the maternal instinct reinforced the belief in the necessary spiritual connection for women between sex and reproduction, and limited the development of birth-control ideas. But the limits were set by the entire social context of women's lives, not by the intellectual timidity of their ideas. For women's "control over their own bodies" to lead to a rejection of motherhood as the *primary* vocation and measure of social worth required the existence of alternative vocations and sources of worthiness. The women's rights advocates of the 1870s and 1880s were fighting for those other opportunities, but a significant change had come only to a few privileged women, and most women faced essentially the same options that existed fifty years earlier. Thus voluntary motherhood in this period remained almost exclusively a tool for women to strengthen their positions within conventional marriages and families, not to reject them.

19

FOOTNOTES

[1] There is no space here to compensate for the unfortunate general lack of information about the Free Lovers. The book-in-progress from which this paper is taken includes a fuller discussion of who they were, the content of their ideology and practice. The interested reader may refer to the following major works of the Free Love cause:

R. D. Chapman, *Free love a Law of Nature* (New York: author 1881).

Tennessee Claflin, *The Ethics of Sexual Equality* (New York: Woodhull & Claflin, 1873).

———, *Virtue. What Is It and What It Isn't; Seduction, What It Is and What It Is Not* (New York: Woodhull & Claflin, 1872).

Ezra Heywood, *Cupid's Yokes: or, The Binding Forces of Conjugal Life* (Princeton, Mass.: Co-operative Publishing Co., n.d., probably 1876).

———, *Uncivil Liberty: An Essay to Show the Injustice and Impolicy of Ruling Woman Without Her Consent* (Princeton, Mass.: Co-operative Publishing Co., 1872).

C. L. James, *The Future Relation of the Sexes* (St. Louis: author, 1872).

Juliet Severance, *Marriage* (Chicago: M. Harman, 1901).

Victoria Claflin Woodhull, *The Scare-Crows of Sexual Slavery* (New York: Woodhull & Claflin, 1874).

———, *A Speech on the Principles of Social Freedom* (New York: Woodhull & Claflin, 1872).

———, *Tried as by Fire; or, the True and the False Socially* (New York: Woodhull & Claflin, 1874).

[2] Heywood, *Cupid's Yokes, p. 20.*

[3] Claflin, *The Ethics of Sexual Equality,* pp. 9-10.

[4] *Woodhull & Claflin's Weekly* 1, no. 6 (1870): 5.

[5] Heywood, *Cupid's Yokes,* pp. 17-18.

[6] Letter to her daughter Alice, 1874, in the Isabella Beecher Hooker Collection. Beecher Stowe Mass. This reference was brought to my attention by Ellen Dubois of SUNY-Buffalo.

[7] Alice B. Stockham, M.D., *Karezza, Ethics of Marriage* (Chicago: Alice B. Stockham & Co., 1898), pp. 84, 91-92.

[8] Theodore Stanton and Harriot Stanton Blatch, eds., *Elizabeth Cady Stanton as Revealed in Her Letters, Diary and Reminiscences* (New York: Harper & Bros., 1922), 2:210 (Diary, 9-6-1883).

[9] Ben Barker-Benfield, "The Spermatic Economy: A Nineteenth Century View of Sexuality," *Feminist Studies* 1, no. 1 (Summer 1972): 53.

[10] J.J. Rousseau, *Emile* (New York: Columbia University Teachers College, 1967), p. 132. Rousseau was, after all, a chief author of the Victorian revision of the image of woman.

[11] Dora Forster, *Sex Radicalism as Seen by an Emancipated Woman of the New Time* (Chicago: M. Harman, 1905), p. 40.

[12] Norman E. Himes, *Medical History of Contraception* (New York: Gamut Press, 1963).

[13] Heywood, *Cupid's Yokes,* pp. 19-20, 16.

[14] Ibid., pp. 19-20; *Woodhull & Claflin's Weekly* 1, no, 18 (September 10, 1870): 5.

[15] Heywood, *Cupid's Yoke,* pp. 14-15.

[16] Stockham, *Karezza,* pp. 82-83, 53.

[17] See for example, *Free Enquirer,* ed. Robert Owen and Frances Wright, (May 22, 1830), pp. 235-236.

[18] Stockham, *Karezza,* p. 86.

[19] Heywood, *Cupid's Yoke,* p. 19.

[20] Charlotte Perkins Gilman, *Women and Economics* (New York: Harper Torchbooks, 1966), pp. 38-39, 43-44, 42, 47-48, 209.

[21] In England, for example, it was not until 1891 that the courts first held against a man who forcibly kidnapped and imprisoned his wife when she left him.

[22] Ezra Heywood, *Free Speech: Report of Ezra H. Heywood's Defense before the United States Court, in Boston, April 10, 11, and 12, 1883* (Princeton, Mass.: Co-operative Publishing Co., n.d.), p. 16.

[23] Quoted in Nelson Manfred Blake, *The Road to Reno, A History of Divorce in the United States* (New York: Macmillan, 1962), p. 108, from the *New York Tribune,* May 12, 1871 and July 20, 1871.

[24] Letter of August 29, 1869, in Hooker Collection, Beecher-Stowe Mss. This reference was brought to my attention by Ellen Dubois of SUNY-Buffalo.

[25] Isabella Beecker Hooker, *Womanhood: its Sanctities and Fidelities* (Boston: Lee and Shepard, 1873), p. 26.

20

26 Elizabeth Cady Stanton Mss. No. 11, Library of Congress, undated. This reference was brought to my attention by Ellen Dubois of SUNY-Buffalo.

27 See for example, *Lucifer, The Light-Bearer*, ed. Moses Harman (Valley Falls, Kansas: 1894-1907) 18, no. 6 (October 1889): 3.

28 Victoria Woodhull, *The Scare-Crows*, p. 21. Her mention of the YMCA is a reference to the fact that Anthony Comstock, author and chief enforcer for the U.S. Post Office of the anti-obscenity laws, had begun his career in the YMCA.

29 *The Word* (Princeton, Mass.) 20, no. 9 (March 1893): 2-3. Emphasis in original.

30 See for example, the National Purity Congress of 1895, sponsored by the American Purity Alliance.

31 *Lucifer* (April 26, 1890), pp. 1-2.

32 N. a. *The Next Revolution: or Woman's Emancipation from Sex Slavery* (Valley Falls, Kansas: Lucifer Publishing Co., 1890), p. 49.

33 Ibid., pp. 8-9.

34 Linda Gordon et al., "Sexism in American Historical Writing," *Women's Studies* 1, no. 1 (Fall 1972).

35 *Lucifer* 15, no. 2 (September 1886): 3.

36 *The Word* 20 (1892-1893).

37 (Slenker) *Lucifer*, May 23, 1907; *Cyclopedia of American Boigraphy* 8: 488.

38 See for example *Lucifer* 18, no. 8 (December 1889): 3; 18, no. 6 (October 1889): 3; 18, no. 8 (December 1889): 3.

39 Heywood, *Free Speech*, pp. 17, 16.

40 Ibid., pp. 3-6. "Comstockism" also is a reference to Anthony Comstock. Noting the irony that the syringe was called by Comstock's name, Heywood continued: "To name a really good thing 'Comstock' has a sly, sinister, wily look, indicating vicious purpose; in deference to its N.Y. venders, who gave that name, the Publishers of *The Word* inserted an advertisement . . . which will hereafter appear as 'the Vaginal Syringe'; for its intelligent, humane and worthy mission should no longer be libelled by forced association with the pious scamp who thinks Congress gives him legal right of way to and control over every American Woman's Womb." At this trial, Heywood's second, he was acquitted. At his first trial, in 1877, he had been convicted, sentenced to two years, and served six months; at his third, in 1890, he was sentenced to and served two years at hard labor, an ordeal which probably caused his death a year later.

41 *The Word* 10, no. 9 (March 1893): 2-3.

42 Ibid.

43 See for example Horatio Robinson Storer, M.D., *Why Not? A Book for Every Woman* (Boston: Lee and Shepard, 1868). Note that this was the prize essay in a contest run by the A.M.A. in 1865 for the best anti-abortion tract.

44 Claflin, *Ethics;* Emanie Sachs, *The Terrible Siren, Victoria Woodhull, 1838-1927* (New York: Harper & Bros., 1928), p. 139.

45 Elizabeth Cady Stanton, Susan Anthony, Matilda Gage, eds., *History of Woman Suffrage,* 1:597-598.

46 Heywood, *Cupid's Yokes*, p. 20; see also *American Journal of Eugenics,* ed. M. Harman 1, no. 2 (September 1907); *Lucifer* (February 15, 1906; June 7, 1906; March 28, 1907; and May 11, 1905).

47 I will deal with early feminists' ideas concerning eugenics in my book.

48 Elizabeth Cady Stanton to Martha Wright, June 19, 1871, Stanton Mss. This reference was brought to my attention by Ellen Dubois of SUNY-Buffalo; see also Stanton, *Eight Years After, Reminiscences 1815-1897* (New York: Schoeken, 1971), pp. 262,297.

49 Stanton and Blatch, *Stanton as Revealed in Her Letters*, pp. 132-133.

50 *Papers and Letters,* Association for the Advancement of Women, 1873. The AAW was a conservative group formed in opposition to the Stanton-Anthony tendency. Nevertheless Chandler, a frequent contributor to Free Love journals, spoke here against undesired maternity and the identification of woman with her maternal fuction.

51 *Woodhull & Claflin's Weekly* 1, no. 20 (October 1, 1870): 10.

52 Woodhull, *Tried as by Fire*, p. 37; Lillian Harman, *The Regeneration of Society.* Speech before Manhattan Liberal Club, March 31, 1898 (Chicago: Light Bearer Library, 1900).

53 *Woodhull & Claflin's Weekly* 1, no. 20 (October 1, 1870): 10.

54 Ibid.

55 Harriot Stanton Blatch, "Voluntary Motherhood," *Transactions,* National Council of Women of 1891, ed. Rachel Foster Avery (Philadelphia: J. B. Lippincott, 1891), p. 280.

56 Rachel Campbell, *The Prodigal Daughter, or, the Price of Virtue* (Grass Valley, California, 1885), p. 3. An essay read to the New England Free Love League, 1881.

21

57 *Woodhull & Claflin's Weekly* 1, no. 14 (August 13, 1870): 4.

58 In addition to the biography by Sachs mentioned above, see also Johanna Johnston, *Mrs. Satan* (New York: G. P. Putnam's Sons, 1967), and M. M. Marberry, *Vicky, A Biography of Victoria C. Woodhull* (New York: Funk & Wagnalls, 1967).

59 From an advertisement for her novel, *Perfect Motherhood; Or, Mabel Raymond's Resolve* (New York: Murray Hill, 1890), in *The Next Revolution*.

60 Helen Hamilton Gardener, *Pulpit, Pew and Cradle* (New York: Truth Seeker Library, 1891), p. 22.

61 Even the most outspoken of the Free Lovers had conventional, role-differentiated images of sexual relations. Here is Angela Heywood, for example: "Men must not emasculate themselves for the sake of 'virtue,' they must, they will, recognize manliness and the life element of manliness as the fountain source of good manners. Women and girls demand strong, well-bred generative, vitalizing sex ability. Potency, virility, is the grand basic principle of man, and it holds him clean, sweet and elegant, to the delicacy of his counterpart." From *The Word* 14, no. 2 (June 1885): 3.

62 Woodhull, *The Scare-Crows;* Charlotte Perkins Gilman, *Concerning Children.*

63 See for example Blatch, "Voluntary Motherhood," pp. 283-284.

64 *The Word* 20, no. 8 (February 1893): 3.

65 Ibid.

22

THE PREVENTION OF CONCEPTION.

BY ISAAC PEIRCE, M.D.,

TAZEWELL, VA.

There are questions which perhaps it is better for men never to discuss, questions the discussion of which will teach nothing and which should be weighed by the individual conscience and decided within the secret workings of the individual mind. But since such a journal as the REPORTER has opened the question of preventing conception, we can but believe that good will follow its discussion, and thinking men should enter the argument with minds unbiased with what has heretofore been regarded as the moral or religious aspect of the question. Moreover, if we enter this subject at all, we should do so with perfect candor, calling a spade a spade, trying to further the good of our science, and not seeking for something about which to multiply words, never reaching a conclusion. The appearance in the REPORTER of any communication on a subject so fraught with interest to physician and patient alike, invests it in the natural order of things with much weight; therefore those who undertake such communications should see that they have considered their subject well before expressing decided opinions.

At the date when I write, there has appeared but one paper on the subject since the Editorial which prompted it—that of Dr. Blackwood, who enters directly into the argument of the question and deals out facts and opinions with a fearless hand. His courage and candor deserve admiration; but I think he is too hasty in some of his conclusions and in some respects misses the spirit of the discussion, as well as the end to be accomplished. As I see it, the question suggested by the Editor was not how to prevent criminal abortion, but conception. Every man knows the horrors of illicit love and the suffering of misguided though patient and confiding women; no man is insensible to the lifelong shame of a child thrown upon the world without knowing a father; and no man denies the wickedness of criminal abortion. No medical man doubts the suffering, and in many cases permanent injury, of the woman who practises abortion that she may escape the shame of her own wickedness, sinning doubly that she may shield herself and her destroyer from the condemnation of the world. No, it takes no one to tell us this; we all know it too well; the poor creatures come to us almost every day, broken down in health, asking for treatment, asking that their secret be protected, and are willing to undergo anything to relieve their physical suffering and escape the consequences of their error. There are others known to us who are in a worse condition, those who yield to the seducer's embrace, bear their children, and, losing all respect for self, plunge into the depths of misery and disease, becoming of those to whom Mr. Leckey thus alludes in his History of Morals in Europe: "That unhappy being whose very name is a shame to speak, who counterfeits, with a cold heart, the transports of affection; scorned and insulted as the vilest of her sex, and doomed, for the most part, to abject wretchedness and an early grave, she is in every age the perpetual symbol of the degeneration and sinfulness of man; she remains while creeds and civilizations rise and fall, the eternal priestess of humanity blasted for the sins of the people."

Now, knowing the state of things as the members of the medical profession do, and having exerted every effort toward a remedy, is there a possibility of bettering these conditions by preventing conception, as Dr. Blackwood intimates? I say No! I think it can be made plain to every man that we would but bring about a worse state of affairs, if it is possible to imagine anything worse. It is true we may prevent criminal abortion, we may lessen suffering, pauperism, and neglect among children the issue of illicit love; but we do more, and we should look at this side of the question. Is it the woman, who in the hour of her fall thinks of conception? Does a preventive suggest itself to her? Does she dream that this act will crown the love and trust she has given her seducer before the hour of its perpetration? Or is it the man, who, knowing his vows of love and marriage to be false, plans her fall and studies the question of a preventive? In eight cases out of ten, I think it is the man; and, with a sure preventive in his hands, how much stronger will be his argument, how much oftener will his persuasion meet compliance, and we all know how the falsity of his promise of marriage will increase. In the other two cases, it will require no offer of marriage, no vows of love; the desire and the preventive will be all that is required to carry them from the arms of one lover to those of another.

If Dr. Blackwood is right in teaching that a man is not culpable before the laws of God and man who interferes to prevent gestation *before* conception has occurred, as I see it, he will institute a new moral law which will be entirely different from that taught by any period of the world's history. He would make adultery a matter of no consequence so long as conception is prevented with its attendant ills. I think I am just in this view, as I think I have ground for my fear when I shudder at what would be the result of the promiscuous prevention of conception, even with the idea of stopping criminal abortion, pauperism, and suffering of illegitimate children. I believe that if we put into the hands of men a ready and sure means of preventing conception, there will be more prostitutes, fewer marriages, and more disease among women. Syphilis, gonorrhœa, and chancroid will be the next in order requiring a preventive.

I think this is the idea which has kept the question in the dark so long, and not the fear of how moralists and church people would receive it. It has been the fear of instituting a process which we could never hope to control, and which would in a short time become so universal and popular as to admit of no check, that has made cowards of the medical profession. I think we have been cowards heretofore, and the REPORTER will deserve much credit if this discussion should determine for us clearly the questions: Should we ever prevent conception? and, if so, when should we interfere? What is a ready and sure means of effecting the object? and how shall we keep the matter in our own hands? No physician should give a preventive to be used outside of the married state, any more than he should produce abortion for the sake of shielding either mother or father. And if we confine ourselves to married life, a vast number of men, both in and out of our profession, will take the ground that, child-bearing being the natural consequence of matrimony, it would always be wrong to interfere; that a woman who incurs the risk of pregnancy and childbirth should bear it as best she may, never crying out, no matter how great her suffering, always allowing her husband his "rights"; and if she fill a premature grave they would consider it a sad fate, but a fate from which there is no escape. Then, a large number of women who have safely passed their menopause are prepared to laugh at the suffering of their younger sisters, and would look upon any interference as criminal. The younger ones would show a majority on the other side, and some, as in Dr. Blackwood's case, will cry out that suicide is preferable to another pregnancy and delivery.

Assuming that there are cases—and I think few men in the medical profession will not admit that there are cases—where the prevention of conception would restore the happiness of a family, give peace to the wife, turn the husband again to the path of virtue, and prevent the life-long suffering of offspring; then the question arises, what are these cases? Dr. Blackwood is right in claiming that there is no justice in condemning children to an inheritance of syphilis, scrofula, tuberculosis, epilepsy, or imbecility, and he might have added a host of others. Where this is to be avoided, I agree with him that prevention is *imperatively demanded*. Again, women who are deformed, women whose lives are made one constant and heavy burden by pregnancy so often repeated that they seem never to have an end, women who are made miserable by the so-called "habit" of abortion, and women whose former gestation and delivery have brought them almost to the door of eternity and who are now in constant dread lest another such ordeal will complete the work and leave a motherless child or children—yes, all these women who are crying out to us from the depths of their fear and misery for help should be heard and their prayers answered. There is no other way. Men *will not* abstain from sexual intercourse, though they see their trusting wives gradually slipping from them, and if here and there one is found who yields to the entreaties of his wife to spare her from a horrible death and a premature grave, he will most often be one who is damning his own soul, ruining his life, and saving the woman from one evil only to render her more unhappy by gratifying his lust among harlots.

Recognizing the fact that there are cases in which prevention is a necessity, and that our duty is to work a reform that will prevent much of the physical suffering and mental distress among women, I am at a loss to know what are the means best suited to accomplish this and how to apply them. The number of preventive measures which have been proposed is not small, but the selection of one which we can control and its restriction to those alone who really need it are not easy. Let it become generally known that the medical profession countenances a preventive even in a few cases, and there is reason to fear this will be stretched to a license which will work much mischief to women who are already experimenting in this direction, who have no reason why they should not fulfill the God-given function which makes happy homes, and who are now only held in check by the judgment of the world. Will it not also place in the hands of men a ready argument with which to destroy the purity of loving, trusting girls? As I see it, this is the reason why to the deserving as well as the undeserving we have for so long given the old woman's advice, "to take a glass of cold water before going to bed and *nothing else.*" This is the reason why we have allowed women to lead lives of misery and seemed not to heed their cries for relief. Thoroughly awake to the fact that the prevention of conception is sometimes right, that it is sometimes a plain duty, it is to be hoped that we may yet find some means whereby we can overcome the difficulty of the problem and confer a blessing on those who will appreciate it most.

"BIRTH CONTROL."
WHAT SHALL BE THE ATTITUDE OF THE MEDICAL PROFESSION TOWARDS THE PRESENT DAY PROPAGANDA?

BY

GEO. W. KOSMAK, M.D.,
Attending Surgeon.

DURING recent years there has occurred a more or less widespread interest and agitation in the matter of regulating the size of the family by methods dealing with the prevention of conception. It would seem appropriate that the members of the medical profession could with justice partake of the discussion on this subject and give their special attention to a question which apparently concerns them directly. Moreover, through their knowledge of what is involved it might be proper that they should take the initiative in presenting this topic in its various aspects to the community. The claims which have been brought forward largely by lay persons to limit the size of the family appear very plausible and a large number of recruits have been obtained and enrolled in this wave of agitation which now seems to be sweeping the country. Lawyers, ministers, social reformers, ordinary citizens and even some physicians have been gathered into the fold and made to lend their voices to the demand for remedial legislation, among other things, in the attempt to bring about a Utopian condition. Efforts have also been made to have representative organized medical bodies lend their official endorsement to the propaganda of birth control as witnessed in a recent discussion in the New York County Medical Society. Whether the question of family limitation is a medical or a lay problem is still undecided. There are undoubtedly certain phases of the same that can be justifiably discussed from either standpoint but in many ways the two aspects may be said to overlap and care must be taken that neither side confuses the issue. The strictly medical aspects of family limitation cannot be satisfactorily discussed without considering to some extent the social side of the matter and in this way many medical men have been led to support the claims

*Read at a meeting of the New York Obstetrical Society, January 9, 1917. Appeared also in the *Medical Record*, February 17, 1917.

and contentions of lay and clerical agitators without fully realizing the import of their action.

What is the cause of this widespread propaganda relative to "birth control," so-called, which has so greatly agitated portions of Europe and has now invaded our own country.   Is it to be ascribed to an honest intent to better the world, is it another expression of the spread of the feministic movement, is it scientifically demanded, or is it merely another instance of those hysterical waves with which our civilization is so frequently assaulted?   In this connection we hear frequent references to Malthusians and Neo-Malthusians and as a matter of interest it may be well to recall that Thomas R. Malthus was an English political economist who was born in 1766 and died in 1834.   His reputation rests almost exclusively upon the views advanced in a book published in 1826, which is entitled "An Essay on the Principle of Population, or a View of Its Past and Present Effects on Human Happiness, with an Inquiry into our Prospects Respecting the Future Removal or Mitigation of the Evils which it Occasions." He held that population unchecked increases in a geometrical ratio, while food can be made to increase at furthest only in an arithmetical ratio.   He believed that powerful checks on population must be constantly in action, which may be resolved into vice, misery, and moral or prudential restraint.   Methods of preventing conception for the purpose of family limitation have agitated a considerable number of persons, both male and female, who with the best of intentions perhaps have united themselves into a variety of organizations to disseminate knowledge broadcast of contraceptive methods. Malthus has been made a sort of patron saint to serve the purposes of these bodies, which however do not seem to take into account that economic principles that may have been true in his day are now to be accepted with certain reservations. The relation of food production to increase of population is a very complex problem which depends at the present time largely on the abnormal increase in the number of consumers as compared with the producers of food.   Nevertheless the original argument sounds well and is unhesitatingly adopted to serve the desired end. The law is in fact to be appealed to in order to regulate the ultimate effects of a certain natural instinct, which in man as well as in brute creation is one of the strongest of physiological impulses.   Every desire emanating from an individual may require to be curbed at certain periods and this may perhaps be made to include the sexual impulse as well as the others.   Methods for accomplishing this end have formed the subject of discussion for layman, priest, and physician for an almost limitless period of time, and although the growth of civilization may have exercised an influence which cannot be denied there is still much to be accomplished before any unanimity of thought and action is possible.

At this point one might be led to inquire whether a widespread propaganda of birth control is called for by an assumed degeneracy of our race. On examining the contentions of those who support those movements one might be led to believe that the millenium would be reached and that poverty, crime, disease and many other ills would immediately be eradicated as soon as contraceptive measures were more readily employed. Holland is generally taken as the country in which the greatest improvement in this respect is supposed to have occurred from the introduction of contraceptive measures The birth and death records are seized upon and made to serve the purposes of the Neo-Malthusians who claim as the results of their efforts during the last generation that Dutch infantile mortality has decreased and the birth rate actually increased. Thus the birth and death rates in Amsterdam during the period 1881-85 were 37.1 and 25.1 per 1000 of population. The activities of the Neo-Malthusians began in the latter year and during the period 1906-1910 the birth rate in Amsterdam had decreased to 24.7 and the death rate to 13.1 per 1000. The infant mortality including deaths during the first year decreased from 203 per 1000 in the first period to 90 per 1000 in the second. Similar figures are presented for the Hague and Rotterdam. It is a matter of interest what reply will be made to these contentions by those lay and medical workers who believe in the prevention of infant mortality by appropriate pre- or postnatal care, by improved housing conditions, by a supervised milk supply, and similar labors which have occupied their attention in recent decades. It is also interesting in this connection to take note of the mortality among legitimate and illegitimate births presented in this same table. In the period 1880-1882 the legitimate fertility per 1000 married women was 306.4; in 1890-92 it was 296.5 and in 1900 it was 252.7. During these same periods the illegitimate fertility was 16.1, 16.3, and 11.3 per 1000 unmarried women, the same figures holding good in all large Dutch cities. The slight differences here noted can be interpreted in various ways. The declining birth date, however, might very well be used as an argument to study the subject in a very much more careful manner.

The fact that certain families, particularly among the poorer classes, are blessed with a correspondingly larger number of children than those who are better off in the world's goods is seized upon as a text for a sermon in which the biblical instruction to "go forth and multiply" is sought to be negatived. The statement may perhaps be accepted that quality and not quantity should count in the building up of a family and while this is a desideratum not to be gainsaid, the question is how can such an ideal condition be attained? Shall it be by giving sexual license to a married couple by providing them with the knowledge of methods that will prevent conception? or shall it be by teaching them moderation in this as well as in the exercise of all their physical functions? For it must be acknowledged that if we are dealing solely with husband and wife in this discussion, the prevention of conception is a matter that lies entirely within their own personal responsibilities and desires. That complete or temporary continence, however, will ever be attained is to hope for the inevitable, for the expression of the natural sexual impulse in normal individuals is one that cannot be curbed by any laws that man may make. It remains, therefore, merely to discuss measures for negativing the possible results that follow the same. Our good friends who believe in the birth control movement think that they can bring an ideal marital relationship in which conception will result from deliberate and planned effort. If they were better acquainted with the physiology and psychology of this process they would realize that this is practically impossible of attainment. Conception in all but a few instances is an accident and we know very little about the time when it takes place. The phenomena of menstruation are generally regarded as the preparation for conception but whether insemination results before or after the menstrual period is a matter that has never been fully determined or proven. There are many factors concerned with the physiology of conception which could with difficulty be placed under law and regulation. We are told that a married couple should not have children until they are ready for them. If such an important question could be decided for them, would they in the majority of cases ever voluntarily agree that the time was at hand? The maternal instinct is one that is strongly developed in most women and how often do we meet with intense disappointment in cases where after a few years of married life no issue has resulted. Inquiry in such cases has elicited the admission that methods for preventing conception were employed, that children were not desired in the early years of their married life but the mother instinct finally conquers and the desire for offspring is met by disappointment and failure We find this particularly shown in women who are approaching the thirtieth year of their life, who may have married after an active social or business career at the age of 25 or 26, and then after a period of several years' sterility it is often difficult or impossible to find any apparent reason for sterility in either husband or wife. In such cases the previous interference with the normal completion of the sexual act, to my mind, constitutes an important factor in the production of their sterility.

Another class of cases to be considered is that in which one or possibly two children have been born to a couple and in consequence of economic or other factors no further offspring is desired. What physician has not been appealed to by his patients after the normal birth of the first, second, and certainly after the third child, to prevent any further "accidents?" Can he honestly agree to such proposals and what will happen if his guaranteed advice fails, —must he agree to an abortion? What would the population of any country that consisted of one or two-family children amount to after the lapse of a century? Such families would become practically extinct, for the population would be at a standstill, and taking into consideration the relation to the normal death rate there would soon be an excess of deaths over births Notwithstanding all our efforts, however, the families with one or two children will continue to be a part of our social fabric and it is only by the introduction of new blood from families that can spare the same that progress in numbers must be looked for. It is well enough for the birth control enthusiast to talk glibly about small families, to enlarge on the element of quality instead of quantity, but we must remember that the accidents of life take away a large proportion of individuals who never reach maturity and to whom the power of procreation is never given, and who are therefore eliminated as a factor in influencing the birth rate.

Attention must also be called to the fact that society, or to call it in a more popular fashion, the better classes, can only keep up their position by the constant influx of new blood from those strata which are ordinarily regarded as of a lower plane. If the striving for better things was not present among the people what incentive would there be for them to improve the position which they may now hold? This brings me to the point of making a very broad statement that may seem most obnoxious to those whom I would class as my opponents in these views. I sincerely believe that it is necessary for the general welfare and the maintenance of an economic balance that we have a class of population that shall be characterized by *"quantity"* rather than by *"quality."* We need the "hewers of wood and drawers of water" and I would like to ask what our good friends who believe in family limitation would say if the "quantity" factor in our population had diminished as the result of their efforts to such an extent that they would have to perform these laborious tasks themselves. Would the estimable lady who considered it an honor to be arrested as a martyr to the principles advocated by Mrs. Sänger, be willing to deposit her own garbage at the river front rather than have one of the "quantity" delegated to this task for her? I repeat that we need the element of quantity in the make-up of the world's population, for it is from this quantity element that the quality is obtained.

The statement has been made that the poor ought to be afforded the same means of information in regard to the limitation of offspring that are now possessed by the rich or well to do. Is it taken into consideration that other factors may determine the numerical disproportion in the families of the "lower" classes. Laying aside the extraneous features, the carrying out of the sexual impulse is largely the result in most cases of contact, mental, spiritual, social or otherwise between the sexes. The act of conception is rarely the result of deliberation. As already stated it is usually the product of chance and accident. Considering the infrequent personal contact between the male and female members of certain unions among the so-called better classes is it to be wondered at that the opportunities for sexual stimulation are so infrequent, for we find the husband busy with commercial affairs, clubs, sports and his own amusements, while the wife has her household, her clubs, her suffrage interests, her church, her social duties, and perhaps a little time for her family, if there

330

is any, to occupy her time. Aside from a meeting at the dinner table or on the way to a social function many husbands and wives rarely come into that close personal contact which is so common among those to whom the "great light" is to be extended. It must be quite evident to any one who studies the situation from this standpoint that among the latter families, the living conditions are such that husband and wife are necessarily brought into more constant and intimate personal contact. Moreover among such people the gratification of the sex impulse is not regarded in the same light as among those who consider themselves on a higher plane in such respects. It is something accepted by these people as a matter of course and very little thought is given to the same. The fear of conception while undoubtedly present does not seem to concern them as it does those who are supposedly on a higher social level. Such people cannot ordinarily be made to see the value of contraceptive methods nor can they be made to use them in an intelligent manner.

There is another point to be considered in this discussion and that is the size of the ideal family. Shall it consist in any individual case of two, three, four or six children? Shall the size be governed by the ability of the father to support the varying numbers of individuals? Shall his family increase in proportion to his income and what shall happen in case his income declines? Is he to get any sympathy, material or otherwise, in the latter case? These are matters that do not seem to have been considered by the advocates of birth control. Can it be said that the ideal state has been reached when after the desired number in a family has been attained, a physician can be consulted who, by the provisions of the law is entitled to inform this couple as to how further conceptions may be avoided?

Let us examine briefly the various aspects of the question of family limitation or birth control. For the purpose of discussion we may divide the same into the medical, moral, economic and legal. As physicians we might be expected to give especial attention to the medical aspects of the case. That methods of limiting or preventing conception in a large number of married people is a necessity cannot be denied. Cases of tuberculosis, heart disease and other constitutional or mental disturbances may contraindicate conception. It is needless to examine this aspect of the matter further as the contention is admitted. Physicians are, however, now given full power in this class of cases to recommend whatever methods of contraception they may think necessary. To those who advocate a more widespread knowledge of contraceptive methods these facts have been seized upon and made the subject of an argument for complete abolition. A well known phthisiotherapist dramatically recites the evils of pregnancy in tuberculous women and considers that it is an act of the highest charity to inform such a patient of the necessity of preventing conception. Surely no one will deny this contention but should this be made a part of an argument to change the accepted attitude of the profession towards contraceptive methods as they are ordinarily practiced?

Patients with advanced cardiac disease should be similarly restricted in
their procreative faculties, but is it not a fact that sexual intercourse
itself may be a source of danger in such cases? Similar instances might be
cited in other diseases mental and physical and the necessity of contraceptive measures in such cases is acknowledged but should this serve
as an argument in that larger class where no physical or mental contraindications exist, to whom such methods are to be given simply that
they may limit or entirely avoid the birth of children, or for other purposes? It is claimed in this country the frequency of induced abortions is
entirely due to the lack of knowledge in these important matters. Let us
examine this statement somewhat more closely. I believe that if statistics could be gathered from the so-called professional abortionists, whether
doctors or midwives, it would be found that their clientele is not necessarily recruited from among women of large families. It is more probably
limited to women who are either unmarried or who have no desire for
children and that only a small proportion of cases comes from mothers
with large families. It is also claimed that the bearing of four children
is about all that the average woman can stand, but it is a matter of common experience that many women who have had but one child are more
seriously damaged as far as their procreative faculties are concerned than
many women who have given birth to a half dozen children. We find,
moreover, that with the advance of the obstetric art and the proper care
of the patient before and after labor that the complications of pregnancy
and labor are greatly diminished. We do not, for example, meet with
the extensive lacerations of the cervix and perineal floor at the present
time that were prevalent even a decade ago, and in the field of puerperal
sepsis the number of cases has decidedly diminished, although here there
is still room for improvement. The growth of our maternity hospitals
and the constantly increasing number of women who look for expert
obstetrical care will have its reward in the decreased mortality and
morbidity of childbed. The number of children which any woman can
bear is entirely a question of individuality in so far as the medical aspects
of this question are concerned. If the medical profession lends its aid and
support to any movement which tends to inculcate in the minds of the
people the necessity for artifically controlling the birth rate of the nation,
the final result will be a deplorable one and can only reflect disparagingly
on those who have made this unfortunate move. The physician must
carefully divorce his activities from any social or economic movements
that have been developed from this subject. It is stated that family
limitation results in better children and a lower death rate. Let us
examine the statistics gathered in the State of New York during the
years 1906 and 1916. We find, for example, that the infant mortality
in 1906 was 148 per 1000 births and that during the succeeding ten years
there was a steady decline until 1915 the mortality rate was 99 per 1000.
During this same period the number of births had increased from approxi-

mately 183,000 to 242,000 and the number of deaths at all ages from 141,000 to 146,000. In other words, the rates per 1000 population of births and deaths were in 1906, 21.8 births and 16.8 deaths; whereas in 1915 the births were 24 and the deaths 14.5 per 1000 population. This table which is given in detail below*, show sconclusively that the number of deaths in proportion to the number of births has steadily declined and that, in other words, our birth rate is slightly though steadily increasing and that our infant mortality among children of less than a year, is steadily declining. Now to what can we attribute this favorable change? Is it through any decrease in the number of children in the individual family, as the result of contraceptive knowledge, or can it not be more sensibly ascribed to better food, better housing conditions, better care, and similar factors, which have been the aim of health authorities, social workers and physicians to improve. It certainly cannot be claimed by the most enthusiastic advocates of birth control methods that during the past decade their efforts already brought about this successful issue. The statistical figures referred to above probably apply to every other State in the Union.

In discussing the economic aspects of birth control a physician is not really in a position to judge the matter satisfactorily. His opinion is merely guided by the impressions of his environment and it is necessary to examine very closely a subject of this kind in order not to be misled. There is no doubt that a couple in poor or moderate circumstances with a

*STATE OF NEW YORK

BIRTHS, DEATHS AND INFANT MORTALITY WITH CORRESPONDING RATES.

Year	Births	Deaths	Rates per 1000 population		Infant Mortality	
			Births	Deaths	Deaths Under 1 year	Rate per 1000 births
1906	183,012	141,099	21.8	16.8	27,114	148
1907	196,020	147,130	22.8	17.1	28,021	143
1908	203,159	138,912	23.1	15.8	26,561	131
1909	202,656	140,261	22.6	15.6	26,077	129
1910	213,235	147,710	23.3	16.1	27,534	129
1911	221,678	145,912	22.7	15.0	25,310	114
1912	227,120	142,377	23.8	14.9	24,061	109
1913	228,713	145,271	23.5	14.9	25,044	109
1914	240,038	145,476	24.2	14.7	23,731	99
1915	242,950	146,892	24.0	14.5	24,079	99

large number of children find themselves engaged in a constant struggle
to make both ends meet, yet who shall say that the general physical
condition of these children of large families is necessarily poorer than
that in which a smaller number have been born? It is a matter of common observation that where poverty holds forth, the mother with one or
two children may be equally pinched as regards the necessities of life as
where a larger number are present. It is frequently stated that poverty
leads to fertility, although I doubt very much whether this can actually be
proved. We are told that among the poor the stork pays an annual visit
and that one pregnancy is scarcely recovered from before another is
present. Although this cannot be denied in many instances, yet it should
not be accepted as a condition that holds good in every poor family.
We find numberless families among the poorer classes in which two or
three children constitute the limit, and very frequently an interval of
several years has elapsed between their arrival and yet no efforts of birth
control has been practised. It is an attempt to reach these particular
people which has developed the so-called birth control clinics that have
been largely supported and controlled by lay persons. Their principal
claim to existence has been to furnish this class of the population with the
means for preventing conception without much question. In few cases
apparently have proper medical arrangements been made to determine
the necessity or safety of their employment. Does it seem fair to promul-
gate such knowledge in an unrestricted manner to those who are usually
too ignorant to make use of the same? Would it not be better to make an
attempt to arrive at a similar end by educational means, means which
remain to be developed in the future as the occasion arises. The teaching
that by means of small families a married couple's economic status will be
improved is one that appeals to many and yet are we going about the thing
in the right way? Is it safe to interfere with natural processes through
the promulgation of knowledge by those ignorant of the results which
may possibly be brought about by their activities? The economic solution of the difficulties attending the production of large families had
better be arrived at in other ways, ways to be developed by men and
women who are guided by higher principles than the attainment of self
glorification and sensationalism.

The moral aspects of the question involved in birth control do not
usually come within the province of the physician. It is, however, a
matter of common observation by physicians that where the sexual relations have been interfered with, trouble between the partners to a marriage contract is sure to result sooner or later. Instances have undoubtedly
come up in the practice of every physician that have convinced him of
the moral dangers that attend a continuous practice of contraceptive
methods, no matter what their character. We do know, however, that
harsh as the statement may seem, that fear of conception has contributed
to the virtue of many unmarried girls, and likewise kept many boys

straight. Will not the indiscriminate distribution of birth control literature such as is certain to follow any relaxation of legal restrictions such
as those proposed, lead to an increase of immorality that will be difficult
to counteract. Moreover the freedom with which such matters are now
discussed by the press and public must have an unfortunate effect on the
morals of all our young people.

As for the legal aspects involved in this agitation attention may be
called to those sections of the penal code which have caused so much
anguish to our propagandist friends. It appears that there are national,
state and local laws which make it a criminal offence, punishable by
fine or imprisonment, or both, to print, publish or impart information
regarding the control of human off-spring by artificial methods of preventing conception. A Birth Control League was organized a few years ago
in this country for the purpose of reforming these laws, which are claimed
to "result in widespread evil and while they do not prevent contraceptive
knowledge of a more or less vague or harmful character being spread
among the people, these repressive laws do actually hinder information
that is reliable and has been ascertained by the most competent medical
and scientific authorities to be proper for dissemination systematically
among those persons who stand in greatest need of it."

The League specifically declares that "to classify purely scientific
information regarding human contraception as obscene, as our present
laws do, is itself an act affording a most disgraceful example of intolerable
indecency." It is claimed that the highest scientific authority constitutes
the basis for the demand that such changes be made. But when we
examine the pamphlets that have been circulated in this campaign we
find that they are a mixture of arrant nonsense, misinformation, false
reports and in some cases seditious libels on the medical profession.
In most instances they have been compiled by women and in some of these
pamphlets which have been circulated very largely, methods are proposed
which are known to be faulty and in some instances can only be characterized as filthy. Moreover, a number of proprietary articles are recommended, and it is understood, sold by these individuals at largely advanced
prices. In view of their character the claims of these non-medical advisors
can safely be dismissed without being dignified by further professional
comment.

In New York State the penal code as expressed in the celebrated
paragraph No. 1142, makes the general dissemination of contraceptive
knowledge a misdemeanor punishable by fine or imprisonment. It is
frequently stated that the penal code places these subjects on the same
basis as obscene articles but there is nothing in the paragraph to sustain
this contention. Moreover, a succeeding section (paragraph No. 1145)
of the penal law grants specifically to a physician the necessary right to
prescribe to his patients whatever articles or instruments are necessary
for the purpose referred to or for the prevention of disease. A Supreme

Court justice (Mr. Justice Kelby of the Second Department, Brooklyn) has, in fact, recently handed down an opinion that section 1145 amounts to a declaration by the legislature that conception shall not be prevented except when physicians in good faith believe it to be inimical to the health of the particular individual under a physician's care. The opinion is a lengthy one and was published in detail in the *New York Law Journal* of December 5, 1916. This would seem to make clear and dismiss the contentions advanced by so many physicians that the law now prohibits them from the proper exercise of their professional functions.

In concluding this paper the writer desires to draw attention to a few cardinal facts. The propaganda of birth control as now developed cannot be ignored by the profession. We must be prepared to assume a certain responsibility in the matter. It is necessary for us to separate and weigh carefully the arguments advanced in favor of legally disseminating knowledge relating to contraceptive measures. Of the necessity or advisability of the latter in some cases there is no denial but advice thus given constitutes a personal and privileged communication between the doctor and patient in which common sense rather than statute-book law should dictate the course to be pursued. How can such a case honestly treated ever come before a judge or jury for trial? On the other hand if such information is scattered broadcast by non-medical persons, often with ulterior motives, will this be for the good of the community? Is knowledge thus imparted not a potential danger to the recipient notwithstanding that it is labelled harmless by its distributors.

As practically all those in favor of birth control seem to consider a change in the penal code necessary to the proper development and exercise of this movement, will not this latitude result, among other things, in the more widespread practice of the professional abortionists who have already assumed the practice of contraceptive measures as a part of their specialty and would profit by any relaxation in the law such as that contemplated. With the aid of the statute as its stands it is possible to control to a large degree such methods of illegal practice. Notwithstanding the argument that the prevention of conception and the production of unlawful abortion should not be considered in the same class, they are practically so as far as the activities of these illegal practitioners is concerned. The profession should give its undivided aid and support to every legal measure that will aid in the restraint and prosecution of these vampires.

What therefore shall be the attitude of the medical profession in regard to this question? Shall we listen to the unrestrained harangue of the reformer, usually a lay person with little conception of the medical aspects involved, who within the narrow bounds of his or her vision sees the solution of all the faults of our social system eliminated through the dissemination of knowledge regarding contraceptive measures? Shall we lend ourselves to the spirit of license which such sentiments naturally must convey? Eminent physicians have declared that sexual continence is not productive of the many disorders which have been attributed to it. Shall we instill into the minds of our patients a negative version of this truth? Would it not be more dignified and appropriate if we as physicians, both individually and collectively, divorce ourselves from this sensational propaganda and when called upon for advice by our patients in these matters, study the conditions carefully in each case and counsel them as the occasion demands, freeing their minds at the same time from the nauseating slush that now characterizes the subject. The lay propagandists seem to have entirely misinterpreted the attitude of those members of the profession who are opposed to their methods? It is to the unrestrained dissemination of so-called information regarding the prevention of conception by irresponsible and equally ignorant individuals that the profession objects, and likewise any change in the penal code which would let down the bars to doubtful and irregular practices. The arguments of the propagandists for the limitation of offspring are perhaps true in some ways, but will the situation be bettered by the procedures they advocate? Would it not be better to develop educational methods either by individual or cooperative instruction by which restraint can be taught to husbands primarily and wives secondarily, by which licence as now advocated. will be superseded by moderation and common sense. The value of continence is not referred to in any of the numerous pamphlets issued by the propagandists— "have intercourse as often as you want," they say to their followers, "we will tell you how to avoid the consequences." That is the sum and substance of their plaint.

Education in a broad sense can and will contribute more to the solution of this problem, as of others, than ill-advised legislative enactments and sensational propagandist movements. No need has yet been shown for the widespread adoption of methods of family limitation nor have the results claimed by the propagandists in other countries been based on statistical evidence which extends over a sufficient length of time to warrant its unqualified acceptance. There are too many pitfalls associated with a movement of this kind to warrant its general adoption and the profession is already fully cognizant of the necessary knowledge and supplied with necessary powers that will allow it to properly advise its clientele in such matters without the assistance of non-medical bodies or a change in the laws specifically relating to this question. The wave of hysteria which is now sweeping over the country and carrying along with it many minds, both weak and strong, will soon dash itself against the rock of truth and common sense and thus let us hope, finally and completely disappear.

TEN GOOD REASONS *for* BIRTH CONTROL

In succeeding numbers of the BIRTH CONTROL REVIEW, we shall give Ten Reasons, one by one why women should be given the power, without interference by the law, to regulate the size of their families, using as the means:

Birth Control

The Use of Harmless and Effective, Mechanical or Chemical Methods of Prevention, called Contraceptives.

Our reasons will be supported by statements of authorities and will be based on personal, social and international considerations.

Reason I—WOMAN'S RIGHT

The first reason is nearest home. It is Woman's Right, in a democracy, to decide whether and when she will be a mother.

Every fair-minded man and woman will see the justice of this claim. A sense of dignity, freedom and responsibility toward motherhood is backed by the support of the representative thinkers of our day.

Here is what a few of them say:—

> *To create a race of well-born children it it essential that the function of motherhood should be elevated to a position of dignity and this is impossible as long as conception remains a matter of chance.*
>
> Declaration of Principles of American Birth Control League.

"WOMEN'S desire for freedom is born of the feminine spirit, which is the absolute, elemental inner urge of womanhood. It is the strongest force in her nature; it cannot be destroyed. The chief obstacles to the normal expression of this force are undesired pregnancy and the burden of unwanted children. Society, in dealing with the feminine spirit . . . can resort to violence in an effort to enslave the elemental urge of womanhood, making of woman a mere instrument of reproduction and punishing her when she revolts. Or, it can permit her to choose whether she shall become a mother and how many children she will have. It can go on crushing what is unc'rushable, or it can recognize woman's claim to freedom, and cease to impose destructive barriers. If we choose the latter course we must not only remove all restrictions on the use of contraceptives, but we must legalize and encourage their use."

MARGARET SANGER.

"THE emancipation of women would be impossible, inconceivable, without the voluntary control of reproduction. The relation of Birth Control to the feminist movement is comparable to the relation which the foundation of a house bears to the superstructure. It is essential, fundamental, not only to the emancipation of women, but to the contemplation of their emancipation. Women cannot be free, cannot develop their potentialities, cannot even begin to plan their lives, as long as they are subject to haphazard pregnancies."

EDITH HOUGHTON HOOKER.

"ARTIFICIAL Birth Control will further increase the independence of women, and their opportunities, besides maternity, of effective self-expression."

J. ARTHUR THOMSON.

"KNOWLEDGE of how to regulate the size of the family is in the United States a class privilege. The organized movement for the emancipation of women does not demand for wives unhindered access to knowledge of the means of limiting the family. The movement is in the hands of women of the classes, who already have such access. And yet, to hosts of hollow-eyed mothers, release from the bearing of unwanted children would have a thousand times the practical value of access to the professions, or the right to vote and hold office. Until she is rid of bondage to the results of the sexual demands lawfully made upon her by her husband, what is called 'emancipation' is a mockery to the wife of the poor man."

EDWARD ALSWORTH ROSS.

INVOLUNTARY STERILIZATION IN THE UNITED STATES: A SURGICAL SOLUTION

PHILIP R. REILLY

Eunice Kennedy Shriver Center,
Waltham, Massachusetts 02254 USA

ABSTRACT

Although the eugenics movement in the United States flourished during the first quarter of the 20th Century, its roots lie in concerns over the cost of caring for "defective" persons, concerns that first became manifest in the 19th Century. The history of state-supported programs of involuntary sterilization indicates that this "surgical solution" persisted until the 1950s. A review of the archives of prominent eugenicists, the records of eugenic organizations, important legal cases, and state reports indicates that public support for the involuntary sterilization of insane and retarded persons was broad and sustained.

During the early 1930s there was a dramatic increase in the number of sterilizations performed upon mildly retarded young women. This change in policy was a product of the Depression. Institutional officials were concerned that such women might bear children for whom they could not provide adequate parental care, and thus would put more demands on strained social services. There is little evidence to suggest that the excesses of the Nazi sterilization program (initiated in 1934) altered American programs. Data are presented here to show that a number of state-supported eugenic sterilization programs were quite active long after scientists had refuted the eugenic thesis.

BACKGROUND

AT THE CLOSE of the 19th Century in the United States several distinct developments coalesced to create a climate favorable to the rise of sterilization programs aimed at criminals, the insane, and feebleminded persons. Evolutionary theory demanded a biological view of man. Francis Galton (Darwin's cousin), an eminent scientist, invented the science of eugenics (Galton, 1869, 1874, 1883) which he fitted to the tenets of Darwinism.

In the United States concern about defective persons could be found in many quarters. Southern whites who opposed miscegenation sought intellectual proof that the Negro was inferior. American criminologists and prison officials (Boies, 1893) were heavily influenced by Lombroso's arguments that most criminal behavior was biologically determined (Lombroso-Ferrero, 1972). Among the physicians and social workers who ran the nation's asy-

lums there was growing despair as the mid-century thesis (Sequin, 1846) that the retarded and insane were educable faded. About 1880, physicians who were doing research into the causes of idiocy and insanity developed the notion of a "neuropathic diathesis" (Kerlin, 1881) that relied on hereditary factors to explain problems as diverse as alcoholism, epilepsy, and crime. As one investigator wrote, "there is every reason to suppose epilepsy in the children may have its hereditary predisposition in some form of habitual crime on the part of the parent" (Clarke, 1879).

The most important event preceding the rise of sterilization programs was probably the publication of Richard Dugdale's (1875) study of the "Jukes," a New York family with a propensity for almshouses, taverns, brothels, and jails. The "Jukes" spawned a new field in sociology: extended field studies of degenerate families, a field that reached its apogee at the Eugenics Record Office (ERO) in Cold

Spring Harbor (Danielson and Davenport, 1912; Estabrook and Davenport, 1912). From 1910 through 1914 more than 120 articles about eugenics appeared in magazines, a volume of print making it one of the nation's favorite topics.

Soon after becoming Director of the Station for Experimental Evolution at Cold Spring Harbor, Charles B. Davenport, one of the earliest American champions of Mendelian genetics, moved to apply the principle of particulate inheritance to humans. Among his most important colleagues was H. H. Goddard, who published an immensely popular study of the "Kallikaks," a family that he claimed had an eminent line and a degenerate line running in parallel over many generations. The book did much to rationalize principles of negative eugenics (Goddard, 1912).

During the period from 1890 to 1917 the United States was washed by a tidal wave of immigrants. Assimilation was painful. The economy was so perturbed by the dramatic expansion of the labor pool that, despite a commitment to internationalism, even the great unions called for restrictions on immigration (Higham, 1965). From 1875 on, proposals to curb immigration were constantly before the Congress. Starting with the "Chinese Exclusion Acts" in the early 1880s, the federal government gradually built its legal seawalls higher. During the 1890s the Boston-based Immigration Restriction League sought to justify legal barriers to entry on the grounds that some races were inferior to the average American stock (Ludmerer, 1972). This argument became a major issue in American eugenics.

The late 19th Century spawned numerous plans to control the "germ plasm" of unfit individuals. In many states young retarded women were institutionalized during their reproductive years. State laws were passed to forbid marriage by alcoholics, epileptics, the retarded, and persons with chronic diseases (Davenport, 1913a). Some legislatures considered proposals to castrate criminals (Daniel, 1907), and a few superintendents of asylums actually engaged in mass castration (Daniel, 1894). It was at this juncture that a technological innovation helped to reorient social policy.

THE SURGICAL SOLUTION

The first American case report of a vasec-

tomy was published by Ochsner, a young Chicago surgeon. Dissatisfied with castration as a therapy for severe prostatic hypertrophy, he guessed that cutting the vasa deferentia might cause the tissue to involute. But when his patients told him that after the vasectomy they noted no impairment of sexual desire or function, he immediately grasped the eugenic implications of the new operation. Ochsner argued that vasectomy could eliminate criminality inherited from the "father's side" and that it "could reasonably be suggested for chronic inebriates, imbeciles, perverts and paupers" (Ochsner, 1899).

Three years later Sharp, a surgeon at the Indiana Reformatory, reported the first large study on the effects of vasectomy. He claimed that his 42 vasectomized patients felt stronger, slept better, performed more satisfactorily in the prison school, and felt less desire to masturbate! Sharp urged physicians to lobby for a law to empower directors of state institutions "to render every male sterile who passes its portals, whether it be almshouse, insane asylum, institute for the feebleminded, reformatory or prison" (Sharp, 1902:414).

During the next few years there was a spate of articles by physicians claiming that vasectomy offered a solution to the problem of limiting the births of defective persons. Some physicians began to lobby vigorously for mass sterilization. For example, a Philadelphia urologist drafted a compulsory sterilization law that passed both legislative bodies in Pennsylvania, but died under a gubernatorial veto (Mears, 1909).

In 1907 the Governor of Indiana signed the nation's first sterilization law. It initiated the involuntary sterilization of any habitual criminal, rapist, idiot, or imbecile committed to a state institution and diagnosed by physicians as "unimprovable." After having operated on 200 Indiana prisoners, Sharp quickly emerged as the national authority on eugenic sterilization. A tireless advocate, he even underwrote the publication of a pamphlet, "Vasectomy" (Sharp, 1909). In it he affixed tear-out post cards so that readers could mail a preprinted statement supporting compulsory sterilization laws to their legislative representatives!

Although the simplicity of vasectomy focused their attention upon defective men, the eugenicists were also concerned about defective women. Salpingectomy, the surgical closure of the Fallopian tubes, or oviducts, was not yet perfected as a surgical operation, and the morbidity after intra-abdominal operations was high. Eugenic theoreticians had little choice but to support long-term segregation of feebleminded women. They were, however, comforted in their belief that most feebleminded women became prostitutes and were frequently rendered sterile by pelvic inflammatory disease (Ochsner, 1899).

Pro-sterilization arguments peaked in the medical literature in 1910, when roughly one-half of the 40 articles published since 1900 on the subject appeared. These articles almost unanimously favored involuntary sterilization of the feebleminded. As time went by, physician advocates suggested casting the eugenic nets more widely. Appeals to colleagues that they lobby for enabling laws were commonly heard at annual meetings of state medical societies (Reilly, 1983a). At the annual meeting of the American Medical Association, Sharp enthralled his listeners with reports on a series of 456 vasectomies performed upon criminals in Indiana. After hearing him, Rosenwasser, a New Jersey official, announced that he would seek a bill for the compulsory sterilization of habitual criminals in his state (Sharp, 1907). New Jersey enacted such a law eighteen months later.

The most successful physician lobbyist was F. W. Hatch, Secretary of the State Lunacy Commission in California. In 1909, he drafted a sterilization law and helped to convince the legislature (which was highly sensitive to eugenic issues because of the influx of "racially inferior" Chinese and Mexicans) to adopt it. After the law was enacted, Hatch was appointed General Superintendent of State Hospitals and was authorized to implement the new law. Until his death in 1924, Hatch directed eugenic sterilization programs in ten state hospitals and approved 3000 sterilizations — nearly half of the nation's total (Popenoe, 1933).

By 1912, support for eugenical sterilization was widespread among physicians. Even eminent professors such as Lewellys Barker (who succeeded Osler as Physician-in-Chief at The Johns Hopkins Hospital), cautiously favored such programs (Barker, 1910). G. F. Lydston, a prominent Chicago surgeon, was an outspoken advocate of radical eugenics policies (Lydston, 1912). The editor of the *Texas Medical Journal* regularly published pro-sterilization articles (Daniel, 1909).

Between 1907 and 1913, 16 legislatures passed sterilization bills, 12 of which became law and 4 of which were vetoed. The evidence is circumstantial but strong (Reilly, 1983a) that a mere handful of activists played a key role in pushing this legislation. Sharp's work in Indiana and Hatch's efforts in California were obviously influential. In New Jersey, Dr. David Weeks, Chief Physician at the Village for Epileptics, lobbied for and later implemented a sterilization law (*Smith v. Board of Examiners*, 1913). In Oregon, an activist woman physician spearheaded a pro-sterilization drive (Owens Adair, 1905).

HARRY HAMILTON LAUGHLIN

The history of American eugenics, especially of involuntary sterilization, is the chronicle of a small, dedicated group of activists whose ideas attracted widespread interest in society. At the center stood Harry Hamilton Laughlin, a Missouri school teacher recruited by Davenport in 1910, and a man who devoted his life to the cause of eugenics (Haller, 1973).

In 1907 Laughlin, after some restless years of high-school teaching, obtained a post as a biology instructor at a small college. Excited by the rediscovery of Mendelian genetics, he plunged into breeding experiments and wrote to Davenport for advice in analyzing his results (Laughlin, 1907). Correspondence led to an invitation to study genetics at Cold Spring Harbor for the summer, and Laughlin returned from his visit to New York with renewed commitment to genetics. Over the next two years he assisted Davenport in gathering data for his studies. In the autumn of 1910 he moved to Cold Spring Harbor to become the first and only Superintendent of the Eugenics Record Office (ERO), a post he held for 29 years (Hassencahl, 1980).

Laughlin worked tirelessly to develop the ERO. His two major tasks were (1) to train an army of field workers (young women who would work at state hospitals and asylums to amass pedigree studies), and (2) to store and index the massive amount of material that this army generated. A meticulous person, Laughlin triple-indexed the pedigrees and stored them in fire-proof vaults. Today these hundreds of thousands of cards slumber peacefully in a cellar at the University of Minnesota. When Laughlin arrived at Cold Spring Harbor, the first field workers were already identifying persons who might be at high risk for bearing feebleminded children. By the close of 1912, when he participated in a committee to study "the best practical means of cutting off the defective germ plasm in the American population" (Hassencahl, 1980: 95), Laughlin strongly favored involuntary sterilization. This blue-ribbon group concluded that "approximately 10 percent of our population, primarily through inherent defect and weakness, are an economic and moral burden on the 90 percent and a constant source of danger to the national and racial life" (Davenport, 1913b: 94). Laughlin reduced the committee's work to publishable form (Laughlin, 1914a,b). The longer of the two monographs, an exhaustive legal study, included a model bill that tried to cast the eugenics net as widely as possible without violating the U.S. Constitution. With this publication Laughlin emerged as a leading figure in the eugenics movement. He began to receive important speaking invitations, such as one to the First National Conference on Race Betterment, in January, 1914 (Hassencahl, 1980).

The war years (1914–1918) slowed the eugenics movement. American eyes turned to the trenches, and the tide of immigrants from southeastern Europe subsided. For Laughlin it was a period of consolidation. Eager to rub shoulders with leading geneticists, he earned a doctorate in biology at Princeton University and published his only really scientific papers (on mitosis in the root tip of the onion, e.g., Laughlin, 1918).

During the early 1920s Laughlin became the nation's leading expert on the twin eugenic policies of restrictive immigration and selective sterilization. In 1920 the Bureau of the Census released the results of a demographic study of 634 "institutions for the care of defective, dependent and delinquent classes" (Hassencahl, 1980: 171) that was based largely on survey work that he had performed for it. His finding that immigrants were over-represented in these institutions came to the attention of the House Committee on Immigration and Naturalization. In April, 1920, Laughlin testified before that body, and warned that the nation's hospitals would soon be filled with immigrants (Laughlin, 1923a). His testimony provided Albert Johnson, the Chairman of the Committee, with the scientific ammunition he needed in his campaign for a restrictive immigration law. In 1921 he appointed Laughlin an "Expert Eugenical Agent" and charged him to study "alien inmates and inmates of recent foreign extraction in the several state institutions for the socially inadequate" (Hassencahl, 1980: 182). The goal was to show irrefutably that the foreign-born were draining American resources.

The plan was simple. Starting with a set of census figures that described the portion of the American population represented by each national group (e.g., Italian, Swedish), Laughlin calculated the number of each group that he expected to find in the institutional population. If a greater-than-expected number appeared in his survey of 93 institutions, then that group was contributing a disproportionate share of socially inadequate citizens to the American melting pot. Laughlin (1923b) found only 91 per cent of the expected number of native whites in the institutional sample, but determined that the institutions housed 125 per cent of the expected number of foreign-born persons and 143 per cent of those to be expected from southeastern Europe. He concluded that while "making all logical allowances for environmental conditions, which may be unfavorable to the immigrant, the recent immigrants (largely from southern and eastern Europe), as a whole, present a higher percentage of inborn socially inadequate qualities than do the older stocks" (Laughlin, 1923b: 755). As the Committee commissioned no other studies, Laughlin's findings helped to build a staunchly pro-restrictionist record.

Laughlin's report may not have swayed many votes in an already restrictionist Congress (the Immigration Act of 1921 preceded his work), but it did help to entrench a national-origins quota system that was much more ambitious than earlier plans. The system, which limited the annual immigration of people from each European country to three per cent of the number of Americans that had claimed that country as place of origin in the Census of 1890, sharply curtailed the flow from southeastern Europe (Garis, 1927). Laughlin's report rationalized that reference base. He showed that people from southeastern Europe were difficult to assimilate. Would it not be unfair to open the gates to those who were unlikely to adapt to new ways? Would it not be unfair to bias immigration flow in favor of newcomers? The succeeding Immigration Act of 1924 was the greatest triumph of the eugenics movement.

In 1920 the indefatigable Laughlin finished his exhaustive study of eugenical sterilization. The 1300-page manuscript, which strongly favored involuntary sterilization of institutionalized persons, could not be published without philanthropic support. Both the Carnegie Institution (Hassencahl, 1980) and the Rockefeller Foundation (Davis, 1921) refused to underwrite the cost. At this juncture, Harry Olson, a Chicago judge and a staunch eugenicist, arranged for the Chicago Municipal Psychopathic Laboratory (devoted to biological studies of crime) to publish the work (Laughlin, 1922). It solidified Laughlin's standing with the inner circle of eugenicists.

Throughout the 1930s Laughlin was a tireless advocate of sterilization, but his stature in the scientific community deteriorated as the science of genetics matured. After Davenport retired in 1934, John C. Merriam, President of the Carnegie Institution of Washington, began to reduce Laughlin's budget at the Eugenics Record Office. Nevertheless, Laughlin remained a prominent figure and was frequently called upon to advise civic groups about eugenic policy. For example, he acted as special consultant to the State of Connecticut in its survey of "human resources." Then, in 1938, President Merriam cut off Laughlin's research funds and guaranteed his salary for only three months longer (Merriam, 1938). In a matter of days Laughlin ended his 29 years at Cold Spring Harbor, and the Eugenics Record Office was closed (Hassencahl, 1980). Laughlin died in 1942.

THE EARLY STERILIZATION LAWS

In studying the rapid rise of the early sterilization legislation, one is hampered by a paucity of state legislative historical materials. Fortunately, a few studies (Rhode Island State Library Legislative Reference Bureau, 1913) shed light on the origin of th legislation. The

writings of Harry H. Laughlin also provide much material.

Four small but influential groups lobbied hard for these laws: physicians (especially those working at state facilities), scientific eugenicists (including prominent biologists like David Starr Jordan, President of Stanford University), lawyers and judges, and a striking number of members of the nation's richest families. There were, of course, opponents as well. But, except for a handful of academic sociologists and social workers, they were less visible and less vocal. Having already mentioned some scientists and physicians, I shall briefly note the other supporters of this legislation.

Among the influential lawyers who pushed for sterilization laws was Eugene Smith, President of the National Prison Association. He viewed sterilization as the only solution to a dangerously spiraling prison budget (Smith, 1908). Warren Foster (1909), Senior Judge of New York County, argued in popular periodicals that recidivists should be sterilized. He assured his readers that scientists had proved that criminality was hereditary, and that a compulsory sterilization law would not violate the U.S. Constitution. His campaign provoked a critical editorial by *The New York Times*, but in 1912 New York did enact a sterilization law.

The enthusiastic support that America's wealthiest families provided to the eugenics movement is a curious feature of its history. First among many was Mrs. E. H. Harriman, who almost single-handedly supported the Eugenics Record Office in its first five years. The second largest financial supporter of the ERO was John D. Rockefeller, who gave it $400 each month (Davenport, 1911). Other famous eugenic philanthropists included Dr. John Harvey Kellogg (brother to the cereal magnate), who organized the First Race Betterment Conference, held in 1914, and Samuel Fels, Philadelphia soap manufacturer. Theodore Roosevelt was an ardent eugenicist, one who urged Americans to have large families in order to avoid racial dilution by the weaker immigrant stock (Roosevelt, 1914).

Of the vocal opponents of the eugenics movement, Alexander Johnson and Franz Boas were among the most important. Johnson, leader of the National Conference of Charities and Correction, thought that sterilization was less humane than institutional segregation. He dreamed of "orderly celibate communities segregated from the body politic," where the feebleminded and insane would be safe and could be largely self-supporting (Johnson, 1909). Boas, a Columbia University anthropologist, conducted a special study for Congress to determine whether immigrants were actually being assimilated into American culture. His findings (Immigration Commission, 1910) argued that Hebrews and Sicilians were easily assimilable — a conclusion that was anathema to most eugenicists.

The extraordinary legislative success of proposals to sterilize defective persons suggests that there was substantial support by the general public for such a plan. Between 1905 and 1917 the legislatures of 17 states passed sterilization laws, usually by a large majority vote. Most were modeled after the Indiana plan, which covered "confirmed criminals, idiots, imbeciles, and rapists." In Indiana, if two outside surgeons agreed with the institution's physician that there was no prognosis for "improvement," such persons could be sterilized without their consent. In California, the focus was on sterilizing the insane. The law permitted authorities to condition a patient's discharge from a state hospital upon a consent to undergo sterilization. As the hospitalization was of indeterminate length, people rarely refused sterilization, so the consent was rendered nugatory (Laughlin, 1922).

How vigorously were these laws implemented? From 1907 to 1921 there were 3233 sterilizations performed under state laws. A total of 1853 men (72 by castration) and 1380 women (100 by castration) were sterilized. About 2700 operations were performed on the insane, 400 on the feebleminded, and 130 on criminals. California's program was by far the largest (Laughlin, 1922).

Sterilization programs ebbed and flowed according to the views of key state and institutional officials. For example, in 1909 the new Governor of Indiana quashed that state's active program. In New York, activity varied from institution to institution. In the State Hospital at Buffalo the superintendent, who believed that pregnancy exacerbated schizophrenia, authorized 12 hysterectomies, but in most other hospitals no sterilizations were permitted, despite the state law. Similar idiosyn-

cratic patterns were documented in other states (Laughlin, 1922).

Although the activities of persons opposed to sterilization are difficult to document, the record of lawsuits attacking the constitutionality of sterilization laws makes it clear that the courts were in general unfriendly to eugenic policy. Between 1912 and 1921 eight laws were challenged, and seven were held to be unconstitutional. The first two cases were brought by convicted rapists, who argued that sterilization violated the Eighth Amendment's prohibition of cruel and unusual punishment. The Supreme Court of Washington, impressed with Dr. Sharp's reports that vasectomy was simple, quick, and painless, upheld its state law (State v. Feilen, 1912). But, a few years later, a federal court in Nevada ruled that vasectomy was an "unusual" punishment and struck down a criminal sterilization law (Mickle v. Henrichs, 1918).

In six states (New Jersey, Iowa, Michigan, New York, Indiana, and Oregon), constitutional attacks were leveled at laws that authorized sterilization of feebleminded or insane persons who resided in state institutions. The plaintiffs argued that laws aimed only at institutionalized persons violated the Equal Protection Clause and that the procedural safeguards were so inadequate that they ran afoul of the Due Process Clause. All six courts invalidated the laws, but they divided in their reasoning. The three that found a violation of the Equal Protection Clause did not clearly oppose eugenical sterilization. Their concern was for uniform treatment of all feebleminded persons. The three that relied on due-process arguments to reject the laws were more antagonistic to the underlying policy. An Iowa judge characterized sterilization as a degrading act that could cause "mental torture" (Davis v. Berry, 1914).

The New York case was especially interesting because prominent eugenicists testified. It grew out of a dispute between the superintendent of the Rome Custodial Asylum, who opposed sterilization, and the institution's Board of Examiners, who favored it. To resolve the matter, the Board voted to sterilize a young feebleminded man named Frank Osborn, who quickly sued. At the trial, the superintendent argued that there were no convincing data to support assertions that the prevalence of feebleminded people was rising; he argued further that vasectomy might harm "high grade" feebleminded men by encouraging immorality. Of the several eugenicists who testitifed, Davenport, the most important, adopted the most conciliatory position. Perhaps sensing defeat, he favored segregation over involuntary sterilization. The court, concerned that the law sacrificed individual rights to "save expense for future generations," struck it down (In re *Thompson*, 1918).

From 1918 to 1921, the years during which these cases were decided, sterilization laws faded as quickly as they had appeared. One reason why the courts were less sympathetic to sterilization laws than the legislatures had been was that sterilization petitions (like commitment orders) touch the judiciary's historic role as protector of the weak. The judges demanded clear proof that the individual would benefit from being sterilized. Another reason may have been that scientific challenges to eugenical theories about crime began to appear. For example, two physicians who studied behavior found "no proof of the existence of hereditary criminalistic traits" (Spaulding and Healy, 1914: 42).

THE 1920s

After World War I arguments that eugenical sterilization programs were needed to protect the nation's "racial strength" resurfaced. The major impetus was the sudden arrival of hundreds of thousands of immigrants from southeastern Europe and Russia — people with languages and cultures that were unfamiliar to Americans (Ludmerer, 1972). The xenophobia triggered by this massive influx reinforced concern for the dangers of miscegenation and helped to renew interest in biological theories of crime. Some eugenicists wrote inflammatory essays. Madison Grant, a wealthy New Yorker and conservation enthusiast, argued that there were a multitude of scientifically distinct races and that admission of "inferior" types threatened the nation (Grant, 1916). Grant's was the most prominent of a whole genre of popular essays warning Americans to beware of diluting their racial vigor (Stoddard, 1920).

The concurrent concern for miscegenation reflected the weakening of Southern white society's control over the lives of Blacks. Of course, anti-miscegenation laws are as old as slavery itself (Wadlington, 1966). After the Civil War, however, the burgeoning "colored" population (largely a product of institutionalized rape before then) stimulated the enactment of new laws that redefined as "Negro" persons with ever smaller fractions of black ancestry (Mencke, 1959). This trend culminated in 1924, when Virginia adopted a law that defined as White "one who has no trace whatsoever of any blood other than Caucasian." It forbade the issuance of marriage licenses until officials had "reasonable assurance" that statements as to the color of both man and woman were correct, it voided all existing interracial marriages (regardless of whether they were contracted legally elsewhere), and it made cohabitation by such couples a felony. The Virginia Racial Integrity Act was enforced by Walter Plecker, Director of the Bureau of Vital Statistics, a zealous eugenicist who corresponded regularly with Laughlin (Reilly, 1983b).

The early 1920s were also marked by an interest in biological theories of criminality somewhat akin to those legitimized in the 19th Century by Lombroso. Orthodox criminologists were not responsible for this development (Parmelee, 1918; Sutherland, 1924). Tabloid journalists did foster a popular interest in hereditary criminality. *World's Work*, a popular monthly, featured five articles on the biological basis of crime. One recounted the innovative efforts of Harry Olson, Chief Justice of the Chicago Municipal Court. Convinced that most criminals were mentally abnormal, Olson started a Psychopathic Laboratory and hired a psychometrician to develop screening tests to identify people with criminal minds (Strother, 1924).

During the 1920s many eugenics clubs and societies sprouted, but only two, the American Eugenics Society (AES) and the Human Betterment Foundation (HBF), exerted a significant influence on sterilization policy. The AES was conceived at the Second International Congress of Eugenics in 1921 by Henry Fairfield Osborn, President of the American Museum of Natural History, together with a small group of colleagues. By 1923 the new society was sufficiently well organized to lobby against a New York bill supporting special education for the handicapped, an idea that it considered to be dysgenic (AES Memorandum, 1923).

In 1925 Irving Fisher, a Yale professor, relocated the AES headquarters in New Haven. For the next few years its major goal was public education. The AES published several pamphlets, sponsored "fitter family" contests and underwrote the cost of several book projects by its members (Whitney, 1928). The Depression of the 1930s caused a great fall in donations to the Society and when Ellsworth Huntington, a Yale geographer, became president of the society in 1934, it was moribund. With the aid of Frederick Osborn, a wealthy relative of the founder, Huntington breathed new life into the organization. Huntington and Osborn redirected the aims of the AES toward "positive" eugenics policies such as family planning and personal hygiene (Huntington, 1935). After World War II, the Society evolved into one for the study of social biology and concerned itself with issues such as population, nutrition, and education. The major influence of the AES on sterilization policy was in its early days, when some of its educational materials did favor sterilization laws.

The wealthiest eugenics organization was the Human Betterment Foundation (HBF), started by a California millionaire named Ezra Gosney. In 1926 Gosney convened a group of experts to study the efficacy of California's sterilization program. This group confirmed the long-term safety of undergoing sterilization and concluded that California had benefitted from the sterilization of prison and other institutional inmates. Gosney came to believe that a massive sterilization program could reduce the number of mentally defective persons by one-half in "three or four generations" (Gosney and Popenoe, 1929). During the 1930s the HBF was the most vocal advocate of eugenical sterilization. It mailed hundreds of thousands of eugenics pamphlets to college teachers across the nation, sponsored a column on "social eugenics" in the *Los Angeles Times*, aired radio programs, and underwrote hundreds of lectures. The HBF also meticulously collected annual sterilization data and published an annual score card. It remained vigorous until

Gosney's death in 1942 (*Pasadena Star News*, 1942).

Despite the constitutional inadequacies of earlier statutes, in the mid-1920s involuntary sterilization became a major legislative issue. During the period from 1923 to 1925 sterilization laws were enacted in twelve states. Drafted with greater concern for constitutional issues than pre-war legislation these laws usually required the assent of parents or guardians and preserved the patient's right to a jury trial on the question of whether he or she was "the potential parent of socially inadequate offspring" (Laughlin, 1926: 65).

Opponents of sterilization quickly attacked the new laws. In June, 1925, the highest Michigan court upheld Michigan's sterilization statute and ruled that the program was "justified by the findings of Biological Science" (*Smith v. Probate*, 1925). The really crucial case involved the constitutionality of a Virginia law that was decided by the United States Supreme Court. In May, 1927, Oliver Wendell Holmes, writing for the majority, upheld the involuntary sterilization of the feebleminded, concluding: "It is better for all the world, if instead of waiting to execute degenerative offspring for crime, or to let them starve for their imbecility, society can prevent those who are manifestly unfit from continuing their kind. The principle that sustains compulsory vaccination is broad enough to cover cutting the Fallopian tubes" (*Buck v. Bell*, 1927: 205). The Supreme Court's decision greatly boosted the pace at which sterilization programs were enacted and implemented. During the next few years the number of states with sterilization laws jumped from 17 to 30, and the number of sterilizations performed on institutionalized persons rose substantially. The years 1927 to 1942 were a triumphant period for persons who believed that sterilization would help solve some pressing social problems.

YEARS OF TRIUMPH

What was the driving force behind the second wave of sterilization laws? As was the case before World War I, a small group of activists from influential quarters persuaded scientifically unsophisticated legislators that sterilization was necessary, humane, and just.

The lobbyists succeeded in part because of

favorable views expressed in the medical profession. During the period from 1926 to 1936 about 60 articles, the vast majority in favor of eugenical sterilization, appeared. A few physicians played crucial roles. For example, Dr. Robert Dickinson, a noted gynecologist, presented a pro-sterilization exhibit at the 1928 meeting of the American Medical Association (Editorial, 1928). A visitor to the 1929 meeting of the American Association for the Study of the Feeble-Minded reported that his colleagues were convinced that it was "absolutely impossible to cope with the problem of feeblemindedness without judicious use of sterilization" (Editorial, 1929: 136). In 1930, that organization voted 227 to 16 in favor of sterilization of the "mentally defective" (Watkins, 1930). Throughout the 1930s, the American Association for the Study of the Feeble-Minded strongly favored eugenical sterilization. In the general medical community support for it was strong, but not uniform. Only 18 state medical societies officially backed sterilization programs (Whitten, 1935). Yet in some states the support of physicians was extremely strong. For example, one investigator halted his further efforts to survey the views of Indiana physicians on sterilization because he found agreement among more than 400 of them so homogeneous (Harshman, 1934).

The legislative victories were impressive; nevertheless, the crucial measure of whether eugenic notions triumphed is to count the actual number of sterilizations performed. I disagree with Kevles (1985) who has characterized the number of persons who were subjected to involuntary sterilization as relatively small. As he put it, "[f]rom 1907 to 1928 fewer than nine thousand people had been eugenically sterilized in the United States" (p. 106). Although Kevles has enriched our understanding of the prominence of eugenic ideas in American intellectual life, his failure to analyze carefully available statistics on sterilization led him to underestimate the impact of eugenic laws. For example, he noted that "by the mid-thirties some twenty thousand sterilizations had been legally performed in the United States" (p. 112) and that enforcement of the laws pursuant to which these operations were performed "was minuscule by 1950" (p. 169).

My analysis of surveys conducted by the Human Betterment Foundation permit *minimum* estimates of mass sterilization and compel some striking conclusions:

(1) Between 1907 and 1963 there were eugenical sterilization programs in 30 states. More than 60,000 persons were sterilized pursuant to state laws.

(2) Although sterilization reached its zenith during the 1930s, several states vigorously pursued this activity throughout the 1940s and 1950s.

(3) At any particular time, a few programs were much more active than the rest. In the 1920s and 1930s California and a few midwestern states were most active. After World War II, several southern states accounted for more than half of the involuntary sterilizations performed upon institutionalized persons.

(4) Beginning about 1930, there was a dramatic rise in the percentage of women who were sterilized.

(5) No revulsion against Nazi sterilization policy seems to have curtailed American sterilization programs. Indeed, more than one-half of all eugenic sterilizations occurred after the Nazi program was fully operational.

From 1929 to 1941 the Human Betterment Foundation conducted annual surveys of state institutions to chart the progress of sterilization. Letters from hospital officials indicate what factors influenced their programs. The most important determinants of the scope of a program's operation were the complexity of the due-process requirements of the relevant laws, the level of funding, and the attitudes of the superintendents themselves. A West Virginia official complained that his law had so many amendments as to "practically annul it" (Denham, 1933). An Arizona physician reported that there was no money to pay for surgery (Develin, 1933). On the other hand, in Alabama a physician superintendent reported that he had secured funds to sterilize *every* patient before discharge from the state hospital and had operated upon 184 persons in two years (Partlow, 1935).

The HBF surveys strongly suggested that the total number of sterilizations performed upon institutionalized persons was underreported. Respondents frequently indicated that eugenic operations were conducted outside the confines of state hospitals. The Assistant Attorney General of Maine wrote that "many more operations have been performed (than are reported) but I suppose we shall have to go by the records" (Folsom, 1936: 1). An Indiana superintendent asserted that "hundreds of operations have been done in the community" (Dunham, 1936: 1).

Until 1918, only 1422 eugenical sterilizations were reported as performed pursuant to state law. Ironically, the sterilization rate began to rise during the very period when the courts were rejecting the first round of sterilization statutes (1917–1918). During 1918 to 1920 there were 1811 reported sterilizations, a four-fold increase over the rate in the prior decade. During the 1920s annual sterilization figures were stable (Fig. 1), but in 1929 there was a large increase in sterilizations. Throughout the 1930s more than 2000 institutionalized persons were sterilized each year, a rate triple that of the early 1920s.

This rapid increase reflected changing concerns and changing policy. In the Depression years superintendents of many hospitals, strapped by tight budgets, decided to sterilize mildly retarded young women. Before 1929 about 53 per cent of all eugenical sterilizations had been performed on men. Between 1929 and 1935 there were 14,651 reported operations, 9327 upon women and 5324 on men. In several states (e.g., Minnesota, Wisconsin)

virtually all the sterilized persons were women. This fact becomes even more impressive when one considers that the salpingectomy operation incurred a relatively high morbidity and a much higher cost than did vasectomy. In California, at least five women died after undergoing eugenical sterilization (Gosney and Popenoe, 1929).

During the 1930s institutionalized men were also being sterilized in unprecedented numbers. This was largely a consequence of the great increase in the total number of state programs. Unlike the "menace of the feeble-minded" that haunted policy before World War I, the new concern was to cope with harsh economic realities. As many superintendents saw it, fewer babies born to incompetent parents might mean fewer state wards in the future. Sterilizing and paroling mildly retarded women eased overcrowding and, it was argued, permitted them to live more successful lives than if they were burdened with children (State Board of Control of Minnesota, 1934).

The triumph of eugenic sterilization programs in the United States during the 1930s influenced other nations. Canada, Germany, Sweden, Norway, Finland, France, and Japan enacted sterilization laws. In England, sterilization was ultimately rejected, but in Germany the Nazis sterilized more than 50,000 "unfit" persons within one year after enacting a eugenics law.

Persons Sterilized

FIG. 1. STERILIZATIONS PERFORMED EACH YEAR UPON INSTITUTIONALIZED PERSONS, 1919–1959

Proposals favoring eugenical sterilization were common in England early in this century. In 1907 Galton's colleagues organized the Eugenics Education Society (EES), which soon included hundreds of prominent academicians. Although committed to eugenic ideals, the EES pursued a moderate course, stopping short of a policy that included involuntary sterilization (Searle, 1976). The major legislative action during this era was the Mental Deficiency Act of 1913, a law that clearly favored educational programs for the feebleminded—a decidedly dysgenic policy.

During the late twenties the EES, troubled by the rising welfare budget, did lobby for voluntary eugenic sterilization. In 1931 the House of Commons rejected such a bill, but it created a Committee to study the question. Three years later it filed a report that roundly criticized programs like those in the United States as being inadequately supported by genetic evidence (Brock, 1934).

The German interest in eugenics had roots that were entwined with 19th Century European racial thought, a topic beyond the scope of this review. In the early years of this century a spate of books preached the need to protect the Nordic germ plasm. A German eugenics society was formed in 1905, and in 1907 the first sterilization bill was offered in the Reichstag (Lenz, 1934). It failed to pass. The devastation of World War I halted the German eugenic movement, but by 1921 groups were again actively lobbying for eugenics programs. Of particular importance was the publication by three prominent German scientists, Erwin Baur, Eugen Fischer, and Fritz Lenz, in 1923 of a textbook on human heredity and eugenics. The contribution of Baur, an eminent plant geneticist who did not usually stray from his area of expertise, is notable. According to Glass (1981), Baur was deeply troubled by the suffering of Germans during the occupation of the Rhineland and became concerned that it was essential for the more robust German citizens to reproduce vigorously in order to counter the influx of inferior types. According to Popenoe (1935), Lenz influenced Hitler's ideas on racial purity.

When the Nazis swept into power, they quickly implemented a program to encourage larger, healthier families. According to Kopp (1935), Gosney and Popenoe influenced Nazi sterilization policy. The Nazis restructured tax laws to favor childbearing and enacted a law to curb reproduction by "defective" persons. This law created a system of "Hereditary Health Courts" that judged petitions brought by public health officials recommending that certain citizens burdened with any of a long list of disorders (feeblemindedness, schizophrenia, manic depressive insanity, epilepsy, Huntington's chorea, hereditary blindness, hereditary deafness, severe physical deformity, or habitual drunkenness) be subjected to compulsory sterilization. In 1934, the courts heard 64,499 petitions and ordered 56,244 sterilizations, for a "eugenic conviction" rate of 87 per cent (Cook, 1935).

During the middle 1930s the Nazis cast an even larger net. In 1934 the German Supreme Court ruled that the law applied to non-Germans living in Germany, a decision that threatened gypsies. From 1935 through 1939 the annual number of eugenical sterilizations grew rapidly. Unfortunately, key records perished during the war. Yet in 1951 the "Central Association of Sterilized People in West Germany" charged that from 1934 to 1945 the Nazis sterilized 3,500,000 people, often on the flimsiest pretext (*New York Herald Tribune*, 1951). The Nazi program was eugenics run amok. In the United States no program even approached it in scope or daring.

CRITICS OF EUGENIC STERILIZATION

The difficulty of explaining why eugenic ideas appealed so strongly to some Americans extends to the task of understanding why interest faded when it did. Of course, there were always critics. At various times geneticists, social scientists, physicians, and (most effectively) the Catholic Church opposed sterilization programs. After World War II civil libertarians, lawyers, patients' families, and patients themselves repudiated the old notions.

During the heyday of eugenics the science of genetics was also making important strides. After training with Thomas Hunt Morgan in Columbia's fly room, talented graduate students brought *Drosophila* genetics to other universities. By the mid-1920s quite a number of academic geneticists were critical of the unsophisticated ways of the eugenicists. Yet few tried to counter the eugenicist's political activities (Haller, 1963; Ludmerer, 1972).

The first major scientist to take on the eugenicists was Herbert Spencer Jennings, a zoologist at The Johns Hopkins University. He severely criticized the statistical methods used by Laughlin to conduct his immigration studies and offered his arguments both in scientific periodicals (Jennings, 1924) and popular magazines (Jennings, 1923). He consistently argued that sterilizing the feebleminded was a futile gesture, although he quite approved of the use of voluntary sterilization by enlightened couples at high risk for bearing defective children (Jennings, 1930).

A few leading geneticists criticized the various famous pedigree studies. Morgan, a man who shunned political battles, contented himself only with dismissing the studies of the "Jukes" and "Kallikaks" as inadequate investigations of the interaction between genetic and environmental influences (Morgan, 1925). Raymond Pearl was more vocal. He ridiculed the early pedigree work and urged eugenicists to throw away their "old-fashioned rubbish" (Pearl, 1927: 263). Another important critic was Hermann Joseph Muller, renowned for the discovery that radiation induces mutations. At the Third International Congress of Eugenics he shocked the audience by attacking its most dearly held tenets (Carlson, 1982).

During the 1930s developments in Europe greatly increased concern among geneticists that substantial harm might be done in the service of state-supported eugenic policy. No doubt influenced by the economic climate, the English National Council of Labour Women passed a resolution in favor of sterilizing defective persons. Soon after they swept to power in 1933, the Nazis dismissed hundreds of Jewish professors. Events like these provoked the great British geneticist, J. B. S. Haldane, to launch a scathing attack on eugenics.

Later published as *Heredity and Politics* (Haldane, 1938), his book masterfully separated genetic fact from political fantasy. For example, Haldane pointed out that the eugenic idea popular in 1910 (an era of full employment) that pauperism was genetically determined was untenable during a Depression in which 1,500,000 Englishmen were out of work. Not a little of Haldane's book deflates the perfectionist dreams of those who advocated sterilization. After demonstrating the weakness of its scientific foundations, he dismissed "compulsory sterilization" ". . . as a piece of crude

Americanism like the complete prohibition of alcoholic beverages" (Haldane, 1938: 86). His book did much to alert the growing genetics community to the need to refute the quasi-scientific eugenicists (Glass, pers. commun., 1985).

Catholic priests were among the earliest critics of eugenical sterilization. Their attack became intense after World War I, and was part of a larger concern over the efforts of eugenicists to restrict immigration. In the United States, the Catholic Church, a small minority in a Protestant nation, was a church of immigrants. By the mid-1920s eugenicists had recognized the Catholic Church as a major enemy. They attributed the gubernatorial veto of a Colorado sterilization bill to lobbying by the Denver Chapter of the Knights of Columbus (Johnson, 1927).

In 1930 Pope Pius XI issued *Casti Connubi*, the encyclical on Christian marriage, which included the first official condemnation of eugenical sterilization. Pius XI asserted that civil authorities had no right to deprive guiltless persons of a "natural faculty by medical action" (Pius XI, 1939: 96–97). *Casti Cannubi* rallied Catholic organizations to oppose eugenic laws. This action in turn polarized some Protestant groups to argue in favor of eugenical sterilization.

During the 1940s Catholic opposition to both eugenic sterilization and elective sterilization to limit family size was widespread. Leading eugenicists saw Catholic opposition as their "greatest obstacle." According to Olden (1945), a founder of the New Jersey Sterilization League, Roman Catholic priests defeated proposed laws in Wisconsin, Maine, Alabama, and Pennsylvania. Olden reported that in Alabama, priests and nuns had "invaded" the legislature and that in Pennsylvania lobbying against sterilization had been masterminded by the Cardinal's office. Catholic opposition to sterilization played a major role in delaying widespread elective surgery to limit family size.

The first important clinical critique of eugenic sterilization was developed by Dr. Walter Fernald (1919), a Boston psychiatrist. After studying 646 non-sterilized feebleminded persons who had been discharged from institutions, he found that few of them became parents. He concluded that the eugenicists had mistaken the relatively high fertility of persons of low socioeconomic status for that of the mentally defective. Despite his refutation of a major tenet of eugenic thought, the study had no immediate impact on social policy.

Just as a few zealous eugenicists did so much to enhance their cause, a single physician played a crucial role in attacking the beliefs upon which eugenical sterilization was grounded. This was Dr. Abraham Myerson, a neurologist at Tufts University, who began to question eugenics in the mid-1920s (Myerson, 1928). In 1930, he published a study showing that the feebleminded were born in roughly equal proportions in all segments of society, a finding that contrasted sharply with the eugenic thesis that a relatively few persons in the lower classes produced a disproportionate number (Myerson and Elkind, 1930).

In 1934 the American Neurological Association asked Myerson to chair an investigation of eugenic sterilization. The committee's conclusions rejected several major eugenic notions. It determined that the increasing number of institutionalized persons was the consequence of a better system of medical care, and was not due to a rising incidence in births of retarded persons. It also reported that the most common affliction suffered by newly institutionalized persons was cerebral atherosclerosis, a fact explained by the increasing longevity of the population. Overall, the committee found that "the race is not going to the dogs, as has been the favorite assertion for some time" (Myerson, 1936a: 6). It concluded that there was no pressing need for involuntary sterilization programs, but that voluntary sterilization might be a reasonable option for some individuals.

The report had an immediate political impact. The *New York Times* published a letter summarizing the report's findings (Myerson, 1936b). It provided important ammunition with which opponents torpedoed a sterilization bill then under consideration in Albany (Cooper, 1936). From the mid-1930s onward, sterilization bills fared less well than during the preceding decade. Nevertheless, already-entrenched programs continued to sterilize about 2500 institutionalized persons each year.

AFTER WORLD WAR II

During the 1920s behavioral psychologists (Watson, 1927) had advanced views of intelligence that were incompatible with the eugenic

thesis. Nevertheless, prior to World War II, social scientists were not much involved in efforts to halt sterilization programs. After the War, as psychologists replaced physicians in much of the mental health field, some of them challenged established sterilization policy. Also after World War II, in 1950, the families of retarded persons formed their own lobbying group, the National Association for Retarded Children (NARC). By 1960 it had achieved an important political presence, a success that was redoubled when President Kennedy took office. NARC and its allies rejected eugenic sterilization. In 1962 the President's Commission on Mental Retardation reaffirmed this view. By the early 1960s most state sterilization programs had stopped.

With the onset of World War II there had been a sharp decline in the number of eugenic sterilizations in the United States. Although manpower shortages were significant, other factors were also at work. In 1939 the Eugenics Record Office closed its doors, and in 1942 the Human Betterment Foundation ceased its activities. Later in that year the Supreme Court, considering its first sterilization case in 15 years, struck down an Oklahoma law that permitted certain thrice-convicted felons to be sterilized (*Skinner* v. *Oklahoma*, 1942). Although its specific impact is difficult to assess, the postwar civil rights movement also surely contributed to the failure of sterilization programs to return to earlier levels. Despite these changes, some sterilization programs continued, albeit at reduced activity levels.

Between 1942 and 1946 there were only half as many eugenic sterilizations annually as had been performed annually during the 1930s. Reports of institutional officials make it clear that this decline was almost completely owing to a lack of surgeons and nurses (Taromianz, 1944; Perry, 1945). The Supreme Court decision was not important in causing the decline. Avoiding an opportunity to overrule *Buck* v. *Bell* (1927), the Justices instead demanded that involuntary sterilization be practiced in accordance with the Equal Protection Clause. The Oklahoma law was struck down because it spared certain "white collar" criminals from a punitive measure aimed at other thrice-convicted persons, not simply because it involved sterilization.

In charting the sterilization trends during the 1940s and 1950s I have relied primarily on the surveys conducted annually by the Sterilization League of New Jersey, a group founded in 1937 that underwent several changes of name. After some internal feuding in the early 1940s the organization, refueled financially by Clarence Gamble, a physician and a relative of the wealthy soap family, renamed itself Birthright, Inc. Besides being instrumental in maintaining sterilization statistics, Birthright, largely thanks to the zeal of Dr. Gamble, helped stimulate several states to initiate new programs. His greatest victory was in North Carolina, where he obtained government approval to conduct extensive mental testing on grade-school children and then used those data to identify persons who would "benefit" from sterilization. Within a few years North Carolina had one of the nation's largest eugenic sterilization programs (Woodside, 1950).

During the late 1940s there was no definite indication that sterilization programs were about to decline further. After hitting a low of 1183 in 1944, there were 1526 operations in 1950. Slight declines in many states were balanced by rapid increases in North Carolina and Georgia. By 1950, however, there were strong signs that sterilization was in disfavor even among institutional officials. For example, during the 1930s and 1940s 100 persons in San Quentin prison had been sterilized each year. In 1950 the prison surgeon told Birthright that new officials at the Department of Correction were "entirely adverse" to the program (Stanley, 1950: 1). Several institutional superintendents in other states informed Birthright that they no longer believed that heredity was a major factor in mental retardation (Missouri official, 1950). During that year sterilization bills were considered in only four states, and all of them were rejected (Butler, 1950).

There were major changes in state sterilization programs in 1952. The California program, for years the nation's most active, was moribund, dropping from 275 sterilizations in 1950 to only 39 in 1952. In that year Georgia, North Carolina, and Virginia (having together sterilized 673 persons) were responsible for 53 per cent of the national total. General declines in most other states continued throughout the 1950s and by 1958, when Georgia, North Carolina, and Virginia sterilized 574 persons, they accounted for 76 per cent of the reported operations. The data do not suggest that the

southern programs were racially motivated. Only as to South Carolina, where in 1956 all 23 eugenical sterilizations were performed upon "Negro females" (Hall, 1956), might such suspicion ! - entertained. The North Carolina program was unique in that it was directed largely at non-institutionalized, rural young women (Woodside, 1950). As recently as 1963 the state paid for the eugenic sterilization of 193 persons, of whom 183 were young women (Casebolt, 1963). Despite their persistence, the southern programs must be seen as a local eddy in a tide of decline.

INVOLUNTARY STERILIZATION TODAY

During the 1960s the practice of sterilizing retarded persons in state institutions virtually ceased. Still, the laws remained. In 1961 there were eugenic sterilization laws on the books of 28 states (Landman and McIntyre, 1961). Between 1961 and 1976 five laws were repealed, six were amended, and one state (West Virginia) adopted its first sterilization statute. Since 1976 there has been a trend to repeal the laws. Currently, eugenic sterilization of institutionalized retarded persons is permissible in 19 states, but the laws are rarely invoked. A few states have even enacted laws that expressly forbid sterilization of any persons in state institutions. On the other hand, several constitutional attacks upon involuntary sterilization have failed (In re *Cavitt*, 1968, *Cook* v. *State*, 1972; In re *Moore*, 1976).

If the mid-1930s saw the zenith of eugenic sterilization, the mid-1960s saw its nadir. The pendulum of policy continues to swing, however. The late 1960s saw the first lawsuits brought by parents of non-institutionalized retarded females on the basis that sterilization was both economically essential and psychologically beneficial to the family's efforts to maintain the adult daughters at home (*Frazier* v. *Levi*, 1968).

In 1973 the debate over the sterilization of institutionalized persons whom officials had decided were unfit to be parents flared in the media. The mother of a young man whom physicians at the Partlow State School in Alabama wished to sterilize challenged the constitutionality of the enabling statute. When Alabama officials argued that they did not need statutory authority so long as consent was obtained from the retarded person, the federal judge not only overturned the law, but decreed strict guidelines to control the process of performing "voluntary" sterilizations at Partlow. The key feature was the creation of an outside committee to review all the sterilization petitions (*Wyatt* v. *Aderholt*, 1973).

Also in 1973 the Department of Health, Education and Welfare (HEW) became enmeshed in a highly publicized sterilization scandal. That summer it was reported that an Alabama physician working in a family planning clinic funded by HEW had sterilized several young, poor black women without their consent. The National Welfare Rights Organization and two of the women sued to block the use of all federal funds to pay for sterilizations. This prompted HEW to draft strict new regulations; but a federal judge struck them down, holding the HEW could not provide sterilization services to any legally incompetent persons (*Relf* v. *Weinberger*, 1974). Revamped several times, the HEW guidelines were subject to continuous litigation for five years. Late in 1978 "final rules" were issued that prohibited sterilization of some persons (those under 21 and all mentally incompetent persons) and created elaborate consent mechanisms for competent persons who requested sterilization, to be paid for by public funds (Federal Register, 1978).

During the last five years the debate over sterilizing the mentally retarded, although no longer cast in a eugenic context, has reheated. The key issue now is how to resolve the tensions between the society's duty to protect the incompetent person and the right of that person to be sterilized. The court must be convinced that the operation will benefit the patient. More than 20 appellate courts have recently been asked to consider sterilization petitions. This spate of litigation has resulted because physicians are extremely reluctant to run the risk of violating the civil rights of the retarded. The courts have split sharply. In the absence of express statutory authority, six high state courts have refused to authorize sterilization orders (In the Matter of *S.C.E.*, 1977). This abdication of power by the courts was in part stimulated by an unusual lawsuit in which a sterilized woman later sued the judge who approved the surgery. Ultimately, the principle of judicial immunity was upheld by the

United States Supreme Court (*Stump* v. *Sparkman*, 1978).

The majority of appellate courts have ruled that local courts of general jurisdiction do have the power to evaluate petitions to sterilize retarded persons. In a leading case, the highest court in New Jersey held that the parents of an adolescent girl with Down's Syndrome might obtain surgical sterilization for her if they could provide clear and convincing evidence that it was in "her best interests" (In re *Grady*, 1981). Since then, high courts in Colorado, Massachusetts, and Pennsylvania have ruled in a similar manner. These decisions promise that in the future the families of some retarded persons will in appropriate circumstances be able to obtain sterilizations for them, regardless of their institutional status.

The great era of sterilization, however, has passed, although grim reminders of unsophisticated programs that once flourished linger. In Virginia persons sterilized for eugenic reasons decades ago have sued the state, on grounds that it was a violation of their civil rights. Depositions taken before the trial indicated that many of the persons who were sterilized were not retarded (*Poe* v. *Lynchburg*, 1981). Although they lost their argument that the operations were performed pursuant to an unconstitutional law, the plaintiffs did win a settlement that requires Virginia to attempt to locate all persons who were sterilized by the state and inform them of the consequences of the operation. Well into the 1970s a few states (Iowa, North Carolina, Oregon) still operated sterilization programs. Between 1971 and 1977 the Iowa Board of Eugenics considered 215 sterilization petitions and authorized 179 of them (*Howard* v. *Des Moines Register*, 1979).

Is the saga of involuntary sterilization over? Our present knowledge of human genetics makes the return of mass eugenic sterilization very unlikely. However, it is more difficult to predict the future of sterilization programs founded on other arguments. During the 1960s a number of state legislatures considered bills to tie welfare payments to "voluntary" sterilization (Paul, 1968). In 1980 a Texas official made a similar suggestion (*New York Times*, 1980). Unscientific opinion polls conducted by magazines and newspapers in Texas (*The Texas Observer*, 1981) and Massachusetts (*The Boston*

Globe, 1982) have found significant support for involuntary sterilization of the retarded.

The pressing demands of population control in India and China have resulted in social policies that create strong incentives to be sterilized. Since launching the "one child" program in 1979, China has rapidly altered the social fabric of one billion people (Intercom, 1981). What may not happen here in the United States may transpire in other countries, with different legal codes and different population pressures. As resources continue to shrink and the earthly neighborhood becomes more crowded, societal incentives in favor of sterilization may some day be as common as compulsory immunizations, but the eugenic vision is not likely to provide its intellectual rationale.

REFERENCES

AMERICAN EUGENICS SOCIETY MEMORANDUM. 1923. On the matter of the proposed state aid in New York for local classes for the feebleminded. Unpublished. Laughlin archive. Kirksville, Missouri. Used with permission.

BARKER, L. 1910. The importance of the eugenics movement and its relation to social hygiene. *J. Am. Med Assoc.*, 54: 2017-2022.

BOIES, H. M. 1893. *Prisoners and Paupers*. G.P. Putnam's Sons, New York.

BOSTON GLOBE. 1982. Item. *The Boston Globe*. March 31, 1982, p. 1.

BROCK, R. 1934. *Report of the Departmental Committee on Sterilization*. His Majesty's Stationary Office, London.

BUCK v. BELL. 1927. 274 U.S. 200.

BUTLER, F. O. 1950. Report. Association for Voluntary Sterilization Archive, University of Minnesota.

CARLSON, E. A. 1982. *Genes, Radiation and Society*. Cornell University Press, Ithaca.

CASEBOLT, S. L. 1963. Letters to Human Betterment Association of America. Association for Voluntary Sterilization Archive, University of Minnesota.

CLARKE, H. 1879. Heredity and crime in epileptic criminals. *Brain*, 2: 490-527.

COOK v. STATE. 1972. 495 P. 2d 768.

COOK, R. 1935. A year of German sterilization. *J. Hered.*, 26: 485-489.

COOPER, F. 1936. Letter to Dr. Harry Perkins. Ellsworth Huntington Papers, Yale University.

DANIEL, F. E. 1894. Emasculation of masturbators—is it justifiable? *Tx. Med J.*, 10: 239-244.

——. 1907. Emasculation for criminal assaults and incest. *Tx. Med J.*, 22: 347.

——. 1909. Sterilization of male insane. *Tx. Med. J.*, 25: 61-67.

DANIELSON, F., and C. B. DAVENPORT. 1912. *The Hill Folk: Report on a Rural Community of Hereditary Defectives*. Eugenics Record Office, Cold Spring Harbor.

DAVENPORT, C. B. 1911. Letter to H. H. Laughlin. Davenport Papers. American Philosophical Society, Philadelphia.

——. 1913a. *State Laws Limiting Marriage Selection.* Eugenics Record Office, Cold Spring Harbor.

——. 1913b. Report of the committee on eugenics. *Am. Breeder's Assoc.*, 6: 92-94.

DAVIS v. BERRY. 1914. 216 F. 413.

DAVIS, K. B. 1921. Letter to R. D. Fosdick. Rockefeller Archives, Tarrytown.

DENHAM, C. 1933. Letter to E. S. Gosney. Association for Voluntary Sterilization Archive, University of Minnesota.

DEVELIN, A. L. 1933. Letter to E. S. Gosney. Association for Voluntary Sterilization Archive, University of Minnesota.

DUGDALE, R. L. 1877. *The Jukes: A Study in Crime, Pauperism, Disease, and Heredity*. G. P. Putnam's Sons, New York.

DUNHAM W. F. 1936. Letter to E. S. Gosney. Association for Voluntary Sterilization Archive, University of Minnesota.

EDITORIAL. 1928. Eugenical sterilization at the meeting of the A.M.A. *Eugenical News*, 13: 115.

——. 1929. Notes and news. *Eugenical News*, 14: 136.

ESTABROOK A., and C. B. DAVENPORT. 1912. *The Nam Family: A Study in Cacogenics*. Eugenics Record Office, Cold Spring Harbor.

FEDERAL REGISTER. 1978. 43(217): 52146-75.

FERNALD, W. 1919. After-care study of the patients discharged from Waverly for a period of twenty-five years upgraded. *Am. J. Ment. Def.*, 4: 62-81.

FOLSOM, S. B. 1936. Letter to E. S. Gosney. Association for Voluntary Sterilization Archive, University of Minnesota.

FOSTER, W. W. 1909. Hereditary criminality and its certain cure. *Pearson's Magazine*, 1909: 565-572.

FRAZIER v. LEVI. 1968. 440 S.W. 2d 579.

GALTON, F. 1869. *Hereditary Genius: An Inquiry into Its Laws and Consequences*. Macmillan, London.

——. 1874. *English Men of Science: Their Nature and Nurture*. Macmillan, London.

——. 1883. *Inquiries into Human Faculty and Its Development*. Macmillan, London.

GARIS, R. L. 1927. *Immigration Restriction*. Macmillan, New York.

GLASS, B. 1981. A hidden chapter of German eugenics between the two world wars. *Proc. Am.*

Philos. Soc., 125: 357–367.

GODDARD, H. H. 1912. *The Kallikak Family*. Macmillan, New York.

GOSNEY, E. S., and P. POPENOE. 1929. *Sterilization for Human Betterment*. Macmillan, New York.

GRANT, M. 1916. *The Passing of the Great Race*. Charles Scribner's Sons, New York.

HALDANE, J. B. S. 1938. *Heredity and Politics*. W. W. Norton, New York.

HALL, W. S. 1956. Letter to Human Betterment Association of America. Association for Voluntary Sterilization Archive, University of Minnesota.

HALLER, M. H. 1963. *Eugenics: Hereditarian Attitudes in American Thought*. Rutgers University Press, New Brunswick.

———. 1973. Laughlin, Harry Hamilton. In E. T. Janis (ed.), *Dictionary of American Biography* (Supp. 3), p. 445–446. Charles Scribner's Sons, New York.

HARSHMAN, L. P. 1934. Medical and legal aspects of sterilization in Indiana. *J. Psycho-Asthenics*, 39: 183–206.

HASSENCAHL, F. J. 1980. *Harry H. Laughlin, Expert Eugenics Agent for the House Committee on Immigration and Naturalization 1921–1931*. Ph.D. Thesis. Case Western Reserve University. University Microfilms, Ann Arbor.

HIGHAM, J. 1965. *Strangers in the Land*. Atheneum, New York.

HOWARD v. DES MOINES REGISTER. 1979. 283 N.W. 3d. 289.

HUNTINGTON, E. 1935. *Tomorrow's Children*. Macmillan, New York.

IMMIGRATION COMMISSION. 1910. *Changes in Bodily Form of Descendants of Immigrants*. U.S. Government Printing Office, Washington, D.C.

IN RE CAVITT. 1968. 159 N.W. 2d. 566.

IN RE GRADY. 1981. 426 N.W. 2d. 467.

IN RE MOORE. 1976. 221 S.E. 2d. 307.

IN RE THOMPSON. 1918. 169 N.Y.S. 638.

IN THE MATTER OF S.C.E. 1977. 378 A. 2d. 144.

INTERCOM. 1981. China's one-child population future. *Intercom*, 9(8): 1: 12–14.

JENNINGS, H. S. 1923. Undesirable aliens. *Survey*, December 15, 309–312.

———. 1924. Proportion of defectives from the southwest and from the northeast of Europe. *Science*, 59: 256–257.

———. 1930. *The Biological Basis of Human Nature*. Norton, New York.

JOHNSON, A. 1909. Race improvement by control of defectives (negative eugenics). *Ann Am. Acad. Penal Soc. Sci.*, 34: 22–29.

JOHNSON, R. H. 1927. Legislation. *Eugenics*, 2(4): 64.

KERLIN, I. 1881. The epileptic change and its appearance among feeble-minded children. *Proc. Assoc. Med. Officers*, 1881: 202–210.

KEVLES, D. J. 1985. *In the Name of Eugenics*. Alfred A. Knopf, New York.

KOPP, M. 1936. The German sterilization program. Association for Voluntary Sterilization Archive, University of Minnesota.

LANDMAN, F. T., and D. M. McINTYRE. 1961. *The Mentally Disabled and the Law*. University of Chicago Press, Chicago.

LAUGHLIN, H. H. 1907. Letter to C. B. Davenport. Davenport Papers. American Philosophical Society, Philadelphia.

———. 1914a. *The Scope of the Committee's Work. Bull. No. 10A*. Eugenics Record Office, Cold Spring Harbor, New York.

———. 1914b. *The Legal, Legislative and Administrative Aspects of Sterilization. Bull. No. 10B*. Eugenics Record Office, Cold Spring Harbor.

———. 1918. The dynamics of cell division. *Proc. Soc. Exp. Biol. Med.*, 15(8): 32–44.

———. 1922. *Eugenical Sterilization in the United States*. Chicago Psychopathic Laboratory of the Municipal Court, Chicago.

———. 1923a. Nativity of institutional inmates. In C. B. Davenport, H. F. Osborn, C. Wissler, and H. H. Laughlin (eds.), *Eugenics in Race and State*, p. 402–406. Williams and Wilkins, Baltimore.

———. 1923b. Analysis of America's modern melting pot. *Hearings before the House Committee on Immigration and Naturalization*. 67th Cong. 3rd Session. Government Printing Office, Washington, D.C.

———. 1926. *Historical, Legal, and Statistical Review of Eugenical Sterilization in the United States*. American Eugenics Society, New Haven.

LENZ, F. 1934. Eugenics in Germany. *J. Hered.*, 15: 223–231.

LOMBROSO-FERRERO, G. 1972. *Lombroso's Criminal Man*. Patterson-Smith, Montclair.

LUDMERER, K. 1972. *Genetics and American Society*. Johns Hopkins University Press, Baltimore.

LYDSTON, G. F. 1912. Sex mutilations in social therapeutics. *N.Y. Med J.*, 95: 677–685.

MEARS, J. E. 1909. Asexualization as a remedial measure in the relief of certain forms of mental, moral, and physical degeneration. *Boston Med. Surg. J.*, 161: 584–586.

MENCKE, J. G. 1959. *Mulattoes and Race Mixture: American Attitudes and Images 1865–1918*. UMI Research Press, Ann Arbor.

MERRIAM, J. C. 1938. Letter to H. H. Laughlin, Davenport Papers. American Philosophical Society, Philadelphia.

MICKLE v. HENRICHS. 1918. 262 F. 687.

MISSOURI OFFICIAL. 1950. Letter to New Jersey Sterilization League. Association for Voluntary Sterilization Archive, University of Minnesota.

MORGAN, T. H. 1925. *Evolution and Genetics*. Princeton University Press, Princeton.

MYERSON, A. 1928. Some objections to sterilization. *Birth Control Rev.*, 12:81-84.
——. 1936a. *Eugenical Sterilization.* Macmillan, New
York.
——. 1936b. Research urged. *New York Times*,
March 15, 1936, p. 18.
MYERSON, A., and R. ELKIND. 1930. Researches
in feeblemindness. *Bull. Mass. Dept. of Ment. Dis.*,
16: 108-229.
NEW YORK HERALD TRIBUNE. 1951. Germans made
sterile by Nazis seek pensions. *New York Herald
Tribune*, January 14, 1951, p. 12.
NEW YORK TIMES. 1980. Official urges sterilization
of Texas welfare recipients. *New York Times*,
February 28, 1980, p. A16.
OCHSNER, A. 1899. Surgical treatment of habitual
criminals. *J. Am. Med. Assoc.*, 53: 867-868.
OLDEN, M. S. 1945. Present status of sterilization
legislation in the United States. *Eugen. News*, 30:
3-14.
OWENS-ADAIR, B. 1905. *Human Sterilization*. Pri-
vately printed. Portland, Oregon.
PARMELEE, M. 1918. *Criminology.* Macmillan, New
York.
PARTLOW, W. D. 1935. Letter to E. S. Gosney. As-
sociation for Voluntary Sterilization Archive,
University of Minnesota.
PASADENA STAR NEWS. 1942. E. S. Gosney passes
at 86 years. *Pasadena Star News*, September 15,
1942, p. 6.
PAUL, J. 1968. The return of punitive sterilization
laws. *Law Soc. Rev.*, 4: 77-110.
PEARL, R. 1927. The biology of superiority. *Am.
Mercury*, 47: 257-266.
PERRY, M. L. 1945. Letter to New Jersey Ster-
ilization League. Association for Voluntary Ster-
ilization Archive, University of Minnesota.
PIUS XI. 1939. On Christian marriage. In *Five Great
Encyclicals*, p. 96-97. Paulist Press, New York.
POE v. LYNCHBURG. 1981. 518 F. Supp. 789.
POPENOE, P. 1933. The progress of eugenical steril-
ization. *J. Hered.*, 28: 19-25.
——. 1935. The German sterilization law. *J. Hered.*,
26: 257-260.
REILLY, P. R. 1983a. The surgical solution: the writ-
ings of activist physicians in the early days of
eugenical sterilization. *Persp. Biol. Med.*, 26:
637-656.
——. 1983b. The Virginia Racial Integrity Act
Revisited. The Plecker-Laughlin correspon-
dence: 1928-1930. *Am. J. Med. Genet.*, 16: 483-
492.
RELF v. WEINBERGER. 1974. 372 F. Supp 1196.
RHODE ISLAND STATE LIBRARY LEGISLATIVE REF-
ERENCE BUREAU. 1913. *Sterilization of the Unfit.*
State Printing Office, Providence.
ROOSEVELT, T. 1914. Twisted eugenics. *Outlook*, 106:

30-34.
SEARLE, G. R. 1976. *Eugenics and Politics in Britain
1900-1914.* Nordhoff, Leyden.
SEQUIN, E. 1846. *Traitment moral, hygiène et education
des idiots et des autres enfants arriérés.* J. B. Bailliere,
Paris.
SHARP, H. C. 1902. The severing of the vasa
differentia and its relation to the neuropsy-
chiatric constitution. *N. Y. Med. J.*, 1902: 411-414.
——. 1907. Vasectomy as a means of preventing
procreation of defectives. *J. Am. Med. Assoc.*, 51:
1897-1902.
——. 1909. *Vasectomy.* Privately printed. Indianapolis.
SKINNER v. OKLAHOMA. 1942. 316 U.S. 535.
SMITH v. BOARD OF EXAMINERS. 1913. 85 N.J.L. 46.
SMITH v. PROBATE. 1925. 231 Mich. 409.
SMITH, E. 1908. Heredity as a factor in producing
the criminal. *J. Crim. Law Criminol.*, 4: 321-322.
SPAULDING, E. R., and W. HEALY. 1914. In-
heritance as a factor in criminality. In *Physical
Basis of Crime*, p. 19-42. American Academy of
Medicine Press, Easton.
STANLEY, L. L. 1950. Letter to New Jersey Steril-
ization League. Association for Voluntary Steril-
ization Archive, University of Minnesota.
STATE v. FEILEN. 1912. 70 Wash. 65.
STATE BOARD OF CONTROL OF MINNESOTA. 1934.
Eighteenth Biennial Report. State Prison Print-
ing Department, Stillwater.
STODDARD, L. 1920. *The Rising Tide of Color against
White World Supremacy.* Charles Scribner's Sons,
New York.
STROTHER, F. 1924. The cause of crime: defective
brain. *World's Work*, 48: 275-281.
STUMP v. SPARKMAN. 1978. 435 U.S. 349.
SUTHERLAND, E. H. 1924. *Criminology.* Lippincott,
Philadelphia.
TAROMIANZ, M. A. 1944. Letter to New Jersey
Sterilization League. Association for Voluntary
Sterilization Archive, University of Minnesota.
TEXAS OBSERVER. 1981. Item. *The Texas Observer*,
March 20, 1981, p. 7.
WADLINGTON, W. 1966. The Loving case: Virginia's
anti-miscegenation statute in historical perspec-
tive. *Va. Law Rev.*, 52: 1189-1223.
WATKINS, H. A. 1930. Selective sterilization.
J. Psycho-Asthenics, 35: 51-67.
WATSON, J. 1927. *The Way of Behaviorism.* Charles
Scribner's Sons, New York.
WHITNEY, L. 1928. *The Basis of Breeding.* Earle C.
Fowler, New Haven.
WHITTEN, B. D. 1935. Sterilization. *J. Psycho-
Asthenics*, 40: 58-68.
WOODSIDE, M. 1950. *Sterilization in North Carolina.*
University of North Carolina, Chapel Hill.
WYATT v. ADERHOLT. 1973. 368 F. Supp. 1382.

Rubber Wars: Struggles over the Condom in the United States

JOSHUA GAMSON

Department of Sociology
University of California, Berkeley

IN THE EARLY 1940s, Dr. Woodbridge Morris, general director of the Birth Control Federation of America, complained to the American Social Hygiene Association about its overwhelming silence on the matter of condoms. "I appeal to you to take a positive stand in this matter," he argued, "before the public finds out that you are, in fact, permitting the spread of venereal disease because the most effective method to control it *happens to be* a method of contraception."[1] Meanwhile, Catholics arguing against birth control continued to treat the condom exclusively as a contraceptive method, arguing that in using such devices to "positively frustrate the procreative purpose of sexual intercourse" couples "pervert the order of nature and thus directly oppose the designs of nature's Creator."[2]

Over forty years later, the condom is again being pushed and pulled, molded into a variety of meanings. Jerry Della Femina, chairman of the advertising agency hired in the mid-1980s to promote LifeStyles condoms, complained in 1986 of the network ban on condom advertising, echoing Dr. Morris: "What the networks are saying is, 'We don't care if people die. We have our policy.'"[3] Testifying before the House Subcom-

---

I am grateful for comments on an earlier version of this paper to Kristin Luker, William Gamson, and two anonymous reviewers.

[1] Quoted in Allan M. Brandt, *No Magic Bullet: A Social History of Venereal Disease in America since 1880* (New York, 1985), p. 159; my emphasis.

[2] Father Francis J. Connell, "Birth Control: The Case for the Catholic," *Atlantic Monthly* 164 (October 1935): 469.

[3] Quoted in Joanne Lipman, "Controversial Product Isn't an Easy Subject for Ad Copywriters," *Wall Street Journal* (December 8, 1986), p. 1.

[*Journal of the History of Sexuality* 1990, vol. 1, no. 2]

mittee on Health and the Environment in 1987, Alfred Schneider, vice president for policy and standards at Capital Cities/ABC, argued for network policy on the grounds that "it is impossible to separate this product from the original and long-standing use of the product, which is for birth control purposes." Therefore, he argued, ABC could not advertise condoms without violating "standards of good taste and community acceptability."[4]

The condom, then, is a disease preventive that happens to also block conception and a conception preventive that happens to block viruses. Meanings proliferate in popular discourse: the condom is a sign of sexual license, a sign of sexual maturity, a sign of AIDS awareness, a cumbersome piece of rubber, and so on. Oddly, bitter fights are and have been waged—in courts, in congressional hearing rooms, on television, in scientific journals, and presumably in bedrooms as well—to make the condom mean certain things and not others, to associate it with some uses and behaviors and dissociate it from others. It is given a power afforded few inanimate objects.

Why and how do people fight so hard to make their meaning of the condom dominate? Two heated periods in the public history of the condom in the United States (court cases of the 1930s and 1940s and publicity debates of the 1980s) provide an interesting base from which to formulate an answer. In the process of examining the theoretical and historical aspects of condom disputes in the United States, I hope to raise and suggest answers to more general questions: How do objects become endowed with meaning? How do meanings shift? What conditions underlie and affect the construction of these meanings? What difference does an object's uses—for example, a sexual use—make for this process?

THEORETICAL ISSUES

The theoretical issues with which this essay begins and ends rest in the realm of cultural theory. The assumption of a "social construction" of the condom obviously defines the approach taken here. I am clearly opposed to an approach that treats the meaning of an object as inherent in the object itself. Among those who do argue that cultural meanings are actively constructed, though, the question of *who* in fact does the constructing is too often avoided.

On the one hand, many theorists of culture operate under a model in which texts and artifacts are read as "mirrors" of collective sentiment or values, symptoms of the goings-on in a society's "psyche." Here, by relying

---

[4]U.S. Congress, House, Committee on Energy and Commerce, Subcommittee on Health and the Environment, *Condom Advertising and AIDS,* 100th Cong., 1st sess., February 10, 1987, pp. 48–49.

on the notion of a homogeneous national psyche—itself an ideological assertion—"reflection theory" generally sidesteps the agents of meaning construction.[5]

A competing model sets ideology and those constructing it at the center. Here, the audience becomes largely irrelevant, injected with the repressive ideology of powerful leaders and cultural producers. (Theodor Adorno and Max Horkheimer, and the subsequent writings from the Frankfurt School, articulated this "hypodermic" model in its purest form.)[6] A more sophisticated version of this view, in discussions of ideological hegemony, conceives of cultural products not as unproblematically imposed but as constantly renewed through the incorporation of challenges. Still, cultural products within this model are generally analyzed to reveal the successful operation of "ruling-class" hegemony.[7]

Overemphasizing unity, then, a model of cultural artifacts-as-mirrors underemphasizes the concrete actions taken to *produce* them as pieces of culture, to infuse them with particular meanings. Overemphasizing ideological production, the model of a hegemonic culture is ill equipped to explain when and how certain pieces of a culture become *disputed:* it does not take seriously the possibility of cultural construction by a wide variety of interested groups, nor does it go far in explaining how new agents might enter the game and under what conditions, nor when and how interests collide and align.

The approach taken in this paper is nearer to this second model in that it pays close attention to the mobilization of particular meanings to serve particular political and material interests; however, it sees the construction process as at times a more open one than "hegemony" implies. The evidence presented here suggests that the process does involve a wide range of players, that new players and new strategies and new constraints arise, and that shifts in meaning occur in a process more complex than an inevitable incorporation of subversive alternatives.

Examining the condom at disputed moments means recognizing that

---

[5]For an early examination of "reflection theory," see Milton Albrecht, "The Relationship between Literature and Society," *American Journal of Sociology* 59 (March 1954): 425–36. For examples of this sort of argument, see Leslie Fiedler, *Love and Death in the American Novel* (New York, 1966); Martha Wolfenstein and Nathan Leites, *Movies: A Psychological Study* (New York, 1970); or nearly any popular magazine analysis of cultural trends. For a helpful review of sociological approaches to culture, see Richard Peterson, "Revitalizing the Culture Concept," *Annual Review of Sociology* 5 (1979): 137–66.

[6]Theodor Adorno and Max Horkheimer, "The Culture Industry: Enlightenment as Mass Deception," in *Mass Communication and Society,* ed. James Curran et al. (Beverly Hills, CA, 1977), pp. 349–83.

[7]For further discussions of this model, see Raymond Williams, "Base and Superstructure in Marxist Cultural Theory," *New Left Review* 82 (November–December 1973): 3–16; Todd Gitlin, "Prime Time Ideology," *Social Problems* 26 (February 1979): 251–66.

asserting what the condom "is"—the meaning of these pieces of latex or skin—becomes a central part of actors' strategies to affect the distribution and use behaviors surrounding it. Drawing from recent studies of social movements, it seems most fruitful to view these battles as "framing contests," in which actors "assign meaning to and interpret relevant events and conditions"—and, I would add, relevant *objects*—"in ways that are intended to mobilize potential adherents and constituents, to garner bystander support, and to demobilize antagonists."[8] Viewed as definitional disputes, debates over the condom can illuminate the process of cultural construction, its dynamics and limits, and its link to concrete actors.[9]

THE OBJECT ITSELF: THE MALLEABLE CONDOM

Condom-like sheaths are known to have been used for disease prevention as early as the sixteenth century (some writers date their origins as early as ancient Egypt) when the Italian anatomist Gabriello Falloppio recommended lubricated linen condoms as protection against venereal disease. Condoms are first known to have been used against conception in the eighteenth century; by then, sheaths were widespread.[10] With the vulcanization of rubber in 1844 came a major boost, the mass production and distribution of condoms (they had previously been made from animal skins). With the introduction of latex in the early 1930s came a second revolution. Sales increased enormously, and prices dropped. By the mid-1930s, the fifteen major condom manufacturers were producing one and a half million a day at an average price of a dollar per dozen.[11]

The naming of the condom is somewhat more mysterious, attributed variously to Daniel Turner but—revealingly—called the "French letter" by the English and "la capote anglaise" ("the English cape") by the French.[12]

Built into this history is a malleability born of the condom's dual function. This distinguishes it from other contraceptives and allows it to more easily become the focal point of framing debates. Most contraceptives have

[8]David Snow and Robert Benford, "Ideology, Frame Resonance, and Participant Mobilization," *International Social Movement Research* 1 (1988): 198. See also William Gamson and Andre Modigliani, "Media Discourse and Public Opinion on Nuclear Power: A Constructionist Approach," *American Journal of Sociology* 95 (July 1989): 1–37.

[9]Jerome Himmelstein has examined *The Strange Career of Marihuana* (Westport, CT, 1983) from "killer weed" to "drop-out drug" through a similar lens, tracing "how the public discussion of marihuana has been framed or structured; how this conceptual framework has been socially shaped by moral entrepreneurs, the social locus of use, and broader social conflicts; and how it in turn has helped to determine the nature of marihuana laws" (p. 146).

[10]Linda Gordon, *Woman's Body, Woman's Right: A Social History of Birth Control* (New York, 1976), p. 44.

[11]Ibid., p. 317.

[12]Norman E. Himes, *Medical History of Contraception* (New York, 1936), pp. 187, 192.

a "fixed" overall meaning—they are used before or during sexual inter-course to block conception—and disputes have generally involved groups asserting individual "rights" opposed to groups asserting the need for state involvement in the regulation of "morality." Disputes over distribution of and access to contraceptives do not generally revolve around definitions of the objects themselves.[13] Because condoms have been used to serve an additional prophylactic function, however, the *definition* of the condom itself becomes a tool.

Definitional strategies center primarily around these two possible meanings: actors first of all attempt to push the condom toward the end of a contraceptive-prophylactic continuum that serves their needs. They push it along a second continuum, as well. The condom, as an external barrier method, always carries with it—even more strongly than do other devices, such as the birth control pill—some connotation of sexual intercourse. Thus, running alongside and intersecting with the contraceptive-prophylactic dimension is a sexual one: actors try to push the condom toward or away from its place in sexual intercourse—wrapping it, for example, in medicinal garb or playing up its image as the friend of the prostitute. At times, the usefulness of locations on these continua conflict, and putting together a successful definition is especially tricky: a group may want to define the condom as a prophylactic, for example, while denying an association of condom use with disease itself.

How do these fights over meaning play themselves out? Who are the actors in the struggles and what are their interests? The condom has brought together institutions and actors guided largely by political interests (increasing or limiting women's collective power and individual control, increasing or limiting the group's power, furthering political careers) and material interests (building and defending professional expertise, expanding and controlling a market share). These actors in turn operate within a state for which two self-defined interests are triggered by the condom: (1) an interest in monitoring and protecting public health; and (2) an interest in monitoring and regulating the sexual "morality" of its citizens.[14] During two periods in this century, to which we turn now, these actors have been especially visible fighting over the condom.

[13] See, for example, the October 1935 arguments in *Atlantic Monthly* (pp. 463–73); the arguments of the Supreme Court justices in Griswold et al. v. Connecticut, 381 U.S. (1965); David M. Kennedy, *Birth Control in America* (New Haven, CT, 1970); John D'Emilio and Estelle B. Freedman, *Intimate Matters: A History of Sexuality in America* (New York, 1988); and Gordon.

[14] Much has been written recently about state involvement in the regulation of sexuality and sexual behavior, and the (mostly economic) interests guiding this regulation. See, for example, many of the essays in *Powers of Desire: The Politics of Sexuality*, ed. Ann Snitow, Christine Stansell, and Sharon Thompson (New York, 1983); and Catherine MacKinnon, "Feminism, Marxism, Method and the State," *Signs* 7 (Spring 1982): 515–44.

CONDITIONS OF THE EARLY DEBATES

In the first three-quarters of the nineteenth century, information about contraceptive devices circulated widely in the United States, through journals and advertisements in newspapers and almanacs. In 1873, however, the federal government became heavily involved in the regulation of contraceptive information and devices through the Comstock Act, which forbade the mailing, interstate transportation, and importation of contraceptive materials and information. Many state legislatures followed suit with their own "little Comstocks."[15]

It was not until the 1930s and 1940s that successful battles to revise state control over access to contraception and contraceptive information were waged. Even then, it was not through direct industry confrontation of the laws themselves that reforms were achieved, but indirectly, through court rulings on issues in which definitions of the condom—which assumed a central place in liberalization of the Comstock statutes—were necessary.

Why did the industry or others fail to confront the laws directly? There is little direct data available to explain inaction; when we look closely at the actions taken, though, clues begin to surface.[16] In the 1920s, birth control activist Margaret Sanger became convinced by a New York ruling to shift her strategy from repeal to reform of birth control legislation; a limited attack was necessary. New York Judge Frederick Crane had construed a state law to allow physicians the right to prescribe contraception "for the cure and prevention of disease," while birth control in general would continue to be classified as obscene and unlawful. Sanger, convinced "that her best prospect lay in a legislative amendment that would keep control in medical hands,"[17] spent nearly a decade pursuing "doctors-only" bills, with no success. This pursuit of an amendment allowed Sanger to gather medical support behind her and "appealed also to the cautiously limited liberalism emerging in the Protestant churches." She failed, David Kennedy argues, because she had to settle for "impotent political allies" in her fight: "Birth control, in spite of quietly growing public acceptance, was still such an explosive subject that most politicians preferred to avoid it."[18]

Sanger's activity, however, set the stage for understanding the activity of the 1930s and 1940s: it highlights both the constraints and the key ac-

[15] D'Emilio and Freedman, pp. 159–60.

[16] And why did change take so long? Here, although I do not examine the period between the 1870s and the 1930s, one again finds clues by noticing key actors in the later activities. The rise of women's activism, birth control activism in particular (see Gordon), and the solidification of medical professionalism (see Peter Conrad and Joseph Schneider, *Deviance and Medicalization* [St. Louis, 1980]) may have been necessary conditions for a challenge to the state's control.

[17] Kennedy, p. 220.

[18] Ibid., pp. 227, 231.

tors. What was apparently necessary was a battle in which the interests of "most politicians," who depend in their careers on "public acceptance," were not central. Moreover, a strategy was necessary that would not provoke mobilization on the part of the strongest opponent of birth control, the Catholic church. Contraception had long been opposed by religious leaders, and a "vigorous campaign" against it had been conducted by the Catholic hierarchy since the end of the nineteenth century, reaching its climax in Pope Pius XI's 1930 statement that those who used contraception violated "the law of God and nature" and were "stained by a grave and mortal flaw."[19] The context was such that direct changes through the legislature were unlikely and politically unfeasible. The site of activity necessarily became the judiciary, and the route for change paralleled Sanger's attempts: limited reforms through an embracing of the medical profession.[20]

CONDOMS IN COURTS: THE DEBATES OF THE 1930s AND 1940s

In the court cases of the 1930s and 1940s, the condom was consistently treated as an *exemption* from a state regulation of sexual behavior, which itself went unchallenged. This exemption turned on a *reframing* of the condom in medical terms.

Interestingly, the series of court cases surrounding condoms began with an internal industry dispute: in 1930 the Youngs Rubber Corporation charged C. I. Lee and Company with a trademark infringement in interstate commerce. In resolving the dispute, the Court of Appeals essentially redefined federal statutes. Condoms, the judge argued, "may be used for

---

[19]Quoted in J. T. Noonan, "Contraception," in *New Catholic Encyclopedia,* 16 vols. (New York, 1967), 4:274. We see examples of a strategy that does not take the Catholic hierarchy and constituency into account later: birth control advocates in Massachusetts twice forced a referendum amending state statutes, in 1942 and 1948. Both cases aroused "bitter campaigns in which the Protestant and Catholic Churches aligned against each other," and in both cases the referendum lost by approximately a seven to five margin (see Jack Hudson, "Birth Control Legislation," *Cleveland-Marshall Law Review* 9 [May 1960]: 248).

[20] The medical profession, it should be noted, was particularly well-positioned for this alliance. Since the mid-nineteenth century, groups of physicians had been working to professionalize medicine, striving for both prestige and "for exclusive professional and economic rights to the medical turf," which had until then been quite open. Much of the striving involved "medical crusading," assertions of expertise and calls for licensing and regulation, the translation of "social goals of cultural and professional dominance into moral and medical language." As a contraceptive, control of the condom would appear to be in the domain of the state; as a medical device, control would switch to medical professionals (see Conrad and Schneider, pp. 10–14). The strategy by doctors of avoiding the sexual dimension in their public actions, and sticking close to disease, is illustrated nicely by the American Medical Association. The AMA, "seeking not to appear to sanction promiscuity," refused to discuss condoms as preventive techniques in their 1960s campaign for the control of syphilis and gonorrhea (see Brandt, *No Magic Bullet* [n. 1 above], p. 176).

either legal or illegal purposes. If, for example, they are prescribed by a physician for the prevention of disease . . . their use may be legitimate; but, if they are used to promote illicit sexual intercourse, the reverse is true. . . . A manufacturer of drugs or instruments for medical use may in good faith sell them to druggists or other reputable dealers in medical supplies, or to jobbers for distribution to such trade."[21] Here, federal law is reinterpreted through the lens of medicine and the condom exempted as a legitimate medical instrument—that is, as a prophylactic. The "legitimate use" argument, and the Youngs Rubber case itself, became central in later rulings.[22] In 1936, the liberalization of federal birth control law was solidified in the One Package case, in which the Second Circuit Court of Appeals literally defined the federal statutes to allow "the importation, sale or carriage by mail of things which might intelligently be employed by conscientious and competent physicians for the purpose of saving life and promoting the well being of their patients."[23] State rulings followed this path; the Massachusetts Supreme Court (which two years earlier had upheld state regulations under the "police powers of the state to control the morals of its people") ruled that "where an appliance, in this case a sheath-type rubber condom, had the *dual capacity* of being a contraceptive and at the same time preventing venereal disease, the *prosecution must show that the seller is aware that the buyer intends to use the device for a contraceptive purpose.*"[24] In this case, the court explicitly directed condom advocates toward the disease frame.

It is important to notice that, while the liberalization of state regulation of birth control information and distribution came about through a definition of a legitimate (that is, medical) use of *any* device, the cases themselves all involved condoms. The dual function of the condom provided openings allowed few other sexual objects. By activating the disease-prevention meaning to which condoms lend themselves and disassociating them from sex, manufacturers could align with the medical profession and the pharmacists to exempt themselves from federal statutes. Moreover, the conflicting interests of the state are themselves triggered. On the one hand, the state has a long history of protecting the institution of monogamous marriage and reproduction via sanctions on sexual behaviors that depart from it; on the other hand, the state is institutionally invested in protecting the public health of its citizens.[25] In the condom, these two interests conflict— a conflict on which interested groups, quite literally, capitalize.

---

[21]Youngs Rubber Corporation v. C. I. Lee & Co., 45 F.2d 103 (1930).

[22]In, for example, Davis et al. v. United States, 62 F.2d 473 (1933).

[23]United States v. One Package of Japanese Pessaries, 86 F.2d 737 (1936).

[24]Hudson, p. 247; my emphasis.

[25]The tug of different state interests may have become especially acute during the world wars, when protecting soldiers from venereal disease through condom distribution became

Just how important the disease frame was—and how important its distance from the contraception frame—becomes apparent in a test case brought by a licensed physician in Connecticut in 1942. In this case, it was to be determined if a physician would be exempted from the prohibition of the right to disseminate contraceptive advice or prescribe a device when another pregnancy would likely result in the death of the mother. The court upheld the statute, arguing that "sexual abstinence might be practiced where life was endangered."[26] For disease prevention, information about and provision of devices tends to be allowed, and this is a function exclusive to condoms; for the prevention of conception, even to save a life, abstinence is recommended.

It is clear that change took place here in the judicial arena, and I have suggested why this was the necessary site and why a limited strategy in which the condom was central was necessary. What was gained by the alliances that produced the changes? Birth control activists, who were not in fact actively involved in the court battles, saw some of their goals achieved: the Comstock Act and many of the "little Comstocks" suffered a partial defeat. Doctors, also involved peripherally, won a certain symbolic victory, a furthering of their status as professionals, in that exceptions read into the statutes often turn on insuring that control of information and dissemination was in the hands of medical professionals. The greatest material gains, though, were for manufacturers (who could now expand marketing so long as it was for "legitimate use") and pharmacists (who took their places as the main gatekeepers of disease-preventative condom distribution).

In part the gains for manufacturers and pharmacists were a result of their ability to mobilize around the disease frame, which allowed them to defuse active resistance (by moving the condom away from "religious"—that is, intercourse-related—issues) and call up alliances (from birth control reformers and professional doctors). These strategies were taken within the limits set by state interests. In part, I would add, they also benefited from a situation set up by the state regulation they opposed. Given that *dissemination* of condoms (rather than use or publicity) was the focus of state regulation, and that the judiciary was the logical site of dispute, manufacturers and pharmacists became the key actors directly involved and arguing their cases in the judicial process. Other actors—condom users or activists for or against—were essentially excluded from direct participation.

---

policy. The success of the disease frame may well have been strengthened not only by the fact that the government had actually positively sanctioned condom use but also by the interest of the state in dissociating this policy from a positive sanctioning of sexual promiscuity (see Brandt, *No Magic Bullet*).

[26]Tileston v. Ullman, 129 Conn. 84 (1942).

CHANGED CONDITIONS

Between the 1940s and the 1980s, several major developments significantly altered the context in which discussions over the condom took place. Building on the legal openings provided by earlier condom disputes and supported by a strong and growing women's movement, birth control activists continued to challenge the legal status of contraceptives. In 1965, with the Supreme Court ruling in *Griswold et al. v. Connecticut* removing the ban on contraceptive use by married persons, the government effectively withdrew from this area of moral policing. And in 1977, the Supreme Court ruled that the advertisement and display of contraceptives could no longer be prohibited and that the arguments that "advertisements of contraceptive products would offend and embarrass those exposed to them" and would "legitimize sexual activity of young people" were not cause for restricting First Amendment rights to expression.[27]

While the state was withdrawing from regulation, the condom was withdrawing from popularity. With the introduction of the birth control pill in 1960 and the broadened use of both the pill and the intrauterine device (IUD) in the 1960s and 1970s, the condom went into a dramatic decline. Between 1965 and 1970, condom use declined 22 percent.[28] From the mid-1970s to the mid-1980s, U.S. condom sales fell by half.[29] This all changed, of course, with the spread of AIDS from the early 1980s onward and the discovery that the use of latex condoms helps prevent the transmission of the HIV virus. The promotion of "safer sex" through condom use, diligently pursued by former Surgeon General C. Everett Koop, other public health officials, and AIDS activists,[30] had immediate effects on the condom industry. Retail condom sales in the United States rose 25 percent from 1982 to 1984.[31] In 1986 alone, sales increased by 10 percent, and stock prices in Carter-Wallace, Incorporated (the manufacturer of Trojans, which has over half of the U.S. market share) shot up by 55 percent. From 1987 to 1989, sales increased by more than 60 percent.[32]

These changes have meant a shift in the site of disputes over the condom. The condom has very rarely appeared as a subject of judicial or

[27]Carey, Governor of New York, et al. v. Population Services International et al., 431 U.S. 678 (June 9, 1977).

[28]Charles F. Westoff and Norman B. Ryder, *The Contraceptive Revolution* (Princeton, NJ, 1977), pp. 21–22.

[29]Colin Leinster, "The Rubber Barons," *Fortune* 114 (November 24, 1986): 106.

[30]For a well-documented, and overly sensationalist, history of the epidemiological and political history of AIDS, see Randy Shilts, *And the Band Played On* (New York, 1987). For a representative activist response to AIDS and sex—a response consistently ignored in Shilts's writing—see Douglas Crimp, "How to Have Promiscuity in an Epidemic," in *AIDS: Cultural Analysis/Cultural Activism*, ed. Douglas Crimp (Cambridge, MA, 1987), pp. 237–71.

[31]Robert Hatcher et al., eds., *Contraceptive Technology, 1988–1989* (Atlanta, 1989), p. 345.

[32]"Can You Rely on Condoms?" *Consumer Reports* 54 (March 1989): 135–42.

legislative inquiry since the 1960s. Instead, interested parties take to the realm of publicity, as witnessed in the continuing debate over condom advertising. Condom use declined, one physician lamented in a medical journal in 1979, due to "bad press" during the 1960s and 1970s. The solution he proposed to revive use was revealing: "Persuading the National Association of Broadcasters to allow condom advertising on radio and television and . . . convincing the copy-review boards of newspapers and magazines in general circulation to do the same. . . . Organized medicine and public health can take the lead here by bringing their collective weight to bear on the media to reverse this policy. . . . Medical journals can also help by encouraging condom manufacturers to advertise their products . . . in their pages."[33] In the contemporary site of condom dispute, different constraints and some new actors operate. Essentially, it is media practices that determine the rules, and new alliances are forged, now with the advertising industry.

CONDOMS IN THE MASS MEDIA: THE DEBATES OF THE 1980s

The sudden boom in sales provided by the "free publicity" of AIDS—tastelessly appearing as one of *Adweek's* "Hottest Markets of 1987"[34]—has triggered a marketing war between the established manufacturers. (Two manufacturers, Carter-Wallace and Schmid's, which makes Ramses and Sheik, account for approximately 90 percent of the market.)[35] It has also triggered the entry of a number of new contenders. In 1986, Carter-Wallace launched a multimillion-dollar advertising campaign, the largest expenditure on a single advertising campaign ever undertaken in the industry, and a new company planned to spend between two and three million dollars marketing its Mentor condoms.[36] In this war, manufacturers have faced two related problems: first, how to break down media-industry opposition to condom advertising; and second, how to target condoms to particular markets. In both cases, not surprisingly, condom manufacturers have steered away from contraception and toward prophylactic use.

Until recently, condom advertising was found only in heterosexual men's sex magazines. In 1985, the first billboard condom advertisements

---

[33]Yehudi Felman, "A Plea for the Condom," *Journal of the American Medical Association* 241 (June 8, 1979): 2517–18.

[34]"The Business of AIDS," *Adweek* 37 (April 16, 1987): 4–11.

[35]The division of the market share has changed little in recent years, with Carter-Wallace at 56 percent and Schmid's at 34 percent, and the remaining 10 percent divided about equally between Circle Rubber and Ansell-Americas (see Leinster).

[36]Gail Appleson, "Women Taking Condom Initiative," *Oakland Tribune* (October 19, 1986).

appeared (for Ramses, in Atlanta). By 1986, many mainstream magazines—including women's magazines such as *Modern Bride, Vogue,* and *Family Circle*—had dropped their long-standing bans on contraceptive advertising and accepted condom ads. Over the course of 1987, nearly all bans were dropped, including those by holdouts such as *Newsweek, U.S. News and World Report,* and the publications of Time, Incorporated.[37]

How has such a dramatic change come about? Primarily, it is a result of aggressive and careful marketing shifts on the part of manufacturers. The underlying factor is the revision by advertisers of what it is that is being sold, so that publications (like the courts in the 1930s) can essentially *maintain* their ban on contraception: condoms are simply not contraceptives. Manufacturers and their marketers do not attempt to repeal mass media "policy" but accept its logic and work around it.

Companies actively target new markets with new marketing strategies. The key here is the basic logic of marketing: try to determine the tastes, desires, and fears of the audience in order to play to them. "We're not going after the *Playboy* and *Penthouse* market," says Mentor vice president Al Mannino. Instead, they are going after (heterosexual) women, who, according to an industry spokesperson, now make up 40 percent of the condom market, and in particular after unmarried women who, according to a recent study, used almost twice as many condoms in 1987 as they did in 1982.[38] To reach women, as well as the non-*Playboy* male markets, strategies have been adapted to give the condom a makeover, from sleazy to "smart."

A major part of this strategy involves desexualizing the condom. Advertising, says *Family Circle*'s advertising director, is "more clinical, nonsuggestive, and informative," a "service" to readers, as opposed to the earlier "suggestive" ads, with their "emphasis on the *pleasure* of the product."[39] The condom, Mentor's health-care products manager asserts, "is a personal hygiene item."[40] Manufacturers and advertisers link themselves directly to public health. "I say God bless him," said John Silverman, the president of Ansell-Americas, makers of LifeStyles, after the surgeon general announced that using condoms is the best protection against HIV-infection

[37] Patricia Winters, "TV Stations Embrace Condom Spots," *Advertising Age* 58 (February 2, 1987): 4.

[38] See "Packing Protection in a Purse," *Time* 132 (August 15, 1988): 65. Another major new market is the gay male population. Carter-Wallace recently decided against advertising in gay media, after "reviewing a cross section of the regional and national gay publications," arguing that "we advertise in a lot of general media and feel gay men aren't excluded as readers" (quoted in *Adweek* [December 14, 1987], p. 26)—a logic belied by its selective application. Here, again, the minimum-risk decision-making process seems to lead companies to avoid associating their products too closely with controversial activities and populations.

[39] Quoted in Leinster; my emphasis.

[40] Quoted in Patricia Winters, "Condom Ads Aim at AIDS," *Advertising Age* 57 (December 1, 1986): 99.

barring abstinence. Silverman quickly called for adjusting advertisements to "support [the surgeon general's statements] with public information advertising" and wrote to drugstores that "Ansell has accepted the challenge and responsibility for taking this message to the public."[41] The LifeStyles ads made no mention of contraception.

Packaging has been similarly adapted both to reach women and to detach the condom from its sexual-player meaning. Schmid's Koromex brand uses a simple box with a businesslike holder. Mentor condoms are sold in individually wrapped, small plastic cups. Carter-Wallace's Trojans for Women are sold in pink and lavender boxes picturing a woman; according to the company's vice president of marketing, this will help the product make the move to "display space in the feminine hygiene section of stores."[42]

In a variety of ways, then, the condom has now been marketed as not-a-contraceptive: it is a beauty aid, a personal hygiene item, a public service— all, however euphemistically, based on the prevention-of-disease frame and a desexualization of the condom.

None of this, however, has yet broken the ban by the three major television networks. (Many local affiliates have broken over the last three years with network policies, agreeing to air the commercials, usually under the conditions that they not run during children's hours, not promote condoms as birth control, and are in "good taste." In addition, many cable stations accept the ads.)[43] The code of the National Association of Broadcasters has held that broadcasters should refuse any advertisement "which the station has good reason to believe would be objectionable to a substantial and responsible segment of the community."[44] This is, of course, the same basic logic guiding newspapers and magazines: do not endanger your sales by offending your audience.

Since 1986, industry outsiders have stepped up pressure on television advertising. Planned Parenthood placed coupons addressed to network executives in print ads in 1986 and in 1987 charged that the networks were "out of step with the great majority of the American people," producing a Harris survey showing support for condom advertising.[45] House subcommittee hearings in February 1987, including testimony by Surgeon

[41]Lipman (n. 3 above), p. 19.

[42]Quoted in "Condom Makers Aim at Women Buyers," *Wall Street Journal* (June 21, 1988), p. 41.

[43]See the testimony in U.S. Congress, House, Committee on Energy and Commerce (n. 4 above).

[44]Myron Redford, Gordon Duncan, and Denis Prager, eds., *The Condom* (San Francisco, 1974), p. 152.

[45]"Harris Survey Shows Viewers Think Contraceptive Ads OK," *Broadcasting* 112 (March 30, 1987): 178.

General Koop, other public health officials, and network executives, focused directly on condom advertising. The outcome was a nonbinding "urging" of network cooperation. Despite charges of "media malpractice" and pointed arguments between congressmen, the hearings give testimony to the recognition that the networks, not the government, determine their own policy. Again, it becomes clear that the arena of dispute, and those setting the rules in that arena, has shifted.

Ralph Daniels, NBC's vice president of broadcast standards, defended NBC's policy of refusing condom commercials at Congressional hearings: "Our network television standards reflect the fact that we provide program service to over 200 individual television stations serving local communities across the nation. The audience served by these stations includes a wide range of religious beliefs, social attitudes and mores, as well as local and regional concepts of propriety and acceptability. . . . *Broadcast network or local stations cannot ignore the fact that condoms are also contraceptive devices.*"[46] While noting that broadcast standards "evolve," Daniels and other network representatives argued that, despite its function as a disease preventative, the condom continues to be seen as a contraceptive and thus to trigger antagonistic beliefs, mores, and concepts of propriety that the networks literally cannot afford to arouse. "At some point," said an NBC spokesperson after the hearings, "there may be such a health problem that we would have to overlook the morality question."[47] Here, like the courts before them, the change-makers instruct interested parties in how to become exempt from the rules under which they operate: define the product in terms of "health problems" (disease) rather than "morality questions" (contraception).

OLD AND NEW VOICES

It is not entirely clear why the networks are so resistant, though one might speculate that their gradual decline in the television market with the advent of cable television and video cassette recorders leaves them in a position particularly disposed against taking advertising risks. The fear of offense and backlash is not unreasonable. Once again, arousing the Catholic church and the religious Right, which hold that "it is not responsible action to use the devastating disease of AIDS as an excuse to promote artificial birth control,"[48] does pose a threat.

[46]U.S. Congress, House, Committee on Energy and Commerce, pp. 36–37; my emphasis.

[47]Quoted in Steven Colford, "Nets Take Flak on Condom Ads, but Congress Won't Force Change," *Advertising Age* 58 (February 16, 1987): 79.

[48]Bishop Bevilacqua, "The Questions Raised by School-based Health Clinics," *Origins* 17 (September 3, 1987): 188.

The Catholic discussion of AIDS itself pushes those wishing Catholic support or fearing Catholic opposition toward a birth control version of the condom. In 1987, the U.S. Catholic Administrative Board stated that AIDS educational efforts "could include accurate information about prophylactic devices or other practices proposed by some medical experts as potential means of preventing AIDS. . . . Such a factual presentation should indicate that abstinence outside of marriage and fidelity within marriage as well as the avoidance of intravenous drug abuse are the only morally correct and medically sure ways to prevent the spread of AIDS."[49] This qualified approval, briefly ratifying the disease-only approach, set off immediate controversy within the church. Bishops responded that "it is never morally permissible to employ an intrinsically evil means to achieve a good purpose"[50]—that is, they reasserted the contraception frame to reinforce, in the pope's phrasing, the church as "sole interpreter of the law of God and 'expert in humanity.'"[51]

This controversy led American bishops to develop a further document in November 1989, which emphasized abstinence and insisted more explicitly that "it is not condom use that is the solution to this health problem," reunifying in a 219–4 vote in favor of the new statement.[52] What is important to note in this controversy is more than how the bishops reasserted the contraceptive frame in the service of moral arguments, a clue to those needing their support. Significantly, their action was also heavily focused on *publicity*: Archbishop John May of St. Louis complained of "a misunderstanding in media reports" of the original document; Archbishop Roger Mahony of Los Angeles complained that "many media headlines and stories have seriously confused our Catholic people." Not to address the issue, May argued, "would leave people to learn of them from factually misleading campaigns designed to sell certain products or to advocate safe sex without reference to a moral perspective."[53]

Opponents of *any* condom publicity also enter the debate on occasion, again articulating the limits of mass media risk-taking. Representing an organization called the Committee on the Status of Women on an ABC "Nightline" report devoted to condom advertising, Kathleen Sullivan argued for abstinence publicity rather than condom publicity: "I'm really appalled at you gentlemen," she said to liberal San Francisco supervisor

[49]U.S. Catholic Conference Administrative Board, "The Many Faces of AIDS," *Origins* 17 (December 24, 1987): 482–89.

[50]"Reaction to AIDS Statement," *Origins* 17 (December 24, 1987): 489–92.

[51]Pope John Paul II, "Address to Vatican AIDS Conference," *Origins* 19 (November 30, 1989): 435.

[52]U.S. Bishops' Meeting, "Called to Compassion and Responsibility: A Response to the HIV/AIDS Crisis," *Origins* 19 (November 30, 1989): 429.

[53]"Reaction to AIDS Statement," pp. 490, 491.

Harry Britt (arguing for condom ads) and WCVB-Boston general manager Jim Coppersmith (arguing for television's autonomy). "You ought to be really debating how much you are going to stress the how and why of abstinence. . . . [Children] want to know the reasons how and why it's healthier for them not to be sexually active."[54] Coppersmith quickly responded that his station "would never do anything . . . that would encourage promiscuity. I recognize clearly that abstinence is the very best defense against AIDS."[55] This exchange captures these two players' acknowledgment of each other's power: the conservative moralist recognizing that television is a key to achieving her goals and the station manager recognizing that the conservative constituency must be reassured of television's conservatism.

Condoms also take a place in the actions of the AIDS activist movement, which, while not powerful enough to set severe limits on the actions of policymakers, does add a grassroots voice to condom debates.[56] AIDS activists make explicit attempts to "campaign and organize in order to enter the amphitheater of AIDS commentary effectively and unapologetically on our own terms."[57] The position taken by groups such as the AIDS Coalition to Unleash Power (ACT UP), and other predominantly gay-run AIDS organizations, is similar to the public health position but broadened to include larger goals and strategies than safer sex education. For these groups, of course, the condom almost never appears as a contraceptive and is instead entirely a life-saving protection against HIV transmission. It becomes a symbol not only of *safe* sex, though, but of safe *sex:* ACT UP, for example, calls for "explicit" sex education[58] that recognizes sexual diversity. Condoms take their place within an overall strategy that includes, for example, same-sex "kiss-ins" and "fuck me safe" T-shirts. These groups, while sharing the disease frame with public health officials and mainstream AIDS service organizations, *resexualize* the condom to bring it into a broader challenge to sexual norms. Implicit in this is a recognition that AIDS, like venereal disease,[59] often serves to stigmatize; since much discourse associates AIDS with "immoral" behavior, the condom method of prevention is also tied into questions of morality. Their strategy for waging

[54]"Nightline," ABC News (January 21, 1987). Representative William Dannemeyer took a similar position during the Congressional hearings. See U.S. Congress, House, Committee on Energy and Commerce.

[55]"Nightline," ABC News (January 21, 1987).

[56]See Josh Gamson, "Silence, Death, and the Invisible Enemy: AIDS Activism and Social Movement 'Newness,'" *Social Problems* 36 (October 1989): 351–67.

[57]Simon Watney, *Policing Desire: Pornography, AIDS and the Media* (Minneapolis, 1987), p. 54.

[58]ACT UP/San Francisco, "Our Goals and Demands" (1988, informational flyer).

[59]See Allan Brandt, "AIDS: From Social History to Social Policy," in *AIDS: The Burdens of History,* ed. Elizabeth Fee and Daniel M. Fox (Berkeley, 1988), pp. 147–71.

a challenge to stigmatization and sexual hierarchy aims directly for "the amphitheater," largely through events staged for the media—and through the free publicity of street postering.[60] (See fig. 1.)

In the case of current condom advertising, then, the shift in the site of activity has brought new players (advertisers, marketers) to the forefront and others more deeply into the game. Significantly, the essential definitional strategy remains the same and has had a similar effect: by defining the condom as a health *exception* to an overall accepted "morality," those who stand to benefit achieve a liberalization of (in this case informal) policy surrounding it. Obviously, the manufacturer-advertiser alliances most successful at breaking down barriers stand to gain the most financially, through access to targeted markets, and this is the crux of the battle. The limits of their strategies are set primarily by media practices—guided by low-risk decision making—which, in turn, are limited by the threat of backlash from organized religious and conservative groups, on the one hand, and pressure from public health officials and birth control and AIDS activists on the other.

CONTINUITIES AND DISCONTINUITIES

A number of continuities are visible through these two moments of contestation over the condom. First of all, what remains at the center of both contests are the material interests of condom manufacturers; what shifts are the alliances made by manufacturers and the constraints on their tactics. (Among these constraints, the fear of religious-group backlash also seems to remain constant.) Second, both periods clearly involve a liberalization of regulation of the condom: in the earlier period, a liberalization of laws regulating distribution and access; in the later one, a liberalization of publicity policies. These changes were achieved through the successful placement of the condom in a "disease and prevention" interpretive package. This package, moreover, involves in both cases a desexualization of the condom, a disassociation of the condom from pleasure (through its centralization in medical hands in the 1940s and through a pulling-back from the "pleasure sell" in the 1980s) and, of course, from reproduction (through its severing from contraceptive functions).

The fundamental use of "meaning" in these cases has a certain irony: actors whose primary interest is in increasing profits liberalize policies and democratize access to condoms by distancing themselves from liberal mod-

---

[60]See J. Gamson. For a discussion of the relationship of social movements to the mass media, see Todd Gitlin, *The Whole World Is Watching* (Berkeley, 1980).

Prevent AIDS. **Practice safer sex.** Use a condom. **It could save your life.**

JUST SAY KNOW

FIG. 1 "Just Say Know": Poster produced by Boy With Arms Akimbo, an anonymous group of "cultural activists," for an ACT UP–sponsored benefit, San Francisco, January 1990. According to a recent flyer (March 1990), Boy With Arms Akimbo approaches social change through "intellectual subversion and visual intervention," the appropriation and reappropriation of images to challenge conventional discourse; thus, when asked for permission to reproduce this poster, one member responded, "Appropriate away. That's our point."

279

els (that is, models involving individual choice in contraception), claiming instead an exemption from the conservative model (that is, sexual behavior can and should be state-regulated), thereby affirming that model.

Even more instructive, though, are the discontinuities. The most important shift is in the *site* of activity from the courts to the media, from legality to publicity. Additionally, state involvement has become much less central and the marketplace much more central. This shift brings with it a change in agents of control: the gatekeepers are no longer judges but media executives. New actors are also thus brought onto the scene. Publicity battles are in fact more open to a broad range of actors than are legal ones. Second, the *point of intervention*—that is, at what stage control is attempted and contested—has shifted. In the 1930s and 1940s, availability was at issue (through the control of actual distribution by the state); in the 1980s, visibility and knowledge are at stake.

How do we make sense of these shifts? Clearly, accidents of history (the rise of diseases such as venereal disease and AIDS) provide opportunities for the mobilization of certain interpretive frames. But this does not explain a shift in site, the entry of new actors, and a shift in the point of intervention.

Broadly put, it seems that recognizing the increased centrality of consumption, and a rise in the industry of publicity, are necessary for understanding these changes. Consumerism, the "mass participation in the values of the mass-industrial market," argues Stuart Ewen, developed from the 1920s on as "an aggressive device of corporate survival"—essentially to educate a nation of buyers to fit the needs of growing mass production. Advertising, whose impulse was one of "actively channeling social impulses towards a support of corporation capitalism and its productive and distributive priorities," became the "key apparatus for the stimulation and creation of mass consumption."[61] In this context, it is a shaping of consciousness, the shaping of a particular relationship toward goods and buying, that becomes central.

Without delving into the complex history of mass culture and advertising (and the growth of radio and television technologies immediately drafted into commercial services), I would suggest that this historical shift, while not altering the strategy of frame construction, accounts for the shift in site and the balance of opportunities and constraints. "A consumption economy," argues Daniel Bell, ". . . finds its reality in appearances."[62] In a market- and publicity-centered society, spreading or restricting public knowledge of the condom, controlling its *image,* becomes more central

---

[61]Stuart Ewen, *Captains of Consciousness: Advertising and the Social Roots of the Consumer Culture* (New York, 1976), pp. 54, 81.

[62]Daniel Bell, *The Cultural Contradictions of Capitalism* (New York, 1976), p. 68.

than opening or restricting manufacture and distribution of condoms; *knowledge* of sexual behavior logically becomes a focal point of attempts to control that behavior. This in turn means that more players can enter the dispute—witness, for example, the use of the media by AIDS activists and others to publicize condoms for safe-sex use. Finally, it means that the constraints on manipulating the condom's definition turn on media rather than judicial processes, dependent on perceptions of "public sentiments," the maximum profits with the minimum risk. This opens up new avenues for those attempting to effect change: by showing through polling, for example, or through social movement activity that the public does or does not accept a particular interpretive package. At the same time, it sets financial considerations firmly at the center of the struggle over the meaning of the condom.

CONCLUSION: WHAT ABOUT THE REST OF US?

I have argued that the condom has been made to mean many different things; that its meaning is in fact determined by the actions taken by interested parties involved in "framing" the condom to help achieve their interests; that broad historical shifts have changed the arena of these debates from legality to publicity; and that, despite a continuity in overall strategies, these shifts provide different opportunities and constraints for actors and, thus, changes in what the condom may come to mean. This is not a process particular to the condom, although the condom is particularly easy to manipulate, but a model for understanding how cultural artifacts are generally imbued with meanings.

Questions remain, however. What does "the public" actually think? And what is the relationship between their interpretation of what the condom is and their use of the object? These are questions that examining attempts to frame the condom in particular ways do not answer. Even superficial evidence suggests that, while different framing strategies explain how it has been possible to open access to condoms and knowledge about them, the shifts in public meaning do not translate into shifts in private understandings. The source of most people's definitions—and the control over their knowledge—may be influenced very little by maneuverings in courts and advertising boardrooms. Cultural understandings differ across social groupings (race, class, gender, age) as well as across time—clearly subject to other influences.[63] Lee Rainwater, for example, found in the

---

[63]For discussions of the complexities of audience reception, see Fred Fejes, "Critical Mass Communications Research and Media Effects: The Problem of the Disappearing Audience," *Mass Communications Review Yearbook* 5 (1985): 517–30; David Morley, *The "Nationwide" Audience* (London, 1980); and Janice Radway, *Reading the Romance* (Chapel Hill, NC, 1984).

early 1960s that "working class women. . . generally have less information about sex and contraception than do the men . . . [and] have learned what they do know (usually about condoms) from their husbands since their marriage."[64] Similar discrepancies between the definitional packages discussed in this paper and the attitudes toward the condom by young men and teenagers are readily seen.[65] Even if one does assume reception by the public of the condom in the public frame that dominates, it is not at all clear that "knowledge" of "what it is" translates into use behaviors. Other considerations and definitions (comfort, who the partner is, and so on) may very well override the publicly asserted definitions of the condom.

This, of course, has implications for the theoretical issues with which I began. Asking in whose interest certain definitions of the condom serve and how these interested parties achieve particular arrangements is necessary for understanding the operation of cultural "meaning" and its relation to access both to the condom itself and to knowledge about it. The approach, though, suggests its own limitations. Looking at the framing process alone also leaves many key questions unanswered—in particular, the reception of interpretive packages by individuals and groups of condom users. How similar are public frame and private? How are they linked? Without explaining discrepancies between construction and reception— the relationship between interest groups shaping public definitions of the condom and the privately held definitions of citizens—questions of impact on everyday lives are left blurry.

[64]Lee Rainwater, *And the Poor Get Children* (Chicago, 1960), p. 64.
[65]Redford, Duncan, and Prager, eds. (n. 44 above), pp. 182, 172.

Contraceptives for Males

William J. Bremner and David M. de Kretser

In the last twenty years, new contraceptive techniques for females have received wide publicity and use. Over 20 percent of American women in the reproductive age group used oral contraceptives in 1973.[1] Other measures, particularly the intrauterine device, are also very popular. With increasing use of these techniques has come increasing awareness of their hazards. Although the risks for an individual woman using either oral contraceptives or an intrauterine device are extremely small,[2] they are certainly present and have contributed to heightened pressure for alternative means of contraception. An increasingly voiced feeling has been that, if there are to be definite health risks associated with adequate contraception, these risks should be shared between the male and female partners.

Research concerning sexual reproduction in general and possible methods of controlling it has been spurred by awareness of the worldwide threat posed by unchecked population growth. Contraceptive

We appreciate the help of Mrs. Vimy Wilhelm of the Australian Federation of Family Planning Associations in obtaining statistics about female contraceptive use, and the typing of Mrs. Jill Volfsbergs.

1. George Washington University Medical Center, Department of Medical and Public Affairs, "Oral Contraceptives," in *Population Report*, ser. A. no. 1 (Washington, D.C., April 1974). The figure quoted is a minimum, since it includes only contraceptives sold in pharmacies and omits those obtained from hospitals, family planning clinics, subsidized programs, and samples. In Australia, Canada, and northern Europe, this figure is approximately 30 percent.

2. The most serious risk of the oral contraceptives for females relates to excessive blood clotting. For healthy women in the reproductive age group not taking the pill, the risk of dying from abnormal blood clotting is approximately 1 in 500,000 per year. For similar women taking oral contraceptives, the risk is approximately 1 in 60,000 and can probably be decreased by using pills with low estrogen content. The risk of acquiring a clot serious enough to require hospitalization while ingesting the pill is approximately 1 in 2,000.

[*Signs: Journal of Women in Culture and Society* 1975, vol. 1, no. 2] techniques that have been effective on a wide scale in the United States and other developed countries have often not been successful when introduced into different cultural and political climates. The desirability of having a wider range of contraceptive techniques available has become apparent and has led to funding for research along these lines from such organizations as the Population Council, the World Health Organization, the Ford Foundation, the International Planned Parenthood Federation, and the American National Institutes of Health, among others. A significant part of the work that has resulted has been concerned with developing techniques for contraception in the male.[3]

The main methods of contraception that have been available to the male have been coitus interruptus, condoms, and vasectomy. Each has obvious disadvantages in terms of reliability in preventing pregnancy and of acceptability. Vasectomy, while generally quite reliable and increasingly acceptable, has the very significant disadvantage of being irreversible with present techniques in the majority of subjects. The present article will review areas of recent work directed at extending our capabilities for male contraception and consider which methods may become available for general use. We will begin with a description of some of the major relevant aspects of male reproductive physiology.

Normal Male Reproductive Physiology

The pituitary (fig. 1) is a gland about the size of a large pea and is attached to the base of the brain. It secretes many hormones, among which two are directly concerned with testicular function. These are follicle-stimulating hormone (FSH) and luteinizing hormone (LH), the names deriving from their biologic functions in females. The secretion of LH and FSH by the pituitary is stimulated by at least one substance normally formed in the hypothalamus, an area of the brain immediately above the pituitary. This substance, called LH-releasing hormone (LHRH), is necessary for normal secretion of LH and FSH and, if deficient owing to hypothalamic disease or chemical inhibition, will result in inadequate LH and FSH production.

The blood stream carries LH and FSH to the testes, where LH is chiefly responsible for stimulating the production of testosterone, the main male sex hormone, and FSH is responsible for the production of sperm.[4] Adequate amounts of both LH and FSH are necessary for nor-

3. D. M. de Kretser, "The Regulation of Male Fertility: The State of the Art and Future Possibilities," *Contraception* 9 (June 1974): 561–600.

4. Sperm are the cells with mobile tails that, following intercourse, ascend the female reproductive tract and fertilize the ovum. "Seminal fluid" is composed of sperm plus seminal plasma, the fluid secreted by the testes and glands along the male reproductive tract, including the prostate.

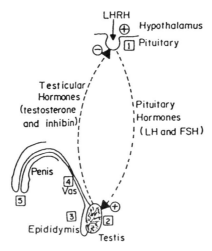

FIG. 1.—A highly schematic diagram of some of the important anatomic and hormonal factors in human male reproduction. Numbers in squares denote various sites where normal processes theoretically could be blocked, leading to infertility. Symbols: + implies that the hormone(s) exert(s) a stimulatory effect on the organ to which the arrow is directed; − implies an inhibitory effect.

mal spermatogenesis; if they are low, sperm production will be decreased or absent. Normal secretion of LH is necessary to maintain normal testosterone production by the testes. If testosterone levels are low, male libido and sexual potency decrease, and many other metabolic functions are adversely affected.

Sperm are produced in the testes from precursor cells by processes of cell division and maturation requiring about seventy days. They are transported through a series of channels into the epididymis, a sacklike structure on the outside of the testis. The epididymis forms the beginning of the vas deferens, which carries sperm to the penis to be expelled during ejaculation. Maturation of sperm, which is necessary for their fertility, occurs during their transit through the epididymis and vas. Abnormalities of any of these structures can lead to deficient or absent production of normal sperm.

Possible Male Contraceptives

The main sites at which a male contraceptive might work are numbered in figure 1.

1. Potential male contraceptives may be effective through inhibition of FSH and LH production by the pituitary, either by a direct action on the pituitary itself or indirectly through suppression of LHRH secre-

tion. Many steroid hormones,[5] including the male sex hormone testosterone, the female hormones estrogen and progesterone, and numerous similar substances, are known to inhibit the production of both LH and FSH from the pituitary in males and females. Indeed, probably the major mechanism through which most of the oral contraceptives (which are various combinations of female sex hormones) are effective in females is their inhibition of the pituitary hormone secretion necessary to produce eggs from the ovary. Many steroids are available that could be taken orally by the male and would be expected to cause infertility through inhibition of pituitary hormone secretion. A major potential difficulty with this form of therapy is that, if LH secretion is inadequate, the amount of testosterone produced by the testes would be low, causing decreased libido and impotence. A possible solution to this problem is to use testosterone itself as the agent to suppress LH and FSH. High dosages of testosterone taken orally or in shots will suppress LH and FSH, causing infertility, and also will replace the testosterone not produced by the testes. Dosages of testosterone in this range, however, have been found to cause an increase in the red cells of the blood in some men. They may have other harmful metabolic effects. A compromise solution being evaluated in several laboratories, including our own and that of Dr. C. A. Paulsen in Seattle, is that lower dosages of testosterone in combination with one of the progesterone-like hormones or some other derivative of one of the sex steroids would simultaneously induce infertility through suppression of LH and FSH, replace the deficit in endogenous testosterone production, and avoid the undesirable side effects of excessive testosterone. Preliminary results suggest that this type of therapy does indeed work without, to date, any unacceptable side effects.[6] One disadvantage already recognized is that, because of the length of time necessary to produce mature sperm in the testes, there is a period of time (two to three months) between beginning to ingest the drugs and achieving infertility. A similar period intervenes between discontinuing the medication and regaining fertility. Even if such a form of contraception is found to be effective, it will probably be at least five years before it is available for general use, owing to governmental requirements for rigorous demonstration of efficacy and lack of toxicity.

Another class of contraceptive agents under intensive investigation in many laboratories relates to the recent discovery, in the laboratory of Andrew Schally in New Orleans, of the structure and method of synthe-

5. "Steroid" denotes compounds of a certain chemical structure. Many human hormones, particularly those produced by the adrenal glands, testes, and ovaries, are of this structure. Relatively minor chemical variations superimposed on the basic steroid structure cause the markedly different biological effects of the various steroid hormones. Chemists have added additional variations to produce compounds that in many cases have biological effects different from those of steroids produced in the body.

6. R. Skoglund and C. A. Paulsen, "Danazol-Testosterone Combination, a Potentially Effective Means for Reversible Male Contraception: A Preliminary Report," *Contraception* 7 (May 1973): 357–65.

sis of LHRH, the hypothalamic hormone that causes the release of LH and FSH from the pituitary.[7] Many attempts are under way to develop a chemical analogue of LHRH or an antibody directed against LHRH that would inhibit its action on the pituitary. This would cause decreased production of LH and FSH and infertility in either the male or the female. These lines of investigation have so far met with only partial success, and, if such an agent is effective as a contraceptive, it will be limited by the necessity to replace the testosterone not produced because of decreased secretion of LH.

Perhaps an ideal male contraceptive would be one that blocked pituitary FSH secretion without affecting LH. This would presumably cause deficient spermatogenesis, to the point of infertility, without impairment of testosterone secretion. Work in our laboratories[8] and in those of Dr. Paul Franchimont in Belgium and Dr. Brian Setchell in England has shown that such a substance (a protein, not a steroid) is present in extracts of testicular tissue and seminal fluid of animals. If this protein can be purified in sufficient amounts and is not harmful when given to humans, it will have considerable importance in future contraceptive techniques. It is clear, however, that at least several years' work remains before this substance, named inhibin, reaches the stage of trial in humans.

2. The testis itself is the second major site that could be affected by male contraceptives. Many drugs, such as those used to treat some human tumors, are known to inhibit sperm formation by a direct effect on the testis but are accompanied by unacceptable side effects on other parts of the body, such as the bone marrow and the gastrointestinal tract. Several different classes of drugs have similarly been shown to cause infertility but also other toxicity.[9] Perhaps the most promising was the bis (dichloroacetyl) diamine class, which was developed to the stage of clinical trials in humans in the early 1960s before it was noted that ingestion of alcohol while taking the contraceptive drug caused severe flushing and irregular heartbeats. Therefore, further trials were abandoned.[10] It remains possible that an analogue of one of these drugs or a new chemical may be found which is capable of causing reversible infertility in men without unacceptable toxicity to some other tissue, but no such compound is available to testing at the present time.

3. The epididymis functions as a site for storage and maturation of

7. A. Schally, A. Arimura, and A. Kastin, "Hypothalamic Regulatory Hormones," *Science* 179 (January 1973): 341–49.

8. V. W. K. Lee, E. J. Keogh, H. G. Burger, B. Hudson, and D. M. de Kretser, "Studies on the Relationship between FSH and Germ Cells: Evidence for Selective Suppression of FSH by Testicular Extracts," *Journal of Reproduction and Fertility* (1975), in press.

9. H. Jackson, "Antispermatogenic Agents," *British Medical Bulletin* 26 (January 1970): 79–86.

10. C. G. Heller, D. J. Moore, and C. A. Paulsen, "Suppression of Spermatogenesis and Chronic Toxicity in Men by a New Series of Bis (Dichloroacetyl) Diamines," *Toxicology and Applied Pharmacology* 3 (January 1961): 1–11.

sperm as well as a channel through which they pass to the vas. If this function is impaired, sperm may be produced in normal numbers but not be able to fertilize eggs. Two drugs have been used, with limited success, in attempts to block normal function of the epididymis. Studies in rats have suggested that cyproterone acetate will impair epididymal function, causing nonfertile sperm without affecting the function of the testis. Some investigators have reported similar results in humans,[11] but others have found that the infertility is associated with decreased libido and potency,[12] which may be related to the drug's known property of antagonizing the effectiveness of testosterone and, at least in high dosage, of inhibiting the secretion of LH. More trials of this compound in humans are under way at the present time.

Another drug thought to affect the epididymis is α-chlorohydrin. It has been shown to produce sterility in laboratory animals, but also (at a dosage two to three times that necessary to cause sterility) to cause permanent anatomic changes in the epididymis and at higher dosages to cause death from bone marrow toxicity.

Although satisfactory agents for affecting epididymal function are not available at present, this form of contraception has several theoretic advantages and will be pursued actively in further studies. Among these advantages is the fact that, if epididymal function alone were affected, libido and potency would be normal. Second, the onset and termination of the effect on fertility would be much more prompt (about two weeks) than in the case of agents affecting the pituitary and testis (two to three months).

4. The vasa deferentia are small tubes passing from each testis to meet just prior to entering the penis. Because all sperm have to travel through them to get to the penis and because they are easily accessible to the surgeon as they pass through the scrotum, the vasa have been a common site of surgical intervention to impair fertility. The familiar procedure of "vasectomy" involves taking out a small section of each vas and tying off or cauterizing the blind ends in an attempt to induce permanent sterility without affecting the function of the testis.[13] This has been a remarkably successful procedure, both in preventing pregnancy (over 99 percent effective in most series of operated subjects) and in absence of side effects.[14] Its outstanding disadvantage continues to be that it is essentially irreversible with present surgical techniques and so is

11. J. Hammerstein, personal communication.
12. N. Laschet and L. Laschet, "Adrenocortical Function, Corticotrophic Responsiveness and Fertility of Men during Longterm Treatment with Cyproterone Acetate," in *Third International Congress on Hormonal Steroids*, ed. V. H. T. James (Amsterdam: Excerpta Medica, 1970).
13. R. Hackett and K. Waterhouse, "Vasectomy Reviewed," *American Journal of Obstetrics and Gynecology* 116 (June 1973): 438–55.
14. Several studies have found no evidence of hormonal abnormalities in subjects up to two years after vasectomy. Most studies have found a very high proportion of positive psychological adjustments to the procedure. Development of antibodies to sperm occurs in

not practicable for a man who wants to father children at some time in the future. It is quite commonly possible (50–80 percent of cases) to reattach the two ends of the vas so that sperm can actually get through from the testis to the penis, but these sperm are usually incapable of fertilizing eggs. Only about 20–40 percent of men who have had vasectomies can be made fertile again.

Many variations in surgical technique have been introduced in an attempt to improve the reversibility of vasectomy.[15] In recent years, various devices have been placed in the vas that operate essentially like a faucet that can be opened and closed by a minor surgical procedure. Unfortunately, the rate of ability to induce pregnancy when the devices are open has so far not been significantly greater than that when a vasectomy is repaired.

A technique that theoretically could be used in conjunction with vasectomy in order to guarantee future fertility is freezing sperm, or "sperm banking." Seminal fluid, collected by masturbation, could be frozen prior to a man's undergoing a vasectomy, then thawed and artificially inseminated into a woman at the time pregnancy is desired. The use of sperm banking in animals, particularly cattle, has met with enormous success. In humans, however, the technology of freezing and thawing sperm without killing them has so far not progressed far enough to make this a reliable method of guaranteeing future fertility. The most optimistic recent reports suggest that approximately 70 percent of samples so treated will be capable of inducing fertility.[16] Some men's sperm are able to survive freezing much better than others, but there is no way as yet of predicting which sperm have this characteristic. Although the procedure of sperm banking has severe limitations at present, it is quite possible that improvements in the methods of freezing and thawing may lead to a reliable way of guaranteeing future fertility. If these improvements are forthcoming, sperm banking in combination with vasectomy would be an effective contraceptive technique for many men. Obviously, if sperm banking becomes a practical reality, it will also introduce many sociological and moral problems, for example, the possibility of using frozen sperm from outstanding men for eugenic purposes in women of succeeding generations.

5. Interference with sperm progression from the penis into the vagina by a device such as a condom or by a sexual practice such as coitus interruptus is a well-known contraceptive technique. Coitus interruptus has limitations in terms of acceptability and, even if performed opti-

the blood of about 50 percent of men who have undergone vasectomy. This is not known to cause any adverse reaction in men so affected.

15. J. Sciarra, C. Markland, and J. Speidel, eds., *Control of Male Fertility* (New York: Harper & Row, 1975).

16. E. Steinberger and K. Smith, "Artificial Insemination with Fresh or Frozen Semen: A Comparative Study," *Journal of the American Medical Association* 223 (February 1973): 778–83.

mally, suffers from the fact that the first few drops of semen ejaculated contain a very high concentration of sperm, which may be produced in some cases before the subjective feeling of an orgasm occurs. Condoms, if used correctly, have a good rate of preventing unwanted pregnancies and are associated with no known health hazard. The necessity of having to employ a condom in each act of intercourse, however, is enough of a burden to cause many men to use them only intermittently, with a subsequent increase in the induction of pregnancy.

A class of techniques for contraception that theoretically might be adapted to act at several of the above sites is that of immunizing a man against some component of his own reproductive system, such as his sperm or his FSH. An effective antibody to one of these components might induce absence of spermatogenesis (FSH antibodies) or production of infertile sperm (sperm antibodies).[17] Efforts have been under way for at least fifty years to develop contraceptives of this type for both males and females but have so far not produced a satisfactory technique.

Antibodies can, in general, be produced in one of two ways. Most commonly they are produced by injecting a person with a substance (the "antigen") that is not normally produced in the body, thereby stimulating the body's immune system to produce its own antibodies. This is the technique employed in vaccination against infectious diseases and is called "active immunization." Alternatively, antibodies can be produced by injecting the antigen into another person or animal and then collecting the antibodies made by that organism and injecting these antibodies into the person whom one desires to make immune. This is called "passive immunization" and has been employed to make antitoxins against diseases such as tetanus.

Either mechanism of antibody production could theoretically be used to produce sterility in humans, but each has serious problems. If FSH or LHRH is employed in active immunization, the man's immune system will produce antibodies to these materials, thereby probably causing infertility.[18] However, the immune system may well continue to produce antibodies for months to years after the injection, since it may be continually restimulated by the hormones produced in the body. This would produce prolonged and possibly permanent sterility and therefore offer no advantage over other methods.

If a sufficiently specific antigen can be prepared from sperm, antibodies to this antigen might be produced in men. Since the antigens associated with sperm normally produced in the body do not enter the

17. T. S. Li, "Sperm Immunology, Infertility and Fertility Control," *Obstetrics and Gynecology* 44 (October 1974): 607–23.

18. The hormones would have to be chemically attached to a large protein prior to injection, thereby making a complex that the body's immune system would regard as foreign.

blood stream, they would presumably not constitute a continuing source of antigenic stimulation, and the infertility induced by sperm antibodies may be reversible. This series of deductions, however, involves several steps that have not yet been experimentally verified. The technique, although promising, cannot be considered an imminent possibility for effective male contraception.

Passive immunization entails even more difficulties. It involves injecting an extract of blood from another organism, usually an animal, which contains many substances that are foreign to the man being injected. This often induces a disease called serum sickness and may, over a long period of time, cause kidney damage and other serious ill effects. It seems unlikely that passive immunization, at least as presently practiced, will ever be an acceptable contraceptive technique.

To summarize the prospects for male contraception, it is unlikely that new techniques will be available on a wide scale for at least five years. The method that seems likely to be available first is the use of some combination of sex steroids to inhibit the production of LH and FSH. Depending upon the rate of technological improvements, sperm banking in combination with vasectomy may also become a useful procedure. The possibilities of a specific inhibitor of FSH production, of drugs causing infertility through a direct testicular or epididymal mechanism, and of antibody therapy all seem very worthy of further investigation but are not likely to be available for general use in the near future.

As a general comment on the history of research in fertility control, it is of interest that contraceptive techniques for men seem to have lagged somewhat behind those available for women, particularly in the area of oral preparations. Questions about the reason for this discrepancy are often asked of workers in the area of male contraception. The questions are occasionally asked with a certain degree of hostility, with the implication that scientists in the area of reproduction research have been responsive to male opinion, which is held to regard the female as the sex responsible for contraception, and therefore have concentrated on developing techniques for female contraception.

There are undoubtedly many answers to these questions. Certainly one answer relates to the fact that biologists have felt that it is technically easier to disrupt the mechanisms responsible for ovulation in the female than it is to disrupt spermatogenesis in the male. It was pointed out in many early papers on the topic that it was necessary to prevent the production of only one egg each month in a female, while in a male it was necessary to prevent the production of billions of sperm, and that an 80 or 90 percent reduction in sperm production was not sufficient to cause infertility in many cases. In addition, the processes of sperm migration from the vagina to the egg, fertilization of the egg, and implantation of the fertilized egg in the uterus all occur in the female only.

Therefore, measures such as intrauterine devices can be used in females but not in males.

Another answer may be that there has, in fact, been some degree of prejudice among workers in the area, or in agencies awarding money for research, as to which sex should bear the responsibility for contraception. If so, this has not been mentioned in the scientific literature on the subject, nor has it been part of the personal experience of the authors.

One interesting, and perhaps less widely appreciated answer, is to be found in the writings of two pioneers in the development of oral contraceptives for females, Gregory Pincus and Joseph Goldzieher.[19] Both have stressed the crucial roles played by the feminist and birth control advocate Margaret Sanger and her friend Mrs. Stanley McCormack in enlisting the aid of the scientist Pincus at the Worcester Foundation in Massachusetts and the gynecologist John Rock at Harvard in early studies directed toward an effective contraceptive for women. Both women supplied considerable moral support and social pressure, and Mrs. McCormack furnished large amounts of personal money to finance contraceptive development. Dr. Pincus's book *The Control of Fertility* was dedicated to Mrs. McCormack. As exemplified by the work of those women, a significant factor in the rapid early development of the female oral contraceptive was feminist pressure and support for a method by which a woman could control her own fertility. It is only at the present time that techniques for male contraception may be starting to approach those already available for the female.

As this review has indicated, the traditional male contraceptive techniques of coitus interruptus, condoms, and vasectomy have many limitations in terms of reliability, acceptability, and reversibility. Many of the possibilities for improvement in the technology of male contraception offer promise of significant advancement toward these objectives. We have tried to make some assessment of the potential value of each new technique and the time it may take before the innovations, if successful, are generally available. While it is apparent that it will be at least several years before any of the new techniques has a significant impact on human reproductive practices, it seems quite possible that some of them may eventually have a profound impact. It is, of course, very probable that techniques not considered here will appear. Such new techniques may well surpass those presently envisaged and, with luck, may help give future generations less cause for worry about the related problems of overpopulation and the health hazards of adequate contraception.

Prince Henry's Hospital Medical Research Centre, Melbourne

19. G. Pincus, *The Control of Fertility* (New York: Academic Press, 1965); and J. Goldzieher and H. Rudel, "How the Oral Contraceptives Came to Be Developed," *Journal of the American Medical Association* 230 (October 1974): 421–25.

DISCOURSE PROCESSES 7, 171–200

The Prescription of Contraception: Negotiations Between Doctors and Patients*

ALEXANDRA DUNDAS TODD

Suffolk University

Maria, an Anglo woman, age 26, married, with medium length brown hair, slightly plump, has a tired air about her which can quickly turn to animated interest if her lively sense of humor is aroused. She has four children ranging in age from 11 years to 6 months, and a varied contraceptive history. This history includes successful use of the birth control pill for several years, followed by a period of 1 year when she would forget to take the pill regularly regardless of how hard she tried. This problem arose at a time when her two oldest (and at that time only) children were starting school and she wanted another child. Her husband was against the idea. She changed contraceptive method to an IUD (intrauterine device) which was inserted painlessly but caused her intense pain with menstruation. After several months she had it removed. She and her husband are presently using condoms which Maria feels interfere with their sexual life as well as creating anxiety for her. She is presently anxious to avoid impregnation, whereas her husband would like her to be pregnant again.

Based on Maria's unsuccessful experience with the birth control pill, her dissatisfaction with the condom, and her mistaken understanding (and thus rejection) that the diaphram needs to be inserted right before intercourse, similarly to the condom, she feels her options for birth control are limited. Given her current fears of pregnancy she has come to her local community clinic to once again try the IUD. While this decision makes her nervous— she previously needed prescription pain pills with this method—she feels resigned to this choice.[1]

Maria came to the gynecologist with a lot of information and misinformation regarding her reproductive needs and choices. The complexity of her past experiences, contraceptive method, state of health, family relations, did not surface in the doctor-patient interaction. Throughout this interview Maria's overriding concern was whether or not the IUD was the right choice. She fluctuated between a

*Grateful acknowledgement is made to Sue Fisher, Rich Frankel, Hugh Mehan, Ron Ryno, Thomas Scheff and Will Wright for their helpful suggestions and comments.

Correspondence and requests for reprints should be addressed to Alexandra Todd, Department of Sociology, 8 Ashburton Place, Suffolk University, Boston, MA. 02108.

Footnotes appear in a "Reference Notes" section at the end of this article.

definite yes, "I want it, I'm gonna do it . . . ." " to a not so sure "ummmmmmmmm, yeah (weak)". Following the actual exam, Maria raised her first and only question during this medical interaction. The question inquired about pain and the IUD.

> Maria: It won't hurt will it?
> Doctor: Oh, I doubt it.
> Maria: I'm taking your word (laugh).
> Doctor: I haven't had anybody pass out from one yet.
> Maria: The last time/
> Doctor (cuts patient off with a joke, both laugh)
> Maria: The last time when I had that Lippes Loop, oh God/
> Doctor: (interrupts patient) / You won't even know what's going on, we'll just slip that in and you'll be so busy talking and you won't know it.[2]

In this sequence Maria attempts to raise the topic of pain 3 times. She is referring to her past experience of pain *after* insertion of the IUD. The doctor interrupts two of her attempts to question him, providing her with information based on his assumption that she is concerned with pain *during* insertion of the IUD. Since this exchange took place as the doctor was preparing to leave, he ended the interview unaware that Maria's question went unanswered and her contraceptive concerns and options unexplored.

Throughout the United States, women regularly visit gynecologists to ask questions about their reproductive systems. Because the two individuals, doctor and patient, have access to different information on the subject, the interaction becomes a negotiated process where needs are developed, discussed, and diagnosed. In medical interactions, patients, such as Maria, ask the original question by initiating the process of seeking health care, yet ironically, the doctor dominates the dialogue in the interview; not simply its diagnostic outcome. Scheff discusses this negotiation in the framework of responsibility in the psychiatric interview, acknowledging the construction of the social reality in the interaction but emphasizing the inequality of power between the participants in this construction.

> The interrogator's definition of the situation plays an important part in the joint definition of the situation which is finally negotiated. Moreover, his definition is more important than the client's in determining the final outcome of the negotiation, principally because he is well trained, secure, and self-confident in his role in the transaction, whereas the client is untutored, anxious, and uncertain about his role. Stated simply, the subject, because of these conditions is likely to be susceptible to the influence of the interrogator. (Scheff, 1968, p. 6)

The doctor-patient relationship has been discussed widely in the literature on medical sociology and the recent literature on women's health. While a number of theoretical frameworks for understanding this relationship have been proposed, the actual sequences of interactive dialogue between participants in the medical setting have received less sociological attention. The systematic study of language, until recently, has been left to other disciplines. Several notable exceptions are Frankel (in press), Paget (1983), West (in press), Fisher (1980), and Cicourel (1975), whose analyses treat speech exchange as the negotiated achievement of language users. My purpose here is to contribute to this literature by examining the discourse between doctors and patients to demonstrate how power is manifested in the interaction and how expression of this power in face-to-face communication is linked to a scientific medical model.

The data for this study are derived from in-depth observations and audiotaped recordings of doctor-patient interactions gathered in two gynecology services during a 2½-year field study conducted in a large metropolitan community in southern California. The first setting was a women's health clinic in a general community health center. This clinic was staffed by medical residents in gynecology from local teaching hospitals. The three doctors I observed in the clinic, Drs. Smith, Jones, and Long,* rotated their work hours, and were paid by the hour. They were assisted by volunteers who took patient histories, dispensed medications, and performed similar work. The second setting was the office of Dr. Masters,* a gynecologist in private practice who conducted all medical interactions with the patients. Dr. Masters is a prominent physician of long standing in the area and a member of the clinical faculty of a nearby medical school.

My data include 20 audiotaped medical interviews, each about 10 minutes long, ten from each setting. Discussion of these data reflects my observations in the settings as well as the taped information broadening the context for interpretation. The taped interactions provide a cross-sectional view of a wider observational context.

My methodological focus should be understood as a spiral, moving from interactional situation to the organizational setting and theoretical structural milieu and back to the interactional situation. While I propose no causal link between the audio taped data and more general observational contexts, I base my analysis on the assumption that interaction, understood here as discourse, does not occur in a vacuum. It is micropolitical as communication is influenced by, and is a reinforcer of, the cultural values of the society in which it takes place.

I will examine the organization of the doctor-patient interviews in three ways. First, distributional analysis is used to present an overview of all of the interactions. Speech act theory offers a means to break the flow of talk into discrete parts. The second way provides insight into the interaction between doctor and patient by focusing on the sequential properties of the medical interviews. Turn-taking analysis combined with consideration of the distribution of speech acts contributes to a broader view of the data than was permitted by the distributional analysis alone. The third way introduces a topical analysis, placing the talk in

---

*pseudonyms

two "frames," one social and contextual and the other technical and medical. Some of the general patterns found in the interaction described by the distributional, sequential, and topical analyses are discussed in the final section.

DISTRIBUTIONAL ANALYSIS

The examination of speech acts is useful in linking the larger institutional levels of social order to everyday interaction and behavior (see Cicourel, 1980; Streeck, 1980). Speech act theory as outlined by Austin and Searle[3] is helpful theoretically for sociological analysis for it treats speech as a social activity based upon the presuppositions and intentions of natural language users. However, speech act theorists address their topic from a static point of view. The basic unit of analysis for speech act theory is the self-contained action unit rather than the interaction unit, where context and mutual participation of the participants are important (see Frankel, forthcoming; Streeck, 1980).

In attempting to expand the speech act framework to include contextual as well as cultural features, D'Andrade has developed a system for the classification of speech acts derived from a socio-cultural framework. His work belongs in the tradition of several writers who in recent years have applied speech act analysis to actual conversations (Cole, Dore, Hall, & Dowley, 1978; Gelman & Shatz, 1977; Labov & Fanshel, 1977; D'Andrade, n:d.; Maseide, n.d.). Table 1 lists D'Andrade's six categories for speech acts: statements, questions, expressives, directives, reactives, and commissives.

I have adapted these categories to my own data, abridging the system in the process to fit the specific needs of the medical interview.

My revisions of D'Andrade's system exclude expressives and commissives. Although these actions take place in the medical setting, they do so subtly and in the form of statements, questions, and directives. Strong emotion is not considered appropriate in the gynecologist-patient relationship, and actions such as vowing and exasperation tend to be played down and absorbed into other acts. I have further subdivided reactives into two categories—reactives and answers. Speech acts will be considered answers when they provide a substantive response to a question, as shown in the following example from a clinic tape.

1. Doctor: You've just finished your period?
2. Patient: Uh hum.
3. Doctor: Okay.

As the display shows, the doctor first states/asks information about the patient. I have coded the doctor's sentences (1) as a question because of its questioning tone and because the doctor receives an answer from the patient (2) that provides information regarding her period. (I have determined the meaning of these conversational acts on the basis of their function in speech rather than their inherent grammatical form.) Third, (3) the doctor acknowledges the patient's answer with

TABLE 1
D'Andrade's Preliminary Speech Act Category System

A. *Statements* (expositives, representatives, assertions)
reports
quotes
instantiations
claims
stimulations (?)
inferences

B. *Directives* (requests, orders exercitives)
suggest/request/order
request object
  agree as to truth
  expression of approval/sympathy/support, etc.
  commitment
  direct action
direct/indirect

C. *Questions*
wh-form
yes/no form
tag form
intonation only form
information only versus other uses

D. *Reactives* (various kinds of agreement or disagreement with what has previously been stated)
agree as to truth versus disagree as to truth
give attention
accede (agree to commit, or actually do) versus refuse

E. *Expressives*
give approval versus disapproval
sympathy, regret, exasperation, etc.
direct versus indirect (accusation, disagreements, etc.)

F. *Commissives*
promise, offer, vow, etc.

a reactive. The classification system used for the following data in this paper thus consists of statements (S), questions (Q), reactives (R), answers (A), and directives (D) applied to conversations between the doctor (D) and the patient (P).

I have aggregated the speech acts in the 20 interviews using the categorization features of D'Andrade's model. Table 2 displays the types of utterances used by doctors and patients during gynecological care involving the prescription of methods of contraception.

Analysis of the data in Table 2 disclosed several differences between the community clinic and the private practitioner's office.

1. The gap between questions asked by the doctors and the questions asked by the patients was greater in Dr. Masters's office than in the clinic. More questions were asked by the doctors and the patients in the clinic. Very few questions were asked by the patients in Dr. Masters's office.

2. The same was true of answers to these questions. The patients in Dr. Masters's office asked fewer questions and received fewer answers. Further, the patients in Dr. Masters's office received fewer answers to questions asked than did the patients in the clinic.

3. In the clinic the doctors uttered more reactives than the patients. The re-

TABLE 2
Aggregate of Speech Acts

Speech Acts	Statement D	Statement P	Question D	Question P	Answer D	Answer P	Reactive D	Reactive P	Directive D	Directive P	Subtotal D	Subtotal P	Total
Clinic													
Number	210	92	251	45	42	244	220	123	69	—	792	504	1296
Percent	70	30	85	15	15	85	64	36	100	—	61	39	100
Dr. Masters Exam													
Number	123	69	105	8	3	96	66	86	85	1	382	260	642
Percent	64	36	93	7	3	97	43	57	99	1	60	40	100
Consult.													
Number	125	53	80	8	6	71	73	139	73	—	357	271	628
Percent	70	30	91	9	8	92	34	66	100	—	57	43	100
Sub Total													
Number	248	122	185	16	9	167	139	225	158	1	739	531	1270
Percent	67	33	92	8	5	95	38	62	99	1	58	42	100

verse was true in Dr. Masters's office.

4. Directives were used more frequently by Dr. Masters than by the clinic doctors.

The analysis also revealed a number of similarities in the two settings.

1. The doctors asked more questions than the patients did. In the clinic the doctors asked approximately 5.6 times as many questions as the patients. Dr. Masters asked approximately 11.5 times as many questions as the patients.
2. The patients provided more answers than the doctors—in the clinic, approximately 5.6 times more. Dr. Masters's patients provided approximately 19 times as many answers as the doctor.
3. The doctors made more statements than the patients. The clinic doctors and Dr. Masters made approximatly twice as many statements as the patients.
4. Doctors (with one exception) made all the directives.

These findings give us important information about who says what to whom and how (Hymes, 1972). Speech act tabulation provides an overview of the data in the form of a distributional analysis which divides talk into discrete components, showing the similarities and differences in the structure of the discourse in the two settings investigated. Comparisons suggest a preliminary view of the distribution of power in the medical interview. As informative as this view may be.

we must keep in mind that speech act analysis masks the way in which doctors and patients act in concert, influencing each other in the course of the interaction. Although cultural presuppositions are used to assign meaning, the theory in and of itself does not display the interconnection of speakers in conversation and thus does not provide the basis for a thorough "grounded" consideration of power relationships. It is the sequential relationship of speaking activities that provides the basis for a broadened understanding of the medical interview.

SEQUENTIAL PROPERTIES

The doctor-patient discourse takes place in the institutional setting of health care delivery. In this system doctors have power both as experts (Ehrenreich & English, 1979) and often as males (Scully, 1980). The patient participates in an exchange which is importantly structured by the sequencing of questions and answers and by the institutional setting, which constrains the form of the interaction. The discourse and the organization are thus embodied in and through each other.

Discourse analysis has revealed a two-part structure in conversation between equal participants in everyday settings, but researchers have found that dialogue between individuals in institutional settings is organized differently.[4] The difference seems to be a consequence of the asymmetry between participants (see West, in press), which produces a third part to the conversation. In the educational setting this third part is an evaluation act: the teacher, in an institutionalized role, evaluates the work that the students are doing (Mehan, 1979). The utterance of the third pair-part in the doctor-patient interaction links the institutional, socio-political context, and the actual, situated discourse (Todd, 1982; Fisher, 1980.)

In the doctor-patient interactions that I discuss, the third part of the conversational sequences is a reactive used by the doctor to maintain control of the floor. The doctor initiates a request for information, receives information from the patient, and acknowledges the answer with a reactive (see Figure 1).

FIG. 1.   Three-Part Structure of an Interaction

The doctor's reactive serves two purposes: The doctor initiates the interaction with a question, and in so doing also offers the topic to be discussed. Thus the reactive serves to end the prior interactional segment and the topic, and is used in combination with a new topic intiation to maintain control of the floor (see Figure 2).

Initiation	Reply	Acknowledgment
D:Q	P:A	D:R
So you haven't had a period since then?	no	all right

In the sequence shown in Figure 2, the utterance "all right" acknowledges the patient's reply and disjoins it from the topically relevant information contained in the continuation of the utterance. What this amounts to is a change of topic and the structure of the speech act by introducing a statement which interprets the problem. The patient acknowledges the information with a reactive and in turn is acknowledged by a reactive from the doctor. This final reactive allows the doctor to proceed with the questioning (see Figure 3).

Initiation	Reply	Acknowledgment
D:S	P:R	D:R
Chances are if you're not having your period, you're not ovulating. . . .	um hum	okay

FIG. 3.   An Interaction with Two Reactives

My data indicate that the doctor initiates sequence and topic-introducing utterances which effectively direct the content and direction of the encounter. The patient's conversation also contains reactives. However, the patient's reactives differ from those of the doctor in that they occur as single speech act turns. The patient does not usually utter a reactive and then continue with this same turn. Rather, she utters the reactive and nothing more.

My findings concerning the use of reactives in the two medical settings indicate subtle, and not so subtle, organizational and interactional influences on discourse. In both settings, the doctor's conversations exhibit how reactives provide the means for changing topics, effectively maintaining control over the conversation; thereby exhibiting medical-institutional power through speech exchange. The differences in the use of reactives according to setting—Dr. Masters used fewer reactives than the clinic doctors and his patients used more reactives than the clinic patients—exhibits interesting patterns in the discourse. Dr. Masters changed topics abruptly, often "talking over" the patient with a directive.

The more abrupt changes in topic account in part for the lower number of reactives and the higher number of directives displayed in Dr. Masters' interviews. Dr. Masters also used directives at a far higher rate than did the clinic doctors in instructing patients about treatment decisions. Examples 1 and 2, involving Drs. Jones and Masters, respectively, involve two patients, both of whom suffer from amenorrhea (cessation of menstruation). Both are asking for renewals of prescriptions for birth control pills. The clinic patient wants to resume use of the birth control pill after a lapse of several months. The patient seen by Dr. Masters has come for a routine renewal of her prescription. The transcript of the discourse displays the very different styles of the two doctors. In the clinic setting, Dr. Jones attempts to explain the problem to the patient, using statements as the major form for imparting instructions and information. Dr. Masters uses directives to a far greater degree, both to mention new aspects of the problem and to impart instructions.

EXAMPLE 1

Dr. Jones	*Patient*
(S) I discussed your problem with one of the head doctors at General, uhm, he knows more about this than I do, about amenorrhea, not getting your period, and he seemed to think, uhm, he seemed to agree with me that it would probably be a good idea to give you a test dose of this progesterone to see if you have a withdrawal bleeding. I assume that this is probably what you were injected with earlier this summer, that was the injection, progesterone, that's a hormone in your body that makes your uterus shed its lining and you bleed, okay, and it works, progesterone is a normal hormone that works in regular cycles. Now, uhm, since you don't have any strong desires to get pregnant at this point, since you did come here for/ contraception, uhm, we both seem to think that it would be, uhm, the best idea to give you the progesterone for five days, have a withdrawal bleed, and then start you on the pill.	(R) /right
(Q) Okay?	(R) Okay
(S) (D) And if you don't get a withdrawal bleed with progesterone, it's a pill which you'll take for five days, if you don't, call us and then we'll see you and we'll have to give you— we'll have to go through another test before we start you on something.	

397

(Q) Okay?

(S) Now, uhm, just judging from what,
     you know, from your story, you
     told me that you did have a little
     spotting/   (R) /uh hum
     from the injection, it sounds
     like you, you just haven't been
     ovulating so this could be a
     problem when you do try to get
     pregnant, you know, when you do
     want to get pregnant again, we'll
     have to get you on regular cycles
     where you're ovulating in order
     for you to get pregnant.

(Q) Okay?   (Q) Now, this time if I
                                                             take it for five days,
                                                             what happens if I just
                                                             start bleeding?

(A) That's—even the slightest amount
     is okay.

(Q) Okay?

(S) Even if it's just a little spot.
     It doesn't have to be like a real
     period, just, you know, a little
     spot here and there. You might
     want to wear a mini pad or
     something.   (R) Okay.

EXAMPLE 2

Dr. Masters   *Patient*

Exam:

(D) Look,

(S) what I think is happening, that your
     pills are not strong enough,

(Q) you know?

(S) I've put you on a very low pill as you
     know/   (R) /right
     and sometimes this will cause
     (swallows word). Now what I would like
     to do is when you finish your pills.
     Let's see, you're on the 28 days.
     (long pause)   (R) Right
                                                         (S) Uh, 21 day.

(R) 21 day, all right.
(D) When you finish your pills, and if
    you don't menstruate within 5 to 6
    days, come in.
(S) I'll give you an injection to get you
    started.   (R) Okay
(D) If you do menstruate, come in then.
(S) I won't charge you anything if I don't
    have to do anything and I will give
    you a prescription for a different
    kind. I want to see what you're going
    to do. I also want you to report back
    on this.   (R) Right, uh hum.
(Q) Okay?   (R) Oh, okay

Consultation:

(D) Now, look,
(S) what I want to do, you understand now,
    all right, you have X number of tablets,
(D) then you finish the package,
(Q) right?   (R) Right
(R) Right.
(D) Finish them wait four to five days.
    If you flow, come in then while you're
    flowing and I'll, uh, uh, and I'll give
    you more pills. (long pause)   (R) Okay
(S) I won't charge you.
(D) If you don't flow, call me,
(S) then I will give you an injection,
(D) don't take any more tablets then.   (R) Uh hum.
(S) I'll give you an injection and I'll,
    uh, get you started with your
    menstruation and I'll give you a
    different type of pill. (long pause)   (R) Okay
(Q) Okay?   (R) All right.
(D) But meanwhile, stay on the pills.
(D) Don't you get into trouble   (R) Right.

Note. The clinic was organized around a single doctor–patient interaction in the examination room; in contrast Dr.Masters' interview started with the exam and was followed by a consultation in the office after the patient had dressed.

Examples 1 and 2 indicate, first, that Dr. Masters' directives provide him with control of the floor much as reactives provide control for the clinic doctors, and second, that the doctor in each setting did most of the talking about the patient's concerns (also see Table 2); the doctors uttered multiple speech acts per turn in

399

both examples, while the patients uttered one or occasionally two speech acts per turn. The distribution of questions and answers in the doctor-patient interviews in both settings shows the doctor asking more questions with the patient providing more answers. In addition, Dr. Masters more often solicits a response from patients. While the clinic doctors did use tag questions and expectant looks to elicit reactives from patients, they never, in my observations, insisted on a response. In fact, they so often rushed through tag questions in their talk that the patient was scarcely able, let alone required, to respond. Dr. Masters, however, used tag questions and/or long pauses accompanied by a piercing, direct gaze to accomplish patient acknowledgement, in some cases repeating this procedure for emphasis. In Example 2, Dr. Masters states, "I also want you to report back on this," emphasizing the words with very direct eye contact. The patient responds with a double reactive, "right, uh hum," to reassure him that she will do her part. Dr. Masters responds with a tag question, "Okay?", for further emphasis, eliciting acknowledgement from the patient in the form of another reactive, "oh, okay."

These examples show the asymmetric distribution of speech acts used by doctor and patient in the medical interview. The sequencing is controlled by the doctor, thus the topics are initiated by the doctor, not the patient. Dr. Masters exemplified this tendency even more than Dr. Jones in that he used many more directives in constructing his turns. Generally, across my data set, the patient brings the problem to the doctor's office with some knowledge about her body and her situation, but the doctor does most of the talking which initiates sequences and topics; the patient primarily seems to respond to the doctor's management of the interview.

The difference in power between doctor and patient can also be seen in how Dr. Masters demanded attention from patients using his social, often jocular, comments as well as responses based on responsibilities with regard to medical treatment. The following example was taken from field notes.

> The doctor turned his attention to the patient and started examining her breasts, which were quite large. As he was looking at her breasts, she was flat on her back staring at the ceiling. Dr. Masters said, "Yes, this is all girl," and smiled at the patient. No one in the room—the patient, the medical assistant, or myself—acknowledged the doctor's remark. The room was unusually quiet as the doctor started checking the patient's other breast. He slowly and measuredly started talking while looking around at each one of us, "I *said*, this is all girl." At this point the patient smiled wanly, the medical assistant chuckled, and I smiled. With our responses the doctor's good humor returned and he told the patient to "get dressed like a good girl" and he would give her "some more happy pills."

In this example, Dr. Masters exerted power in two ways. First he made a statement, which, in the absence of a response, he reiterated. While speakers routinely have rights to reassert that which is not responded to, it seems quite clear that the lack of response was deliberate, rather than not having been heard or understood. The fact that the doctor demanded and received attention from everyone present regarding a topic independent of the patient's health needs constitutes some type of power. Second, while performing the breast exam the physician reinforced gender-based stereotypes of female anatomy in a way that seems incongruous with the purpose of screening the patient for cancer.[5]

The differences in speech composition in the settings (and their similarities) can in part be attributed to differences in the organizational features of a community clinic and the office of a private practitioner. Dr. Masters, a private practitioner, has and exerts more power in his interactions with his patients than do the clinic doctors. Nothing happens in this office that Dr. Masters does not know about, organize, and control. The patient interacts solely with Dr. Masters concerning her reason for visiting a doctor. Her power lies in her freedom to choose not to come, not to return, or to ignore the doctor's instructions. In my observations the strongest reaction that women showed against Dr. Masters in the interaction was silence.

The clinic doctors in contrast, display no possessiveness toward the patients. Clinic patients are seen by rotating doctors in an atmosphere where it is acknowledged that the patient has other resources, such as health educators and counselors, if she is dissatisfied with or confused by the doctor's information. Whereas Dr. Masters exercised solo control, the clinic doctor is only one possible step of a number. Interestingly, Dr. Masters's patients asked fewer questions than did patients in the clinic; it is possible that patients felt less active and more dominated in Dr. Masters's office.

A large sociological literature discusses the institutional level of health care organization and delivery as it is structurally defined and observable, across settings, in traditional medical care in our society (Conrad & Kern, 1981). The discourse that I recorded between doctors and patients can be seen to reflect the institutional power of the physician as it has been discussed and criticized theoretically. The doctor dominates the conversation in an active manner, as discussed above; the patient, whose reproductive life and health are the topic of the interview, remains passive. Lazare and Eisenthal (1979) have found, in their work with psychiatric outpatients, that in the active doctor/passive patient interview the patient's requests often do not clearly emerge in the initial visit. Their research stresses attending to the "patient's perspective" which includes the "patient's definition of the problem, goals of treatment, and requests or desired methods of treatment" (1979, p. 156). They advocate eliciting the patient's interpretation and requests from the very onset of the interview, so that the interaction becomes a negotiation between two active participants. Interactions such as those which I recorded display irony: the patient has come to question the doctor, but the doctor asks all the questions.

SOCIAL AND MEDICAL FRAMES

An analysis of the sequential properties of the doctor-patient interview provides a framework for expanding the speech act distribution to reveal institutional dimensions in the conversation between doctors and patients and in the structure of the interaction. Examination of this interface between the micro level of discourse and the macro level of institutionally designated power offers insights useful for further study of the doctor-patient interview and health care delivery.

In conversations that take place in everyday life and in institutional situations, information is arranged by topic. Neither the distributional nor the sequential properties analyzed this order in the medical interview. My data suggest that the doctors and the patients framed their conversations in different ways—the former in medical, technical terms, and the latter in more social, biographical terms. The doctor-patient interviews that I observed and taped present primarily technical conversations centered around a biologically treated and defined issue—for example, the need for prescription renewal or amenorrhea and its possible connection to use of the birth control pill.

Birth control, however, has more than technical significance for women (Luker, 1975; Todd, 1982). The choice and use of contraceptives interrelate with sexual relationships, contextual circumstances, and life options. While the doctor speaks from a technical, biological standpoint, the patient's speech is often social and contextual. As I noted earlier, doctors exert more control in the interaction than do patients, so that the exchange of information is primarily technical in nature. When patients do take control of the floor, their topics center on the wider social context of their health and their bodies, as exhibited in example 3.

EXAMPLE 3

Dr. Masters	Patient
	(S) Uh, I haven't been here lately because I had to, uh, switch to Kaiser for financial reasons and I've been on a leave of absence and they can't take me back for a while.
(R) Uh huh.	
(Q) What pills are you on now?	

The patient has visited the clinic for birth control pill renewal and has introduced several topics: (a) failure to visit the doctor; (b) an alternative health plan; (c) financial matters; (d) leave of absence; and (e) loss of work. The doctor provides a token acknowledgement of the patient's comments and then immediately changes the topic to the birth control pill that the patient is presently using. The conversation in Example 3 fails to strike a balance between the topics of concern to the patient and the doctor's more technical response (see Halliday & Hasan, 1976). The patient is talking in a social/contextual mode about her health care and her life. The doctor responds with a medical, prescription-oriented question. The patient considers her social/contextual understandings of her life circumstances relevant to her health care; financial matters and the change of health care facilities, combined with a leave of absence and loss of work, are important concerns affecting her selection of birth control. Since he does not probe, the doctor leaves the patient's concerns unacknowledged and unaddressed, a situation which effectively disregards and perhaps discounts the issues.

The interchange in Example 4 takes place during a routine pelvic exam and gynecological checkup for renewal of a prescription for birth control pills.

EXAMPLE 4

Dr. Smith	*Patient*
(R) Okay,	
(S) I'm going to take a little bit of your secretion to look at it to make sure you don't have an infection, too.	
(R) Okay,	
(S) coming out, you're doing fine.	
(D) Stay right there, now.	
(Q) Doing all right, Norma?	(A) I feel uneasy tonight,
	(S) I don't know why.
(R) Well,	
(S) we're almost done.	
(D) Relax.	
(R) Okay,	
(S) your cervix's right there. Right behind your pelvic bone's your bladder.	

The patient has reported a vaginal discharge. Throughout the interaction she has appeared depressed and has expressed dissatisfaction with the birth control pill. In this brief excerpt the patient tentatively makes a social statement about her mood. The doctor's response incorporates this statement into a technical model relating to the exam and directs the patient to relax, changing the topic in turn back to the physiological explanation.

The insertion of a social topic into the medical discourse occurs frequently enough in the data to require explanation and infrequently enough to constitute a breach in the normal interactional flow (Table 3).

403

TABLE 3
Aggregate of Speech Acts Divided Into Medical and Social Frames

Speech Acts	Statement		Question		Answer		Reactive		Directive		Subtotal		Total
	D	P	D	P	D	P	D	P	D	P	D	P	
Clinic													
Frame dominant													
Medical	181	62	229	39	35	214	204	108	67	—	716	423	1139
Social	29	30	22	6	7	30	16	15	2	—	76	81	157
Percentage of speech acts													
Medical frame	16	5	20	3	3	19	18	10	6	—	63	37	100
Social frame	19	19	14	4	4	19	10	10	1	—	48	52	100
Dr. Masters													
Frame dominant													
Medical	197	67	141	12	5	126	119	194	146	1	608	400	1008
Social	51	55	44	4	4	41	20	31	12	—	131	131	262
Percentage of speech acts													
Medical frame	19	7	14	1	1	13	12	19	14	—	60	40	100
Social frame	19	21	17	2	2	16	7	12	4	—	49	51	100

In some instances the social utterance produced an exchange between the doctor and patient based on contextual knowledge, apparently an example of "particularistic co-membership" (Erickson, 1975). In one instance, Dr. Jones' comment on a patient's New England accent triggered a discussion of the doctor's medical school years in Massachusetts. For the most part, however, social comments were isolated remarks embedded in or eliciting a change back to a medical, technical frame. Sequences in the discourse show a weaving back and forth between social and medical topics, with social topics in the minority.

The sequences also display a difference in the way doctors and patients talk about social issues. Occasionally, especially in Dr. Masters' office, the social talk was an equal exchange of information regarding a patient's trip or a subject unrelated to the current reason for the patient's visit. However, the majority of the social utterances in both settings were distinctly different for the doctor and the patient.

When the patient talked of social issues, they were particularistic and contextual, impinging on her life in ways she considered relevant to the discussion of her body, as in Example 5.

In this sequence the doctor presents general medical knowledge to explain why menstrual periods cease after a woman stops taking birth control pills. The pa-

EXAMPLE 5

Dr. Jones   *Patient*

Okay, push back and sit up.
  Sometimes after you go off the
  pill, you cannot have your period
  for a while and that's not
  abnormal, to not have a period.   Uhh, before I had my daughter
                                          my husband was stationed, or no,
                                          it was after I had Sally, he was
                                          stationed overseas and I went off
                                          the pill and my period started
                                          like six weeks after.

---

tient responds with a particularistic, social statement indexing her own body's cycles with contextual, family information. The sequence in Example 6 displays a similar medical/social distinction between the doctor's questions and the patient's answer, and Example 7 similarly exhibits the patient's contextual understanding of her body and reproductive functions during a visit to Dr. Masters' office. This patient's "happiness" is embedded in her information regarding her sexual life.

EXAMPLE 6

Dr. Jones   *Patient*

Yeah. When was the last time that
  it did something like this, was
  late [menstruation]?
                                          I missed one once right before my
                                          wedding. That was in
                                          September, but that was because
                                          I was so nervous, I mean we
                                          were moving all around so/

EXAMPLE 7

Dr. Masters   *Patient*

Your pregnancy test is negative.
                                          Okay, doctor.

That makes us happy.
                                          It makes me happy,
                                          especially since I
                                          broke up with him
                                          three months ago.

405

When the doctors did make social comments regarding the patients' reproductive circumstances, they reflected stereotypical attitudes. The doctors voiced an abstract, social understanding, triggered by the specific situation of the topic being discussed. The sequence presented in Example 8 follows Dr. Masters' direction of the patient not to have intercourse for 2 or 3 weeks, as she is still recovering from childbirth.

EXAMPLE 8

Dr. Masters   *Patient*

                                                Don't worry, I keep telling
                                                you people I could take
                                                it or leave it, preferably
                                                leave it.

Look, honey, you have a husband
    don't you.   Yeah, I know (resignedly).

The patient expressed her disinclination for sexual activity at the present, and Dr. Masters volunteered a stereotypical comment on woman's responsibilities as a sexual partner in marriage. (He did not probe for information about her current sexual problems of initiate discussion of ways for her to understand her present situation.) In Example 9 Dr. Masters, in lecturing to a woman on taking her pills properly and the risks of missing them, exhibited a traditional attitude toward marriage and pregnancy.

EXAMPLE 9

Dr. Masters   *Patient*

You know, and you're going to, uh,
    particularly since you're not
    married, you're going to take
    care of it [baby] and you're
    going to have all the problems.   [nods head]

Dr. Masters here displayed the assumption that marriage comes before pregnancy and drew on it when he urged the patient to use birth control properly.

Similar views of women's reproductive roles can be seen in the clinic interactions. As shown in Example 10, Dr. Smith discouraged a patient from using an intrauterine device (IUD) for several reasons, including the possibility that she would jeopardize her future fertility, and encouraged her to use the birth control pill. (All of the doctors observed in my study generally opposed the IUD.)

EXAMPLE 10

Dr. Smith	Patient

I, I have a very negative opinion of
the IUD, *particularly for young women
who haven't had their family yet,*
because if/

        /Uhm, they give it to girls who
        have abortions and miscarriages.
        Why, why is that?

I think that's just because it's, the,
the women don't feel that they could
take the pill, and it's some, it's some
form of birth control, at least. For
women who just can't remember to take
the pill or won't, that need protection,
then the IUD is some (pause), you know,
is second best. But you're really
taking a big chance of infection.
We're seeing at least I'd say five
in a hundred IUDs that we put in are
coming back with some, some sort of
infection, often not serious, but it
can be very serious. It can mean
hospitalization and antibiotics into
your veins, and some of them even have
their organs operated on or removed
because they get so infected, it can
even result in death, then, you know,
so it's, again, a remote possibility
the same as blood clots are with
birth control pills, although it's
not as remote as that. It happens,
it really, you know, much more
frequently. *Uhm, I think the
scariest thing, even if, you know,
you don't get an overwhelming
infection, with the IUD is that we
don't know what we, what your
future fertility would be like.*
The IUD works by causing a little
infection inside your uterus and it
can climb up inside the tubes and
it may destroy the normal structure
of the tubes enough so that the egg,
which is very small, and the tube,
which is also very small, don't fit,
and they can get hung up. There's a/   /I see.
higher incidence of ectopic pregnancy,
tubal pregnancies, when the
egg stops in the tube and then

407

tries to grow into a baby there, with women who've had IUDs.	Oh, I see.
So I, I really think, unless you're really adament and you, you're willing to take all those chances, I really wouldn't tend, wouldn't recommend it.	(laugh) Okay, that answers/
/So it sounds like maybe the pill really is the right thing.	

Note. Emphasis added.

Here the doctor voiced a primarily technical, but emphatic, warning about the dangers of the IUD and concomitantly promoted the birth control pill.[6] This patient and doctor would assert that the patient had chosen her birth control method herself. This is true, but she had help in the decision-making process. The social assumptions of the doctor in this example center around the "future" fertility of the young, single patient who seeks contraception. There is an assumption, first, that she will one day want children, and second, that despite possible infection, surgery, and death, the IUD is most to be feared as a potential cause of infertility. The discussion implies that for a young woman who already had a family, the potential dangers of infection, surgery, and death, although still problems, would not be quite so serious. In Example 11 (the same speakers as in Example 1), Dr. Jones comments on women's reproductive roles when discussing amenorrhea with a recently divorced, single mother who, in this interview, has shown no interest in having more children and has stated that she is not presently sexually active.

EXAMPLE 11

Dr. Jones	*Patient*
You know, when you do want to get pregnant again, we'll have to get you on regular cycles, where you're ovulating, in order for you to get pregnant, okay?	
	(nods head and changes topic)

Note. Emphasis added.

Doctors from both settings assumed the role of protector of reproductive function (Examples 10 and 11) and discourager of reproductive carelessness (Example 12). In the social talk, the doctors also engaged in "keeping the moral order" (Fisher, 1980). To a white, middle class, married woman Dr. Masters said in a friendly, chatty manner, "Now look Susan, uh, this is the third time (abortion). I'm talking as a friend to you as well as your doctor, okay?" In Example 12 Dr. Masters spoke with a young, single black woman receiving state medical aid.

EXAMPLE 12

Dr. Masters	Patient
Look, you already had two abortions, at seventeen.	
	I know this.
Well!?!	But the next time I become pregnant, I'm gonna keep it.
Yeah, but I mean at seventeen and being single, do you want to be pregnant?	

The doctor was far sterner in talking to her about her reproductive history and future than he had been with the married woman. Dr. Masters strongly disapproved of pregnancies among young, single women and in conversations with me referred to women in this group as "stupid" and generally beyond rationality.

The clinic doctors were rarely so explicit in their advice. Dr. Long, in treating a 21-year-old single woman for postabortion pain, did not discuss her failure to use birth control until the end of the interaction. As he was concluding the exam and preparing to leave the room he rather grimly directed her: "Okay, well, use your diaphragm. Use it every time, remember to take it out every time."

Throughout the exam Dr. Long's patient had evidenced intense pain. The doctor had ignored her signals and moans. As the doctor and I left the examining room, he rolled his eyes at me and said, "She acts like she's fourteen years old." Both settings include such social interpretations. In the office of Dr. Masters they form part of the discourse between doctor and patient. In the clinic they tend to receive subtler expression in the doctor-patient discourse and to be voiced among staff members.

To distinguish between the social/contextual and medical/biological modes of discourse, I drew in my analysis on the concept of "frames"—forms in speech which reference conceptual levels of understanding in an interaction. Frames indicate the parameters for each social or medical topic, and since topics discussed in the interviews invariably had both social and medical components, I used a dominant/residual continuum to avoid the need for a sharp break between the two. My data indicated that patients talk *within* the medical frames but do not initiate medical topics (Todd, 1982). When a patient did initiate a topic, it was generally social. The doctors' social talk, in contrast, tended to be more abstract and embedded in the medical talk.

Like the distributional and sequential analyses, analysis of the frames showed that the doctor exerted more power and control in the interaction than the patient. The doctors' power was used, in part, to reinforce traditional stereotypes of women and their life options; stereotypes that have been increasingly questioned during the past decade. In addition, the doctors' focus on technical information and relative exclusion of contextual information reveals more general conceptions of the purposes and proper domain of science, a matter that I shall consider at greater length in the next section.

PATTERNS IN THE DATA

In the data I collected, several patterns were consistently evident: (a) the doctor's reasoning or interpretation was more in evidence than the patient's; (b) this interactional edge allowed the doctor to express attitudes and views often independent of the patient's concerns and sometimes to display condescension on the part of the doctor toward the patient; and (c) social and medical topics introduced by the doctor generally prevailed while the doctor either ignored the patient's socially embedded topic or used it to return the discourse to a medical topic. Let us consider the three patterns in sequence.

1. In Example 13, from the clinic, a woman complains of dizziness, bloating, tiredness, and general dissatisfaction that she feels is caused by a 5-year course of birth control pills. Dr. Smith is skeptical that the birth control pill is the cause of these problems and dominates the ensuring interchange.

EXAMPLE 13

Dr. Smith	Patient
Do, do your legs swell, is that part of your problem or do you just feel kinda bloated?	
	My, I don't know, my stomach just feels really, like it's out here, you know, very/
/uh hum. Sometimes things get blamed on the pills that aren't always the pills' fault so, like/	/what?
women who say they gained weight because they're taking the pills when ordinarily they're expecting to gain weight, so maybe they eat a little bit more, and, uh, then they gain weight and say, oh, look at the pill made me gain weight.	
	But it makes me hungry. When I'm not on the pill, I don't feel hungry.
I've never heard that one before (laughs).	

Although the doctor laughs off the patient's concerns about the birth control pill, the patient is not unreasonable to question this method of contraception. Seaman (1969) and Seaman and Seaman (1977) stressed the numerous potential health hazards of hormonal tampering. Congress has conducted hearings on the dangers of oral contraceptives. Ehrenreich and English (1979) discussed how doctors, as experts, control women's reproduction despite growing evidence that many of the prescribed medical treatments have negative effects. While the birth control pill may in fact be useful to the patient in Example 13, depending on her contextual needs to avoid pregnancy, the concerns she expressed are legitimate and deserve careful consideration. In Example 14,[7] an interchange between Dr. Masters and a patient of several years, recommendations about medical coverage which disregard the patient's worries and needs are made.

EXAMPLE 14

Dr. Masters	Patient
Uh, Nancy, anything you need, you know that from previous times, you just see me, you know, anything medical, you know. And what I would suggest, now you have a choice between Kaiser and Blue Cross, I think (pause)—take Blue Cross. You don't need anything for office calls. You're not going to go broke on that; you know that. (pause)	Uh hum. Uh hum.
Uh, but you should have some hospital and surgery, so if you want to stay with me, you know, and, uh, I would suggest that you just take Blue Cross on the hospital and surgical, you know, uh, then you can come to me. You know, since you feel that strongly about it, you know.	Okay. But does Blue Cross cover like examinations when you have a physical?
Honey, this is what I'm trying to tell you. Look, yes. Some do, some don't. But that basically is not what you need. Now, you're not going to go broke by coming once or twice a year for a pelvic examination.	I don't think they do. Yeah.

2. The doctor's condescension to patients was observable in sequences throughout the interviews. The doctors in the clinic were more likely than Dr. Masters to explain to a patient what they were doing, but still they often seemed to be talking down to the patient, as if to a child. The exchange in Example 15 was particularly common in the clinic when the patient observed her cervix by using a mirror during the exam.

411

EXAMPLE 15

Dr. Jones	Patient
You can hold the mirror and you can just kind of angle it in and see it [cervix]. Can you see it?	Oh, yeah.
It's like a little pink doughnut?	Uh huh.
It's got a little hole.	Uh huh.
That's where the baby comes out.	

The doctors used the diminutive "little" in many of their explanations—in such statements as the one in example 15; in initiating an exam ("I'm just going to do a little exam"); in applying medication to vaginal warts, ("a little bit on that one and a little bit on this one"); and in teaching women breast self-examination ("just march your little fingers. . .").

Dr. Masters always examined patients without explaining his actions. He often showed a condescending attitude toward patients, however, in demanding a response while revealing a suspicion that they would not follow his instructions. Statements to a Mexican-American woman regarding her usage of birth control: "Okay. I've given it to you in English and Spanish. Now you better mind it, okay?" In Example 16, Dr. Masters addresses a woman who has run out of birth control pills and must wait until her cycle to resume taking them.

EXAMPLE 16

Dr. Masters	Patient
Okay, now, honey, look, so that you don't get pregnant again, I want you to get this foam, okay?	Uh hum.
You go to the drugstore and get it, and then here are the instructions. Read it. Now it's very simple.	
[This woman has not been pregnant recently.]	

Dr. Masters's "honey" and the step-by-step instructions were typically uttered slowly and carefully, as if to a small child, with favorite expressions that included "Have you been a good girl?" meaning "Have you used contraception with intercourse?" and "Here are your happy pills" meaning birth control pills. These phrases generally elicited nervous laughs and twitches from the patients.

3. It should be noted that doctors' control of social and medical topics may jeopardize the outcome of medical treatment. If the patient is in a passive position, she will often refrain from making requests (Lazarre &

Eisenthal 1979). If topics related to contraception are not discussed, as a result, the consequences for her health and life options might be serious. In Example 3 and 4, the doctor shifted the patient's social topic to a medical one. When the doctor wanted to pursue social topics, however, the transition was accomplished with ease. (In demanding a reactive from the patient and staff while performing a breast exam, Dr. Masters required compliance from the patient even when she was reluctant.)

The question of immediate importance from the sociological point of view, then, is why such doctor-patient encounters are so systematically repeated. What could hold such a system together? Why are women compelled to allow entire life experiences and biographies to be erased from the arena of consideration? And why should doctors efface them in the first place? Sexism and elitism are certainly involved, and feminist theories invite consideration of reproduction as a political and social phenomenon. But a full description of oppressive and exploitative aspects of the encounter does not explain the structure of doctor-patient interaction. Why does Dr. Jones, in Example 6, with the best of intentions, consider the patient's marriage irrelevant to her reproductive concerns? Why does Dr. Smith, in Example 4, assume in a kindly way that the patient is uneasy about the pelvic exam, when the patient has specifically stated she is uneasy "tonight"—a night, in fact, when she is considering a change of contraceptive method? Why does Dr. Masters, in Example 3, ignore the patient's economic situation when prescribing contraception?

The discussion that I have presented incorporates three foci in current literature on medical interaction. Critical medical social theory provides a wide range of analyses for understanding the power relationships in the present American medical system (see Conrad & Kern, 1981; Ehrenreich, 1978; Freidson, 1970; Navarro, 1976; Waitzkin & Waterman, 1974). The literature emphasizes study of the organization and institution of health care for understanding medical inequality and dominance of the doctor-patient relationship by physicians. Recent feminist scholarship has also contributed similar writings on women's health care needs in modern medical treatment, concentrating on the effects of patriarchy on reproduction. A third group of writers use conceptual analysis of the influence of a scientific-biological model for understanding issues of health (see McKeown, 1976; Mishler et al., 1982; Todd, 1982; Wright, 1982). Such analysis, which delves into the worldview arising out of the Scientific Revolution of the sixteenth and seventeenth centuries, has received less attention than the other two, but in my opinion makes an essential contribution to an understanding of interactions such as those that I recorded.

The doctors' exclusion of the realms of life to which the patients refer has origins that predate the sixteenth and seventeenth centuries, but advances associated with the Scientific Revolution are probably chiefly responsible for this exclusion. Certainly the Scientific Revolution institutionalized the "scientific" notion that the conscious mind can be understood as separate from the mechanical body; an idea that has particularly influenced medical theory and practice (McKeown, 1976). The medical model assumes a health care delivery service based on the understanding that illness is a biological function of individuals or individual organs. Merchant (1980), in her historical work on the rise of modern science, discusses how the image and control of nature and reproduction changed from female to male, active to passive, holistic to mechanical parts, contextual to context free, and subjective to objective with the discovery of scientific laws of external forces and a natural order beyond human control. Shifts in human understanding of nature are also reflected in institutional attitudes toward human beings. Medicine defines patients as passive entities divided into mechanical bodily parts. The contextual aspects of patients' lives and their subjective understandings are negated in favor of objective, technical, medical control. This separation, until recently, has gone unexamined by modern medicine and by the doctor and the patient in the interaction, as well as by most social scientists who study the health care delivery system.

CONCLUSION

The analysis of conversations between the doctor and the patient in the medical setting provides significant evidence that linguistic data can be useful for sociological inquiry and as a consequence for affecting the conduct of medicine. In the actual discourse the sociologist can observe situated interaction, organizational features, institutional structures, conceptual definitions, and cultural assumptions of reproduction.

The doctors' position in the interaction is reflected in the distribution and use of speech acts. It is clear that medical knowledge can be shaped and used as an interactional tool for wielding socially defined and accepted power. Medical institutional power in both settings I examined was reflected in domination by the questioning behaviors of doctors and the directive strategies they used. Furthermore, the doctors controlled the sequencing of turns in the interaction.

The topical analysis of social and medical frames and their use by the doctor and the patient in conversation affords additional insights. The doctor, while conforming to acceptable standards within the medical model, truncates the patient's social understandings with clinical, technical definitions, and with stereotypical social definitions of women's proper roles—when and how to be sexually active, when and how to be reproductive, and when and how to use birth control. The doctor and the patient are potentially in conflict in this interaction. The patient asks the doctor for help in understanding how to adjust her body to her social life. The doctor's technical answer assumes that the patient should adjust her social life to her body, because the doctor does not consider information about the patient's social life theoretically relevant to health care delivery. Biological issues of health and illness have been defined as treatable separate from contextual concerns, but such a division is problematic for reproductive cycles, which normally involve social and contextual—and not merely biological—activities in women's lives.

As I have noted, doctors do use social information and interpretations in these interactions. The social frames display views of women, sexuality, and reproductive function—attitudes which have been questioned, criticized, and broadened during the past decade. While traditional roles are related to larger societal and cultural values, expression of them by the medical profession is important because it is the primary institution in control of reproductive processes in this society; doctors' assumptions can play a powerful role, defining women's definitions of self as well as influencing their health (Ehrenreich & English, 1979; Fisher, 1980; Todd, 1982, 1983).

Analysis of medical discourse can lead to improved health care. One might argue that the present system is adequately arranged. Some findings indicate, however (Fisher, in press, Todd, 1982), that when women patients do not have access to interactional and communicational channels in the medical interview, they often receive inappropriate reproductive care which affects their life options. Shuy (1974) has also discussed the inadequacies in doctor-patient relationships involving highly technical physicians and cross-cultural and/or minority patients.

Some evidence suggests, then, that health care delivery today is inadequate. The benefits of further research, whether for medicine or for sociology, will require increased awareness of the usefulness of empirical data in theorizing about the relationship of knowledge and action. Empirical data enlighten theory and, thereby, lead to improved hypotheses. In addition, the data, in this case doctor-patient interviews, are illuminated by theory—critical medical writings, feminist scholarship, and research into the evolution of the scientific world view—and, in fact, cannot be understood without it. Theory and empirical data thus operate in tandem. Together they can suggest potential strategies for improved medical interactions between doctors and patients. It is in the interests of doctors, patients—and of all concerned with improved health care—to broaden current definitions of health and the role of communicative structure in medical settings.

REFERENCE NOTES

[1]This summary is excerpted from information gathered in an indepth interview between Maria and myself after her doctor/patient visit regarding insertion of a contraceptive IUD.

[2]This verbatim sequence of discourse is taken from an audio taped interview between the doctor and Maria during her visit regarding insertion of a contraceptive IUD. I was present during the interview.

[3]Chomsky's (1968) view of linguistics addressed the relationship between linguistic form and meaning in syntactic terms, with the sentence as the unit of analysis. Searle (1969) differs from

Ehrenreich, J. (1978). *The cultural crisis of modern medicine*. New York: Monthly Review Press.
Erickson, F. (1975). Gatekeeping and the melting pot: Interaction in counselling encounters. *Harvard Educational Review*, 45, 44–70.
Fisher, S. (1979). *Negotiation of medical decisions in doctor-patient interaction and their consequences on the identities of women patients*. Unpublished doctoral dissertation, University of California, San Diego.
Fisher, S. (1980). The context of medical decision-making: An analysis of practitioner/patient communication. *Working Papers in Sociolinguistics, 75*.
Fisher, S. (in press). The negotiation of treatment decisions in doctor/patient communication. *Language and Society*.
Frankel, R. M. (in press). Talking in interviews: A dispreference for patient-initiated questions in physician-patient encounters. In G. Psathas (Ed.), *Interaction Competence, New York: Irvington Press*.
Freidson, E. (1970). *Professional dominance*. Chicago, IL: Aldine.
Gelman, R. & Shatz, M. (1977). Appropriate speech adjustments: The operation of conversational constraints on talk to two-year olds. In M. Lewis & L. Rosenbaum (Eds.), *Interaction, conversation and the development of language*. New York: Wiley.
Halliday, M. & Hasan, R. (1976). *Cohesion in english*. London: Longmans.
Hymes, D. (1972). Models of the interaction of language and social life. In J. Gumperz & D. Hymes (Eds.), *Directions in sociolinguistics: The ethnography of communication*. New York: Holt, Rinehart and Winston.
Labov, W. & Fanshel, D. (1977). *Therapeutic discourse: Psychotherapy as conversation*. New York: Academic Press.
Lazare, A. & Eisenthal, S. (1979). A negotiated approach to the clinical encounter. In A. Lazare (Ed.), *Outpatient psychiatry: Diagnosis and treatment*. Baltimore, MD: Williams and Wilkins.
Luker, K. (1976). *Taking chances. Abortion and the decision not to contracept*. Berkeley, CA: University of California Press.
Maseide, P. Cognitive-linguistic approaches to analyses of social interaction: An analysis of clinical talk. Department of Sociology, University of California, San Diego.
McKeown, T. (1976). *Medicine: Dream, mirage, or nemesis*? London: Nuffield Provincial Hospitals Trust.
Mehan, H. (1979). *Learning lessons*. Cambridge, MA: Harvard University Press.
Merchant, C. (1980). *The death of nature: Women, ecology, and the scientific revolution*. San Francisco, CA: Harper and Row.
Mishler, E. et al. (1981). *Social contexts of health, illness and patient care*. London: Cambridge University Press.
Navarro, V. (1976). *Medicine under capitalism*. New York: Prodist.
Paget, M. A. (1983). On the work of talk: Studies in misunderstanding. In S. Fisher & A. D. Todd (Eds.), *The social organization of doctor-patient communication*. Washington, DC: Center for Applied Linguistics.
Sacks, H., Schegloff, E., & Jefferson, G. (1974). A simplest systematics for the analysis of turn taking in conversation. *Language, 50*, 696–735.
Scheff, T. (1968). Negotiating reality: Notes on power in the assessment of responsibility. *Social Problems, 16*, 3–17.
Schiefelbein, S. (1980). The female patient: Heeded? hustled? healed? *Saturday Review, 3*, 29.
Scully, D. (1980). *Men who control women's health: The miseducation of obstetrician-gynecologists*. Boston, MA: Houghton-Mifflin.
Seaman, B. (1969). *The doctors' case against the pill*. New York: Avon.
Seaman, B., & Seaman, G. (1977). *Women and the crisis in sex hormones*. New York: Rawson Associates.
Searle, J. R. (1969). *Speech acts: An essay in the philosophy of language*. Cambridge: Cambridge University Press.

Chomsky by distinguishing between the propositional content of utterances and elocutionary force, or intention to act on the world. Austin (1962) and Searle (1969) shifted the focus from the linguistic emphasis on syntax to the relationship between language and action (also see Streeck, 1980). For example, the utterances "I command" display *and* accomplish the act of commanding.

⁴Sacks, Schegloff, and Jefferson's (1974) turn-taking analysis assumes talk occurring in natural, everyday interactions between equals in contexts where turn taking is spontaneous and turn allocation is free to vary. Sacks et al. describe a two-part structure in which a question receives an answer or a greeting receives a return greeting, and when it is not interrupted by side or insertion sequences, this response takes place in the next immediate turn at talk, providing a sequential organization of conversation. Labov and Fanshel (1977) have provided a structural outline of the psychotherapeutic interview as an interactional event involving the exchange of information in a routine manner with defined boundaries and expected behaviors.

⁵His remark implies that having large breasts made the patient "all girl" as small breasts could not. If this patient had been a man, would such an anatomy-based remark have been applied to his manhood by a doctor checking his genitals for potentially cancerous lumps? The growing literature on doctors' differential treatment of female and male patients suggests that it would not (Schiefelbein, 1980).

⁶Fisher (1979) discusses how gynecological residents promote hysterectomies to women patients. She asserts that the doctors' selling techniques reflect self-interest rather than concern for the patient. Scully (1980) outlines the ways in which gynecological obstetric residents learn to persuade patients to accept treatment procedures.

⁷Interestingly, when I interviewed the patients quoted in Examples 13 and 14, I learned that both had accepted the advice of the doctor. The clinic patient explained to me that her swelling and dizziness were attributable to her diet rather than to the birth control pill. Scheff discusses a similar situation where the psychiatrist asks the questions; the patient answers. Due to the psychiatrist's control of the interview he is in the position to accept, reject, ignore, etc., the patient's answers, while concomitantly "leading her to define the situation as one in which she is at fault" (1968, p. 14). Dr. Master's patient was nervous about possible medical expense for office visits, but had decided to adopt Blue Cross medical insurance.

REFERENCES

Austin, J. L. (1962). *How to do things with words.* Cambridge, MA: Harvard University Press.

Chomsky, N. (1968). *Language and mind.* New York: Harcourt Brace Jovanovich.

Cicourel, A. V. (1975). Discourse and text: Cognitive and linguistic processes in studies of social structure. *Versus: Quaderni de Studi Semiotici,* September-December, 33–84.

Cicourel, A. V. (1980). Three models of discourse analysis: The role of social structure. *Discourse Processes, 3,* 101–132.

Cole, M., Dore, J., Hall, W. S., & Dowley, G. (1978). Situation and task in young children's talk. *Discourse Processes, 1,* 119–176.

Conrad, P., & Kern R. (1981). *The sociology of health and illness: Critical perspectives.* New York: St. Martin's Press.

D'Andrade, R. A. *A tentative cultural classification of speech acts.* Unpublished manuscript, Department of Anthropology, University of California, San Diego.

Ehrenreich, B. & English, D. (1979). *For her own good: 150 years of the experts advice to women.* Garden City, NY: Doubleday, Anchor Books.

Shuy, R. (1974). Problems of communication in the cross-cultural medical interview. *Working Papers in Sociolinguistics, 19*.

Streeck, J. (1980). Speech acts in interaction: A critique of Searle. *Discourse Processes, 3*, 133–154.

Todd, A. D. (1983). Women's bodies diseased and deviant: Historical and contemporary issues. In S. Spitzer (Ed.), *Research in law, deviance and social control, 5*. Greenwich, CT: JAI Press.

Todd, A. D. (1982). *The medicalization of reproduction: Scientific medicine and the diseasing of healthy women*. Unpublished doctoral dissertation, University of California, San Diego.

Waitzkin, H. B., & Waterman, B. (1974). *The exploitation of illness in capitalist society*. Indianapolis, IN: Bobbs-Merril, 1974.

West, C. (in press). When the doctor is a lady: Power, status and gender in physician-patient exchanges. In A. Stromberg (Ed.), *Women, health, and medicine*. Palo Alto, CA: Mayfield.

Wright, W. (1982). *The social logic of health*. New Brunswick, NJ: Rutgers University Press.

ACKNOWLEDGMENTS

"Aristotle." "The Experienced Midwife." *The Works of Aristotle.* (London, 1822): 102–41.

Knox, Robert. "Contributions to the History of the Corpus Luteum, Human and Comparative." *Lancet* (May 9, 1840): 226–29.

Adams, Francis. "On the Construction of the Human Placenta." In *On the Construction of the Human Placenta* (Aberdeen: A. Brown and Co., 1858): 1–42.

Potter, Robert G. Jr., Philip C. Sagi, and Charles F. Westoff. "Knowledge of the Ovulatory Cycle and Coital Frequency as Factors Affecting Conception and Contraception." *Milbank Memorial Fund Quarterly* 40 (1962): 46–58. Reprinted with the permission of the *Milbank Memorial Fund Quarterly.*

Nadler, Henry L. and Albert B. Gerbie. "Role of Amniocentesis in the Intrauterine Detection of Genetic Disorders." *New England Journal of Medicine* 282, No.11 (1970): 596–99. Reprinted with the permission of the Massachusetts Medical Society.

Littlefield, John W. "The Pregnancy at Risk for a Genetic Disorder." *New England Journal of Medicine* 282 (1970): 627–28. Reprinted with the permission of the Massachusetts Medical Society.

Kass, Leon R. "Making Babies: The New Biology and the 'Old' Morality." *Public Interest* 26 (Winter 1972): 18–56. Reprinted with the permission of *The Public Interest.* Copyright (year) by National Affairs, Inc.

Steptoe, P.C. and R.G. Edwards. "Birth after the Reimplantation of a Human Embryo." *Lancet* 2 (1978): 366. Reprinted with the permission of the Lancet Ltd.

Harrison, Michael R. "Unborn: Historical Perspective of the Fetus as a Patient." *Pharos* 45, No.1 (1982): 19–24. Copyright 1982 by Alpha Omega Alpha Honor Medical Society. Reprinted by permission.

Elias, Sherman, and George J. Annas. "Perspectives on Fetal Surgery." *American Journal of Obstetrics and Gynecology* 145 (1983): 807–12. Reprinted with the permission of Mosby Year Book, Inc.

Reed, James. "Public Policy on Human Reproduction and the Historian." *Journal of Social History* 18 (Spring 1985): 383–98. Reprinted with the permission of the *Journal of Social History*, Carnegie Mellon University.

Wertz, Dorothy C., James R. Sorenson, and Timothy C. Heeren. "Clients' Interpretation of Risks Provided in Genetic Counseling." *American Journal of Human Genetics* 39 (1986): 253–64. Reprinted with permission of the University of Chicago Press. Copyright 1986 University of Chicago Press.

Seibel, Machelle M. "A New Era in Reproductive Technology: In Vitro Fertilization, Gamete Intrafallopian Transfer, and Donated Gametes and Embryos." *New England Journal of Medicine* 318, No. 13 (1988): 828–34. Reprinted with the permission of the Massachusetts Medical Society.

Butler, William J. and Paul G. McDonough. "The New Genetics: Molecular Technology and Reproductive Biology." *Fertility and Sterility* 51, No.3 (1989): 375–86. Reprinted with the permission of the American Fertility Society.

Clarke, Adele E. "Controversy and the Development of Reproductive Sciences." *Social Problems* 37, No.1 (1990): 18–37. Reprinted with the permission of the University of California Press. Copyright (1990) by the Society for the Study of Social Problems.

D'Alton, Mary E. and Alan H. DeCherney. "Prenatal Diagnosis." *New England Journal of Medicine* 328, No.2 (1993): 114–20. Reprinted with the permission of the Massachusetts Medical Society.

Neumann, Peter J., Soheyla D. Gharib, and Milton C. Weinstein. "The Cost of a Successful Delivery with In Vitro Fertilization." *New England Journal of Medicine* 331, No.4 (1994): 239–43. Reprinted with the permission of the Massachusetts Medical Society.

Weiner, Nella Fermi. "Of Feminism and Birth Control Propaganda (1790–1840)." *International Journal of Women's Studies* 3, No.5 (1980): 411–30. Reprinted with the permission of the author.

McLaren, Angus. "Contraception and the Working Classes: The Social Ideology of the English Birth Control Movement in Its Early Years." *Comparative Studies in Society and History* 18 (1976): 236–51. Reprinted with the permission of Cambridge University Press.

Anderton, Douglas L. and Lee L. Bean. "Birth Spacing and Fertility Limitation: A Behavioral Analysis of a Nineteenth Century Frontier Population." *Demography* 22, No. 2 (1985): 169–83. Reprinted with the permission of the authors and the Population Association of America.

Gordon, Linda. "Voluntary Motherhood: The Beginnings of Feminist Birth Control Ideas in the United States." *Feminist Studies* 1 (1973): 5–22. Reprinted with the permission of the publisher, *Feminist Studies,* Inc., c/o Women's Studies Program, University of Maryland, College Park, MD 20742.

Peirce, Isaac. "The Prevention of Conception." *Medical and Surgical Reporter* 59 (1888): 614–16.

Kosmak, Geo. W. "'Birth Control:' What Shall Be the Attitude of the Medical Profession Towards the Present Day Propaganda?" *Bulletin of the New York Lying In Hospital* 11 (1918): 88–99.

Sanger, Margaret, et al. "Ten Good Reasons for Birth Control: Reason I: Woman's Right." *Birth Control Review* 12 (1928): 3.

Reilly, Philip R. "Involuntary Sterilization in the United States: A Surgical Solution." *Quarterly Review of Biology* 62, No.2 (1987): 153–70. Reprinted with the permission of the University of Chicago Press. Copyright 1987 University of Chicago Press.

Gamson, Joshua. "Rubber Wars: Struggles over the Condom in the United States." *Journal of the History of Sexuality* 1, No.2 (1990): 262–82. Reprinted with the permission of the University of Chicago Press. Copyright 1990, University of Chicago Press.

Bremner, William J. and D.M. de Kretser. "Contraceptives for Males." *Signs* 1, No.2 (1975): 387–96. Reprinted with the permission of the University of Chicago Press, publisher. Copyright 1975 University of Chicago Press.

Todd, Alexandra Dundas. "The Prescription of Contraception: Negotiations between Doctors and Patients." *Discourse Processes* 7 (1984): 171–200. Reprinted with the permission of the Ablex Publishing Corporation.

EDITORS

Series Editor

Philip K. Wilson, MA, Ph.D., is an assistant professor of the history of science at Truman State University (formerly Northeast Missouri State University) in Kirksville, Missouri. After receiving his undergraduate degree in human biology from the University of Kansas, he pursued work towards an MA in medical history at the William H. Welch Institute for the History of Medicine at The Johns Hopkins School of Medicine and received his Ph.D. in the history of medicine from the University of London. He has held postdoctoral positions at the University of Hawaii-Manoa and Yale University School of Medicine before settling in Missouri.

Wilson has received scholarly support including a Logan Clendening Summer Fellowship, an Owsei Temkin Scholarship, a Folger Shakespeare Library Fellowship, a Wellcome Trust Research Scholarship, and grants from the Hawaii and Missouri Committees for the Humanities for medical and science history projects. He was a founding member of the Hawaii Society for the History of Medicine and Public Health. Wilson has contributed chapters to volumes including *The Popularization of Medicine 1650–1850* (Routledge), *Medicine in the Enlightenment* (Rodopi), and *The Secret Malady: Venereal Disease in Eighteenth-Century Britain and France* (University Press of Kentucky), articles in the *Annals of Science,* the *London Journal,* and the *Journal of the Royal Society of Medicine,* and is a regular contributor of medical and science history entries to many dictionaries and encyclopedias. Currently, Wilson is pursuing research on women's diseases, osteopathy, and eugenics in Kirksville, Missouri, where he lives with his wife, Janice, and son, James.

Assistant Editors

Ann Dally, MA, MD, received her Master's degree from Oxford University, having been an exhibitioner in modern history at Somerville College. She then studied medicine at St. Thomas' Hospital, London, qualifying in 1953. After some years of general medical practice, she specialized in psychiatry, a specialty she practiced until her retirement in 1994. Meanwhile she pursued her interests in the history of medicine, receiving her doctorate in that subject in 1993. The book based on her doctoral thesis, *Fantasy Surgery, 1880–1930,* will shortly be published as part of the Wellcome Institute for the History of Medicine (London) series. Her most recent book, *Women Under the Knife. A History of Surgery* (Routledge), follows a long publishing history of books including *The Morbid Streak, Why Women Fail, Mothers: Their Power and Influence, Inventing Motherhood: The Consequences of an Ideal,* and a book of memoirs, *A Doctor's Story.* Currently a Research Fellow at the Wellcome Institute for the History of Medicine (London), she lives with her husband Philip Egerton in West Sussex, England and has four children and seven grandchildren.

Charles R. King, MD, MA, is a professor of obstetrics and gynecology at the Medical College of Ohio. He received his BA from Kansas State University, an MD from the University of Kansas, and has completed post graduate medical training at the University of Kansas and the University of Oregon. He has since received an MA in medical history from the University of Kansas. King has been the recipient of Rockefeller Foundation, National Endowment for the Humanities, American College of Obstetricians and Gynecologists-Ortho, and Newberry Library Fellowships for projects in medical history. He is the author of numerous publications regarding women's health, including articles in the *Bulletin of the History of Medicine, Kansas History,* and the *Great Plains Quarterly,* and has recently completed *Child Health in America* (Twain). He currently lives with his wife, Lynn, in Temperance, Michigan.

Milton Keynes UK
Ingram Content Group UK Ltd.
UKHW040711141024
449569UK00005B/97